Kuhlenbeck: The Central Nervous System of Vertebrates

Hartwig Kuhlenbeck

The Central Nervous System of Vertebrates

A General Survey of its Comparative Anatomy with an Introduction to the Pertinent Fundamental Biologic and Logical Concepts

S. Karger · Basel · München · Paris · London · New York · Sydney

Vol. 5, Part II

Mammalian Telencephalon
Surface Morphology and Cerebral Cortex

The Vertebrate Neuraxis as a Whole

Subject and Authors Index to Vols. 1–5
by J. Gerlach and M. Gaab, Würzburg

155 figures comprising 295 illustrations

S. Karger · Basel · München · Paris · London · New York · Sydney

Cataloging in Publication
Kuhlenbeck, Hartwig, 1897–
The central nervous system of vertebrates: a general survey
of its comparative anatomy with an introduction to the pertinent
fundamental biologic and logical concepts/Hartwig Kuhlenbeck. –
Basel; New York: Karger, [etc.], 1967–1978
5 v.: ill.
1. Anatomy, Comparative 2. Vertebrates – anatomy & histology
I. Title
QL 937 K96c
ISBN 3–8055–2645–8

Printed in Switzerland by Pochon-Jent AG, Bern
ISBN 3–8055–2645–8

«Die vergleichende Anatomie, Histologie, Architektonik und Embryologie des Zentralnervensystems bildet ferner einen umfangreichen Zweig und zugleich eine unentbehrliche Methode der neurobiologischen Forschung. Sie verrät die zahlreichen Wege, durch welche die Evolution der Nervensysteme der verschiedenen Tiersorten im phylogenetischen Zusammenhang ihre heutige Verschiedenartigkeit zustande gebracht hat. Vertieft man sich dabei genügend in den Zusammenhang von Form und Funktion, so gelangt man in eine wunderbare Welt der Harmonie zwischen Geist und lebendem Nervensystem...

Wer vergleichende Anatomie des Nervensystems sagt, sagt also auch vergleichende Physiologie – Psychologie und – Biologie, und das ist ein Gebiet, aus welchem die künftige Forschung mit vollen Zügen schöpfen kann...

Dass der Mensch für den Menschen sich zunächst interessiert, ist verzeihlich und naheliegend. Hat er aber einmal erkannt, dass er nur ein Glied in der Tierreihe bildet und dass sein Hirn, das Organ seiner Seele, aus dem Tiergehirn und somit aus der Tierseele stammt, so muss er doch zur Erkenntnis gelangen, dass das Studium der Neurobiologie dieser seiner Verwandten das grösste Licht auf sein eigenes Nerven- und Seelenleben werfen muss.»

August Forel
(«*Die Aufgaben der Neurobiologie*»)

Preface

The lectures on the central nervous system of vertebrates, given by the author during his first sojourn in Japan, 1924–1927 (Taishô 13 to Shôwa 2), intended to foster the interest in comparative neurologic studies based upon the morphologic principles established by the *Gegenbaur* or *Jena-Heidelberg School of Comparative Anatomy*. Notwithstanding their introductory and elementary nature, these lectures, published by Gustav Fischer, Jena, in 1927, included a number of advanced as well as independent concepts, and represented, as it were, the outline of a further program.

Despite various vicissitudes, and although I found the prevailing intellectual climate in the realm of biologic sciences rather unfavorable to the pursuit of investigations related to the domain of classical morphology, I have, *tant bien que mal*, carried on with my studies as originally planned, and propose to summarize my viewpoints in the present series, designed to represent a general survey, and projected to comprise five separate volumes, of which the first two are now completed. It can easily be seen that the present series follows closely the outline of my old 'Vorlesungen', meant to stress 'die grossen Hauptlinien der Hirnarchitektur und die allgemeinen Gesetzmässigkeiten, welche in Bau und Funktion des Nervensystems erkannt werden können'.

Comparative anatomy of the vertebrate central nervous system requires a very broad and comprehensive background of biological data, evaluated by means of a rational, consistent, and appropriate logical procedure. Without the relevant unifying concepts, comparative neurology becomes no more than a trivial description of apparently unrelated miscellaneous and bewildering configurational varieties, loosely held together by a string of hazy 'functional' notions. A perusal of the multitudinous literature dealing with matters involving the morphologic aspects of neurobiology reveals, to the critical observer, considerable confusion as regards many fundamental questions.

For this reason, the present attempt at an integrated overall presentation includes a somewhat detailed scrutiny of problems concerning the significance of configuration and configurational variety with respect to evolution and to correlated reasonably 'natural' taxonomic classifications. Because comparative anatomy of the central nervous system embodies the morphological clues required to infer the presumable phylogenetic evolution of the brain, a number of general questions referring to ontogenetic evolution are critically considered: it is evident that both the inferred phylogenetic sequences and the observable ontogenetic sequences represent evolutionary processes suitable for a comparison outlining the obtaining invariants.

Moreover, the comparison of organic forms involves procedures closely related to *analysis situs*. Thus, a simplified and elementary discussion of the here relevant principles of *topology* was deemed necessary.

Finally, since vertebrate comparative anatomy and vertebrate evolution, including the origin of vertebrates, cannot be properly assessed in default of an at least moderately adequate familiarity with the vast array of invertebrate organic forms, a general and elementary survey of invertebrate comparative neurology from the vertebrate neurobiologist's viewpoint, that is as seen by an 'outsider' with a modicum of first-hand acquaintance, has been included as volume two of this series. The approximately 20 pages and 12 figures dealing with this matter in my 1927 'Vorlesungen' have thus, of necessity, become rather expanded.

US N.I.H. Grant NB 4999, which is acknowledged with due appreciation, made possible the completion of Volumes 1 and 2 of this series, and, for the time being, the continuation of these studies, by supporting a 'Research Professorship' established to that effect, following my superannuation, at the Woman's Medical College of Pennsylvania.

Concluding this preamble to the present series, I may state with CICERO (*De oratore*, III, 61, 228): '*Edidi quae potui, non ut volui sed ut me temporis angustiae coegerunt; scitum est enim causam conferre in tempus, cum afferre plura si cupias non queas.*'

H.K.

Foreword to Volume 5, Part II

The present part-volume, containing chapters XIV, XV, and XVI, completes the whole work's *Special or Systematic Subdivision (Spezieller oder Systematischer Teil)*.

Chapter XIV, dealing with the surface morphology of the Mammalian telencephalon and with thereto related topics pertaining to the purview of 'anthropology', includes, in this connection, some general comments concerning the Human mind in social, cultural, and historical perspective. This *excursus* seems appropriate on the basis of AUGUST FOREL's broad concept of neurobiology and because of the evident fact, stressed by SCHOPENHAUER, that our entire phenomenal world, as experienced in consciousness, merely represents a brain phenomenon.

Chapter XV deals with morphologic, structural, architectural and functional aspects of the Mammalian cerebral cortex, including engraphy and consciousness. The epistemological significance of consciousness is pointed out and critically discussed.

The final chapter XVI reviews aspects of the Vertebrate neuraxis as a whole, with emphasis on typological features in the taxonomic series, followed by some comments on phylogeny.

The bibliographies appended to these chapters are, of necessity, meant to be selective, but should easily enable those interested in further particulars to find the required additional references. For my old '*Vorlesungen*', published 50 years ago, I had chosen the subtitle: '*Eine Einführung in die Gehirnanatomie auf vergleichender Grundlage*'. Although, in comparison, the present series has assumed the size which, before midcentury, would have been that of a detailed '*Handbuch*', the present work should likewise be considered as representing merely a somewhat elaborate '*Introduction*' to the overall domain of neurobiology and its interdisciplinary relationships, with particular emphasis on the significance of morphology.

As in the preceding volumes, numerous duly credited illustrations were taken from the public domain of published scientific literature also including contributions by my collaborators and myself. Illustrations without credit reference are previously unpublished originals from my own studies. Professor J. GERLACH *(Würzburg-Lützelbach)* and Dr. M. GAAB (*Würzburg)* have kindly taken over the task to prepare the appended *Subject and Author Index*.

As before, I am obliged to the *Medical College of Pennsylvania* for the facilities of my '*Laboratory of Morphological Brain Research*', and particularly grateful to the *Alumnae Association* of the whilom *Woman's Medical College of Pennsylvania* which includes my many former students, for the continued generous contributions to the funds necessary for my work.

In concluding the *Special or Systematic Subdivision* of this Series, I also wish to express my thanks and appreciation to those who, for many years, substantially helped me in the preparation of this work: to my expert Chief Technician and trusted Secretary in the former Department of Anatomy at the *Woman's Medical College of Pennsylvania*, the late MISS VERA MENOUGH, to my efficient Secretary at my '*Laboratory of Morphologic Brain Research*', MRS. DOLORES BRENNAN, and to MESSRS. GILBERT and RING, Philadelphia, for expert photographic and illustrative work. I am also particularly thankful to MR. THOMAS KARGER, *Basel*, for his cooperation and understanding in undertaking to publish this extensive Series, and I greatly appreciate the careful work performed by the staff of S. KARGER A.G.

It is also a pleasure to express my thanks to my esteemed old friend, colleague and former student Prof. J. GERLACH for many helpful discussions in Germany and in Philadelphia, also for his and MRS. GERLACH's repeated assistance with work at my Philadelphia Laboratory, moreover to Prof. GERLACH and Dr. GAAB for the abovementioned preparation of the Index. I am likewise grateful to my esteemed old friend the late Professor HUGO SPATZ, former Director of the *Kaiser-Wilhelm* (now *Max-Planck) Institute of Brain Research*, to his successor Professor R. HASSLER, and to the members of that Institute, for numerous relevant discussions and for the hospitality extended to me during my repeated visits. Last but indeed not least, I owe thanks to my wife, who for more than half a century, travelling with me the road of life, gave me encouragement and moral support in my endeavors.

H.K.

Table of Contents of the Present Volume

Volume 5 Part II: Mammalian Telencephalon: Surface Morphology and Cerebral Cortex The Vertebrate Neuraxis as a Whole

Table of Contents of the Complete Work

Volume 1 Propaedeutics to Comparative Neurology

Volume 2 Invertebrates and Origin of Vertebrates

Volume 3 Part I: Structural Elements: Biology of Nervous Tissue

Volume 3 Part II: Overall Morphologic Pattern

VI. Morphologic Pattern of the Vertebrate Neuraxis

VII. The Vertebrate Peripheral Nervous System and its Morphological Relationship to the Central Neuraxis

Volume 4 Spinal Cord and Deuterencephalon

VIII. The Spinal Cord

Volume 5 Part I: Derivatives of the Prosencephalon: Diencephalon and Telencephalon

XII. The Diencephalon

XIII. The Telencephalon

Volume 5 Part II: Mammalian Telencephalon: Surface Morphology and Cerebral Cortex The Vertebrate Neuraxis as a Whole

XIV. Surface Morphology of the Mammalian Telencephalon

XV. The Mammalian Cerebral Cortex

XVI. The Central Nervous System as a Whole: Typologic Features in the Vertebrate Series

XIV. Surface Morphology of the Mammalian Telencephalon

1. General Remarks: Lissencephalic and Gyrencephalic Brains, Volume-Surface and Size Relations; the Domestication Problem

As regards the surface morphology of the Mammalian lobus hemisphaericus, it is customary to distinguish *lissencephalic* (smooth) and *gyrencephalic* (convoluted) brains. This distinction, however, refers essentially to the neopallial surface in the wider sense, i.e. including the parahippocampal surface.

All Mammalian brains, both lissencephalic and gyrencephalic ones, display, however, more or less distinctive surface grooves approximately indicating the boundaries (1) between neopallium and piriform lobe, (2) between piriform lobe and basal cortex, as well as (3) between parahippocampal pallium and hippocampal formation. In the precommissural region, moreover, there occurs rather frequently (4) a sulcus between precommissural hippocampus and paraterminal grisea. Again, in diverse essentially lissencephalic brains, (5) a groove between neopallium *sensu strictiori* and parahippocampal pallium may be present (cf. Fig. 223B, vol. 5/I).

The so-called *choroid fissure* is represented by the attachment of plexus chorioideus ventriculi lateralis to the edge (fimbria and crus fornicis) of the postcommissural hippocampal formation. It is not a surface groove in the aspect here under consideration. The term 'fissure' moreover, is here an evident misnomer[1], since the fold of roof-plate lamina epithelialis, invaded by leptomeningeal tissue, and providing the choroid plexus villi, becomes, as it were, a narrow mesenterium-like duplicature, which may be single or double (cf. e.g. Figs. 226, p. 451, and 234A, p. 458, vol. 3/II), being sealed by the mesodermal tela chorioidea (cf. p. 336, vol. 3/I). One folded leaf of the duplicature, if single, is attached to the fornix system, and the other to the lamina

---

[1] On the other hand, a 'fissure' evidently results if, as e.g. in routine dissection of the Human brain, the choroid plexus is torn away, leaving the fringes known as taenia fornicis and taenia chorioidea. *Gray's Anatomy* (29th ed., Lea & Febiger, Philadelphia 1959) thus correctly defines on p. 904 the choroid fissure as a cleft-like space which remains after removal of the lateral ventricle's choroid plexus.

affixa (cf. e.g. Fig. 241, p. 467, vol. 3/II). More caudally, both laminae may become attached to the concavity of the curved fornix edges (cf. Fig. 236, p. 461, vol. 3/II).

Discounting thus the so-called choroid fissure, the aforementioned sulci, which KAPPERS (1920) designated as *primitive grooves (Primitivfurchen)* are the following: On the lateral aspect, (1) *sulcus rhinalis lateralis* with an anterior (rostral) and a posterior (caudal) subdivision and (2) *sulcus endorhinalis*[1a], likewise with an anterior and a posterior (periamygdalar) extension. On the medial aspect runs (3) the *sulcus sive fissura hippocampi* which, caudally to corpus callosum, follows the curvature of the hemispheric bend. In Mammals with a reduced supracommissural hippocampus (cf. Fig. 223C, vol 5/I) this sulcus disappears along the body of corpus callosum, being more or less absorbed by sulcus corporis callosi. The *sulcus limitans hippocampi* (4) likewise becomes either continuous with sulcus corporis callosi or disappears on the dorsal callosal surface. If present in lissencephalic brains, (5) a groove between neopallium and parahippocampal pallium is found on the dorsomedial surface of the hemisphere. It is presumably a forerunner of the *sulcus cinguli system* as seen in various gyrencephalic brains. The primitive grooves (1) to (5) of the Mammalian brain are indicated or may be easily identified in several Figures of chapter XIII, volume 5/I (e.g. Figs. 154C, 158, 175A–D, 221B, 222A). The sulcus rhinalis lateralis, either *in toto* or in part, may be barely suggested or even not noticeable in the brain of very small Mammals (cf. e.g. STEPHAN, 1975, p. 875).

These primitive sulci can be conceived to result from boundary interactions in the ontogenetic development of distinctive neural tube regions such as *grundbestandteile* or *formbestandteile* representing, as it were, morphogenetic units, and, in this respect, could be likened to the ventricular grooves, such as the deuterencephalic sulcus limitans, the longitudinal diencephalic sulci, or the telencephalic ventricular sulci. All these grooves are related to the ontogenetic development of the various longitudinal zonal systems characterizing the vertebral neuraxial bauplan. The evolution respectively the transformations of these

[1a] This sulcus, along the dorsal aspect of tractus olfactorius lateralis, indicates the boundary between (pallial) piriform lobe cortex and basal cortex. In macrosmatic Mammals, an additional sulcus (endorhinalis) arcuatus may run along the basal aspect of lateral olfactory tract, delimiting a lateral boundary of the bulge of tuberculum olfactorium proper (cf. e.g. Fig. 221A, chapter XIII of volume 5/I). A '*solco diagonale*' of BECCARI may also indicate a caudal boundary of said tuberculum (cf. Figs. 175A–C, chapter XIII of volume 5/I).

zones were dealt with in volume 3/II. To a variable degree, these primitive grooves of the Mammalian telencephalic surface are also displayed or at least suggested in some representatives of all submammalian gnathostome Vertebrates with inverted telencephalon. Thus, Figure 186E (vol. 5/I) shows in a Selachian an unlabelled shallow groove between B1 and B2 neighborhoods, which may be evaluated as an accessory endorhinal sulcus.

The everted telencephalon of many Ganoids and Teleosts displays a sulcus externus, which was also designated as sulcus endorhinalis (cf. Fig. 194C, vol. 5/I), and may roughly correspond to a boundary between pallial (D) and basal (B) components. However, since this groove is related to the folding obtaining in eversion, its comparison with external grooves of the inverted telencephalon involves debatable questions of morphologic interpretation.

As regards the inverted hemispheric wall of Dipnoans, Figure 202A in volume 5/I shows a lateral pallio-basal sulcus, kathomologous[2] to sulcus endorhinalis, in Protopterus; it is likewise recognizable in Lepidosiren, and shown by Figure 203 in volume 5/I. Upon inspection of this latter Figure it will also be seen that a faint accessory endorhinal groove is suggested between B_1 and B_2 on the lateral side, while a barely recognizable shallow groove on the dorsomedial side corresponds to the boundary between hippocampus and paraterminal body. In addition, a somewhat more pronounced groove corresponds to the boundary between B_4 and B_3. In diverse Amphibians, shallow surface grooves either between D_1 and B_1 or B_1 and B_2 ('sulcus endorhinalis accessorius') are not infrequently seen (cf. e.g. Fig. 205B of vol. 5/I and, in vol. 3/I, Figs. 270B, p. 517, and 275, p. 523).

With regard to submammalian Amniota, sulcus rhinalis lateralis and sulcus endorhinalis are shown in Figure 209B (vol. 5/I) as displayed by a Tortoise. Figure 210B in volume 5/I likewise shows sulcus endorhinalis and sulcus limitans hippocampi, which latter can also be identified in Figure 210A (I) of volume 5/I. This groove, commonly designated as 'sulcus hippocampi' by various authors, must not be confused with the true sulcus hippocampi (3) between parahippocampal cortex and hippocampal formation.

---

[2] Kathomologous but not orthohomologous since it corresponds to a boundary between B_1 and D_1. This latter zone D_1 is not homologous to the Mammalian piriform lobe which essentially derives from a lateral neighborhood of D_2. An external groove between B_1 and B_2 can be designated as 'accessory endorhinal sulcus'.

In Birds (Figs. 214A, C, vol. 5/I), whose peculiar epibasal development results in a considerable pattern transformation, the vallecula corresponds to a modified sulcus rhinalis lateralis, and the sulcus basalis lateralis externus to a modified sulcus endorhinalis, whose rostral extension (u) may protrude dorsally to a basal portion of prepiriform cortex. Although not commonly present, a true hippocampal sulcus or 'fissure' at the boundary between parahippocampal and hippocampal cortex was identified and depicted in the Kiwi by CRAIGIE (1930)[3].

Reverting now to the gyrencephalic Mammalian brains, it could be said that their neopallial (and parahippocampal) surface grooves represent distinctive boundary sulci in a much lesser degree than the primitive sulci.

Although some neopallial grooves doubtless indicate boundaries of cortical fields, many of these sulci do not. ELLIOT SMITH (1907a, b) and others distinguish, with regard to the relation of sulci to cortical areas, *limiting sulci* approximately corresponding to boundaries, and *axial sulci*, extending within a given 'field'. To this, however, one should add a substantial number of sulci whose path does not display any significant relation to distinctive cortical areas. Again, certain sulci which appear as limiting sulci in some forms, appear to have lost this relationship in others. As it were, the convolutional pattern and the cortical 'fields' may display mutually independent displacements, that is to say cortical 'fields' could 'slide' or 'shift' beneath the surface pattern of gyri and sulci.

In toto, the grooves characterizing the gyrencephalic brains are essentially related to the increasing surface expansion of the neopallium *sensu latiori* with its isocortex and parahippocampal cortex.

Since the lissencephalic hemisphere may roughly be compared to a spherical body, it is evident that the surface area, represented by the neopallium, will increase in proportion to the second power of the radius, while the volume will increase in proportion to the third power (sphere: area $4\ \pi r^2$; volume: $\frac{4}{3}\pi r^3$). Considerable expansion of the cortex would thereby produce an excessively bulky volume. Because the depth of the cortex, the amount and length of medullary fibers, the

[3] Cf. Figure 14, p. 242 of CRAIGIE's paper. It is there doubtless the true 'fissura *sive* sulcus hippocampi', although HUMPHREY (1967, p. 672) contests this interpretation. On the other hand, HUMPHREY (loc.cit.) justly points out that the 'fissura hippocampi' depicted by HINES (1923), SHANKLIN (1930) and others in Reptiles is not the true 'hippocampal fissure' (it is, in fact, the sulcus limitans hippocampi between hippocampus and paraterminal grisea, i.e. between D_3 and B_4).

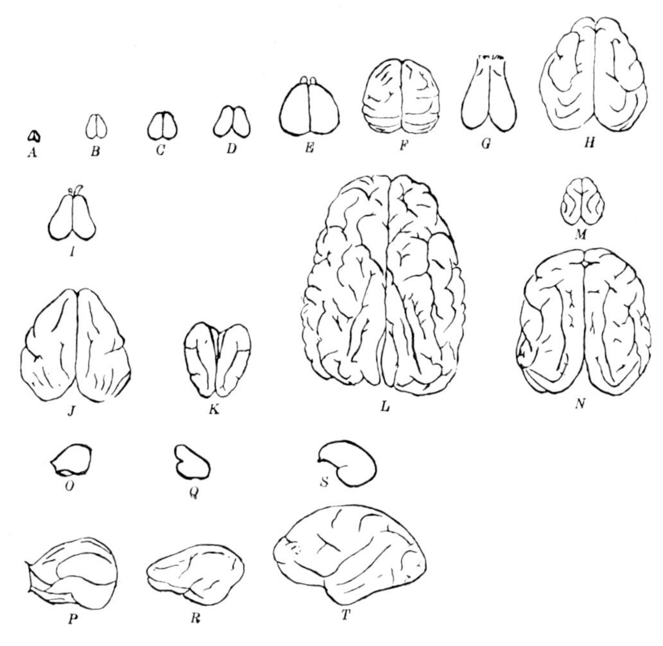

Figure 1. Lissencephalic and gyrencephalic brains of different Mammals drawn at identical reduced scale (from KAPPERS *et al.*, 1936). A–D: smallest Mammalian brains (Chiroptera and Insectivores); A: Vespertilio; B: Peromyscus; C: Tomomys; D: Talpa. E–T: brains of small and large animals of same order; E: Ornithorhynchus; F: Echidna (Monotremata); G: Didelphys; H: Macropus (Marsupialia); I: Cynomys; J: Hydrochaerus (Rodents); K: Tragulus; L: Alces (Ungulata); M: Mustela; N: Felis concolor; (Carnivora); O: Tatusia; P: Choloepus (Edentata); Q: Tarsius; R: Indris; S: Hapale; T: Ateles (Primates).

size of internal grisea and of the ventricular spaces, all of which contribute to the volume, remain limited, it seems obvious that surface expansion beyond a certain range must necessarily result in the formation of folds.

It is well known that, within one and the same Mammalian taxonomic group, the forms with small body and brain size may be lissencephalic and the larger ones gyrencephalic.[4] The folding of the neopallium in these latter is said to counteract an otherwise resulting disproportionate decrease in the surface-volume ratio *(Dareste's and Baillar-*

[4] If the smaller forms are already gyrencephalic, the convolutional pattern is then commonly more complex in the larger ones.

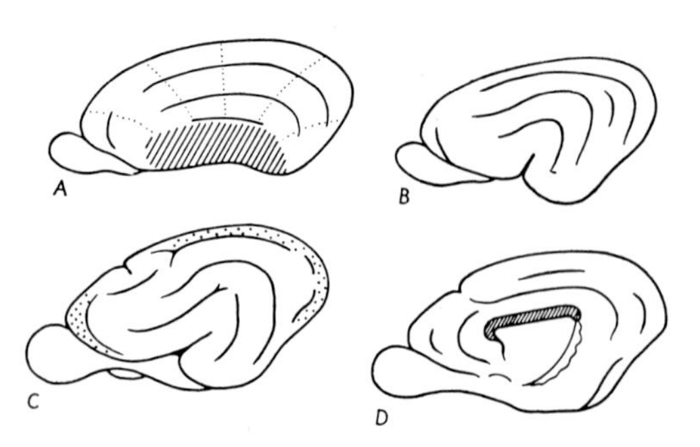

Figure 2. Sketches illustrating LE GROS CLARK's concept of neopallial folding (after LE GROS CLARK, 1945, from YOUNG 1957). A: folds developing at right angle to the lines of stress shown by dotted lines in the thinner pallial expansion dorsally to region of thicker basal ganglia (hatched); B: effect of hemispheric bend in producing arcuate sulci; C, D: lateral and medial aspect of Carnivore (Cat) cerebral hemisphere, showing arcuate sulci (rostral and visual pallial areas are stippled).

ger's law).[5] Figure 1 illustrates this particular aspect of diverse small lissencephalic and larger gyrencephalic Mammalian brains.

Several authors have attempted to elucidate the specific mechanical factors involved in the telencephalic sulcus formation of gyrencephalic brains. LE GROS CLARK (1945) suggested that the pattern according to which these sulci are formed is determined by the occurrence of mechanical stresses originating as the hemispheres expand. Constraining nondeformable structures in the hemisphere itself are presumed to be the basal ganglia ventrally and the corpus callosum dorsally. Externally to the hemisphere, the cranial basis in particular, and the cranial cavity in general represent an enclosed space. Longitudinal folds at right angles to the lines of stress thereby develop and run parallel to the non-compressible constraining structures.[6] Such longitudinal folds are commonly found in subprimate Mammalian brains (Fig. 2). Additional furrows of lesser order form when the brain becomes still larger, and often assume triradiate patterns, closely corresponding to the folds formed on the surface of a collapsing balloon (Fig. 3).

Other authors, e.g. SCHAFFER (1918), LANDAU (1923), CONNOLLY (1950), BARRON (1950), and others, however, consider convolutional

---

[5] Cf. KAPPERS (1921, p.1121), BAILLARGER (1845), DARESTE (1870).

[6] It is here of interest that, in agenesis of the Human corpus callosum, a radial orientation of the sulci on the hemisphere's medial surface replaces the curved longitudinal pattern (cf. Fig. 97 A, p. 266, vol. 3/II).

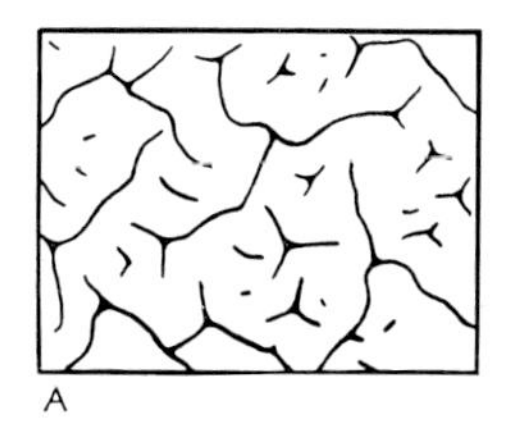

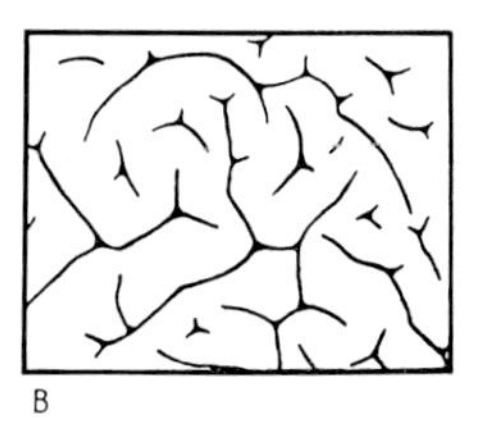

Figure 3. Patterns of folds formed on surface of a contracting balloon (A), compared with sulcal pattern (B) from the parietal lobe of a human brain, showing triradiate patterns, some of them linked as in A (after BULL, from YOUNG, 1957).

development to be the result of purely intrinsic hemispheric growth processes. BARRON (1950) who destroyed selected brain portions of fetal Sheeps *in utero* at a stage when the neopallial sulci are just about to appear, and then allowed the operated fetuses to continue their development *in utero*, found an essential normal pattern of gyri and sulci to be displayed at term or postnatally, by the damaged hemispheres. The cited author reached the conclusion 'that the folding of the cortex of the cerebrum is not due to either a disproportion between (1) the rate of increase in the capacity of the cranial cavity and the mass of the cerebrum or (2) the rate of increase of the cortical surface relative to the underlying structure of the telencephalon and diencephalon'. He then added that his observations 'appear to justify the inference that the forces primarily responsible for cortical folding are resident in the cortex, though the possibility must be recognized that these primary forces may be modified by the growth of cells outside the cortex'.

BARRON's findings, although doubtless supporting the concept of intrinsically cortical growth processes, are not entirely unambiguous and cannot be evaluated as fully convincing. The operations were performed at stages with already well displayed hemispheric bend. Splitting and partial removing of developing corpus callosum would not, *per se*, prevent the effect of remaining and regenerating fibers in centrum semiovale. Destruction of central grisea in the region of the hemispheric stalk would likewise not necessarily prevent the constraining mechanical effect of resulting scar tissue, cysts or of regenerative processes.

A recent review of the problem under consideration was presented by RICHMAN *et al.* (1975) who reach the conclusion that 'the forces responsible for cerebral convolutional development are predominantly intracortical'. The cited authors elaborate a mechanical model of this development based on observations concerning either microgyric or lissencephalic maldevelopment in Man, and postulating the interaction of inner and outer cortical layers. Some authors, e.g. WOOLSEY (1960), WELKER and CAMPOS (1963) and SANIDES (1972), particularly stress the relationships of sulci to specific cortical areas.

A detailed critical review and discussion of the theories concerning sulcus formation and apposite aspects of brain growth is also included in an essay on developmental neurology by DODGSON (1962). Yet the various and in part conflicting hypotheses in this topic remain inconclusive and unsatisfactory. As pointed out in the section on morphogenesis in chapter VI (vol. 3/II, p. 73) nothing of any significance is, up to now, known about the *modus operandi* respectively the relevant components of the 'forces' (physiocochemical interactions) which shape organic configurational patterns.

As regards my own opinion,[7] I would agree with LANDAU (1923) and others who deny an influence of the skull upon brain development and assume that '*die formbildende Kraft für den Schädel liegt im Gehirn*'.

Concerning the intracerebral factors, I believe, however, that the constraint provided by the basal ganglia, as pointed out by LE GROS CLARK (1945), could also be of significance concomitantly with the influence exerted by the hemispheric stalk and the thereto related hemispheric bend. These factors might indeed result in the formation of the curved longitudinal sulci shown in Figure 2. This, of course, does not exclude the factors stressed by BARRON and RICHMAN *et al.*

It is furthermore quite possible that a variety of different events are involved in the development of different sulci in the various regions of the hemisphere. Thus CONNOLLY (1950) assumes the interaction 'of a competitive influence of the cortical areas, thalamo-cortical connections and U-association bundles on the one hand and the genetically determined influence of the group on the other', whatever this may mean with respect to a still not achievable more specific and intelligible formulation.

The terms '*sulcus*' and '*fissure*' *(fissura)* have been used in a rather indiscriminate fashion by the diverse authors, and can now be consid-

---

[7] '*Es kann als sicher gelten, dass die Furchen (Sulci) und die zwischen ihnen liegenden Windungen (Gyri) der Grosshirnoberfläche endogenen Wachstumsvorgängen der Hirnsubstanz ihre Entstehung verdanken. Die Verteilung der Blutgefässe hat auf den Verlauf der Furchen keinen Einfluss, im Gegenteil ist die innere Wachstums- oder Faltungstendenz der Hirnsubstanz eine so starke, dass Wachstum und Relief des Schädels durch das Wachstum des Gehirns beeinflusst werden und nicht umgekehrt.*' (K., 1927, p. 285).

ered as more or less equivalent respectively interchangeable. An attempt was formerly made to distinguish *fissures* as '*Totalfurchen*' from sulci as '*Rindenfurchen*'.[8] The former were supposed to represent folds of the entire brain wall as far as the ventricular surface, while the latter involved only cortex and a neighborhood of medulla. With the exception of fissura hippocampi, however, which, at early developmental stages, is indicated by a very shallow groove corresponding to a minor and rather flat bulge into the ventricular lumen, all other deep 'fissures' (e.g. calcarina, parieto-occipitalis, collateralis, fossa *Sylvii)* do not originate as infoldings of the entire brain wall. The actually obtaining protrusions into the ventricle displayed e.g. in the Human hemisphere as calcar avis, eminentia collateralis, and bulbus cornus posterioris are entirely secondary late growth processes, taking place quite some time after these grooves became distinctly formed (cf. HOCHSTETTER, 1924). Even the hippocampus, bulging into the ventricle, and its very shallow fissura hippocampi, as displayed in the adult brain, are both the result of rather late secondary developmental processes.[9] Further comments on the terms 'sulci' *versus* 'fissures' with regard to the Human hemispheres will be given further below in section 6, dealing with the general evaluation of the telencephalic fissuration pattern.

It is of some interest to compare the formation of telencephalic sulci and gyri with that obtaining in the cerebellum, which likewise displays a cortex. In the large cerebellum of Selachians and Osteichthyes, the sulci or fissures are formed by foldings of the entire wall and thus involve the ventricle.[10] In Birds and Mammals, on the other hand, these grooves are surface folds.[11] In Reptiles, only the Crocodilian cerebellum displays two grooves, with at best very faint ventricular bulge.[12] Generally speaking, the Vertebrate cerebellar sulci are transverse, at right angle to the rostrocaudal cerebellar expansion, but a median longitudinal groove is present in various Fishes, with further complications in Mormyrids. In diverse Mammals, bilateral, more or less longitudinal grooves are likewise added to the transverse ones.[13] The tectum mesencephali with its cortex, *per contra*, does not display sulci and gyri comparable to the telencephalic and cerebellar ones.

---

[8] Cf. the discussion of this topic by HOCHSTETTER (1924).

[9] Cf. K (1927, p. 286).

[10] Cf. volume 4, chapter X, Figures 306, p. 672; 307, p. 673; 312, p. 682; 317, p. 693.

[11] Cf. volume 4, Figures 337B, C, p. 734; 338C, D, p. 736; 344A, B, p. 747.

[12] Cf. volume 4, Figure 332, p. 725.

[13] Cf. volume 4, Figures 348C–E, p. 753.

The significance of telencephalic size respectively mass was mentioned above with reference to lissencephalic and gyrencephalic brains, and various other aspects of the topic concerning brain size, mass or weight, were dealt with in section 8, chapter VI of volume 3/II. A detailed treatise on the evolution of brain size in relation to 'intelligence' with particular emphasis on paleoneurological evidence was recently published by JERISON (1973), who collected the relevant available data in all Vertebrate groups, recent and fossil. HOLLOWAY (1974) has criticized various inconclusive opinions expressed by JERISON, but further comments on these ambiguous questions can be omitted from the present context.[14]

It seems appropriate, however, to deal here briefly with another aspect of the brain size topic, namely with the complex *domestication problem*. More than hundred years ago, about 1868, DARWIN had reported reduction of cranial capacity and brain volume in domestic animals, and pointed out diverse variations displayed by plants and animals under domestication,[15] but his findings were largely neglected until again stressed by KLATT (1912).

At the turn of the century, about 1898, both LAPIQUE and DUBOIS, who were concerned with problems of 'cephalization' (cf. section 8, chapter VI, vol. 3/II) simultaneously and independently stated that in all domestic animals the cephalization coefficient is smaller than in their wild (feral) relatives.[16]

Subsequently KLATT (1912, 1921, 1928, 1932), stressing and confirming DARWINS's findings, investigated the domestication problem. KLATT noted a shortening of various length measurements *(Verkürzung bestimmter Längendimensionen)* in comparison with wild forms, and a substantial, but variable reduction in cranial capacity and brain volume, roughly speaking in an order of magnitude up to between 20 and more than 30 per cent. As regards e.g. Dogs, the reduction is said particularly to affect the hemisphere's sensory areas, being, however, concomitant with a relative increase of the association areas. It should, moreover, be added that the reduction in brain size does not exclusi-

---

[14] A short 'anticritique' directed against HOLLOWAY by GOULD was published in Science *185:* 400–401 (1974). A brief reference to JERISON's previous work can be found on p. 738 of volume 3/II.

[15] Cf. HERRE (1966), with relevant bibliography.

[16] Cf. KAPPERS (1929) for further details and relevant bibliographic data. It should be added that not only domesticated Mammals but also Birds (e.g. Ducks) display said reduction.

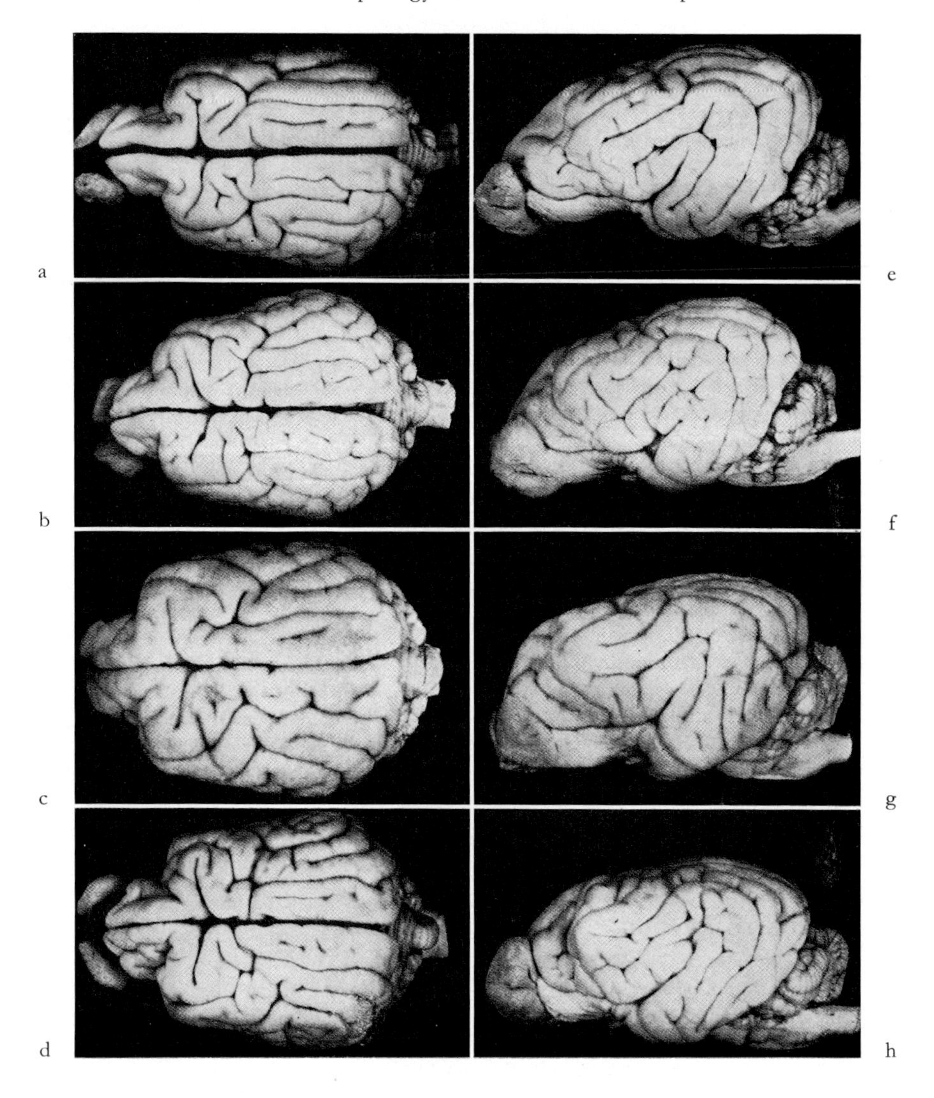

Figure 4. Dorsal and lateral aspects of neopallial gyri and sulci in wild, domesticated, and hybrid Carnivores (from HERRE, 1966). a, b: European Wolf; c, d: large Poodle *(Königspudel)*; e, f: small Poodle *(Zwergpudel)*; g, h: first generation hybrid of male Poodle and female Wolf. For further details concerning the relevant body and brain weights, HERRE's original paper should be consulted.

vely involve the telencephalon, but, to a different and variable degree, also the other main subdivisions of the brain.

Among contemporary authors dealing with this exceedingly complex problem[17] are HERRE (1964, 1966), KRUSKA (1970, 1972, 1973a, b, 1975), KRUSKA and STEPHAN (1973), RÖHRS (1961), RÖHRS and KRUSKA (1969) and STEPHAN (1954).

KRUSKA (1970, 1972, 1973a, b), and his collaborators (KRUSKA and STEPHAN, 1973; KRUSKA and RÖHRS, 1974) have pointed out that although some olfactory and sensory grisea are reduced in domesticated Pigs if compared with feral ones, the greatest reduction is displayed by grisea of the limbic system (hippocampus with parahippocampal cortex). These changes seem related to the behavioral changes obtaining in domestication. Thus, feral Pigs are highly aggressive, and domesticated Pigs become quite docile. It will be recalled that the limbic system most likely represents an important neural mechanism regulating affective tone.

Laboratory Rats display a much better learning capacity than wild Rats, although a decrease in relative brain weight obtains in the former (KRUSKA, 1975). The greater learning ability may well result from greater docility rather than from higher 'intelligence'.

Another interesting feature, pointed out by HERRE (1966) and others, is an apparently more complex fissuration pattern displayed by at least some domesticated forms in comparison to their wild relatives (cf. Fig. 4). This, however, does not mean greater surface expansion, since the relevant convoluted areas were found to be relatively smaller. Certain domesticated animals are said to show fissuration patterns not present in the related wild forms. HERRE (1966) also succeeded in crossing a male domesticated Carnivore form (Poodle) with a female wild one (Wolf). The body and brain weights of the hybrid were of an intermediate order of magnitude, and the Poodle's more pronounced fissuration pattern was retained (Fig. 4g, h).

Some of the poorly understood questions related to the still incompletely investigated domestication problem with its highly diversified manifestations involve the genetic aspects. Present-day theories preclude an effect of environmental factors on the 'genome', but could be compatible with JOLLO's 'lingering modifications' *(Dauermodifikatio-*

[17] Thus, e.g. some domesticated animals such as the Reindeer and the laboratory Mouse are said to make an exception to the relative reduction of brain weight (HERRE, 1966).

nen) as pointed out in volume 1, p. 135. A 'shuffling' or modified combination of obtaining genetic factors was also suggested.[18] The genetics of the relevant structures and the intraspecific variations compared with interspecific differences need further elucidations before plausible theories of evolutionary processes and events may be formulated (cf. HERRE, 1966).

Two important aspects of domestication are designed (artificial, contrived) *constraint*, and *protection*. *Isolation* and *selection* become additional relevant factors. In this respect it is evident that comparable parameters obtain in the evolution of Human populations as 'ethnic' or 'racial' groups, likewise involving constraint, protection, selection and relative isolation, resulting in what might be called vaguely and arbitrarily 'delimitable' 'gene pools'. Thus, Man can be conceived as a '*self-domesticated*' Ape.[19]

2. Remarks on the Non-Primate Fissuration Pattern

Generally speaking and reverting to the longitudinal respectively arcuate sulci on the lateral surface of the neopallium as shown in Figure 2, the Non-Primate (or Subprimate) Eutherian gyrencephalic brains are characterized by a vallecula, fossa, or sulcus of variable

---

[18] In his theoretical discussion of the problem, STEPHAN (1954) enumerates diverse aspects of domestication or captivity with regard to the animal's environment and mode of life. With HERRE he emphasizes '*dass sehr rasch nach der Haustierwerdung die verschiedenen Haustierformen eine erstaunliche Vermannigfaltigung erfahren*'. Concerning genetic aspects STEPHAN states: '*Die veränderte Umwelt scheint also einen mutationsauslösenden Einfluss zu haben. Diese zahlreichen ungerichteten Mutationen erfahren dann unter der Hand des Menschen die gewünschte Auslese.*'

[19] Cf. the comments by SALLER (1930): '*Beim Menschen wie bei den Haustieren wird die natürliche Wirkung einer Auslese vielfach durch menschliche Eingriffe illusorisch gemacht. Wie sich die Haustiere im Zustand der Domestikation befinden und ihre Fortpflanzung durch menschliche Willkür geregelt wird, so befindet sich der Mensch in einem Zustand der Selbstdomestikation* (THURNWALD), *die in seinem Wesen begründet ist. In diesem Zustand werden die Auslesevorgänge vielfach unterbunden, d.h. der Mensch sorgt aus irgendwelchen Motiven – bewusst oder unbewusst – für das Erhaltenbleiben von Varianten, welche im Wildzustand zugrunde gehen würden. So kann die Domestikation die Verteilung der Merkmale innerhalb der Variationsreihe einer Rasse derart verschieben, dass ein zunächst anormales Merkmal im Lauf der Domestikation immer mehr in den Normbereich rückt. Für die Annahme, dass die Domestikation eine Umänderung der Erbmasse und dadurch eine Vergrösserung der primären genotypischen Variabilität des Menschen hervorruft, liegen keine Beweise vor.*' As regards this last point, which remains indeed a moot question, cf. however, the comments by STEPHAN in footnote 18.

depth, the *sulcus Sylvius* or *pseudosylvius*, corresponding to the concavity of the hemispheric bend, whose 'center of rotation' lies within the basal ganglia. Three arcuate sulci *(Bogenfurchen)* may surround that pseudosylvian depression, namely, in basodorsal sequence, *sulcus ectosylvius*, *sulcus suprasylvius*, and *sulcus arcuatus lateralis*.

The gyri delimited by these sulci are, again in basodorsal sequence, the following: (1) *Gyrus arcuatus primus sive Sylvius*, between pseudosylvian sulcus and ectosylvian sulcus; (2) *gyrus arcuatus secundus sive ectosylvius;* (3) *gyrus arcuatus tertius sive suprasylvius;* (4) *gyrus arcuatus quartus sive marginalis aut lateralis* (cf. Fig. 13).

Additional sulci located rostrally to this system will be discussed further below in dealing with diverse Mammalian forms. On the medial aspect, the likewise longitudinal respectively curved sulcus splenialis (cf. Fig. 13) is a rather constant feature in a pattern commonly characterized by further sulci. The behavior and significance of the sulcus hippocampi was pointed out in the preceding section.

On the lateral surface the vallecula of the hemispheric bend, which, in some lissencephalic brains, barely involves the neopallium, but essentially only the piriform lobe (cf. Figs. 221 A, C, vol. 5/I) tends to become progressively deeper, absorbing, as it were, the insula and likewise drawing in the adjacent arcuate sulci and gyri. The insula thereby becomes *opercularized*, being covered by an operculum frontale, parietale, and temporale, particularly developed in certain Primates, and displaying insular sulci and gyri.

Because diverse complexities, such as interruption or fusion of sulci, formation of apparently random accessory or third and fourth order sulci, moreover dubious and thereby controversial relationships to neopallial 'areas', the formulation of a valid overall fissuration pattern upon which homologies can be based, presents substantial intrinsic difficulties. Thus, in diverse instances, the problem of homologization becomes undecidable.[20] To some extent, nevertheless, a reasonably valid comparison of the fissuration pattern in Non-Primate and in Primate gyrencephalic brains seems to be possible. Detailed discussions on the morphology of Mammalian telencephalic sulci and gyri

[20] '*Die Frage, inwieweit Grosshirnfurchen verschiedener Ordnungen miteinander verglichen werden können, berührt die Prinzipien der Homologienlehre selbst*' (K., 1927, p. 287). This involves, e.g., the arbitrary decision whether shifting of neopallial sulci with respect to cortical areas or vice-versa is of significance for homologization. In other words, are fissuration patterns and 'cortical parcellation' (a) two autonomous morphogenetic events, (b) partly interrelated, or (c) strictly correlated.

are included in the treatises by FLATAU and JACOBSOHN (1899), PAPEZ (1929), *Graf* HALLER (1934), KAPPERS *et al.* (1936) and BECCARI (1943). The two latter texts contain numerous bibliographic references to the investigations of the older authors. Of particular interest are also the studies by ELLIOT SMITH (1902, 1903/04a, b, 1907a, b and others). Relevant data concerning domestic Mammals can also be found in the treatises by ELLENBERGER and BAUM (1926) and by SISSON and GROSSMAN (1959). Still more recent pertinent publications are those by BRAUER and SCHOBER (1970) and SCHOBER and BRAUER (1975).

Turning now to the 'aplacental' Mammals,[21] it will, be recalled that in the Prototherian Monotremata, the hemispheres of Ornitorhynchus are lissencephalic (Fig. 1 E), while those of Echidna are gyrencephalic (Fig. 241 A, vol. 5/I, and Fig 1 F of the present volume).

Figure 5 illustrates the lateral aspect of the hemisphere in Echidna. It can be seen that dorsally to the small Sylvian (pseudosylvian) sulcus[22] two roughly arcuate systems of sulci are displayed, the inner one provided by sulcus antesylvius posterior and postsylvius anterior, the outer one by sulcus antesylvius anterior and postsylvius posterior in KAPPERS' nomenclature. ELLIOT SMITH used a notation with Greek letters, likewise indicated in the Figure. Whether these grooves should be considered homologous or analogous to the Eutherian ectosylvian and suprasylvian arcuate sulci remains an open question. On the medial aspect, well dorsally to sulcus hippocampi, there is a longitudinal sulcus (sulcus vallaris) somewhat similar to either sulcus splenialis or genualis of Eutheria, in addition to diverse nondescript transverse (radial) sulci in part continuous with extensions of the aforementioned lateral ones.

Again, on the lateral aspect, rostrally to sulcus antesylvius anterior, there are nondescript variable sulci, whose identification with the rostral lateral fissuration patterns of Eutheria remains uncertain.

Among the Metatherian Marsupials, the brain of the Opossum (Didelphis) is essentially lissencephalic. It nevertheless may display a poorly developed rostral so-called sulcus (or 'fissura') orbitalis in the somatic motor area. On the basis on this relationship, said sulcus

---

[21] The brain morphology of Monotremes and Marsupials was especially investigated by ZIEHEN (1897–1908) who made use of the material collected in Australia by R. SEMON.

[22] As forerunner of the deep fossa Sylvii *sive* fissura cerebri lateralis it could also be called presylvian sulcus, but this latter term has been preempted for a groove located rostrally to the system of arcuate sulci of diverse Eutherian brains.

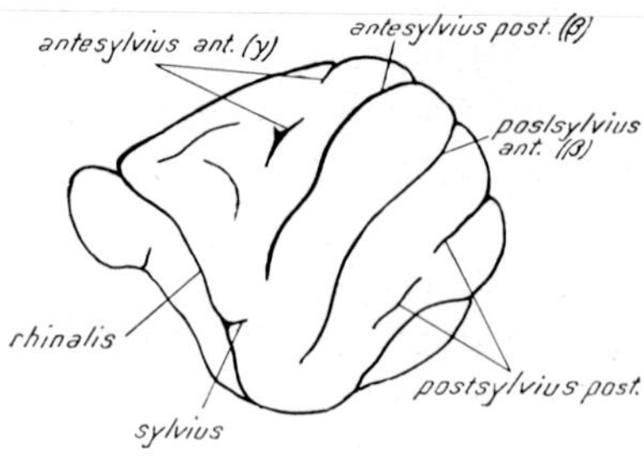

Figure 5. Lateral aspect of the telencephalon in Echidna (after ZIEHEN, from BECCARI 1943). The nomenclature is that of KAPPERS. Greek letters in parenthesis refer to the nomenclature of ELLIOT SMITH.

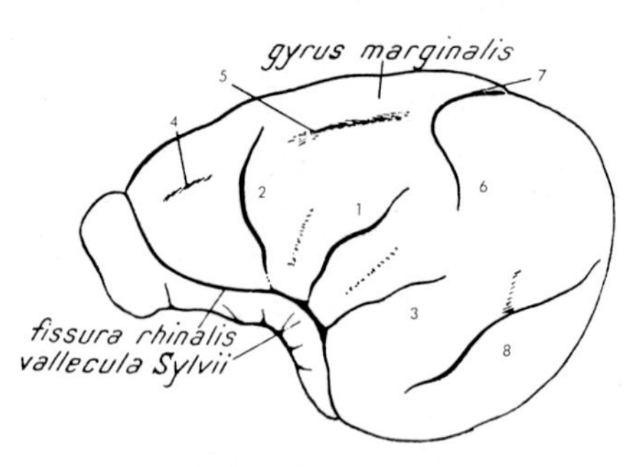

Figure 6. Lateral aspect of the telencephalon in Macropus ualabatus (after ZIEHEN, from BECCARI, 1943). 1: sulcus pseudosylvius; 2: sulcus antesylvius; 3: sulcus postsylvius anterior; 4: rostral sulcus; 5: antero-inferior lateral sulcus; 6: postero-inferior lateral sulcus; 7: superior lateral sulcus; 8: sulcus postsylvius caudalis. The designations have been changed to the present author's simplified nomenclature. Note also two barely suggested (dotted) grooves pertaining to the ante- and postsylvian system.

might be compared with the sulcus cruciatus as seen in Eutherian Carnivores and described further below.

The neopallium of large Marsupials, such as e.g. the Kangaroo Macropus, and also that of some smaller ones such as the Kangaroo-Rat Hypsiprymnus, shows several more or less developed sulci. Since it seems uncertain whether these grooves can be compared in a satisfactory manner with those of gyrencephalic Eutheria, ZIEHEN (1897, 1901) and others, following ELLIOT SMITH, used Greek letters for their designation. It is perhaps preferable to use a more descriptive although non-committal terminology, and to distinguish as here indicated in Figures 6 and 7, a rostral sulcus, a *Sylvian* (or pseudosylvian) sul-

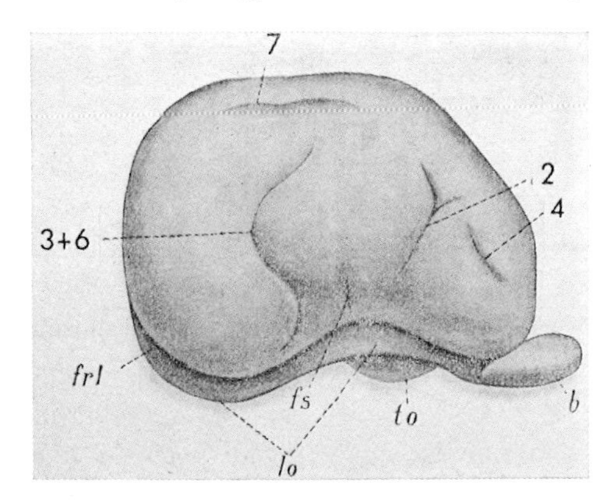

Figure 7. Lateral view of right hemisphere in the Kangaroo-Rat Hypsiprymnus (from LIVINI, 1907). b: olfactory bulb; frl: sulcus rhinalis lateralis; fs: pseudosylvian sulcus; lo: piriform lobe; to: tuberculum olfactorium; 2, 3, 4, 6, 7, as in figure 6.

cus, an antesylvian and one or two postsylvian sulci, and one or more nondescript lateral respectively longitudinal sulci. Data reported by DILLON (1963) seem to indicate substantial taxonomically related variations in Macropodidae.

On the medial aspect of some large brains, as e.g. in Macropus, a longitudinal sulcus with a rostral transverse extension may be present dorsally to sulcus hippocampi. In addition, two or more variable and poorly defined sulci, which could even be shrinkage artefacts, have been depicted in some of the larger Marsupials by LIVINI (1907) and others.

Turning now to the Eutherian Mammals, most Insectivore, Chiropteran, and Rodent brains can be classified as lissencephalic. This generalized statement, nevertheless, requires some qualifications. Thus, in the Chiropteran Pteropus a pseudosylvian sulcus and two arcuate sulci are faintly indicated on the lateral aspect, a somewhat more pronounced longitudinal sulcus, dorsally to sulcus hippocampi, may be seen on the medial surface. In the Rodent Hydrochaerus (Fig. 1) a pattern of sulci, rather similar to that in some Marsupials is recognizable. In Cavia (Fig. 8), a rostrally dorsolateral and caudally more dorsomedial longitudinal groove may be displayed, whose rostral part might correspond to a pre-arcuate or anterior arcuate sulcus, while its posterior part seems to be the groove between isocortex and parahippocampal cortex. The brains of Rats and Mice (Fig. 9 A), on the other hand, are lissencephalic. As regards Lagomorphs, which were once consid-

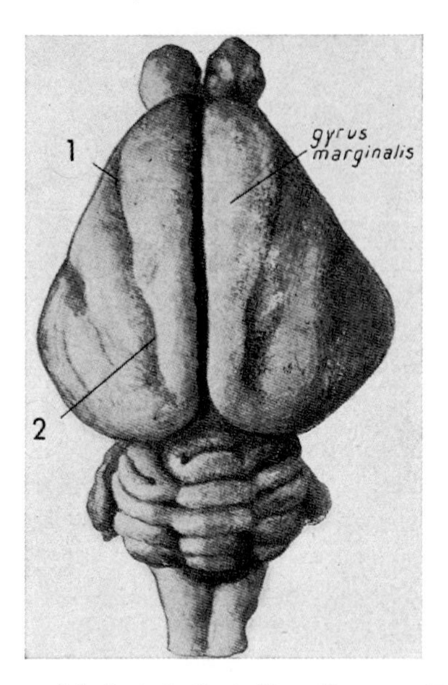

Figure 8. Dorsal aspect of the brain in Cavia (from BECCARI, 1943). 1: probably sulcus lateralis; 2: probably sulcus between isocortex and parahippocampal cortex.

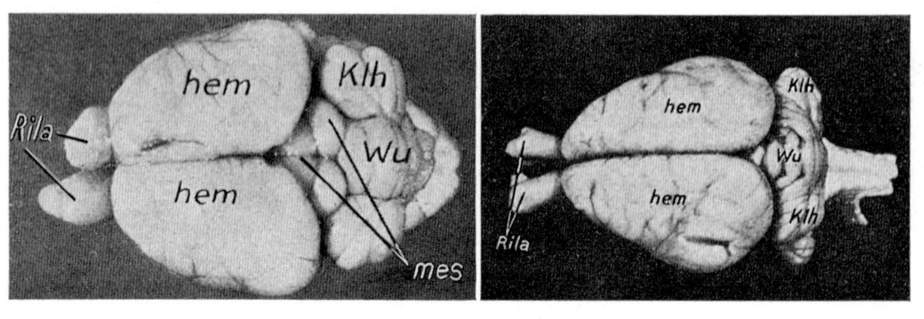

Figure 9. Lissencephalic brains of Rodent Mouse (A) and (B) of Lagomorph Rabbit (from MAURER, 1928). hem: lobus hemisphaericus; Klh: cerebellar hemisphere; Rila: olfactory bulb; Wu: cerebellar vermis. A Magnified about 3×. B Reduced about $^3/_4$. Left hemisphere of Rabbit slightly damaged in caudal region.

ered a suborder of Rodentia,[23] the brain of the Rabbit is frequently described as lissencephalic, and may indeed display a smooth neopallium (Fig. 9B). Yet, rather frequently a dorsolateral faint sulcus is present at the boundary between isocortex and parahippocampal cortex

[23] Cf. volume 1, p. 66.

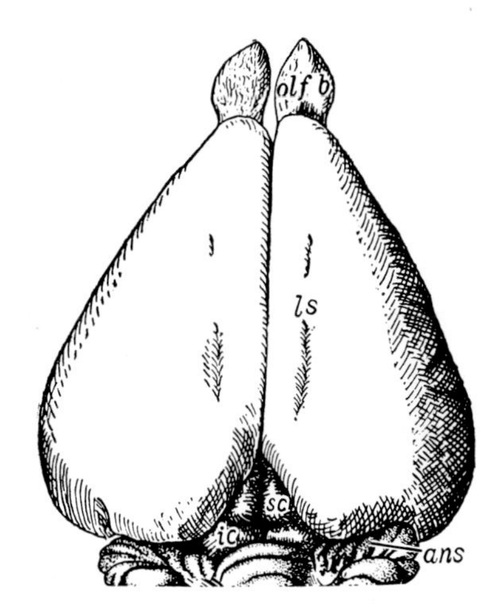

Figure 10. Dorsal aspect of Rabbit hemisphere showing faint longitudinal sulcus (from PAPEZ, 1929). ans: ansiform lobe of cerebellum; ic: inferior colliculus; ls: 'lateral sulcus' (sulcus between isocortex and parahippocampal cortex, cf. Fig. 223 B, vol. 5/I); olf.b.: olfactory bulb; sc: superior colliculus.

(Fig. 223B, vol. 5/I, and Fig. 10 of the present volume). Again, in some instances, nondescript additional pallial sulci, perhaps deformation or shrinkage artefacts may be noted, which in KRAUSE's (1921) interpretation, are shown in Figure 11.

As regards Edentata (Fig. 1O, P), which include lissencephalic and moderately gyrencephalic forms, these latter display a pseudosylvian groove and manifestations of arcuate sulci, which GRAF HALLER (1934) interprets as comparable with those in gyrencephalic Marsupials and Rodents. BECCARI (1943) expresses some doubts concerning this question, but suggests that, of the two sequences of arcuate sulci in Bradypus (Fig. 12), the basal one corresponds to sulcus suprasylvius, and the dorsal one, consisting of two parts, to sulcus arcuatus lateralis of the Carnivore-Ungulate pattern.

The neopallial sulci and gyri of Carnivores have been investigated and described by numerous authors. The fissuration in this Eutherian order appears to be representative of the typical or fundamental pattern for 'intermediate' Mammals, since, despite various modifications most of its essential features can be recognized in Ungulates, Proboscidians and Cetaceans. Some of that pattern's features, moreover, also remain reasonably well identifiable in Primates.

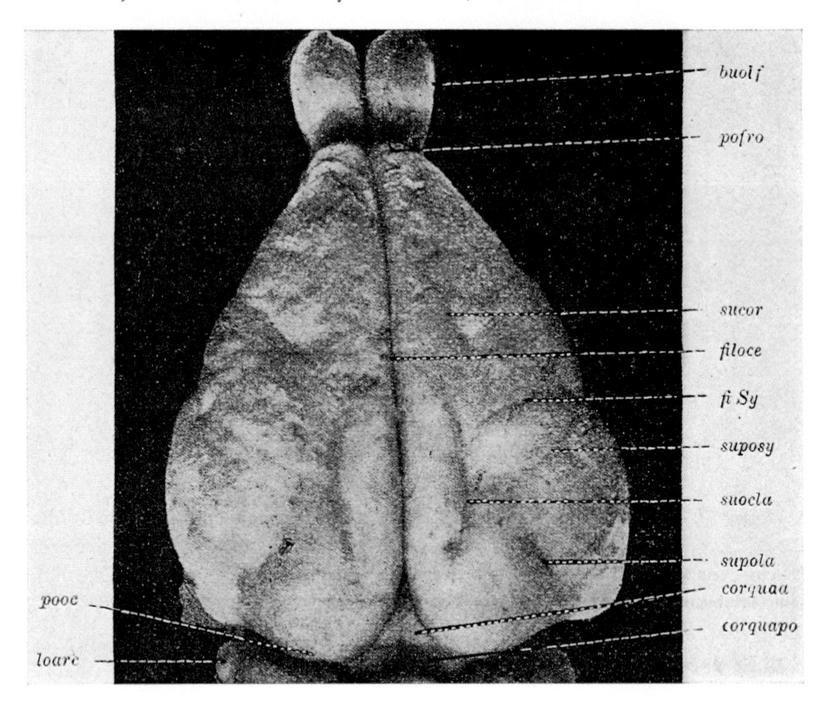

Figure 11. Dorsal aspect of Rabbit hemisphere showing various faint sulci (from Krause, 1921). buolf: bulbus olfactorius; corquaa, corquapo: anterior and posterior colliculus; filoce: fissura longitudinalis cerebri (interhemisphaerica); fi Sy: sulcus pseudosylvius; loarc: ansiform lobe of cerebellum; pofro, pooc: rostral and occipital pole of hemisphere; sucor: 'sulcus coronarius'; suocla: 'sulcus occipitalis lateralis' (groove between isocortex and parahippocampal cortex, cf. Fig. 10); supola: 'sulcus posterior lateralis'; suposy: 'sulcus postsylvicus'.

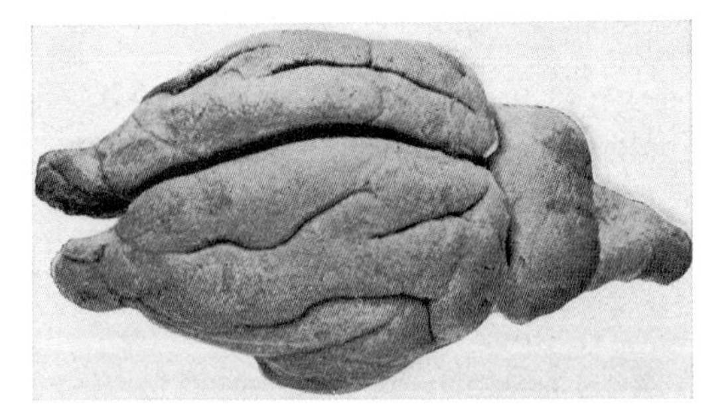

Figure 12. Oblique dorsal view of the brain in the Edentate Bradypus tridactylus, displaying pseudosylvian sulcus (left) and unidentified longitudinal sulci (from Beccari, 1943).

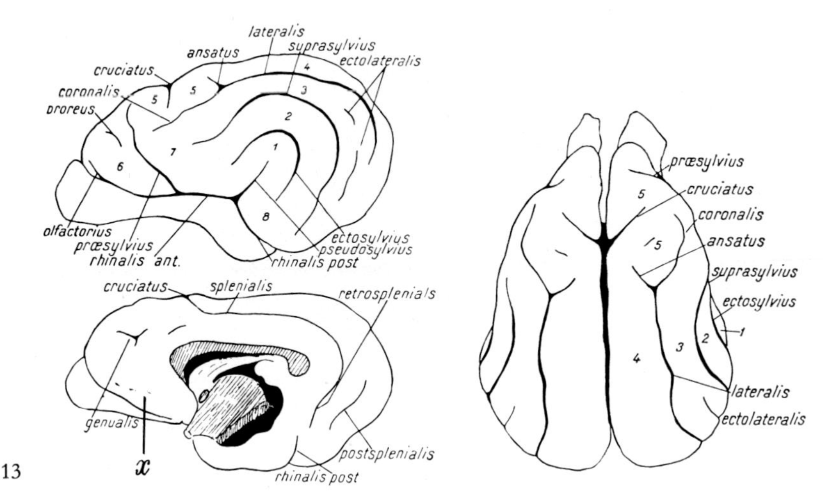

Figure 13. Lateral and medial aspect of telencephalon in the Carnivore Wolf, Canis vulpes (from BECCARI, 1943). 1: first arcuate (Sylvian) gyrus; 2: second (ectosylvian) arcuate gyrus; 3: third (suprasylvian) arcuate gyrus; 4: fourth (marginal) arcuate gyrus; 5: sigmoid gyrus; 6: gyrus proreus sive orbitalis; 7: gyrus reuniens; 8: gyrus polaris. Added x: sulcus rostralis.

Figure 14. Dorsal aspect of hemisphere in the Wolf (from BECCARI, 1943). 1–5 as in Figure 13.

Figures 13 and 14 illustrate the sulci and gyri displayed by Carnivore Canidae (cf. also Fig. 4). On the lateral side, dorsally to the pseudosylvian sulcus, the three arcuate sulci (ectosylvius, suprasylvius, lateralis) with their four arcuate gyri (primus *sive* Sylvius, ectosylvius, suprasylvius, marginalis *sive* lateralis) are conspicuous. Between sulcus arcuatus lateralis and sulcus suprasylvius, a variable sulcus ectolateralis may be present within the caudal bend of the arcuate system, whereby a gyrus ectolateralis between lateral and ectolateral sulci becomes delimited.[24] At the caudobasal end of sulcus ectosylvius, a gyrus polaris is commonly seen.

Rostrally to the system of arcuate sulci runs a sulcus coronalis, which may or may not join the sulcus lateralis, whose rostral end branches into a sulcus ansatus. Rostrally to this latter, the transverse sulcus cruciatus represents an important landmark and lies within the main somatic motor area. A nondescript postcruciate sulcus may be

[24] According to KAPPERS (1921) the ectolateral sulcus as well as an entolateral sulcus found in diverse Ungulates represent merely a 'doubling' of the sulcus arcuatus lateralis.

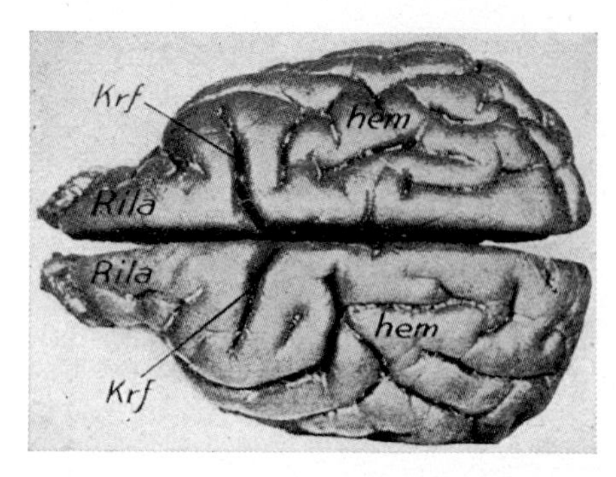

Figure 15 A. Dorsal aspect of the hemisphere of the domestic Dog (from MAURER, 1928). hem: hemisphere; Krf: sulcus cruciatus; Rila: region of gyrus proreus (frontal lobe; partly covered by this lobe, a portion of olfactory bulb can be seen).

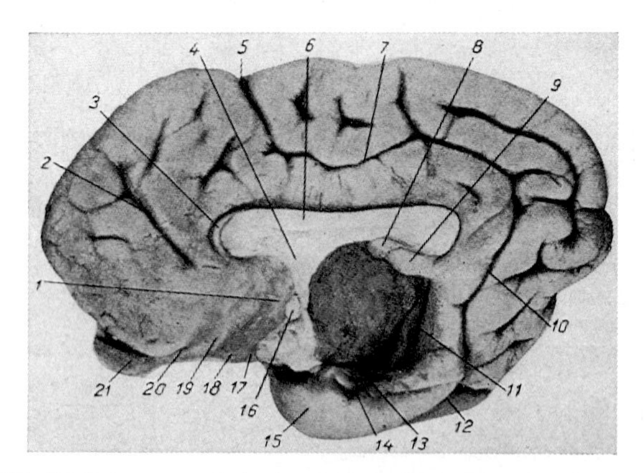

Figure 15 B. Medial aspect of the hemisphere in the Dog (from BECCARI, 1943). 1: paraterminal area; 2: sulcus genualis; 3: supracommissural hippocampus with paraterminal griseum; 4: intraterminal portion of paraterminal grisea ('septum'); 5: sulcus cruciatus; 6: corpus callosum; 7: sulcus splenialis; 8: subcommissural portion of retrocommissural hippocampus; 9: transition of retrocommissural hippocampus to parahippocampal cortex; 10: sulcus retrosplenialis; 11: sulcus (sive fissura) hippocampi; 12: sulcus rhinalis lateralis, pars posterior; 13: fimbria hippocampi; 14: fascia dentata; 15: piriform lobe; 16: anterior commissure; 17, 18: tuberculum olfactorium; 19, 20: olfactory stalk with medial olfactory tract; 21: olfactory bulb. Some minor modifications of BECCARI's terminology have been made.

present, and a variable sulcus proreus lies rostrally to sulcus cruciatus. The oblique sulcus praesylvius *(sive* sulcus orbitalis) runs caudad toward sulcus rhinalis lateralis. A short sulcus olfactorius may extend from the latter sulcus into the rostral neopallium.

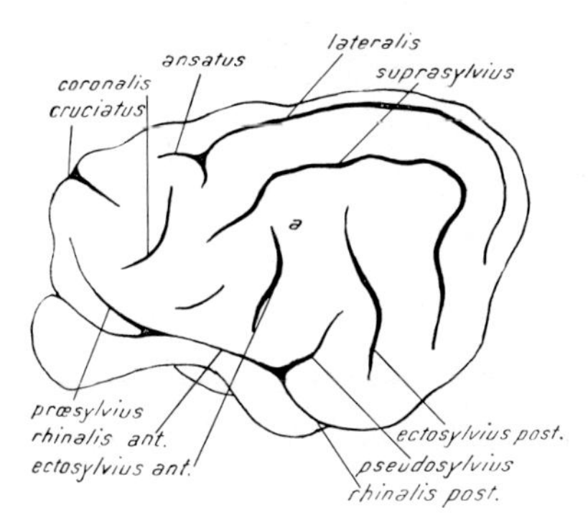

Figure 16. Lateral aspect of hemisphere in the Felid Carnivore domestic Cat (from BECCARI, 1943). a: so-called gyrus felinus.

The Nomenclature of the rostral gyri is not satisfactorily standardized, but one may, with BECCARI (1943) and others, distinguish an anterior and posterior sigmoid gyrus laterally surrounding the sulcus cruciatus, a gyrus reuniens located caudally to presylvian sulcus, and a gyrus proreus *sive* orbitalis rostrally to the presylvian groove.

On the *medial aspect*, there is a longitudinal caudobasalward curving sulcus splenialis, also called sulcus cinguli, whose caudal portion is named sulcus retrosplenialis. The rostral part of sulcus splenialis may or may not join the sulcus cruciatus. The sulcus genualis can be considered a rostral, separate portion of the sulcus splenialis system. This latter, at least to some extent, indicates the boundary between isocortex and parahippocampal cortex, and thereby would correspond to the dorsolateral sulcus mentioned above in Lagomorphs. A sulcus rostralis, basally to sulcus genualis, and, in the medial occipital region, caudally to sulcus retrosplenialis, a postsplenial sulcus are commonly present.

The gyri on the medial aspect are designated in diverse manners, but a gyrus rectus basally to sulcus rostralis, a gyrus cinguli basally to genualis-splenialis system, continuing into gyrus retro- or parasplenialis and hippocampal gyrus[25] around the splenium, are generally recognized. Dorsally to genualis-splenialis system, a gyrus frontalis superior rostrally to cruciate sulcus, a caudally adjacent gyrus parietalis

[25] The gyrus hippocampi, characterized by parahippocampal cortex, should not be confused with the hippocampus (or hippocampal formation), consisting of cornu Ammonis and gyrus dentatus ('fascia dentata').

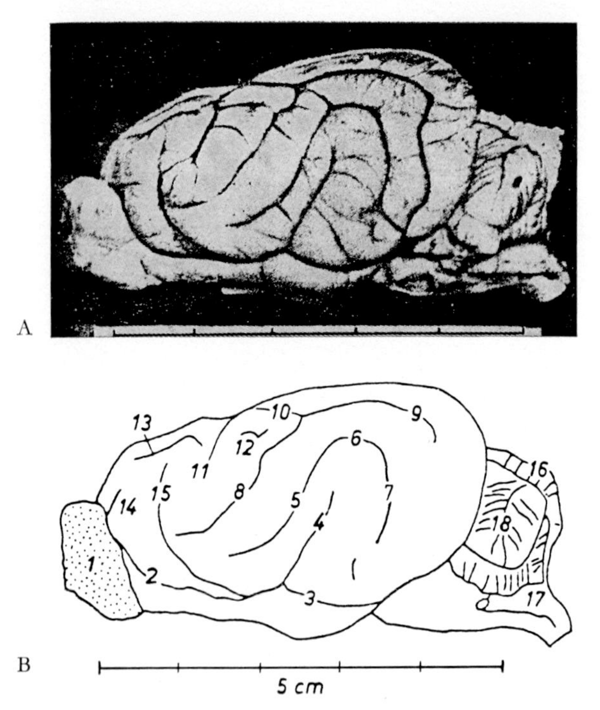

Figure 17 A, B. Lateral aspect of the telencephalon in the Mustelid Carnivore Marten (from SCHUMACHER, 1963). 1: olfactory bulb; 2, 3: sulcus rhinalis lateralis; 4: sulcus pseudosylvius; 5, 6, 7: subdivisions of sulcus suprasylvius; 8: sulcus coronalis; 9: sulcus lateralis; 10: sulcus ansatus; 11: sulcus cruciatus; 12: sulcus postcruciatus; 13: sulcus proreus; 14: sulcus olfactorius; 15: sulcus praesylvius; 16: cerebellar vermis; 17: oblongata; 18: cerebellar hemisphere.

medialis along sulcus splenialis, and a gyrus suprasplenialis dorsally to caudal bend of sulcus splenialis have been named. Rostro-basally the boundaries between isocortex, parahippocampal cortex, precommissural hippocampus, and paraterminal body (diagonal band) are, as a rule, not clearly indicated by unambiguously distinctive surface landmarks. The relevant similarities and very slight dissimilarities between feral and domestic Canidae with regard to the fissuration pattern were mentioned above in dealing with the domestication problem (cf. Fig. 4). Figures 15 A and B depict aspects of the Dog's hemisphere. Inspection of Figure 15 will disclose that in the Dog (as well as in other Carnivore and generally speaking gyrencephalic Mammalian brains) fairly conspicuous asymmetries of the antimeric fissuration pattern may obtain.

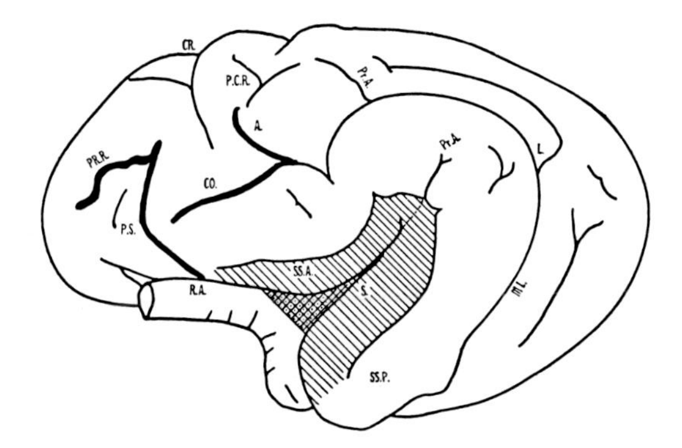

Figure 18. Lateral aspect of the hemisphere in the Ursid Carnivore Ursus malayanus (from KAPPERS *et al.*, 1936). A: sulcus ansatus; CO: sulcus coronalis; CR: sulcus cruciatus; L: sulcus lateralis; ML: sulcus postlateralis; PCR: sulcus postcruciatus; PrA: processus acuminis; PRR: sulcus proreus; RA: sulcus rhinalis lateralis; S: sulcus pseudosylvius; SSA: sulcus suprasylvius anterior; SSP: sulcus suprasylvius posterior. Second arcuate gyrus in oblique hatching, first arcuate gyrus on floor of Sylvian fossa cross-hatched (for other interpretation, cf. Fig. 19).

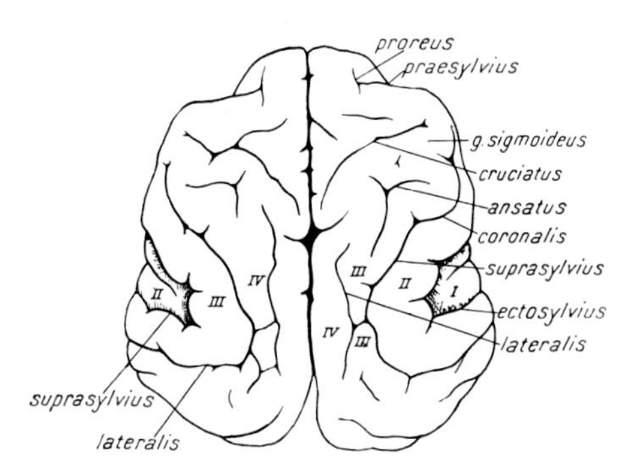

Figure 19. Dorsal aspect of hemisphere in the Ursid Carnivore Ursus maritimus (after GRAF HALLER, 1934, from BECCARI, 1943). I–IV: arcuate gyri, at left in the interpretation of HOLL, at right in that of HALLER v. HALLERSTEIN).

Figures 16 and 17 illustrate the lateral aspect of the hemisphere in the Felid Cat and the Mustelid Marten, whose gyri and sulci show an arrangement differing in some respects from that displayed by Canidae. In Felidae, the ectosylvian sulcus is commonly interrupted by a gyrus felinus, separating a posterior ectosylvian sulcus from an anterior one. In Mustelidae, the sulcus ectosylvius seems to be completely missing.

Ursidae and Pinnipedia display a tendency toward formation of an *insula with opercularization.* There is, however, some doubt as to the in-

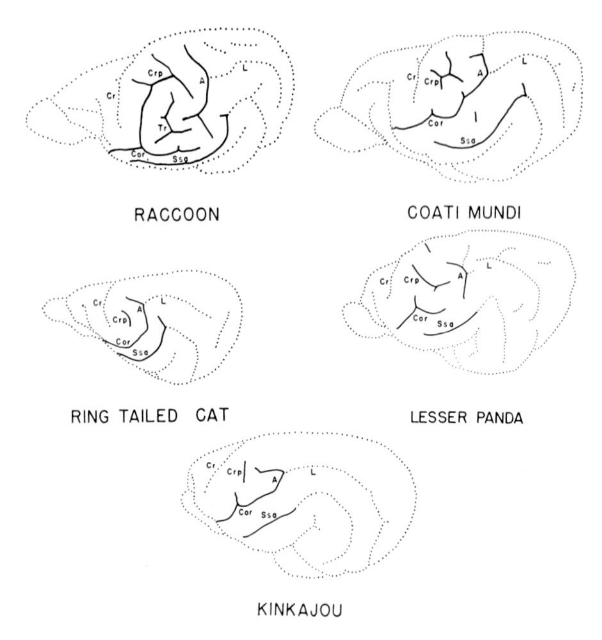

Figure 20. Lateral aspect of hemisphere in five Procyonid Carnivores with emphasis on postcruciate fissuration (from WELKER and CAMPOS, 1963). A: ansate sulcus; Cor: coronal sulcus; Cr: cruciate sulcus; Crp: postcruciate sulcus complex; L: lateral sulcus; Ssa: anterior suprasylvian sulcus; Tr: triradiate sulcus.

terpretation of the actual findings. The entire first arcuate gyrus may form the floor of the insula, the sulcus ectosylvius becoming a sulcus circularis insulae. The pseudosylvian sulcus could now be designated as a true *Sylvian sulcus* (or fissura cerebri lateralis). The groove surrounding this sulcus could be the sulcus suprasylvius with a processus acuminis (cf. Fig. 18). According to GRAF HALLER (1934) however, the groove surrounding the sylvian sulcus would still be the sulcus ectosylvius, in accordance with an incomplete opercularization of the first arcuate gyrus. This difference in interpretation[26] is indicated by Figure 19. The arrangement displayed by Procyonidae, formerly included among the Ursidae, is shown in Figure 20, which should be compared with Figures 18 and 19.

[26] A detailed discussion concerning the problem of the evolution of the Mammalian insula, with particular reference to the investigations by HOLL (1899, 1900, 1902) and other older authors is contained in the treatise by LANDAU (1923). Additional elaborations on this problem with bibliographic references can also be found in the publications by KAPPERS *et al.* (1936), GRAF HALLER (1934) and BECCARI (1943).

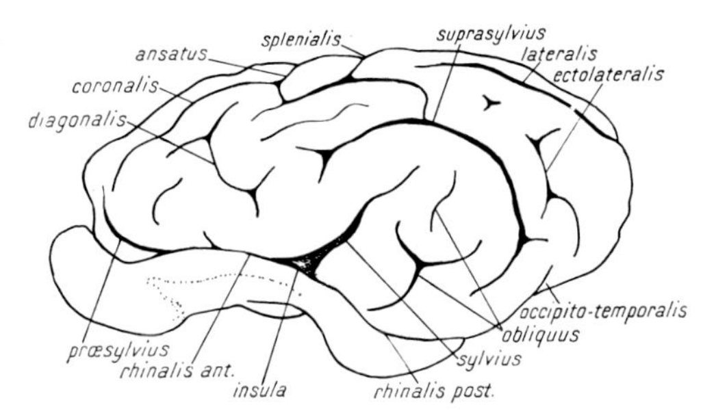

Figure 21. Lateral aspect of telencephalon in the artiodactyle Ungulate domestic Pig (from BECCARI, 1943).

The fissuration pattern in Ungulates[27] is, on the whole, similar to that in Carnivores (cf. Figs. 21–25), but the typical sulcus cruciatus of these latter forms apparently has disappeared. Some authors, e.g. GRAF HALLER (1934) assume that, in diverse forms, a variable, rostro-dorsal, ascending extension of the sulcus splenialis represents a modified sulcus cruciatus, which thus would have joined the splenialis system (cf. Fig. 22C). Sulcus coronalis and ansatus are generally joined as sulcus corono-ansatus, which may display anastomoses with sulcus splenialis medially, and laterally with sulcus lateralis or sulcus suprasylvius. This latter sulcus runs dorsally to sulcus Sylvius, the first arcuate convolution providing the floor of the insula (cf. Fig. 22B). Caudobasally to sulcus suprasylvius, a sulcus obliquus is generally present, and becomes a system of sulci in large Ungulate brains. Rostrally, between sulcus praesylvius and suprasylvius, a variable sulcus diagonalis may likewise expand into a branching system. Additional dorsolateral sulci are sulcus entolateralis and ectolateralis (cf. Fig. 23).

On the medial aspect, there are sulcus rostralis, genualis, splenialis and retrosplenialis with diverse variable and nondescript secondary sulci more or less parallel to this basally concave system. A caudal ramus horizontalis posterior of sulcus retrosplenialis (Fig. 24B) is interpreted, together with part of the former, as a 'forerunner' of, or perhaps better as comparable to, the Primate sulcus calcarinus system. The sulcus retrosplenialis, moreover, may likewise represent a forerunner of the Primate parieto-occipital infolding.

[27] In addition to the descriptions by GRAF HALLER (1964), KAPPERS *et al.* (1936), and BECCARI (1943), it will here be sufficient to point out the papers by KRUEG (1878), concerning Ungulates in general, by MOBILIO (1914/15), dealing with Equidae, and by HARPER and MASER (1976), concerning the American Bison.

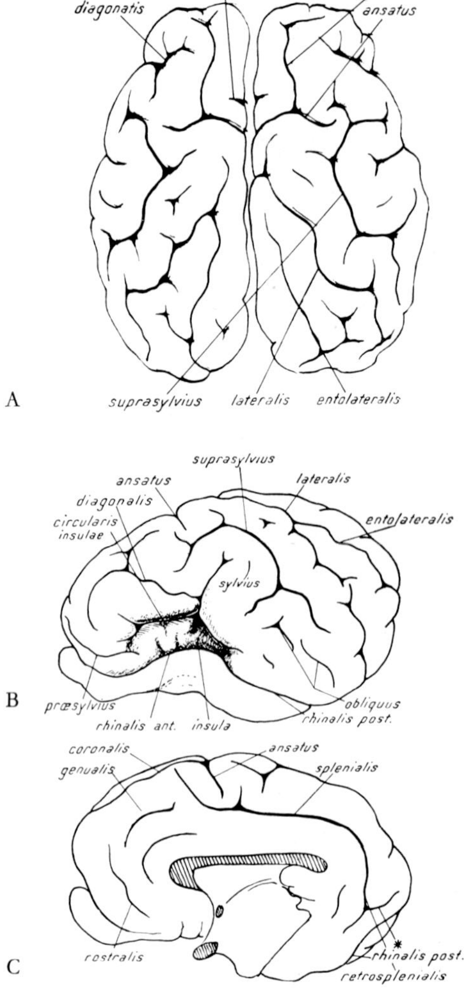

Figure 22 A–C. Dorsal, lateral and medial aspect of the hemisphere in the artiodactyle Ungulate Lamb (from Beccari, 1943).

The large Proboscidian brains[28] of Elephas indicus and Loxodonta africana (Figs. 26–31) are characterized by a conspicuous hemispheric bend resulting in a large temporal lobe. The *sulcus Sylvius* extends caudodorsalward into the system presumably representing the sulcus

[28] Cf. e.g. the older paper by Dexler (1907; Elephas indicus) and the more recent one by Janssen and Stephan (1956; Loxodonta africana).

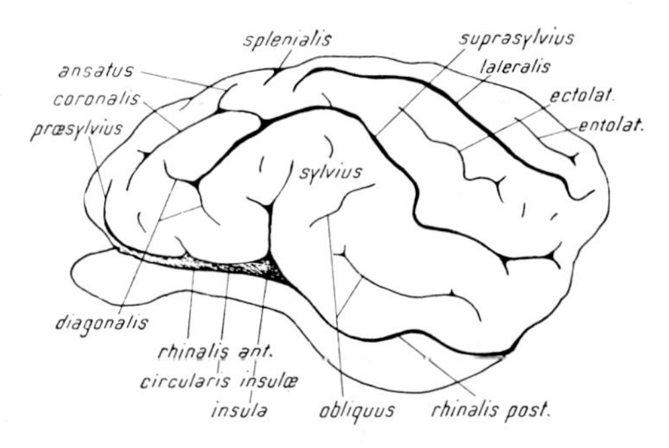

Figure 23. Lateral aspect of the hemisphere in the artiodactyle Ungulate European Deer (from BECCARI, 1943).

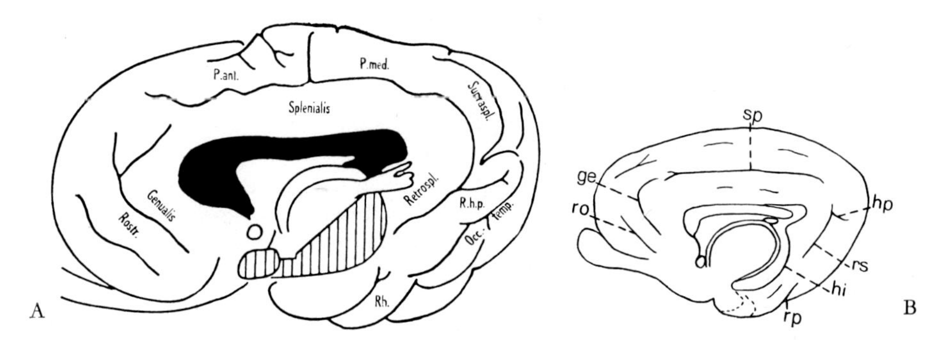

Figure 24A. Medial aspect of the hemisphere in the artiodactyle Ungulate Giraffe (from KAPPERS *et al.*, 1936). Rh: sulcus rhinalis lateralis, pars posterior; Rhp: ramus horizontalis posterior sulci retrosplenialis.

Figure 24B. Medial aspect of a 'generalized' Ungulate brain, illustrating the overall medial fissuration pattern (from K., 1927). ge: sulcus genualis; hi: sulcus *(sive* fissura) hippocampi; hp: ramus horizontalis posterior sulci retrosplenialis; ro: sulcus rostralis; rp: sulcus rhinalis lateralis, pars posterior; rs: sulcus retrosplenialis; sp: sulcus splenialis.

suprasylvius, which is complicated by diverse radiating sulci, and seems to continue as sulcus temporalis nasalis. The so-called third presylvian radial sulcus (Figs. 26B, C) may correspond to either sulcus diagonalis or sulcus praesylvius of Ungulates. The sulcus temporalis caudalis represents a modified sulcus ectolateralis, but these interpretations remain uncertain.[29] On the medial side (Figs. 28, 31) a genualis-splenialis (callosomarginal) system is complicated by an occasional in-

[29] Thus, e.g. the sulcus temporalis nasalis can be interpreted as a sulcus ectosylvius caudalis, the sulcus temporalis caudalis then being conceived as sulcus suprasylvius caudalis.

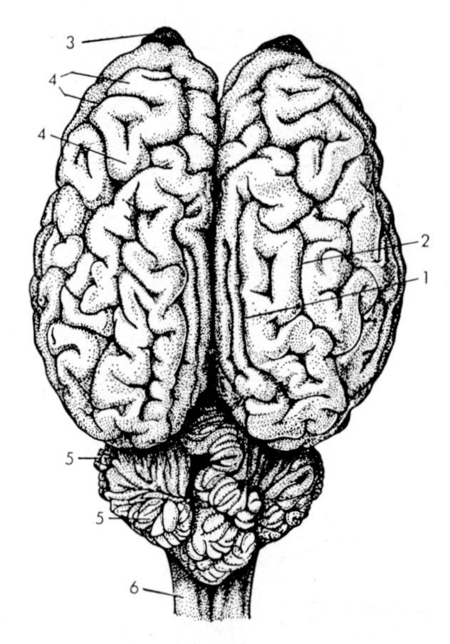

Figure 25. Dorsal aspect of brain in the perissodactyle Ungulate Horse (after SISSON, from ROMER, 1950). Added designations: 1: sulcus lateralis; 2: sulcus suprasylvius (evaluated as s. ectosylvius by some authors); 3: olfactory bulb; 4: rostral gyri and sulci (cf. Figs. 21, 22B, 23); 5: cerebellum; 6: oblongata.

ner (i.e. more basally located) parallel sulcus, and numerous secondary sulci oriented at a right angle to said system. Substantial individual variations with respect to details of the fissuration pattern seem to obtain.

The large telencephalic hemispheres of Cetacea, like those of Proboscidia, are extremely convoluted, due to the presence of numerous randomly arranged tertiary sulci[30] between the secondary main ones, which essentially correspond to the Carnivore-Ungulate pattern (LANGWORTHY, 1932).

The hemispheres appear foreshortened, with wide transverse diameter, and pronounced hemispheric bend. Boundaries between a frontal, parietal, occipital and temporal lobe cannot be defined with any certainty, but the occipitotemporal region displays a voluminous ex-

[30] For purposes of a rough classification, the primitive sulci of the lissencephalic brains (cf. section 1) can be designated as primary, the main sulci of the Carnivore-Ungulate pattern being the secondary sulci. The random duplications of, or additions to, these latter may then be designated as tertiary.

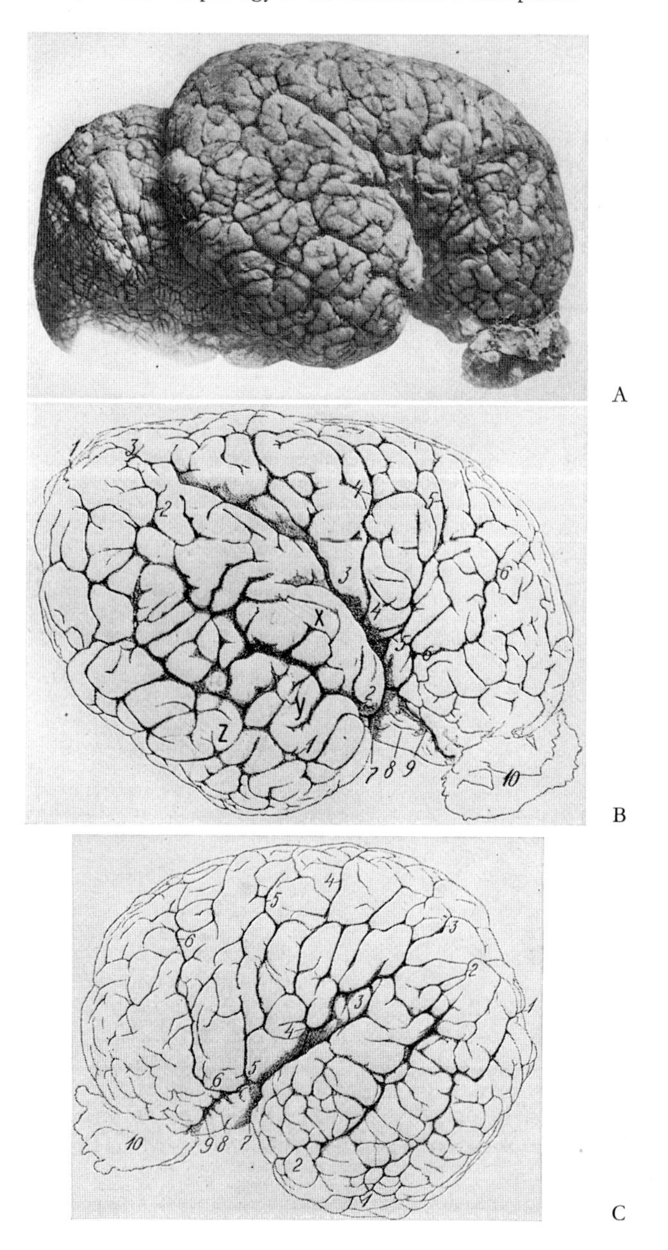

Figure 26 A–C. Lateral aspect of the brain in the Proboscidean Indian Elephant (from DEXLER, 1907). 1: sulcus temporalis caudalis; 2: sulcus temporalis nasalis *sive* rostralis; 3: sulcus *sive* fissura Sylvii; 8: trigonum and tuberculum olfactorium; 9: sulcus rhinalis lateralis, pars anterior; 10: bulbus olfactorius; x, y, z: superior, middle, and inferior temporal gyri; 4, 5, 6: presylvian radial sulci; 7: *fossa Sylvii.*

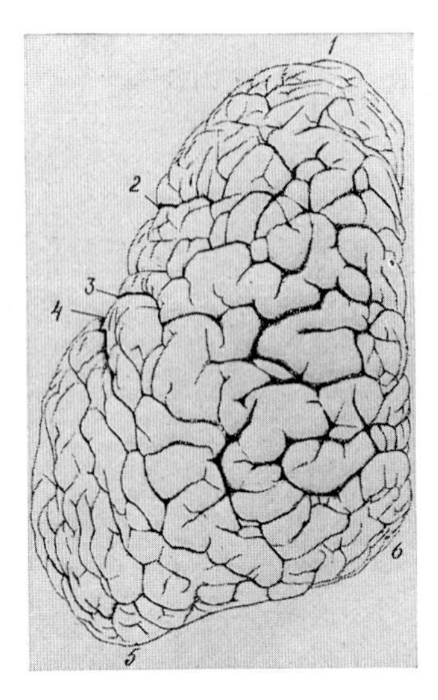

Figure 27. Dorsal aspect of the Indian Elephant's hemisphere (from DEXLER, 1907). 1: rostral pole; 2, 3: dorsal extension of 3rd and 2nd (3) presylvian radial sulci; 4: sulcus Sylvii; 5: caudal pole; 6: angulus dorsocaudalis of hemisphere.

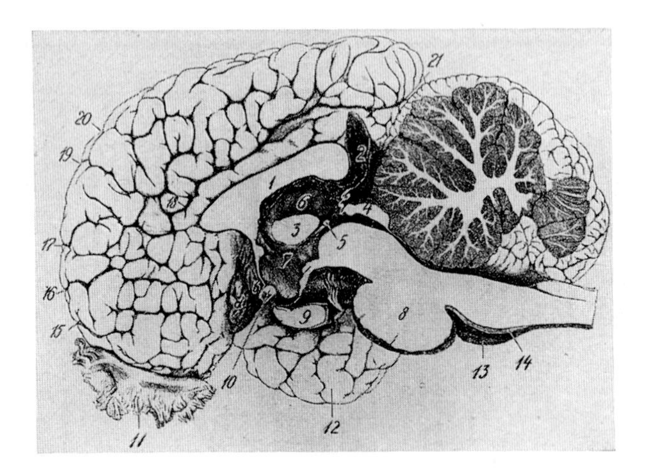

Figure 28. Medial aspect of the Indian Elephant's brain (from DEXLER, 1907). 1: corpus callosum, fornix and septum; 2: 'epiphysial recess' (according to JANSSEN and STEPHAN, 1956, this is a 'suprapineal recess'); an epiphysis was not found to be present by DEXLER in the Indian Elephant, nor by the above mentioned authors in the African Loxodont; 3: massa intermedia; 4: lamina quadrigemina; 5: accessory massa intermedia; 6, 7: third ventricle; 8: pons; 9: hypophysial complex; 10: optic chiasma; 11: olfactory bulb; 12: temporal lobe; 13: inferior olive; 14: fissura mediana oblongatae; 15: 'fissura semicircularis parolfactoria'; 16, 17, 19, 20: radial sulci of medial surface reaching the rostrolateral surface *('die Mantelkante passierende tiefe Radialfurchen');* 18: rostral end of sulcus callosomarginalis (splenialis); 21: caudal end of sulcus callosomarginalis; a: 'trigonum olfactorium'; b: *Broca's diagonal band.*

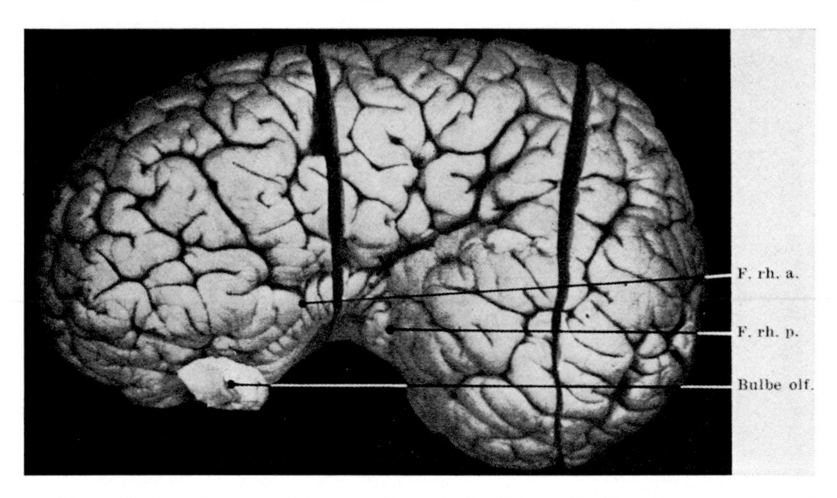

Figure 29. Lateral aspect of the hemisphere in the African Elephant (from JANSSEN and STEPHAN, 1956). F.rh.a., p.: anterior and posterior portion of sulcus rhinalis lateralis; origin of sulcus *sive* fissura lateralis Sylvii at bend between the two parts of sulcus rhinalis lateralis.

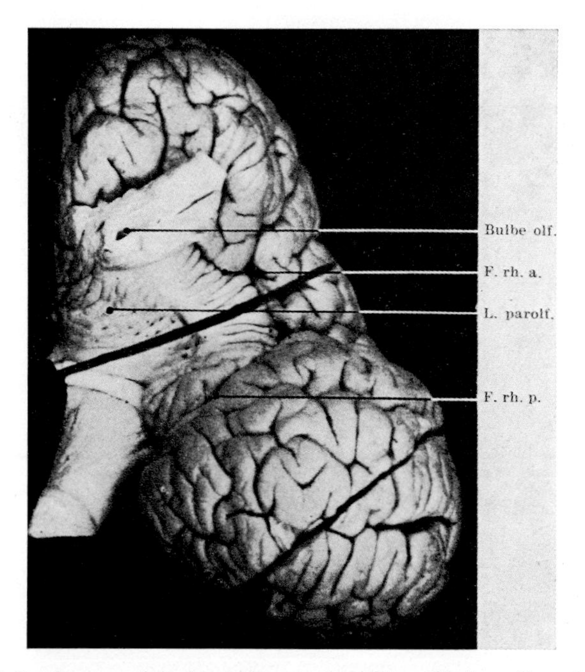

Figure 30. Basal aspect of the hemisphere in the African Elephant (from JANSSEN and STEPHAN, 1956). Bulbe olf: the lead shows the sectioned bulb at region of stalk, pointing to olfactory stalk ventricle; L. parolf.: tuberculum olfactorium *sive* area ventralis anterior (*'lobe parolfactif'*).

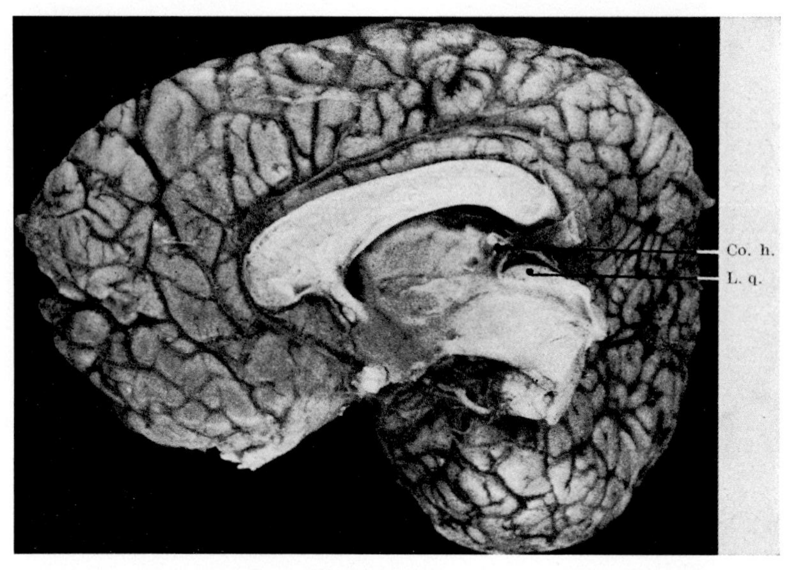

Figure 31. Medial aspect of the hemisphere in the African Elephant (from JANSSEN and STEPHAN, 1956). Co.h.: habenular commissure; L.q.: lamina quadrigemina mesencephali.

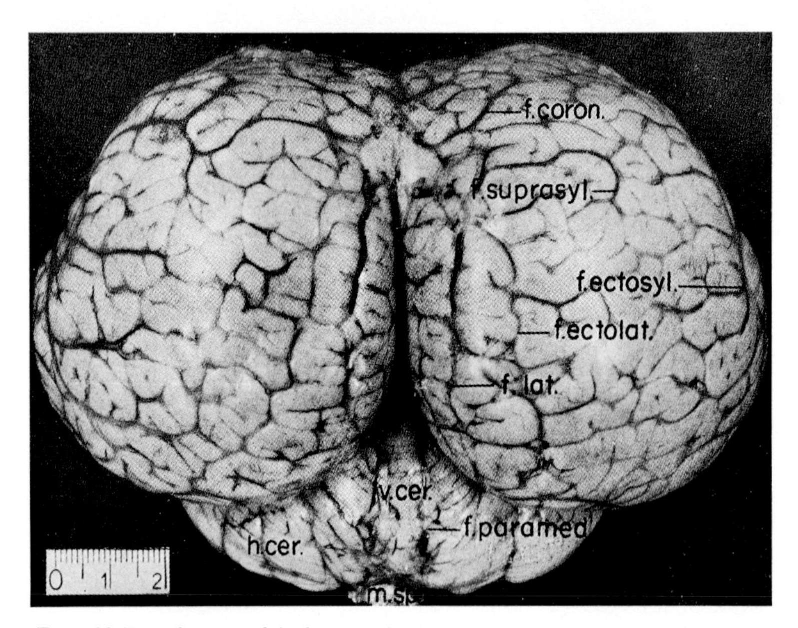

Figure 32. Dorsal aspect of the brain in the Cetacean Porpoise Phocaena (from JANSEN and JANSEN, 1968). Abbreviations for Figures 32–37B. aq.cer.: aquaeductus cerebri; ch.opt.: optic chiasma; col.inf., sup.: inferior and superior colliculus of tectum mesen-

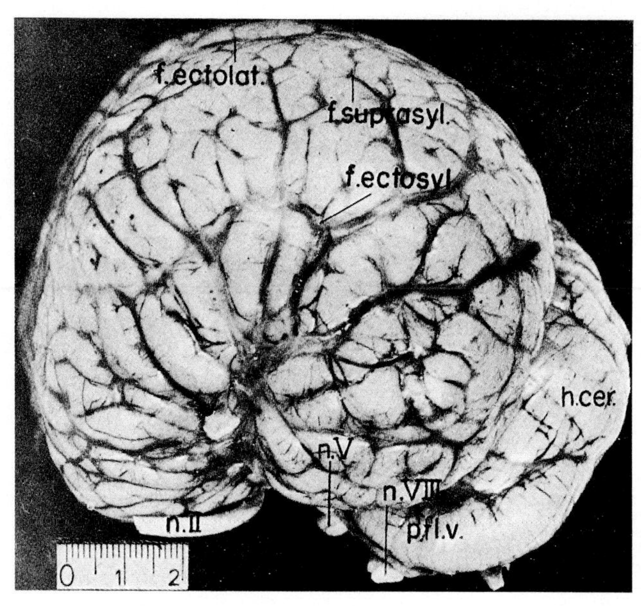

Figure 33. Lateral aspect of brain in the Porpoise Phocaena (from JANSEN and JANSEN, 1968).

pansion (JANSEN and JANSEN, 1968). Data on the Cetacean brain have also been recorded by PILLERI (1962, 1964, 1965/66a, b).

Figures 32–37 illustrate aspects of the hemisphere in an Odontocete, anosmatic Porpoise, and in a Mysticete, microsmatic Whale. It will be seen that, on the lateral and dorsal side, ectosylvian, suprasylvian, ectolateral, lateral and coronal sulci have been identified by JANSEN and JANSEN (1968) in accordance with FLATAU and JACOBSOHN (1899) and other authors. FRIANT (1953, 1954), however, on the basis

---

cephali; com.ant., post.: anterior respectively posterior commissure; corp.call.: corpus callosum; f.: fornix; f.coron.: sulcus coronalis; f.ectolat.: sulcus ectolateralis; f.ectosyl.: sulcus ectosylvius; f.gen.: sulcus genualis; f.hip.: sulcus *sive* fissura hippocampi; f.i.: foramen interventriculare; f.lat.: sulcus (arcuatus) lateralis; f.paramed.: sulcus paramedianus (cerebelli); f.retrosplen.: sulcus retrosplenialis; f.rhin.p.: sulcus rhinalis lateralis, pars posterior; f.suprasyl.: sulcus suprasylvius; g.hip.: gyrus hippocampi; h.cer.: cerebellar hemisphere; l.term.: lamina terminalis; l.pmd.: lobus paramedianus (cerebelli); n. II–XII: cranial nerves II–XII; p.: pons; ped.cer.: pedunculus cerebri (pes pedunculi); p.fl.d., v. (x): paraflocculus dorsalis respectively ventralis (with pars x); sept.pel.: septum pellucidum; tr.olf.: tractus olfactorius; tr.opt.: tractus opticus; tub.olf.: tuberculum olfactorium; v.cer.: vermis cerebelli; V. III.: third ventricle.

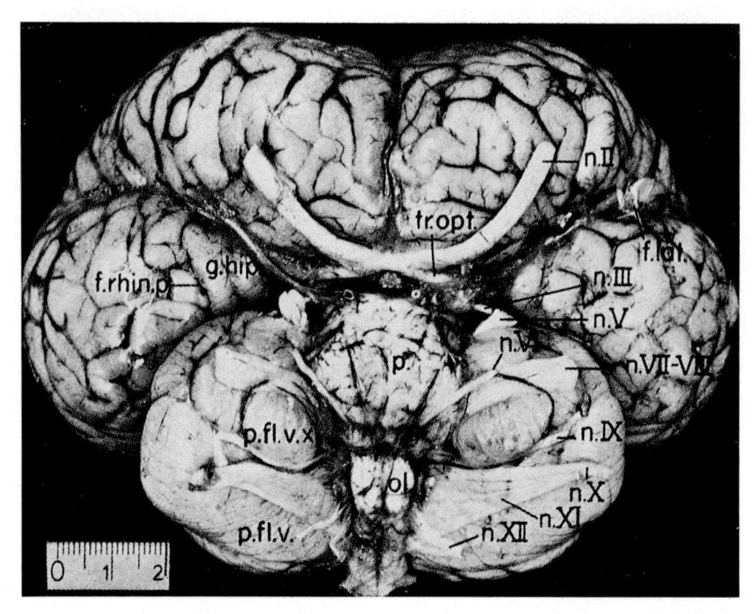

Figure 34. Basal aspect of brain in the Porpoise Phocaena (from JANSEN and JANSEN, 1968). Note absence of olfactory bulb and tract in this anosmatic form.

of embryological data, assumes that the ectosylvian sulcus is buried in the depth of the *Sylvian fossa*, whereby the ectosylvian and suprasylvian sulci of other authors would represent the suprasylvian and ectolateral sulci, respectively.

On the medial aspect in addition to various random sulci a spleniogenual system is displayed (Fig. 37B). JANSEN and JANSEN (1968) designate the outer, more pronounced groove of that system as 'fissura splenialis', and the adjacent roughly parallel sulcus as 'fissura genualis'. In addition to sulcus hippocampi, a sulcus retrosplenialis is also present (Fig. 37A). The fissuration pattern in Cetacea, like that of other Mammalian groups, displays, within a common configurational scheme, numerous taxonomic and individual variations.

Before turning to the Primate fissuration pattern, a short reference to the variable sulci displayed by the posterior part of the piriform lobe in diverse non-Primate as well as Primate Mammals seems perhaps appropriate. These nondescript sulci may occur in relation to the posterior extensions of sulcus endorhinalis and of sulcus rhinalis lateralis, whereby, in addition to the gyrus lunaris (*sive* semilunaris) of basal cortex, and to the gyrus ambiens of piriform lobe cortex, acces-

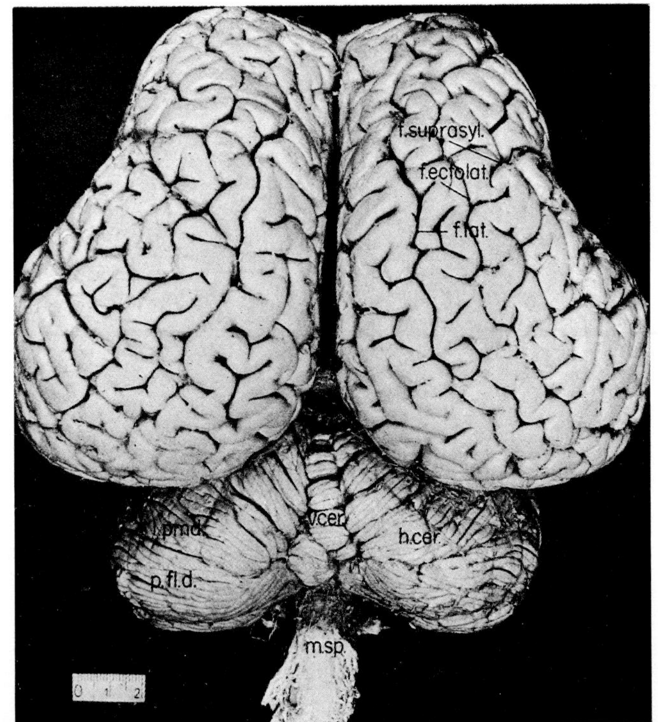

35

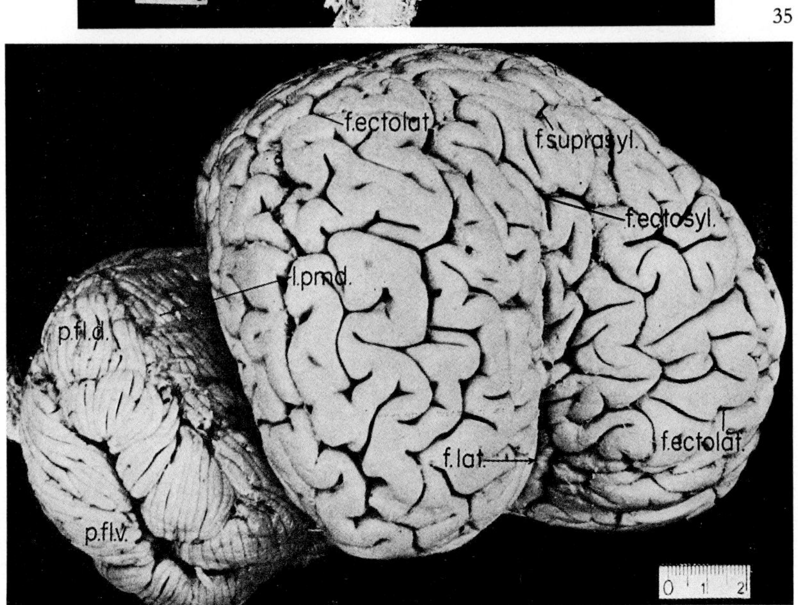

36

Figure 35. Dorsal aspect of the brain in the Cetacean Minke Whale, Balaenoptera acutorostrata (from JANSEN and JANSEN, 1968).

Figure 36. Lateral aspect of brain of Balaenoptera (from JANSEN and JANSEN, 1968).

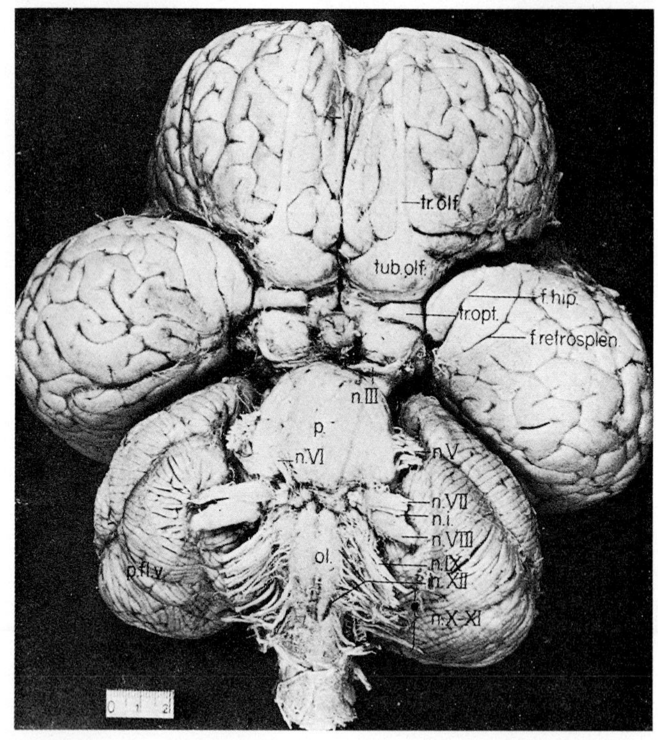

A

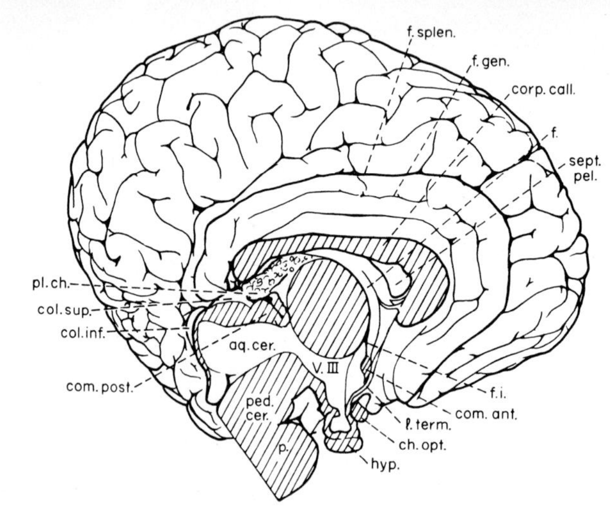

B

Figure 37 A. Basal aspect of the brain of Balaenoptera (from JANSEN and JANSEN, 1968). Note olfactory tract and remnant of olfactory bulb in this microsmatic form with sizeable tuberculum olfactorium (cf. Fig. 34).

Figure 37 B. Medial aspect of brain of Balaenoptera (from JANSEN and JANSEN, 1968).

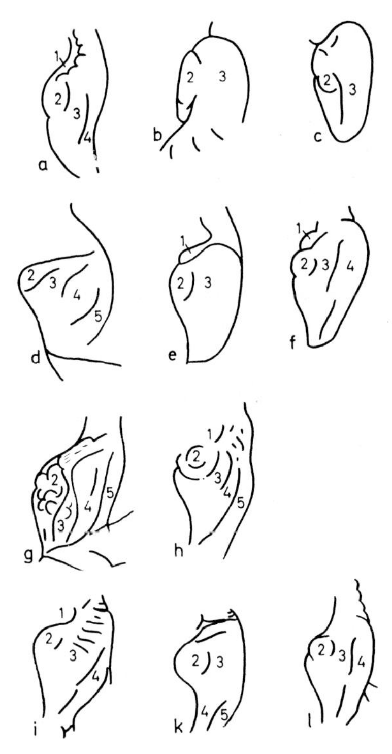

Figure 38. Groove systems displayed by the piriform lobe and its immediate neighborhood in diverse Mammals (after RETZIUS and other authors, from SCHOBER and BRAUER, 1975). a: Macropus rufus (Marsupial); b: Papio cynocephalus (Primate); c: Rattus norvegicus (Rodent); d: Ursus arctos (Carnivore); e: Mustela spec. (Carnivore); f: Viverra spec. (Carnivore); g: Elephant (Proboscidian); h: Horse (Ungulate); i: Camel (Ungulate); k: Llama (Ungulate); l: Antelope (Ungulate); 1: 'gyrus rhinalis intermedius' (essentially bulge of tractus olfactorius); 2: gyrus lunaris (cortical amygdaloid nucleus); 3: gyrus ambiens; 4: 'gyrus sagittalis medius'; 5: 'gyrus sagittalis lateralis' (4 and 5 appear to represent accessory gyri of intermediate and posterior piriform lobe).

sory or supernumerary small gyri become noticeable, as indicated in Figure 38 and named, with the relevant interpretation, in that Figure's legend.

3. The Primate Fissuration Pattern

Several classifications of the order Primates are extant. In the simplified taxonomic survey given in volume 1, the subdivision into three suborders, namely Lemuroidea, Tarsioidea, and Anthropoidea was

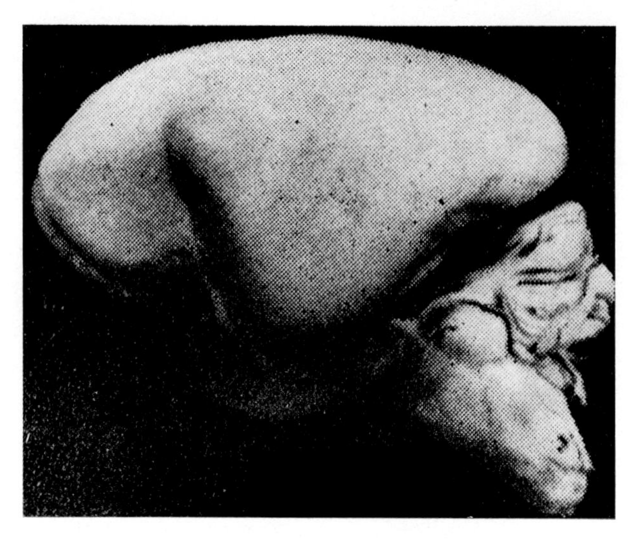

Figure 39. Lateral aspect of the brain in the Tarsioid Primate Tarsius spectrum (from CONOLLY, 1950).

adopted. The last named suborder includes Platyrrhini (South American or New World Monkeys), and Catarrhini (Old World Monkeys), Apes and Hominidae as infraorders.[31]

Small Primates, such as the Tarsioid Tarsius (Figs. 1, 39), the tree shrew Tupaia (if classified as a Primate), and the Anthropoid Hapale (Fig. 1 S) are essentially lissencephalic, although a fairly pronounced Sylvian depression at the hemispheric bend, and an indistinct (superior) temporal sulcus, on the lateral side, moreover faint suggestions of sulci on the medial, particularly occipital aspect in such lissencephalic forms may variably be present.

Detailed data on the external morphology of the Primate telencephalon can be found in a publication by TILNEY (1928), in the contribution by GRAF HALLER (1934), and in the treatises by KAPPERS *et al.* (1936) and BECCARI (1943). Among investigations specifically concerned with that topic, those by KÜKENTHAL and ZIEHEN (1895), ZIEHEN (1896), ZUCKERKANDL (1902, 1903, 1904a, 1905), ELLIOT SMITH (1903/04a, b and others) and CONNOLLY (1950) may be mentioned.

[31] A somewhat different classification into two suborders Prosimii and Anthropoidea, each again with infraorders, was adopted by YOUNG (1955) whose chapter on Primates should be consulted by those interested in the obtaining problems. Additional comments on that topic can also be found on p. 638 of BECCARI's (1943) treatise.

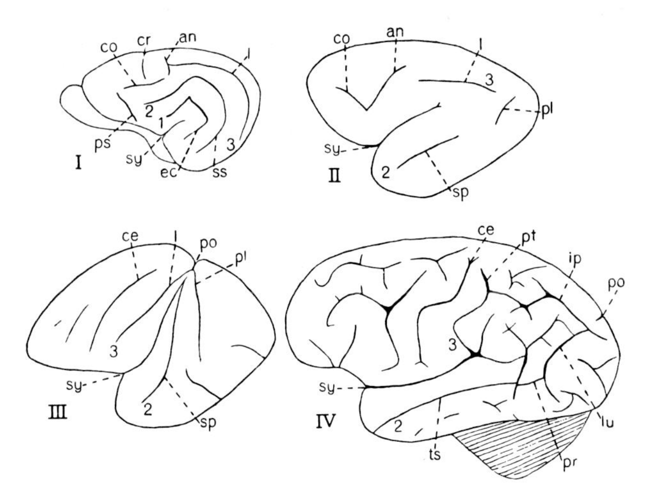

Figure 40. Sketches illustrating neopallial sulci on the lateral aspect of the hemisphere in diverse Mammals (from K., 1927). I: Carnivore Dog; II: Lemuroid Primate Lemur varius (after KAPPERS); III: Catarrhine Primate Macacus; IV: Anthropoid Simiid Chimpanzee (after G. RETZIUS); an: sulcus ansatus; ce: sulcus centralis; co: sulcus coronalis; cr: sulcus cruciatus; ec: sulcus ectosylvius (first arcuate sulcus); ip: sulcus intraparietalis; l: sulcus lateralis (third arcuate sulcus); lu: sulcus lunatus; pl: sulcus postlateralis; po: fissura parieto-occipitalis; pr: sulcus praelunatus; ps: sulcus praesylvius; pt: sulcus postcentralis (presumably third arcuate sulcus); sp: sulcus suprasylvius, pars posterior (presumably posterior part of second arcuate sulcus); ss: sulcus suprasylvius (second arcuate sulcus); sy: fossa sive fissura Sylvii, respectively sulcus pseudosylvius of Carnivores; ts: sulcus temporalis superior (presumably posterior part of second arcuate sulcus).

With respect to the difference in terminology, METTLER (1933) brings a synopsis of the various terms respectively synonyms, and states his preferences. A detailed review of the sulci in the Human cerebral hemisphere, with numerous bibliographic references, is included on pp. 21–60, as chapter III, in BAILEY's and VON BONIN's (1951) treatise on the isocortex of Man.

Generally speaking, the Primate hemisphere differs from that of Carnivores and Ungulates by the expansion of the rostral, occipital, and temporal mantle portions, whereby frontal, parietal, occipital and temporal lobes become more readily distinguishable than in most subprimate forms.[32] Another 'progressive' feature is the transformation of the Sylvian depression or pseudosylvian sulcus, resulting in

[32] As regards the temporal lobe, its considerable expansion in Proboscidea and Cetacea should be kept in mind.

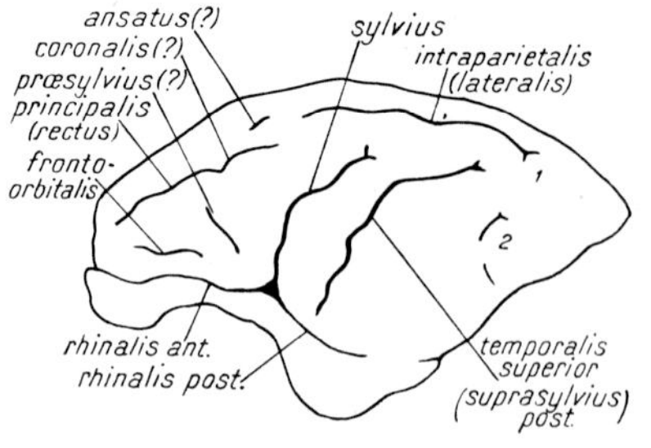

Figure 41. Lateral aspect of the hemisphere in Lemur macaco (slightly modified after GRAF HALLER, 1934, from BECCARI, 1943). 1: sulcus occipitalis transversus (?); 2: sulcus lunatus (?) according to KAPPERS.

opercularization and expansion of the insula in the floor of a deep *fossa Sylvii.* These changes, involving the configuration of the surface area, are correlated with substantial transformations in the arrangement of the sulci.

Nevertheless, despite considerable differences between the Primate and the Carnivore-Ungulate fissuration pattern,[33] some features, related to the Sylvian fossa and its surrounding arcuate sulci on the lateral side and to the '*circumcallosal arch*' on the medial side, are similar in Primates and non-Primates and allow for a comparison in terms of homology.

Although several authors have pointed out that the Sylvian depression respectively the pseudosylvian sulcus in Subprimates differs from the Sylvian fissure of Primates, which latter becomes a complex of sulci and includes an opercularized insula, there can be little doubt that pseudosylvian sulcus and Sylvian fissure display a type of special homology, namely kathomology,[34] augmentative homology *qua* Anthropoids and Hominids, and defective homology *qua* non-Primates.

As regards the concentric arcuate sulci, the ectosylvian groove, which seems to be the least constant, has not been properly identified in Primates. The posterior (postsylvian) portion of the suprasylvian sulcus, on the other hand seems to be preserved and apparently becomes the sulcus temporalis superior. The third arcuate groove or sulcus

---

[33] BECCARI (1943, p. 633) states: '*Il tipo di solcatura degli emisferi dei Primati differisce notevolmente da quello degli altri Mammiferi. Se si eccettua la scissura silvia o silviana e una parte dell'arco circumcalloso, tutti gli altri infossamenti non appariscono uguali a quelli dei C.U.*' (C.U. means here Carnivores and Ungulates).

[34] Cf. volume 1, pp. 194–210, and volume 3/II, pp. 64–65.

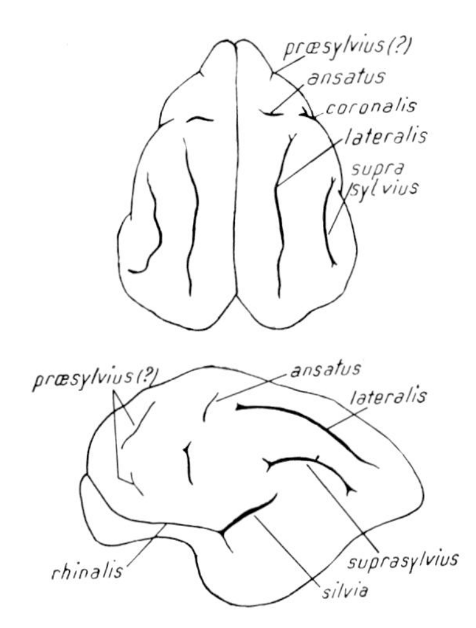

Figure 42 A. Dorsal and lateral aspect of the hemisphere in the Lemuroid Primate Chrysomis madagascariensis (slightly modified after GRAF HALLER, 1934, from BECCARI, 1943).

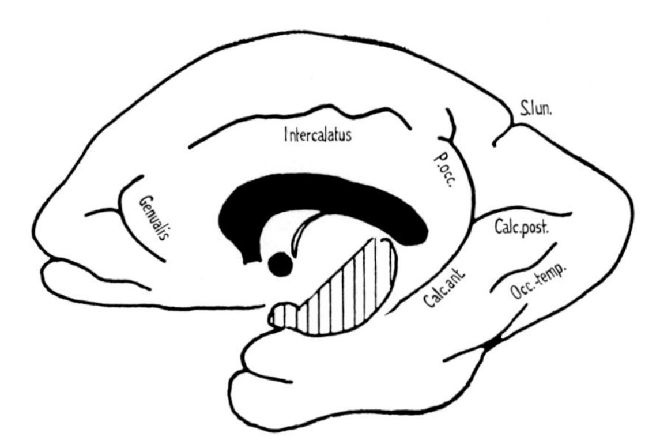

Figure 42B. Medial aspect of the brain in Lemur varius (from KAPPERS *et al.*, 1936). Sulcus 'intercalatus' is the sulcus splenialis.

lateralis is likewise apparently present, its main part presumably becoming postcentral and intraparietal sulci. A caudal part of sulcus lateralis or a corresponding part of sulcus entolateralis (sulcus postlateralis) may perhaps join the sulcus intraparietalis as sulcus occipitalis transversus (Figs. 40, 41), or represent a forerunner of the sulcus lunatus, to be dealt with in more detail further below. A lateral occipitotemporal sulcus (or sulcus occipitalis inferior) can also be seen in the occipital lobe of some forms (cf. Figs. 45, 46 A).

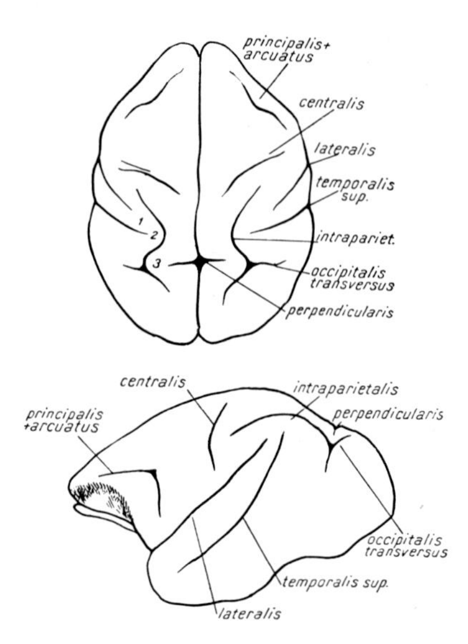

Figure 43. Dorsal and lateral aspect of the hemisphere in the Cebid Pithecia monacha (after GRAF HALLER, 1934, from BECCARI, 1943). 1: supramarginal gyrus; 2: angular gyrus; 3: 'gyrus parieto-occipitalis superior' (I of GRATIOLET).

As regards the sulci of the expanding frontal lobe, it can be assumed, with KAPPERS (1921) and others, that sulcus coronalis and ansatus have joined to form the sulcus centralis. A rostral part of sulcus coronalis is also designated as sulcus rectus *sive* principalis (Fig. 41). It is joined with or separated from a sulcus arcuatus, which latter might become the sulcus precentralis inferior (Figs. 43–46). The cruciate sulcus seems to have disappeared, perhaps by a process of 'amalgamation' comparable to that assumed for Ungulates.

The presylvian (or orbital) sulcus has been regarded by some authors as precursor of sulcus fronto-orbitalis in Anthropoids (Fig. 49), but this is contested by others.

It should be added that these interpretations of the frontal and lateral sulci are by no means certain. Thus, WOOLSEY (1960) and others, who insist on a definite relationship of sulci to functional respectively architectural cortical areas, claim that the primate sulcus centralis could not correspond to the Carnivore-Ungulate ansate and coronal sulci, since these latter do not have the required relationships to the sensory and motor localization patterns. The central sulcus, according to WOOLSEY, is supposed to have been formed between motor and sen-

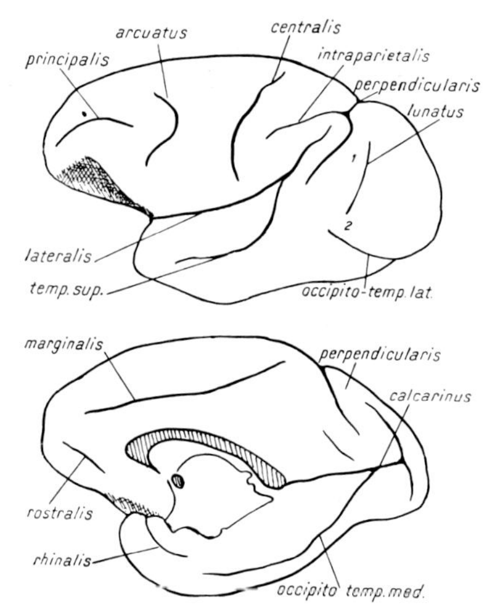

Figure 44. Lateral and medial aspect of the hemisphere in Cebus capucinus (after GRAF HALLER, 1934, from BECCARI, 1943). 1: gyrus parieto-occipitalis medius (II of GRATIOLET); 2: gyrus parieto-occipitalis inferior (III of GRATIOLET).

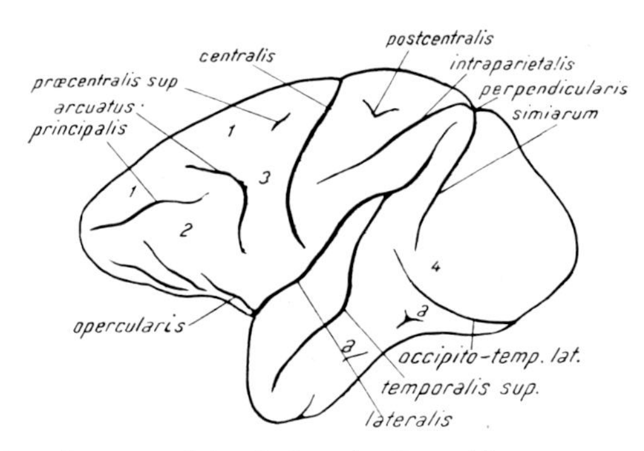

Figure 45. Lateral aspect of hemisphere in Cercopithecus cynosurus (after GRAF HALLER, 1934, from BECCARI, 1943). a: components of sulcus temporalis medius; 1: gyrus frontalis superior primitivus Simiarum; 2: gyrus frontalis inferior primitivus; 3: gyrus praecentralis; 4: gyrus parieto-occipitalis inferior (III of GRATIOLET).

sory regions as a new sulcus, a portion of which is said to be indicated in the Raccoon (cf. Fig. 47A). Woolsey's concept of sulcus centralis evolution is shown in Figure 47B, in accordance with the hypothesis that 'cortical sulci are formed at the boundaries of "physiological" subdivisions'. WOOLSEY's hypothesis does not seem convincing. A comparison of various forms rather clearly seems to indicate shifts of

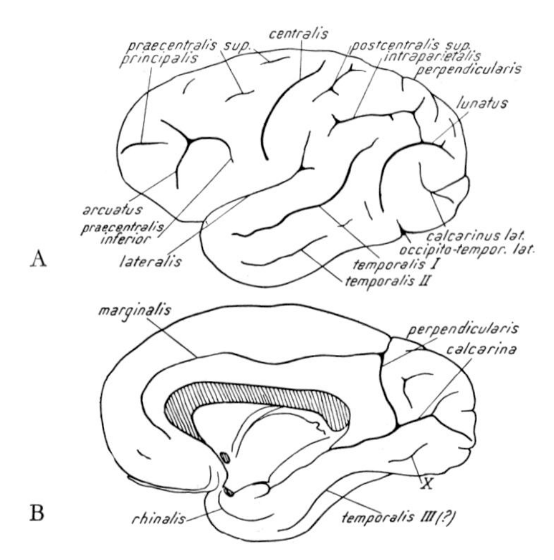

Figure 46 A, B. Lateral and medial aspect of hemisphere in Ateles paniscus (from Beccari, 1943). X: perhaps sulcus collateralis.

cortical areas in relation to sulci. These latter, moreover, often display highly variable, apparently random modifications. Some degree of scepsis concerning the accuracy of the functional and cytoarchitectural mappings 'precisely' elaborated on paper by Woolsey and others seems likewise appropriate. With some reservations, acknowledging the obtaining uncertainties and ambiguities, the interpretation by Kappers and other morphologists still appears reasonably adequate.

The medial aspect of Primate hemispheres is shown in Figures 44 and 50B. A sulcus rostralis and a sulcus marginalis (or sulcus cinguli) can be recognized. The former seems to be identical with the sulcus rostralis in Carnivores and Ungulates. The sulcus marginalis appears to combine sulcus genualis and splenialis of the just mentioned forms, with or without a retrosplenial extension.

The *retrosplenial system* includes a deep sulcus parieto-occipitalis (or perpendicularis), which may reach and joint the lateral system of sulci (cf. Figs. 44, 48 I). Another important groove of the retrosplenial system, usually, but not always joined to the sulcus parieto-occipitalis is the sulcus calcarinus within the visual cortex. The complexities and the divergent interpretations of the sulcus calcarinus are dealt with below in section 4.

Turning now to the great anthropoid Apes and Man, it can be said that Gorilla, Orang, and Chimpanzee display a fissuration pattern very

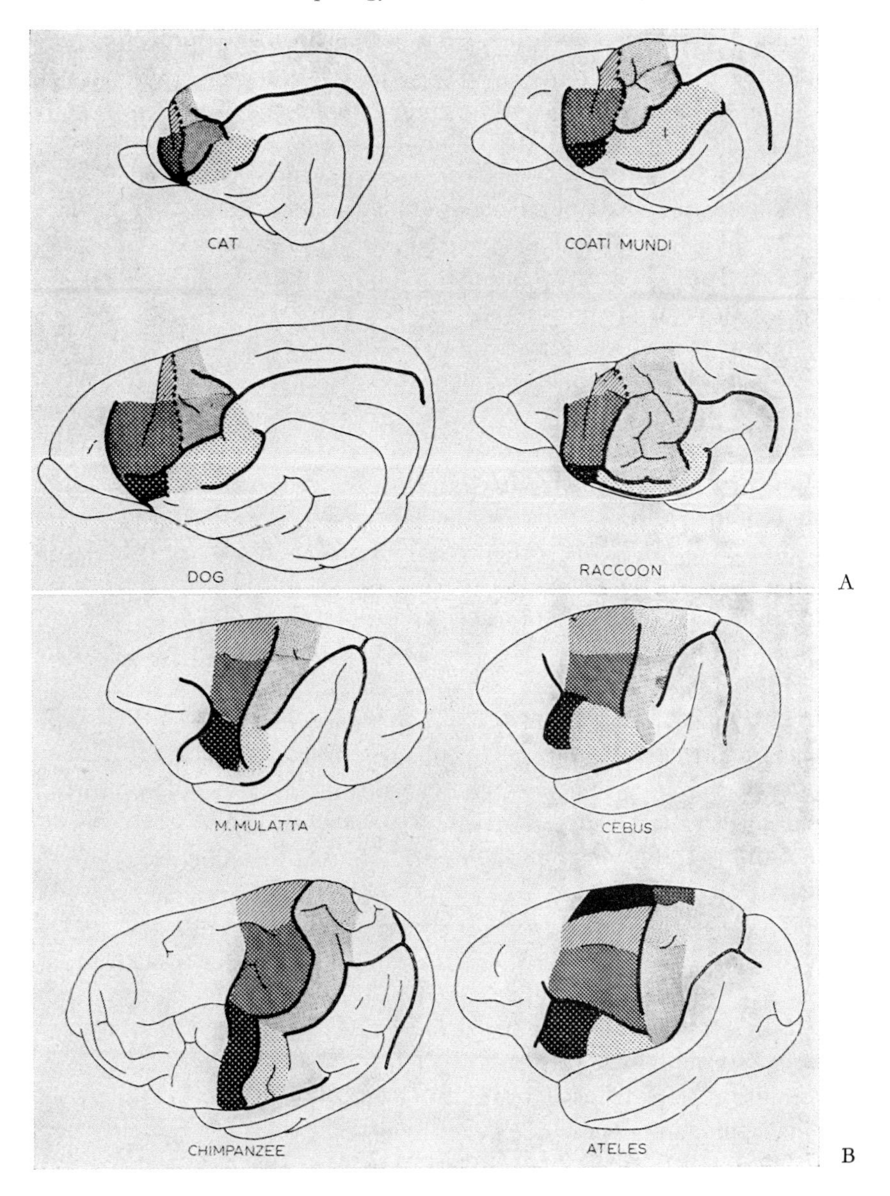

Figure 47 A, B. WOOLSEY's interpretation of homologous sulci (drawn in heavy lines) based on the purported location of motor and sensory cortical areas (from WOOLSEY, 1960). The heavily shaded areae are motor, the lighter shaded ones sensory, both in dorsobasal sequence of leg, arm, face. The dotted line in A is supposed to indicate the location in which the sulcus centralis of Primates (B) subsequently develops in phylogeny.

similar to, and directly comparable with that of Man, while the sulci and gyri of the Gibbon (Hylobates), although likewise rather closely conforming to the overall anthropoid pattern, are considered by some authors, e.g. BECCARI (1943) to be of a type intermediate between the pattern displayed by Monkeys and by the great Apes (Figs. 49–55).

Discounting the numerous individual variations obtaining in all the forms under consideration, it will be sufficient briefly to enumerate the main and most constant gyri and sulci of the Anthropoid pattern in general with particular reference to Man.[35]

The *lateral surface* of the hemisphere displays the complex system of *fissura cerebri lateralis sive fossa Sylvii*, resulting from the phylogenetic and ontogenetic evolution of the primordial Sylvian depression at the hemispheric bend.[36] The basal opening into the fossa Sylvii corresponds to the truncus fissurae lateralis, which generally gives off the short ramus anterior horizontalis and ramus anterior ascendens, the main fissure continuing occipitalward as ramus posterior. By this latter, together with truncus, the temporal lobe is separated from the frontal lobe rostrally, and from the parietal lobe posteriorly.

On the lateral aspect, the *frontal lobe* is caudalward (or occipitalward) bounded by the *sulcus centralis Rolandi*, which corresponds rather closely to the transition between first somatomotor and first somatosensory areas. The basal end of the central sulcus does not, as a rule, reach the fissura Sylvii. The gyrus centralis anterior lies between sulcus centralis and sulcus praecentralis. This latter usually consists of two separate (inferior and superior) portions, each of which is generally connected with a longitudinal groove, namely sulcus frontalis superior and inferior. The gyrus frontalis superior, which extends towards the medial aspect of the hemisphere, is located above sulcus frontalis superior.

The gyrus frontalis medius lies between superior and inferior frontal sulcus und is usually subdivided, by a variable sulcus frontalis medius, which may rostralward branch as so-called sulcus frontomarginalis, into an upper and lower subgyrus.

The gyrus frontalis inferior, basally to inferior frontal sulcus is sub-

---

[35] Reference should also be made to Figures 244 A and B as well as to the comments on various sulci and gyri of the Human brain in connection with questions of function and localization dealt with in section 10 of chapter XIII, volume 5/I.

[36] It should be noted that in some lower Mammalian forms (cf. vol. 5/I, Figs. 221 A, C) the hemispheric bend with its pseudosylvian fossa involves only the piriform lobe and the adjacent basal cortex, without affecting the neopallial isocortical mantle.

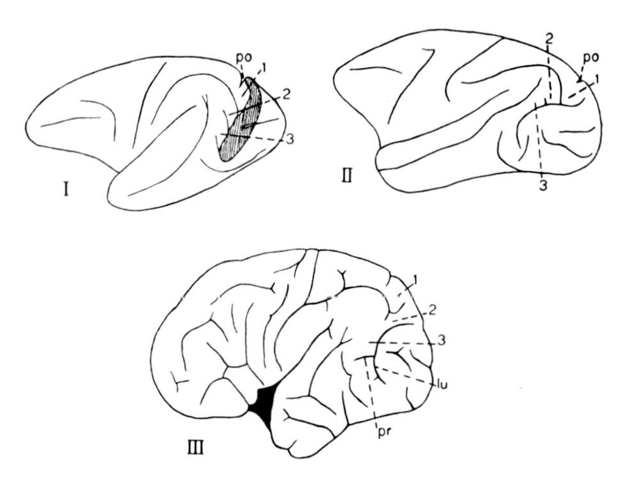

Figure 48. Sketches illustrating the so-called parieto-occipital transition gyri (*Übergangswindungen, plis de passage*) in three Primates (I and II after ZUCKERKANDL, 1904; III after KAPPERS, 1921, from K., 1927). I: Semnopithecus entellus, with cut operculum occipitale (all three *plis de passage* 1, 2, 3 reach the bottom of the *Affenspalte*); II: Cynocephalus (the first *Übergangswindung* is superficial as in Anthropoid Apes); III: newborn Man with sulcus lunatus (gyri 1 and 2 are superficial); lu: sulcus lunatus; po: sulcus parieto-occipitalis; pr: sulcus praelunatus.

divided into pars orbitalis basally to ramus anterior horizontalis of fissura Sylvii, pars triangularis between said ramus and ramus anterior ascendens, and pars opercularis posteriorly to the ramus anterior ascendens.[37] It should be added that the following variations can obtain: ramus anterior horizontalis and ramus anterior ascendens may branch off from a common anterior stem, or three anterior rami (ascendens, intermedius, horizontalis) may be present. Basally to gyrus frontalis inferior are located nondescript gyri orbitales related to variable sulci orbitales. These gyri, which correspond to the 'basal neocortex' emphasized by SPATZ (cf. p. 749 of vol. 3/II), form the frontal lobe's basolateral surface, slanting toward the olfactory tract which runs basally to sulcus olfactorius (Fig. 53D), but is not directly related to that sulcus (MAIR, 1928). To the orbital gyri correspond the impressio-

[37] The relationship of the (usually left) pars triangularis and opercularis of the Human gyrus frontalis inferior (*Broca's convolution*) to the cortical areas concerned with speech was pointed out in section 10 of chapter XIII, volume 5/I (cf. Figs. 260A, B). It is of interest that the configuration of inferior frontal gyrus is displayed as a less clearly defined and less constant far more variable pattern in the subhuman Apes (cf. Figs. 40 IV, 49–50, 51A, B).

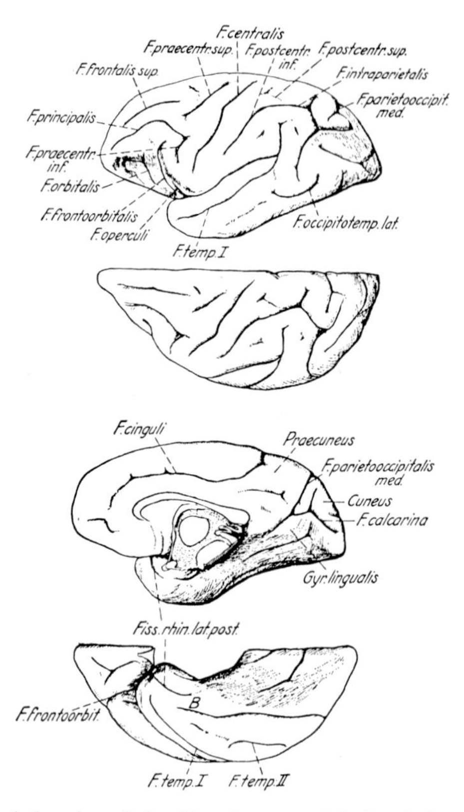

Figure 49. Lateral, dorsal, medial and basal aspects of the hemisphere in the Anthropoid Ape Gibbon (from GRAF HALLER, 1934). The sulcus lunatus, not labelled, and its residual connection with sulcus parieto-occipitalis are easily identified.

nes digitatae and juga cerebralia of the convex facies cerebralis ossis frontalis. The corresponding concavity of the frontal lobe's orbital surface may give a keel-like appearance, particularly conspicuous in many subhuman Primates, but also variably displayed in Human brains. The frontal pole corresponds to an undefined region where neighborhoods of the frontal gyri may merge in a variable pattern.

In the *parietal lobe*, the sulcus postcentralis, frequently consisting of separate inferior and superior parts, runs roughly parallel to the central sulcus. Between this latter and the postcentral sulcus lies gyrus postcentralis. The intraparietal sulcus, often a continuation of the inferior part of postcentral sulcus, runs toward the occipital lobe and may end as transverse occipital sulcus. The intraparietal (or interparietal) sulcus

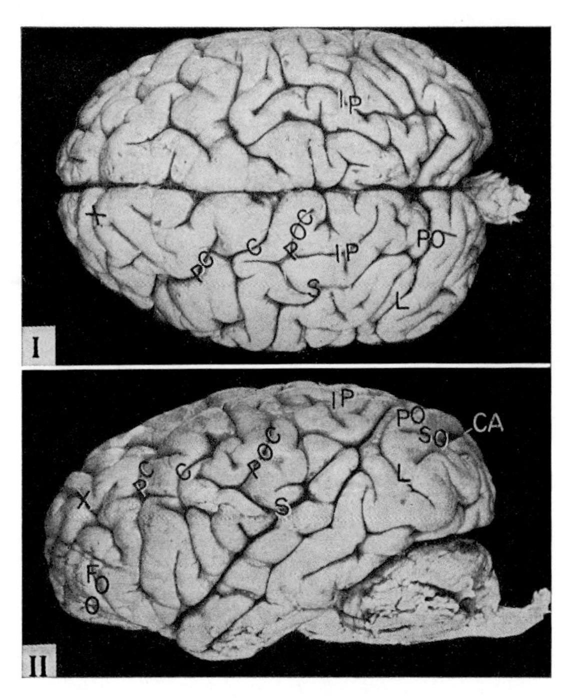

Figure 50 A. Dorsal and lateral aspect of the hemisphere in a Gorilla (from NOBACK and GOSS, 1959). C: sulcus centralis; CA: calcarine sulcus; FO: 'fronto-orbital' sulcus; IP: interparietal sulcus; L: sulcus lunatus; O: 'orbital sulcus'; PC: precentral sulcus; PO: parieto-occipital sulcus; POC: postcentral sulcus; S: Sylvian fissure; SO: sulcus calcarinus externus; X: 'sulcus in frontal lobe'.

separates a lobulus parietalis superior from a lobulus parietalis inferior. The former often displays a variable sulcus parallel to the intraparietal groove. The lobulus parietalis inferior consists of gyrus supramarginalis surrounding the end of fissura Sylvii, and of gyrus angularis, surrounding the superior end of superior temporal sulcus.

The *occipital lobe* is demarcated from the parietal lobe by the dorsolateral extension of the sulcus parieto-occipitalis and, somewhat more vaguely, by the variable sulcus occipitalis transversus. Basally, a nondescript notch or indentation, the incisura praeoccipitalis (preoccipital notch) at the posterior end of sulcus temporalis medius system may be noted. A line drawn from said notch to the end of parieto-occipital sulcus indicates a conventional boundary delimiting the lateral aspect of occipital lobe from the neighboring parietal and temporal lobes. The lateral aspect of the occipital lobe displays, moreover, a variable pat-

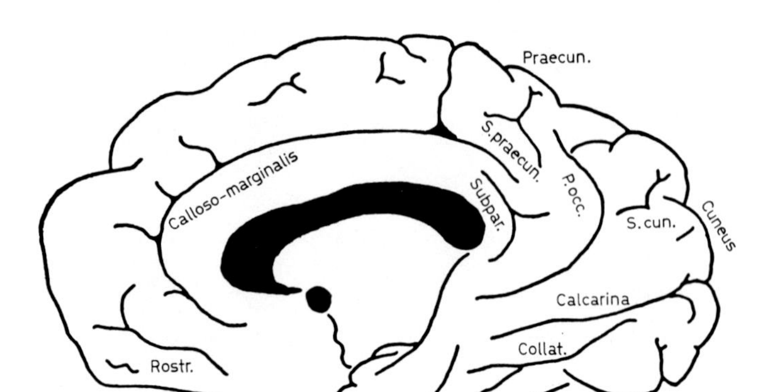

Figure 50 B. Medial aspect of the Gorilla's hemisphere (after ELLIOT SMITH, from KAPPERS *et al.*, 1936).

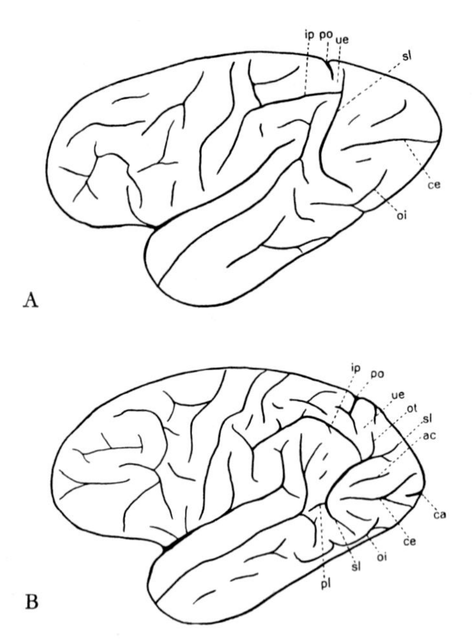

Figure 51. Lateral aspects of the hemisphere in the Anthropoid Apes Chimpanzee (A) and Orang (B) (from K., 1928b). ac: (dorsal) accessory sulcus calcarinus externus; ca: caudal end of main calcarine sulcus; ce: sulcus calcarinus externus (below this: ventral accessory sulcus); ip: sulcus intraparietalis; oi: sulcus occipitalis inferior; ot: sulcus occipitalis transversus; pl: sulcus praelunatus; po: sulcus parieto-occipitalis; sl: sulcus lunatus; ue: first *Übergangswindung*. In this and the following Figures the various unlabelled main sulci and gyri, described in the text, can easily be identified.

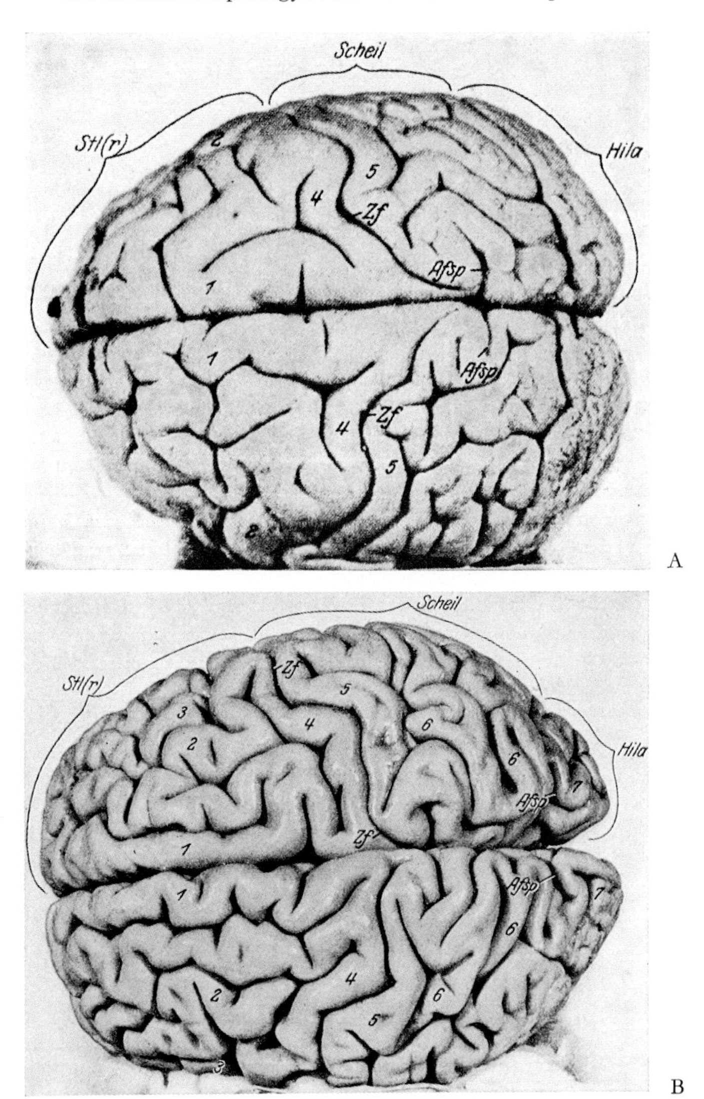

Figure 52. Dorsal aspect of hemispheres in the Chimpanzee (A) and in Man (B) (after MAURER, 1928, from SALLER, 1930). Afsp: parieto-occipital sulcus *('Affenspalte')*; Hila: occipital lobe; Scheil: parietal lobe; Stl (r): right frontal lobe; Zf: sulcus centralis; 1, 2, 3: superior, middle, and inferior frontal gyrus; 4: precentral gyrus; 5: postcentral gyrus; 6: parietal convolutions. A at lesser reduction than B.

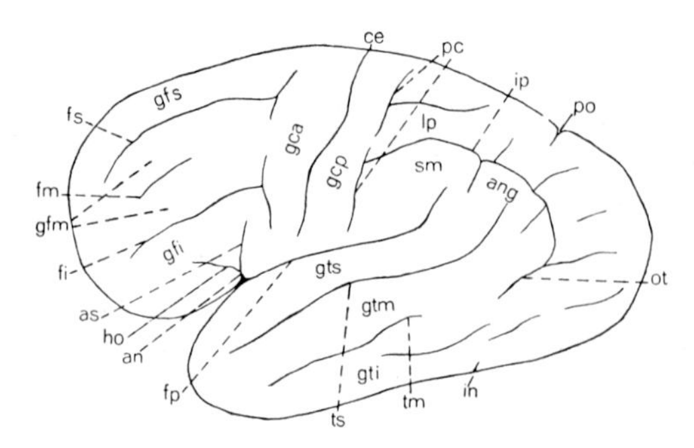

Figure 53 A. Main gyri and sulci of lateral aspect of Human hemisphere (from K., 1927). an: ramus anterior fissurae Sylvii *(sive* f. cerebri lateralis); ang: gyrus angularis; as: ramus anterior ascendens fiss. Sylvii; ce: sulcus centralis Rolandi; fi: sulcus frontalis inferior; fm: sulcus frontalis medius; fp: ramus posterior fissurae cerebri lateralis; fs: sulcus frontalis superior; gca: gyrus centralis anterior *(sive* praecentralis); gcp: gyrus centralis posterior *(sive* postcentralis); gfi: gyrus frontalis inferior; gfm: gyrus frontalis medius (with pars inferior and superior); gfs: gyrus frontalis inferior; gti: gyrus temporalis inferior; gtm: gyrus temporalis medius; gts: gyrus temporalis superior; ho: ramus anterior horizontalis fissurae Sylvii; in: incisura praeoccipitalis; ip: sulcus intraparietalis *(sive* interparietalis); lp: lobulus parietalis superior; ot: sulcus occipitalis transversus; pc: sulcus postcentralis; po: sulcus parieto-occipitalis; sm: gyrus supramarginalis; tm: sulcus temporalis medius; ts: sulcus temporalis superior.

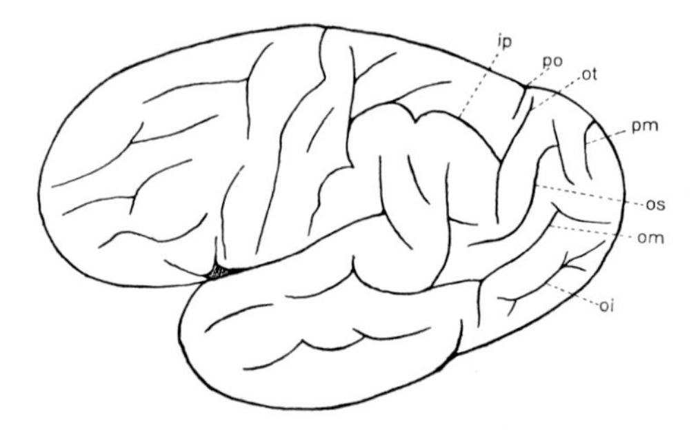

Figure 53 B. Lateral aspect of Human hemisphere with somewhat different, but fairly common fissuration pattern (from K., 1928b). ip, ot, po as in Figure 53A; oi: sulcus occipitalis inferior; om: sulcus occipitalis medius; os: sulcus occipitalis superior; pm: sulcus paramedialis.

tern of sulci, related to the transformations of the sulcus lunatus and sulcus calcarinus systems dealt with further below in section 4. It will here be sufficient to mention, as a fairly common Human pattern, the presence of three more or less longitudinal sulci, namely, in basodorsal sequence, sulcus occipitalis inferior, medius and superior (cf. Fig.

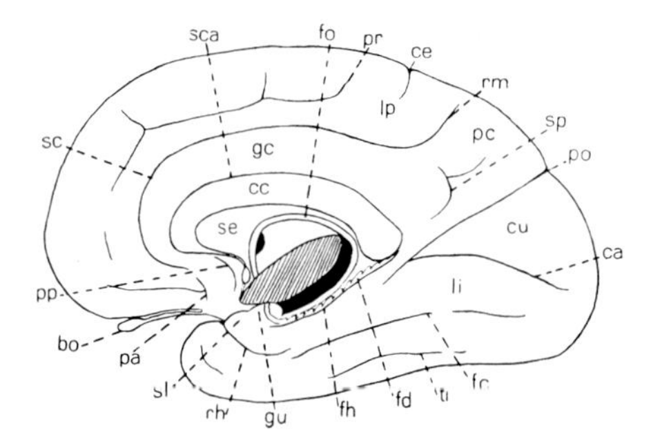

Figure 53C. Main gyri and sulci of medial aspect of Human hemisphere (from K., 1927). bo: olfactory bulb; ca: sulcus calcarinus; cc: corpus callosum; ce: sulcus centralis cu: cuneus; fc: sulcus ('fissura') collateralis; fd: fascia dentata; fh: sulcus ('fissura') hippocampi; fo: fornix; gc: gyrus cinguli; gu: gyrus uncinatus; li: gyrus lingualis; lp: lobulus paracentralis; pa: sulcus parolfactorius anterior; pc: praecuneus; po: sulcus parieto-occipitalis; pp: sulcus parolfactorius posterior; pr: sulcus paracentralis (frequently a branch of sulcus cinguli); rh: sulcus ('fissura') rhinalis (lateralis posterior); rm: ramus marginalis sulci cinguli (this branch rather constantly reaches the dorsal surface caudally to sulcus centralis and either rostrally to, or at same level with, sulcus postcentralis); sc: sulcus cinguli (it may be accompanied by either a dorsal or a basal parallel sulcus, both of which may also be lacking, cf. e.g. Fig. 244B of volume 5/I, and Fig. 55); sca: sulcus corporis callosi; se: septum pellucidum; sl: gyrus semilunaris sive lunaris (cortical nucleus amygdalae, surrounded by here not indicated gyrus ambiens (cf. Figs. 53D and 56A); sp: sulcus subparietalis; ti: sulcus temporalis inferior.

53B). The occipital pole may be reached by the calcarinus system or, as not infrequently in Man, by an extension of sulcus occipitalis medius.

The *lateral aspect of the temporal lobe* is characterized by the rather constant superior temporal sulcus and the more variable middle temporal sulcus, frequently consisting of several parts. The superior temporal gyrus lies between fissura Sylvii and superior temporal sulcus, and the middle temporal gyrus between this latter and the middle sulcus. The inferior temporal gyrus is bounded by the middle and the inferior temporal sulcus, which is located in the basal surface. The inferior temporal gyrus usually displays the *impressio petrosa*, a depression at the rostral end of the temporal lobe's tentorial surface, corresponding to an elevation on the anterior surface of the petrous portion of the temporal bone at the location of superior semicircular canal. The temporal pole is reached by merging neighborhoods of superior, middle and opercular temporopolar gyri.

The operculum temporale, adjacent to, and covering a part of the insula, includes the variable transverse temporal sulci and gyri

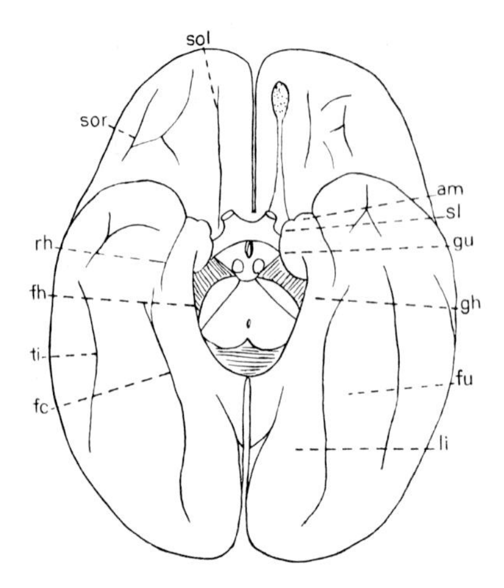

Figure 53 D. Main gyri and sulci of basal aspect of Human hemisphere (from K., 1927). am: gyrus ambiens; fc: sulcus collateralis; fh: sulcus hippocampi; fu: gyrus fusiformis; gh: gyrus hippocampi; li: gyrus lingualis; rh: sulcus rhinalis (lateralis posterior); sl: gyrus semilunaris; sol: sulcus olfactorius; sor: sulci orbitales; ti: sulcus temporalis inferior.

(Heschl's convolutions) corresponding to the auditory projection area. Figure 54 indicates a fairly common pattern of these gyri and sulci in Man.

With regard to the hypothesis that asymmetries of the antimeric Human temporal lobes are associated with the main representation of language function in one (usually the left) hemisphere, Yeni-Komshian and Benson (1976) report on the results of their measurements. The cited authors found, as also noticed by a few previous investigators, that the Human *Sylvian fissure* is generally longer on the left than on the right. To a lesser degree than in Man, the Chimpanzee brains are said to show a similar asymmetry. In the Rhesus brain, however, no significant differences between left and right fissure lengths were detected by the cited authors.

The *insula* represents a hidden lobe within the floor of *fossa Sylvii* ('stem lobe' or '*Stammlappen*' of older authors), covered by the frontal, parietal, and the above-mentioned temporal operculum, and delimited from the opercula by sulcus circularis insulae (Figs. 53E, 54). The basal *limen insulae* represents the Sylvian or hemispheric bend of the sulcus rhinalis lateralis. A sulcus centralis insulae, perhaps likewise corresponding to the convexity of the hemispheric bend (i.e. to the pseudosylvian sulcus) and a variable postcentral and several like-

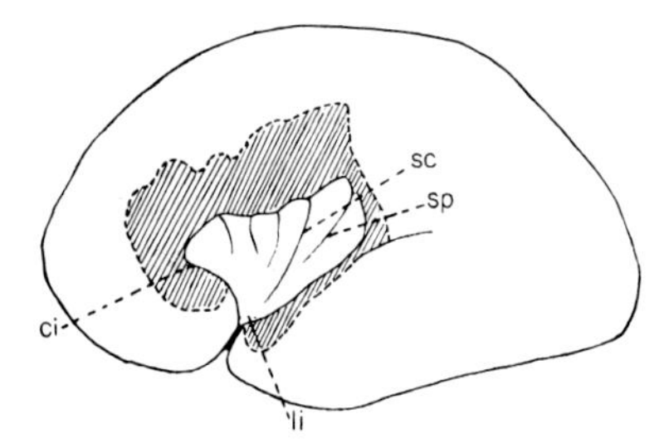

Figure 53 E. Sketch showing aspect of insula in the Human hemisphere (combined after RETZIUS, 1896, and personal observations, from K., 1927). ci: sulcus circularis insulae; li: limen insulae; sc: sulcus centralis insulae; sp: sulcus postcentralis insulae. The oblique hatching represents the cut surface of operculum frontale, parientale and temporale.

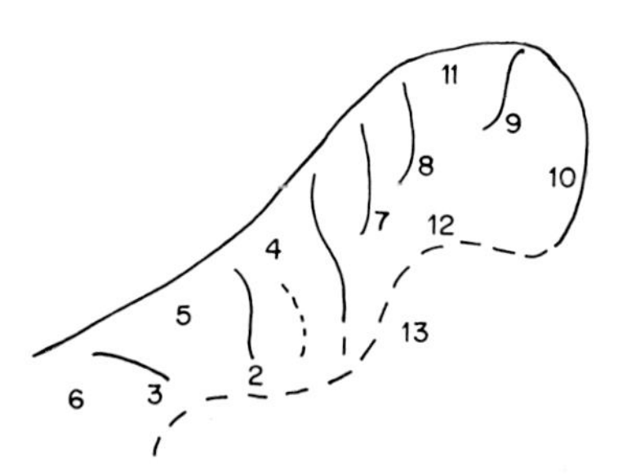

Figure 54. Sketch illustrating a common pattern of *Heschl's transverse temporal gyri and sulci* in Man (based on findings by BECK, 1931, and personal observations). 1: sulcus temporalis transversus primus; 2: s.t.t. secundus; 3: s.t.t. tertius; 4: gyrus temporalis transversus primus; 5: g.t.t. secundus; 6: g.t.t. tertius; 7: sulcus supratemporalis; 8: sulcus temporopolaris lateralis; 9: sulcus temporopolaris medialis; 10: gyrus temporopolaris medialis; 11: gyrus temporopolaris lateralis; 12: gyrus parainsularis; 13: cut region of sulcus circularis insulae. The gyrus around 7 is gyrus supratemporalis, and the dotted sulcus in 4 is the very variable sulcus temporalis transversus intermedius.

wise variable precentral insular sulci are generally present.[38] The superior portion of sulcus circularis may correspond to a remnant of the Mammalian suprasylvian arcuate sulcus. Several questions concerning the presumptive phylogenetic evolution of the Primate insula from insula aperta to insula tecta still remain uncertain and controversial. These problems, with particular reference to the older authors, have been reviewed by LANDAU (1923) and a more recent study on this topic is that by FRIANT (1956).

[38] The gyri rostral to sulcus centralis insulae are the gyri breves, the gyri posterior to said sulcus are the gyri longi. The presence of more than one postcentral insular gyrus is perhaps restricted to man.

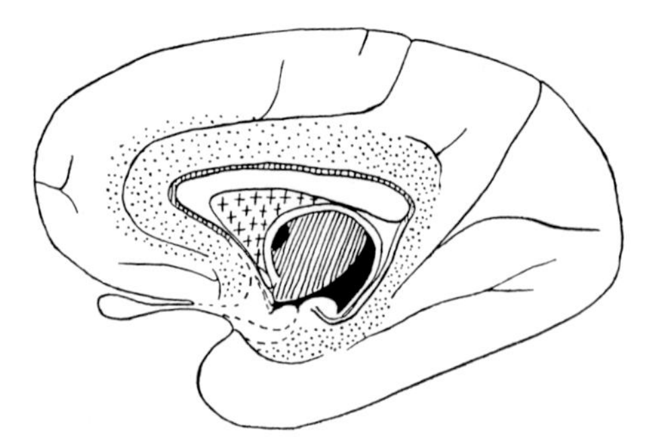

Figure 55. Sketch of medial aspect of Human hemisphere roughly illustrating the '*circonvolution annulaire*' of French authors respectively SCHWALBE's '*Sichellappen*' comprising gyrus fornicatus with adjacent hippocampal formation (from K., 1927). Dots indicate parahippocampal cortex, the precommissural and supracommissural hippocampus is indicated by vertical hatching, and the septum pellucidum by crosses.

The *medial aspect* of the hemisphere includes portions of frontal, parietal, occipital, and temporal lobes, moreover the more or less ring-like *limbic lobe* (Fig. 55) surrounding, as it were, hemispheric stalk and corpus callosum, from which latter it is separated by the *sulcus corporis callosi*. The boundaries of limbic lobe are *sulcus parolfactorius anterior*, *sulcus cinguli*, *sulcus subparietalis*, *rostral end of sulcus parieto-occipitalis*, portions of *sulcus collateralis*, and *sulcus rhinalis*.[39] The portion of limbic lobe between splenium corporis callosi and common stem of parieto-occipital and calcarine sulci includes the so-called isthmus gyri fornicati (cf. below, footnote 43).

Within the frontal lobe, the sulcus rostralis usually ends caudally at the sulcus parolfactorius anterior. Between sulcus rostralis and sulcus olfactorius extends the gyrus rectus, which may reach the frontal pole. The medial portion of gyrus frontalis superior is basally limited by sulcus rostralis and sulcus cinguli,[40] and extends as far posteriorly as the medial portion of sulcus praecentralis or to the variable sulcus paracentralis, which may be connected with an accessory system of sulci running parallel to sulcus cinguli. Precentral and postcentral gyri join around the medial end of sulcus centralis and form the so-called *lobulus*

---

[39] This sulcus rhinalis represents the posterior portion of the Vertebrate sulcus rhinalis lateralis.

[40] The sulcus cinguli, also designated as sulcus callosomarginalis, corresponds to the genualis-splenialis system of the Carnivore-Ungulate fissuration pattern.

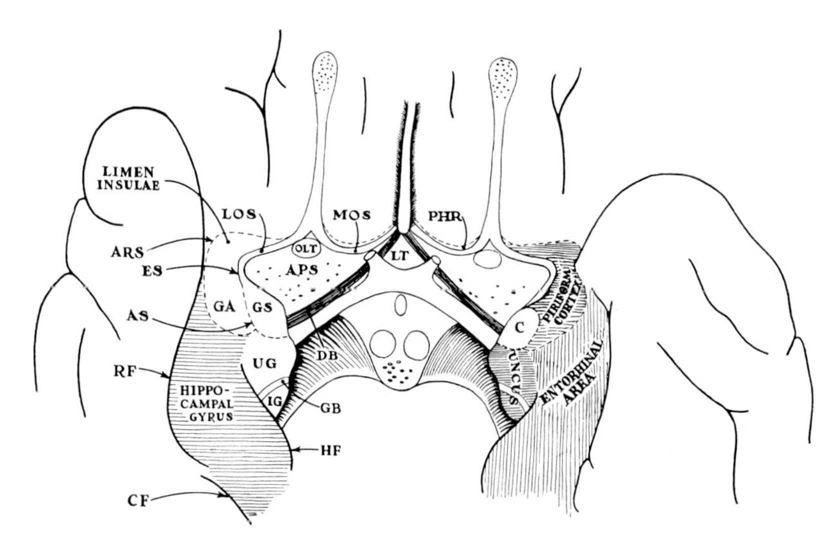

Figure 56 A. Sketch illustrating details of configurations at the base of the Human hemisphere and its surroundings (modified after K., 1927, from NAUTA and HAYMAKER, 1969). APS: substantia perforata anterior; ARS: sulcus rhinalis lateralis anterior; AS: sulcus semiannularis; C: cortical amygdaloid nucleus; CF: sulcus collateralis; DB: *Broca's diagonal band;* ES: sulcus endorhinalis; GA: gyrus ambiens; GB: *Giacomini's band* (end of gyrus dentatus); GS: gyrus lunaris sive semilunaris; HF: sulcus ('fissura') hippocampi; IG: gyrus intralimbicus; LOS: lateral olfactory stria; LT: lamina terminalis (pars hypothalamica); MOS: medial olfactory stria; OLT: 'olfactory tubercle'; PHR: precommissural hippocampal rudiment; RF: sulcus rhinalis lateralis, pars posterior ('rhinal fissure'); UG: uncinate gyrus. The uncus is hippocampal. Cf. Fig. 175 D, vol. 5/I.

paracentralis,[41] whose posterior limit is the ramus marginalis sulci cinguli, often extending toward the dorsal aspect of the hemisphere. The *praecuneus*, pertaining to the parietal lobe, is bounded by ramus marginalis, by sulcus subparietalis and sulcus parieto-occipitalis.

The *occipital lobe* displays the system of sulcus calcarinus, rostrally joining sulcus parieto-occipitalis. The *cuneus* lies between these two converging sulci, and the *gyrus lingualis* between calcarine sulcus and sulcus collateralis. The cuneus, and the above mentioned precuneus of parietal lobe commonly display variable, nondescript sulci cunei respectively praecunei. The likewise variable submerged sulci and gyri

[41] This lobulus thus represents a conjoint neighborhood of frontal and parietal lobe i.e. of precentral and postcentral gyrus.

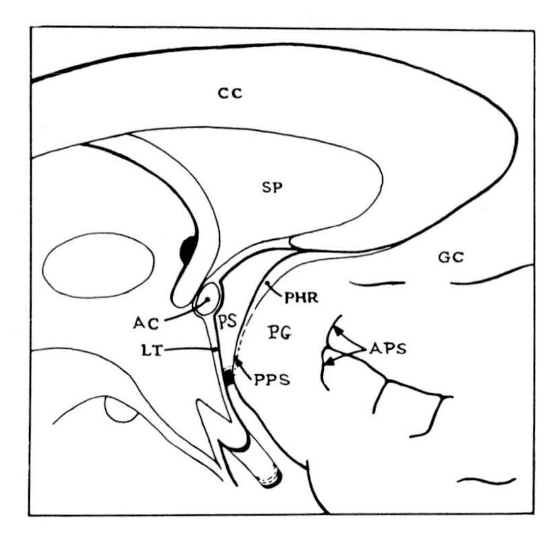

Figure 56B. Sketch illustrating configurations on rostromedial aspect of Human hemisphere (from NAUTA and HAYMAKER, 1969). AC: anterior commissure; APS: sulcus parolfactorius anterior; CC: corpus callosum; GC: gyrus cinguli; LT: hypothalamic part of lamina terminalis (containing the here omitted supraoptic crest); PG: 'parolfactory area or gyrus' (parahippocampal cortex); PHR: rudimentary precommissural hippocampus; PPS: sulcus parolfactorius posterior; PS: 'precommissural septum' (paraterminal body); SP: septum pellucidum. The groove extending frontalward from APS is sulcus rostralis.

in the deep parieto-occipital sulcus, which may also become superficial, have been pointed out by ELLIOT SMITH (1903/04a, b) and were also dealt with by CONNOLLY (1950). At the basal end of sulcus parieto-occipitalis, the gyrus lingualis is continuous with the *gyrus hippocampi* pertaining to a parahippocampal portion of limbic lobe. This transition is also called gyrus occipitotemporalis medialis (PNA). Rostrally, the *sulcus collateralis* may join the sulcus rhinalis. Between sulcus collateralis and sulcus temporalis inferior runs the *gyrus fusiformis*, which apparently corresponds to what the PNA now term gyrus occipitotemporalis lateralis. Nondescript, highly variable accessory sulci in gyrus lingualis and gyrus fusiformis have been designated as sulci sagittales of these gyri. Thus, except for the indistinct preoccipital notch, there is no definite surface marking separating on the medial aspect, occipital and temporal lobes.

Reverting now to the *limbic lobe* and its complex morphologic features, the particular significance of this ring-like configuration was already recognized about 1838 by GERDY, who described a '*circonvolu-*

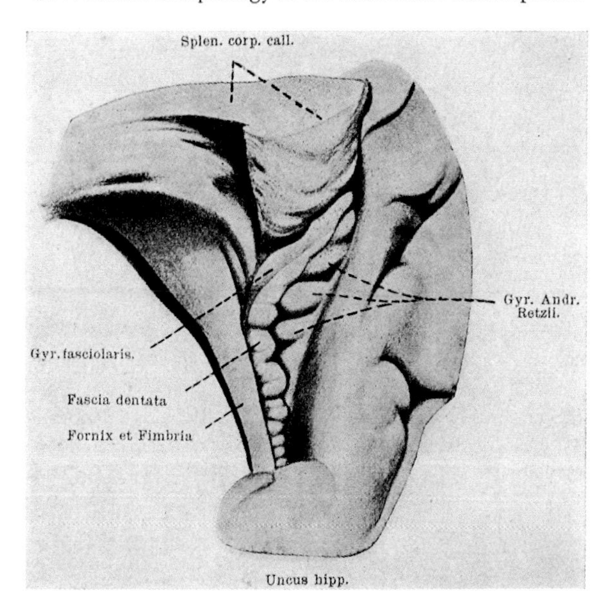

Figure 57 A. Retrosplenial aspect of Human hippocampus, displaying gyrus fasciolaris and *gyri Andreae Retzii* (after RETZIUS, 1898).

tion annulaire'.[42] BROCA (1878) apparently introduced the term '*grand lobe limbique*' and '*limbe de l'hémisphère*'. SCHWALBE (1881) similarly recognized a 'falciform lobe' *(Sichellappen)*. Other authors distinguished concentrically arranged limbic subdivisions, such as *gyrus fornicatus*,[43] *cornu Ammonis*, *gyrus dentatus*, and *fimbria-fornix system*, all of which were included in the so-called rhinencephalon (cf. e.g. VILLIGER, 1920).

On the basis of the presently available data, the configurations of the limbic lobe may be interpreted as follows. Rostrally, near the connection of olfactory tract with the hemisphere's base, the ill-defined stria olfactoria medialis turns dorsomedialward toward the callosal induseum, being adjacent on the rostral side, to a vaguely outlined *precommissural hippocampal rudiment (gyrus geniculatus)* continuous with the

[42] GERDY, *Recherches sur l'encéphale, Journal des connaissances médico-chirurgicales, décembre 1838, réimprimé dans* GERDY, *Mélanges d'anatomie, de physiologie et de chirurgie, Paris 1875* (quoted after K., 1927).

[43] The gyrus fornicatus included gyrus cinguli and gyrus hippocampi, continuous with each other at isthmus gyri fornicati, between parieto-occipital sulcus and begin of postsplenial hippocampal formation.

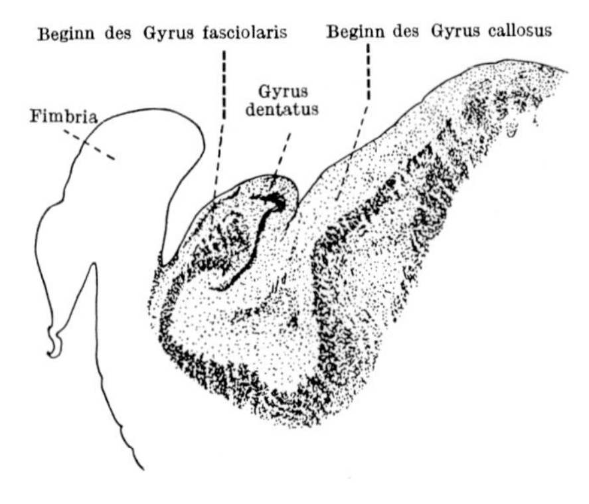

Figure 57B. Section through isthmus gyri fornicati showing relationships of fimbria, gyrus fasciolaris, gyrus dentatus and cornu Ammonis (modified after v. ECONOMO, from GRAF HALLER, 1928).

supracommissural hippocampus running along the sulcus corporis callosi (Fig. 223C, III, vol. 5/I, and Fig. 57C, I, of the present volume). The precommissural hippocampus may or may not be delimited from the rostrally adjacent parahippocampal cortex, which represents the so-called *area parolfactoria*, by a *sulcus parolfactorius medius*. The area parolfactoria (or gyrus parolfactorius)[44] is, in turn, delimited from the frontal isocortex by the *sulcus parolfactorius anterior*.

The stria olfactoria medialis lies within the precommissural paraterminal body, which is usually delimited from the precommissural hippocampal rudiment by the variable and poorly developed *sulcus parolfactorius posterior*. The precommissural paraterminal body, continuous with the induseum verum, represents the *gyrus subcallosus*, which also contains the rostral portion of *Broca's diagonal band*. The groove between gyrus subcallosus and lamina terminalis is the *sulcus parolfactorius postremus*. Vague accessory posterior parolfactory sulci may, within gyrus subcallosus, suggest an outline of stria olfactoria (or 'gyrus olfactorius') medialis. Because of the highly variable and often indistinct development of these configurations, considerable discrepancy concering terminology and interpretation can be found in the literature. Figures 56A and B illustrate the relevant relationships.

At the *transition of supracommissural to postcommissural hippocampus*, where gyrus dentatus and cornu Ammonis begin to become distinctively separated, and sulcus hippocampi joins sulcus corporis callosi,

---

[44] BROCA's *carrefour olfactif*.

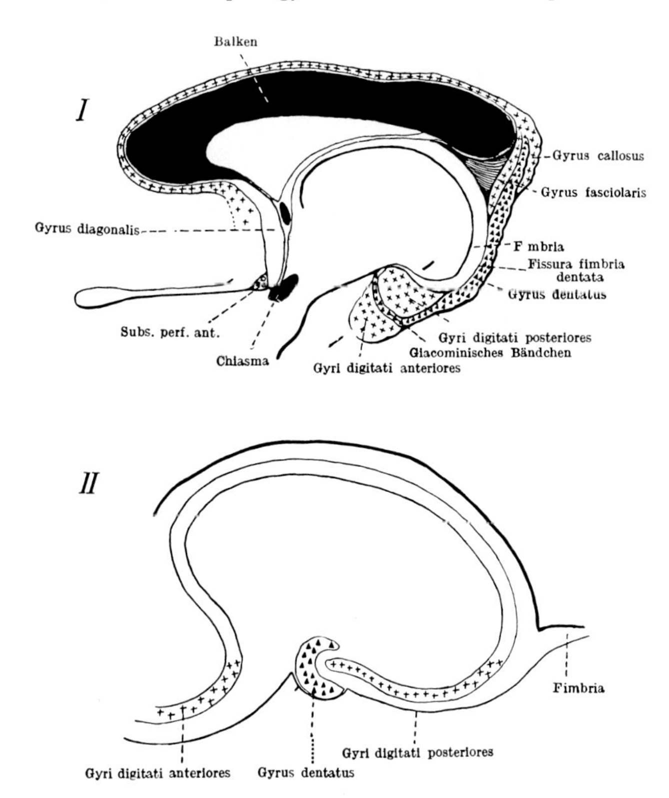

Figure 57C. Graf Haller's concept of pre-, supra-, and postcommissural hippocampal formation (I), and sketch illustrating end of gyrus dentatus as *Giacomini's band* (II) (from Graf Haller, 1928). The plane of section for Figure II is indicated in Figure I. The 'gyri digitati posteriores' are the gyrus intralimbicus, and 'gyri digitati anteriores' are gyrus uncinatus of Figure 56A. The lower part of precommissural hippocampus and the 'gyrus diagonalis' are not quite accurately drawn (cf. Fig. 56B).

this zone of transition, above the isthmus of gyrus dentatus, displays, on both sides of the serrated edge of the emerging fascia dentata, two variable, strip-like small gyri. Caudally to fascia dentata a portion of cornu Ammonis (the so-called gyrus callosus) may be characterized by irregular protrusions designated as *gyri Andreae Retzii*. Rostrally to fascia dentata, the *gyrus fasciolaris* represents here a region of transition between cornu Ammonis and gyrus dentatus. Figures 57 A–C illustrate relevant aspects of these configurations.

In the region of the *uncus hippocampi*, the gyrus dentatus becomes *Giacomini's band*, on both sides of which the cornu Ammonis forms the

58A

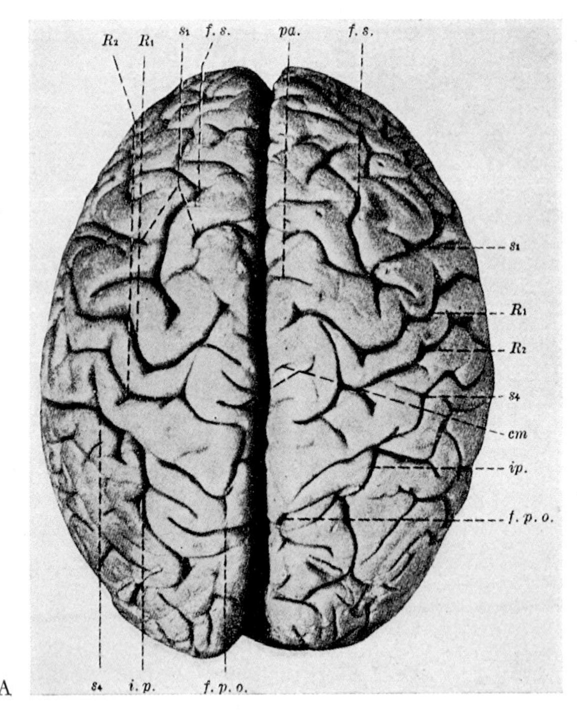

Figure 58. Dorsal (A) and left lateral (B) aspect of Human brain with 'duplication of sulcus centralis' (from GERLACH und WEBER, 1929). cm: pars marginalis sulci cinguli (sive calloso-marginalis); f.i.: sulcus frontalis inferior; f.p.o.: sulcus parieto-occipitalis; f.s.: sulcus frontalis superior; f.S.: fissura Sylvii; i.p.: sulcus intraparietalis; pa.: sulcus paracentralis; R.ant.horiz., R.ant.ax.: ramus anterior horizontalis respectively ramus anterior ascendens fissurae Sylvii; R_1, R_2, s_1, s_4: second, third, first, and fourth sulcus of central region.

configurations designated as *gyrus limbicus* (or *intralimbicus)* caudally, and *gyrus uncinatus* rostrally[45] as shown in Figures 56 A and 57 C. The hippocampal gyrus, respectively the so-called entorhinal area represent parahippocampal cortex. The relationships of the adjacent cortical nucleus amygdalae, of anterior perforated substance, and anterior piriform lobe to the aforementioned limbic lobe configurations are likewise illustrated by Figure 56 A.

In concluding the present section on the overall Primate fissuration pattern, and before dealing in the following section 4 with a specific

[45] Gyri digitati posteriores and gyri digitati anteriores of GRAF HALLER (1928) as indicated in Figure 57 C.

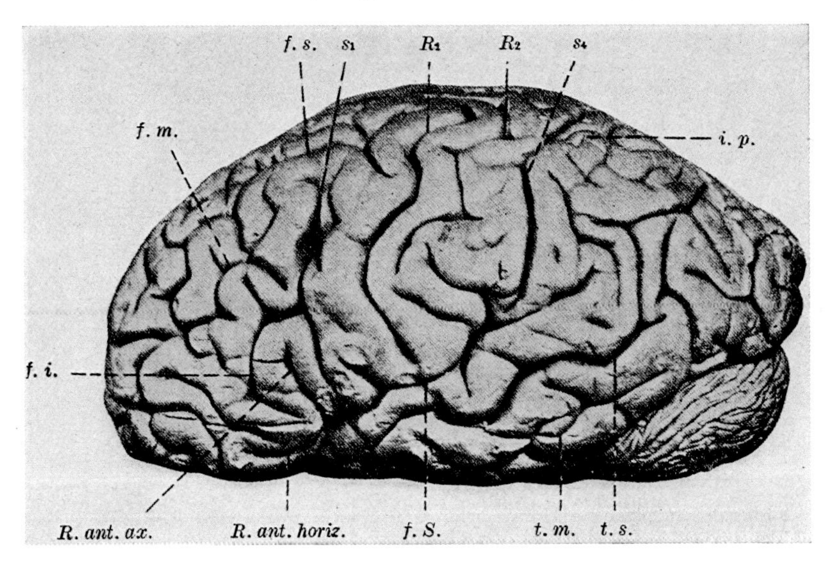

Figure 58 B

problem concerning occipital lobe and sulcus lunatus, brief references to an unusual variation in Man, and to a few aspects of ontogenetic development of sulci in the Human hemisphere are perhaps appropriate.

The *sulcus centralis* is one of the sulci which appear rather early in ontogeny, becoming recognizable at about the 19th week of intrauterine age. It is, moreover, one of the few sulci showing rather constant relationships to distinctive cytoarchitectural and functional cortical areas, since it corresponds to the approximate boundary between somatic motor and somatic sensory cortex. In some rare instances, apparently first reported by GIACOMINI (1882), this sulcus may appear duplicated in one or in both hemispheres.[46] According to GERLACH and WEBER (1929) who reported on a case of bilateral duplication observed in our *Breslau* laboratory (Fig. 58), about 33 cases of sulcus centralis duplication, roughly half of which were bilateral, had been previously recorded in the literature. A cytoarchitectural examination by GERLACH and WEBER as well as in a similar case by ORTON (1911), dis-

[46] GIACOMINI (1882, p. 94–95) remarked: '*Questa duplicità della scissura di Rolando è uno dei fatti piu singolari e piu interessanti que possa presentare la superficie cerebrale. Esso non fu fino ad ora osservato da alcun anatomico.*' It is indeed a strange coincidence that the selfsame rare variety discovered by GIACOMINI was displayed by his own brain, which was studied by SPERINO (1901).

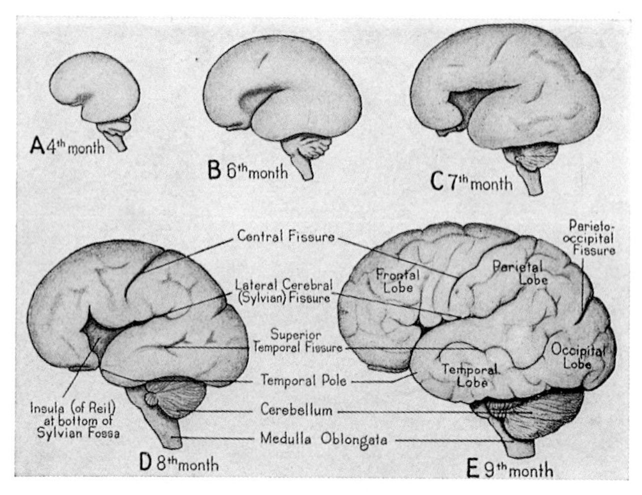

Figure 59. Lateral view of Human hemisphere at successive stages of fetal development (modified after RETZIUS, 1896, from PATTEN, 1953).

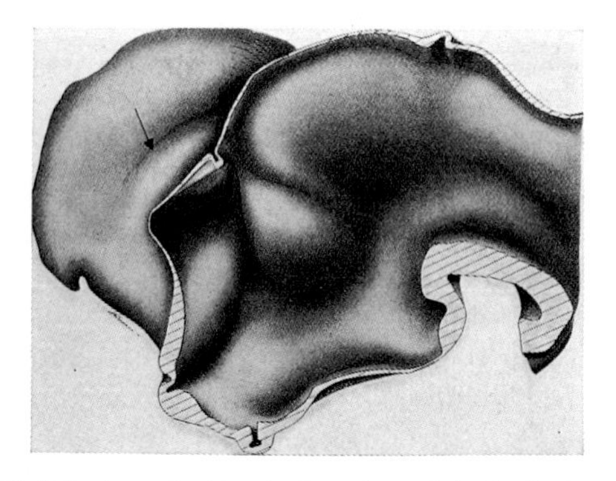

Figure 60 A. Medial view of midsagittal section of the brain in a 19.1 mm Human embryo (about 7th week), showing sulcus hippocampi, indicated by arrow (from HINES, 1922).

closed that the cortex of the supernumerary gyrus between the two 'central' sulci was the koniocortex of the somatic sensory area. On the basis of this criterion, apparently not considered by the previous investigators, the posterior 'central sulcus' in such cases of duplication would actually represent the anterior groove of a duplicated postcentral sulcus, whose posterior unit is connected with the intraparietal sulcus. Variations of this type should therefore not be considered duplications of the central sulcus, but rather of the postcentral sulcus, *simula-*

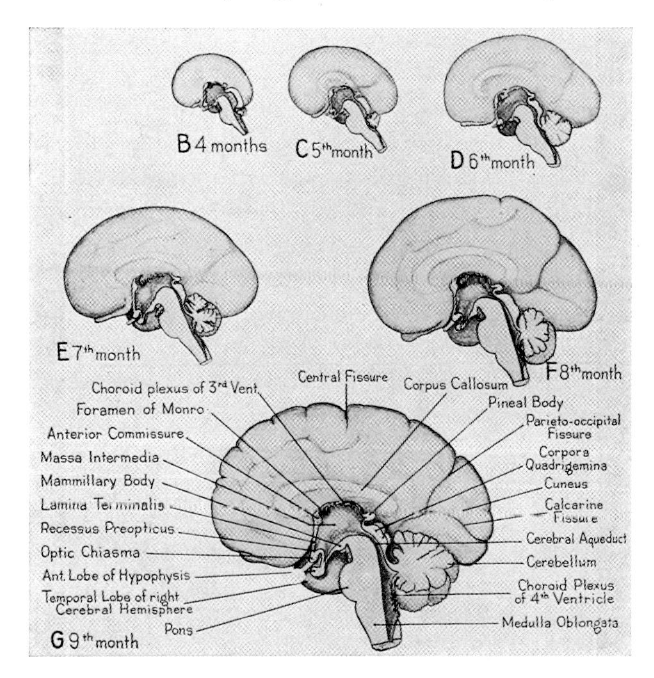

Figure 60 B–G. Medial view of Human hemisphere at successive stages of fetal development (from PATTEN, 1953).

ting a double sulcus centralis Rolandi (ORTON, 1911; GERLACH and WEBER, 1929).

As regards the ontogenetic development of the sulci in the Human hemisphere, and discounting the earlier grooves delimiting olfactory bulb and tract from lobus hemisphaericus, a lateral fovea corresponding to hemispheric bend and pseudosylvian groove becomes well noticeable at 27 mm, in the 9th week. In the 4th lunar month, the hemisphere has still a lissencephalic aspect. The central sulcus appears at about 19 weeks. The subsequent development of sulci and gyri on the hemisphere's lateral aspect is somewhat differently described by various authors,[47] but generally speaking, can be assumed to proceed as roughly illustrated by Figure 59.

On the medial side, the sulcus hippocampi, particularly investigated by HINES (1922) and HUMPHREY (1967), appears as a shallow groove at

[47] A fairly recent embryological study on the development of the Human cerebral sulci, with bibliographic references, was published by MURAKAMI (1955).

about 16 mm (approximately between 6th and 7th week) above the lamina terminalis, before the development of corpus callosum (Fig. 60 A), and subsequently rotates along the hemispheric bend and concomitantly with the formation of the temporal lobe, towards its definitive location while undergoing transformations related to the differentiation of cornu Ammonis and gyrus dentatus. Sulcus cinguli, sulcus parieto-occipitalis, and sulcus calcarinus begin to appear at about the 19th week. Their further development is shown in Figures 60 B–D.

The various controversies related to early so-called transitory sulci and to a medial arcuate sulcus presumably representing artifacts due to poor fixation have been critically reviewed by HOCHSTETTER (1924). Many of the details concerning the ontogenesis of the hemispheric fissuration pattern still remain insufficiently elucidated.

4. The Problem of the so-called Sulcus Lunatus

The individual variations of sulci and gyri in the lateral aspect of the Human occipital lobe are very numerous and diversified, thus precluding the formulation of a generally valid descriptive scheme, such as, despite some qualifications, can be given for frontal, parietal, temporal lobes, insula, and medial hemispheric surface.[48]

It is, moreover, of considerable interest that the Human occipital lobe frequently displays a fissuration pattern typical for the anthropoid Apes and may, on the other hand, show an arrangement of sulci and gyri apparently not found in these latter.

In Catarrhine Cercopithecidae the *sulcus lunatus*, which approximately corresponds to the rostrolateral boundary of the visual cortex (area striata, area 17) commonly joins the dorsolateral extension of sulcus parieto-occipitalis, and forms, with this latter, a fairly deep, opercularized fossa *(operculum occipitale)* into which three so-called parieto-occipital transitional gyri (GRATIOLET's *plis de passage, Übergangswindungen)* can be followed (Figs. 48 I, II, 61). This fossa and the combination of sulcus parieto-occipitalis with sulcus lunatus were designated as *fossa Simiarum (Affengrube)* respectively *fissura Simiana sive sulcus Simianus (Affenspalte)*.

---

[48] GUILLAIN and BERTRAND (1926) remark in their text on topographic anatomy of the Human central nervous system: '*La face externe du lobe occipital représente la portion de l'écorce cérébrale la plus mal systematisée.*'

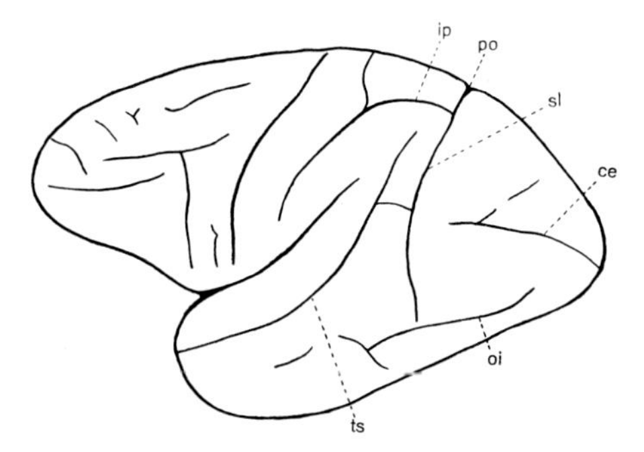

Figure 61. Lateral aspect of the hemisphere in the Catarrhine Primate Cynocephalus babuin (from K., 1928b). ce: sulcus calcarinus externus; ip: sulcus intraparietalis; oi: sulcus occipitalis inferior; po: sulcus parieto-occipitalis; sl: sulcus lunatus; ts: sulcus temporalis superior.

In the anthropoid Apes (Figs. 40 IV, 49–50) the connection between parieto-occipital and lunate sulci tends to disappear, the operculum becoming reduced, and the transitional gyri beeing essentially superficial (Fig. 48). The sulcus lunatus, with a rostral sulcus praelunatus, remains conspicuous and still about corresponds to the limit of area striata.

Within the posterior concavity of the sulcus lunatus, three sorts of grooves, related to the sulcus calcarinus complex, may commonly be seen, as particularly shown in Figure 51 B. These are (1) a rostrolateral extension of the medial sulcus calcarinus system, which may be called *sulcus calcarinus posterior* or pars intermedia of the calcarinus complex; (2) a *sulcus calcarinus externus*, usually but not always separated by a small gyrus from pars intermedia, and frequently triradiate *(sulcus triradiatus)* with a posterior trifurcation and an anterior branch pointing toward sulcus lunatus; (3) two accessory sulci, *sulcus accessorius dorsalis and basalis*, roughly 'parallel' to anterior branch of sulcus calcarinus externus. All these sulci lie within the visual area striata which, as mentioned above, extends approximately as far as sulcus lunatus. Basally, the entire cortical region behind sulcus lunatus is delimited by the sulcus occipitalis inferior, which extends rostrally toward or into the temporal lobe with a wide variety of terminal or branching patterns.

Exactly the same type of fissuration can occasionally be found in the Human brain (Figs. 62 A–C), the sulcus calcarinus externus being either a separate sulcus, continuous with, or a component of, pars intermedia sulci calcarini.

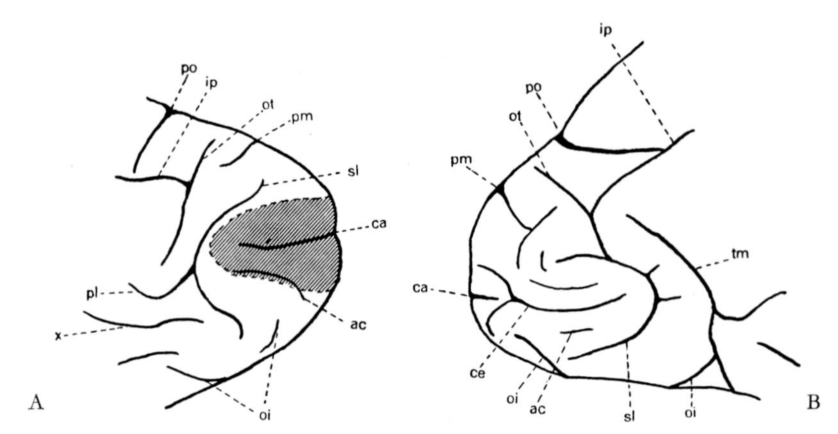

Figure 62 A, B. Human occipital lobes with sulcus lunatus (A) and typical triradiate sulcus calcarinus externus (B) (from K., 1928b). ac: sulcus calcarinus externus accessorius ventralis (in B, an unlabelled dorsal accessory sulcus is present); ca: sulcus calcarinus (in A, continuous with sulcus calcarinus externus); ce: sulcus calcarinus externus; ip: sulcus intraparietalis; oi: sulcus occipitalis inferior; ot: sulcus occipitalis transversus; pl: sulcus praelunatus; pm: sulcus paramesialis; po: sulcus parieto-occipitalis; sl: sulcus lunatus; tm: sulcus temporalis medius; x: sulcus praelunatus inferior (?). The hatched area in A indicates extent of area striata, which does not quite reach the sulcus lunatus.

In the Human brain, however, this Anthropoid pattern tends to be modified, concomitantly with a 'shifting' or 'sliding' of the area striata's outermost posterolateral expansion, which recedes from the caudolateral surface of occipital lobe toward a restricted region in the neighborhood of the occipital pole.[49] This shift may well be related to the increased expansion of FLECHSIG's parieto-occipito-temporal association cortex obtaining in Man. The sulcus calcarinus externus joins pars intermedia sulci calcarini, dorsal and basal accessory sulci may join to form a single U-shaped sulcus rostrally to sulcus calcarinus externus and posteriorly to sulcus lunatus. This latter tends to be dissolved into a dorsal and a basal quadrant which, in various combinations with sulcus praelunatus, assume a longitudinal course, becoming either independent or connected, in a variety of patterns, with neighboring parietal and temporal sulci. Diverse aspects of these transformations are shown in Figures 63A–C and 53B. The 'end result', as it were, is the formation of sulcus occipitalis (lateralis) superior and

[49] The posterolateral respectively posterior region of area striata corresponds to the projection of the macula, as depicted in Figure 145 of volume 5/I.

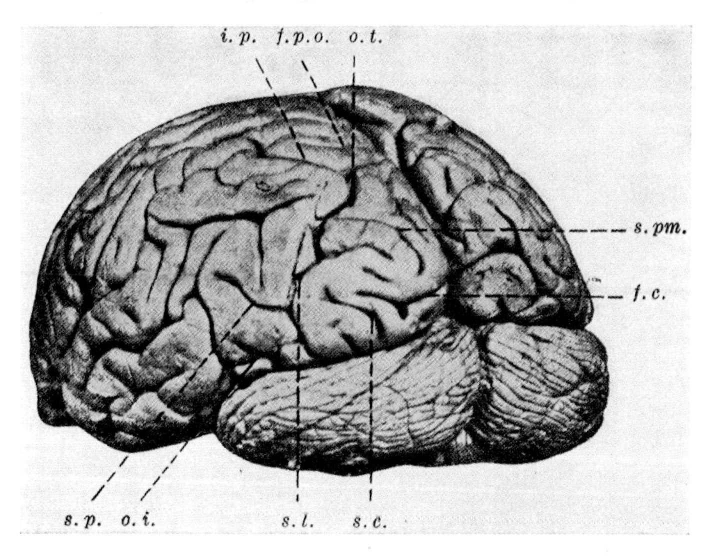

Figure 62C. Left sulcus lunatus (not typically present on right side) in brain with 'reduplication' of central sulcus shown in Figure 58 (from GERLACH and WEBER, 1929). f.c.: sulcus calcarinus externus; f.p.o.: sulcus parieto-occipitalis; i.p.: sulcus intraparietalis; o.i.: sulcus occipitalis inferior; o.t.: sulcus occipitalis transversus; s.c.: sulcus calcarinus externus ventralis accessorius (there is also an unlabelled dorsal accessory sulcus); s.l.: sulcus lunatus; s.p.: sulcus praelunatus; s.pm.: sulcus paramesialis.

medius above the sulcus occipitalis inferior, which latter seems to be a groove not essentially affected by the transformations of the lunatus-praelunatus complex. There is, in addition, a dorsal, frequently somewhat transversely oriented *sulcus occipitalis paramesialis*, pointed out by ELLIOT SMITH (cf. Figs. 53B, 62A, B, 63A, B, 64). This sulcus may extend toward the medial surface and also can be joined to sulcus lunatus or sulcus occipitalis transversus, being rather independent of the lunatus complex. In Anthropoids the sulcus paramesialis appears less well developed and not infrequently is entirely missing.

Figure 63D shows a diagrammatic interpretation of the specifically Human pattern displayed by the lateral aspect of the occipital lobe resulting from a dissolved sulcus lunatus and calcarinus externus complex, with reference to divergences in nomenclature. Nondescript accessory sulci may complicate this pattern. It should be added that the terms 'shifting' and 'sliding' of the area 17, and 'dissolving' respectively 'joining' of sulci have a purely figurative meaning referring to the conceptualized abstraction of a pattern and its topologic transforma-

Figure 63 A. Lateral aspect of Human hemisphere with sulcus lunatus from the author's Breslau neuroanatomy course (from K., 1928b). ce: sulcus calcarinus externus, continuous with pars intermedia (cf. Fig. 65) of calcarine sulcus; ip: sulcus intraparietalis; oi: sulcus occipitalis inferior; ot: sulcus occipitalis transversus; pl: sulcus praelunatus; pm: sulcus paramesialis; po: sulcus parieto-occipitalis; sl: sulcus lunatus. Between ce and sl a semicircular groove, parallel to sulcus lunatus, and interpreted as joined dorsal and ventral accessory external calcarine sulci. The area striata (hatched) extends rostrally beyond that semicircular groove, but does not reach sulcus lunatus.

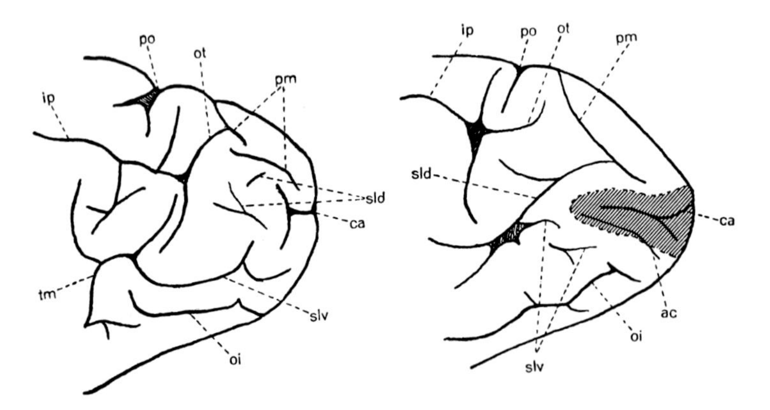

Figure 63 B. Lateral aspect of two Human occipital lobes from the author's Breslau neuroanatomy course, showing patterns of 'dissolved' or 'disrupted' sulcus lunatus (from K., 1928b). ac: sulcus calcarinus externus accessorius ventralis; ca: sulcus calcarinus (continuous with sulcus calcarinus externus); ip: sulcus intraparietalis; oi: sulcus occipitalis inferior; ot: sulcus occipitalis transversus; pm: sulcus paramesialis; po: sulcus parieto-occipitalis; sld: 'dissolved fragments' of dorsal quadrant of sulcus lunatus (in the right Figure: entire dorsal quadrant together with sulcus praelunatus); slv: 'dissolved fragments' of ventral quadrant of sulcus lunatus. The extent of area striata (hatched) is indicated in the right Figure.

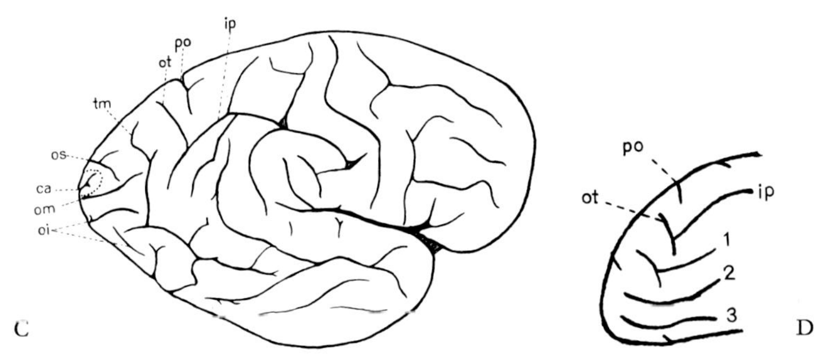

Figure 63 C. Lateral aspect of Human hemisphere from the author's Breslau material, with completely 'dissolved' sulcus lunatus (from K., 1928b). ca: sulcus calcarinus; ip: sulcus interparietalis; oi: sulcus occipitalis inferior; om: sulcus occipitalis medius (interpreted as ventral 'dissolved' portion of sulcus lunatus); os: sulcus occipitalis superior (interpreted as dorsal 'dissolved portion' of sulcus lunatus, the dorsal oblique branch joining said fragment could perhaps represent a rudimentary sulcus paramesialis); ot: sulcus occipitalis transversus; po: sulcus parieto-occipitalis; tm: dorsal expansion of sulcus temporalis medius. The dotted outline around ca indicates extent of area striata.

Figure 63 D. Diagrammatic sketch of specifically 'Human' fissuration pattern on lateral surface of occipital lobe (adapted from LANDAU, 1923, from K., 1928b). 1: sulcus occipitalis lateralis of *Eberstaller;* 2: sulcus occipito-temporalis lateralis of *Zuckerkandl;* 3: sulcus occipito-temporalis lateralis of HOLL, corresponding to sulcus occipitalis inferior of GRATIOLET-BROCA; ip: sulcus intraparietalis; ot: sulcus occipitalis transversus; po: sulcus parieto-occipitalis.

tions. Quite evidently, in the concrete individual case, area 17 does not 'slide', nor do sulci 'dissolve' during ontogenesis in a manner corresponding to the abstract transformations, although some secondary 'joining' of developing sulci might indeed take place.

In other words, an individual typical, atypical, or 'dissolved' sulcus lunatus presumably develops *ab initio* in accordance with genetic factors. The various phenotypic manifestations of the lateral occipital fissuration pattern which, if serially arranged and compared, figuratively suggest disruption or 'dissolving' of said sulcus, can be interpreted as the results of hidden genetic interactions, but quite obviously not as the actual process of disruption (or dissolving) of sulcus lunatus in a given individual developing or adult brain.

VON ECONOMO (1930), moreover, has contested my concept of a 'dissolving' sulcus lunatus by emphasizing its significance as an '*operculum occipitale*' and by stressing the transitional gyri of GRATIOLET, ZUKKERKANDL and others. In v. ECONOMO's interpretation, the 'operculum

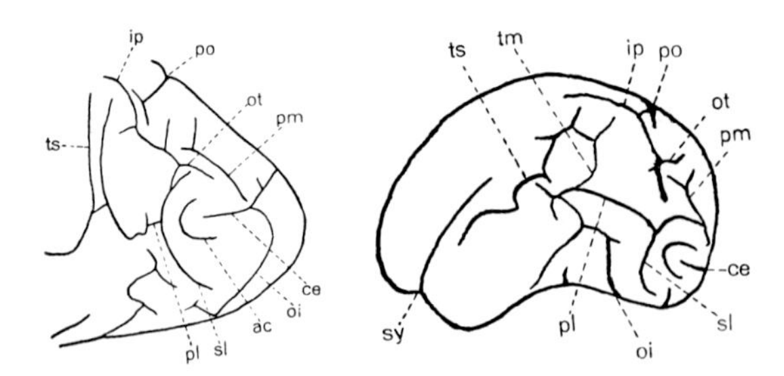

Figure 64. Sulci on lateral aspect of occipital lobe (left) in a native of Celebes and (right) in a Chinese woman (left after KOOY, 1921; right after KURZ, 1924, redrawn and slightly simplified, from K., 1928b). ac: sulcus calcarinus externus accessorius (KOOY interprets said groove as sulcus occipitalis superior, and ELLIOT SMITH as sulcus calcarinus externus joined with calcarine system); ip: sulcus intraparietalis; oi: sulcus occipitalis inferior; ot: sulcus occipitalis transversus; pl: sulcus praelunatus; pm: sulcus paramesialis; po: sulcus parieto-occipitalis; sl: sulcus lunatus; sy: fissura Sylvii; tm: sulcus temporalis medius (*'Rest der lateralen Affenspalte'* according to KURZ); ts: sulcus temporalis superior. The interpretations are those of the present author, for the differing ones of KOOY and of KURZ, their original papers should be consulted.

occipitale', with its sulcus lunatus, recedes as a whole toward the occipital pole and finally disappears, whereby all transitional gyri become superficial, their boundaries being the lateral occipital sulci. The fact, that, even in a pronounced sulcus lunatus such as that of Figure 62A no opercularization whatsoever may obtain, and the occurrence of 'disruption patterns' such as those of Figure 63B are not findings in favor of v. ECONOMO's interpretation which has failed to convince me. However, quite apart from the possibility that both v. ECONOMO's and my own interpretation represent two actually obtaining 'modes of disappearance' of the sulcus lunatus, being thereby not mutually exclusive, the apparent discrepancy between the two views is also partly due to a difference in emphasis on certain configurations, and partly a mere question of semantics.

Reverting to the above-mentioned terms *fossa* respectively *sulcus Simiarum (Affengrube, Affenspalte)*, a number of diverse views have been expressed in the literature with respect to their identification or 'homologization' in the Human brain.

The *sulcus occipitalis transversus* was, at first, considered to represent such homologous groove, but CUNNINGHAM (1890) conclusively demonstrated the fallacy of this interpretation. Subsequently some authors assumed that, for an identification of fossa Simiarum, the decisive criterion should be the opercularization by the combination of sulcus parieto-occipitalis and sulcus lunatus (Fig. 48). This, however, occurs essentially in Cercopithecidae, is rarely and at most barely suggested in Anthropoid Apes, and has not been conclusively recorded in Man.[50] Some authors, e.g. MAURER (1928), evaluate the dorsolateral extension of Human sulcus parieto-occipitalis as *Affenspalte* respectively *Affenspaltenrest* (cf. Fig. 52). Others consider the sulcus lunatus, if present, regardless of depth and opercularization, to represent the sulcus Simiarum in Man.

The *Human sulcus lunatus*, moreover, was not properly recognized until ELLIOT SMITH (1903/04a, b) described it in Egyptian Fellaheens and Sudan Negroes. It was subsequently identified in the brain of Javanese (e.g. KOHLBRUGGE, 1906; KOOY, 1921), Europeans (e.g. ANTONI, 1914; LANDAU, 1923; K., 1928a, b; v. ECONOMO, 1930), Chinese (KURZ, 1924; VAN BORK-FELTKAMP, 1930, 1933), Japanese (HAYASHI and NAKAMURA, 1914; K., 1927, 1928a, b) and can be regarded as a common variation in all races (cf. Figs. 62, 63 A, 64), occurring either bilaterally or unilaterally, being in this case more frequently displayed by the left hemisphere. There are, however, statistical differences with regard to the frequency of this variation in the diverse racial groups.[51] Although the sulcus lunatus represents a distinctively Anthropoid characteristic, doubtless related to the evolution of Man from Simian forms, one would hardly be justified to evaluate its occurrence in a Human brain as necessarily implying 'inferiority' or 'primitive mentality', but rather as an incidental feature of gross morphology quite independent of the multifactorial variables determining 'intelligence', 'behavioral attitudes' or 'mental powers'.[52] Further apposite questions

---

[50] A very deep bilateral dorsolateral extension of sulcus parieto-occipitalis, with some degree of opercularisation on the right, involving sulcus intraparietalis and lobulus parietalis superior, was interpreted as '*Affenspaltenrest*' by LUBOSCH (1929).

[51] Nearly 100 percent in the Fellaheen and Sudan Negroes according to ELLIOT SMITH, about 91 per cent in Chinese, about 40 to 60 per cent in Japanese, about 30 per cent in Europeans (cf. K., 1927). It should be added that these figures are very tentative, since the criteria for the recognition of a 'true' sulcus lunatus, because of its numerous transitional forms, vary in accordance with viewpoints adopted by the various observers. There are, moreover, a number of perhaps not particularly relevant differences in the terminology and the details of interpretation proposed by the diverse authors (cf. e.g. LANDAU, 1923; K., 1928a, b; v. ECONOMO, 1930).

[52] Cf. K. (1927, p.299), also K. (1928a, b). At most, it could perhaps be tentatively assumed that the Human sulcus lunatus is related to a persistent expansion of the macular visual cortex, i.e. of the sensory area mediating greatest visual acuity, which has main-

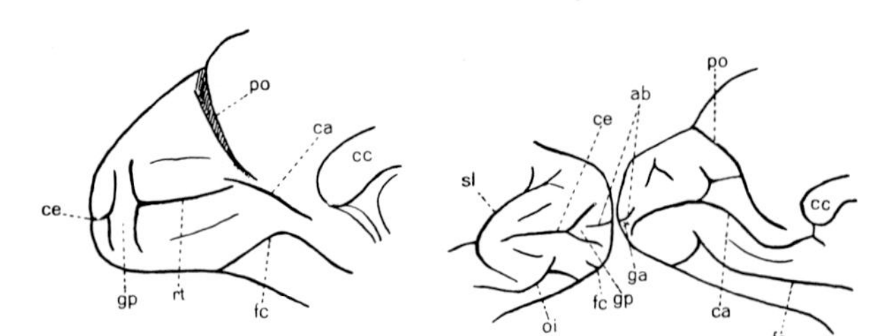

Figure 65. Semischematic representation of calcarine complex on medial surface of Human occipital lobe (left) and calcarine complex in the hemisphere of an Orang (right) (from K., 1928b, left Figure adapted and modified after ELLIOT SMITH, 1903/04b). ab: portion of sulcus calcarinus interrupted by gyrus cuneo-lingualis accessorius; ca: pars anterior of sulcus calcarinus; cc: corpus callosum; ce: sulcus calcarinus externus; fc: sulcus collateralis; ga: gyrus cuneo-lingualis accessorius; gp: gyrus cuneo-lingualis posterior; oi: sulcus occipitalis inferior; po: sulcus parieto-occipitalis; rt: pars intermedia sulci calcarini (sulcus retrocalcarinus sive calcarinus posterior of ELLIOT SMITH); sl: sulcus lunatus. In the left Figure, the narrow gyrus between ca and rt is the gyrus cuneo-lingualis anterior. Many variations obtain in Man, ce may be joined to the rest of calcarinus system, or may be absent. The calcarinus system proper may not display cuneolingual gyri. The parastriate sulci (not labelled) can easily be identified.

concerning individual and racial aspects of brain morphology shall be dealt with further below in section 5 of the present chapter.

Since the configuration of the occipital lobe's lateral aspect may display a sulcus calcarinus externus or triradiate sulcus, which is a portion of the calcarine complex, some additional comments on this latter, whose main components are located on the lobe's medial aspect, seem appropriate.

CUNNINGHAM (1892), ELLIOT SMITH (1907a), LANDAU (1923) and other authors have emphasized that the Human sulcus calcarinus, even if it appears to be a deep single groove, usually consists of three components which may be separated by two '*cuneolingual transitional gyri*' hidden in the floor of the calcarine 'fissure,. The gyrus cuneolingualis anterior separates the pars anterior sulci calcarini, which joins the

---

tained its position despite the changes induced by self-domestication (cf. above, section 1, p. 13. This, however, does not preclude an adequate development or differentiation of the parieto-occipito-temporal association areas. The cytoarchitectural examinations which I undertook in connection with my studies on the sulcus lunatus seemed to indicate that, at least in some instances, the actual surface area of visual cortex (area striata) may be equally extensive in brains with and without sulcus, lunatus, the caudolateral expansion of that area in the former being compensated by greater width on the caudomedial surface in the latter.

parieto-occipital sulcus to form a common stem, from the pars intermedia. This latter, in turn, is separated by gyrus cuneolingualis posterior from the pars posterior sulci calcarini. The two cuneolingual gyri may occasionally become superficial, thereby disrupting the continuity of the calcarinus complex (Fig. 65). An additional accessory gyrus cuneolingualis accessorius may, as a variation, be found within the posterior portion of pars intermedia sulci calcarini. In Anthropoids and other Primates, the gyrus cuneolingualis anterior does not seem to occur, but the accessory gyrus may be present (Fig. 65). The pars posterior sulci calcarini, if separate, is thus the sulcus calcarinus externus *sive* triradiatus on the lateral surface of the occipital lobe. On the medial surface, a variable superior parastriate sulcus cunei, and a likewise variable inferior parastriate sagittal sulcus gyri lingualis frequently run roughly parallel to sulcus calcarinus (cf. Fig. 65) and correspond to the boundary of area striata.

It should finally again be stressed that considerable discrepancies in the terminology and in the interpretation of the occipital gyri and sulci can be noted in the literature. Discussions of the pertinent questions, with attempts at classification, can be found in the contributions by LANDAU (1923), by myself (K., 1927, 1928b), by PAPEZ (1929), v. ECONOMO (1930, GRAF HALLER (1934), KAPPERS *et al.* (1936), BECCARI (1943) and CONNOLLY (1950).

5. Remarks on Attempts at Anthropological Interpretations of Telencephalic Fissuration Pattern and on Some Apposite Topics Concerning the Human Mind in Social, Cultural, and Historical Perspective

Anthropology as a 'science' dealing with Man 'in all his relations' indeed covers a wide field with highly ambiguous 'limits' or boundary zones. It is, in this respect, customary to distinguish *physical* and *cultural anthropology*.[53] The former subdivision deals with morphological as well as physiological aspects, and their similarities or differences

[53] The inherent difficulties in establishing appropriate classifications of 'science' were pointed out in section 7 (p. 34), chapter I of volume 1. These difficulties are related, *inter alia*, to the ambiguities of language, and to the interconnections between rational thought and affectivity ('reason' and 'emotion'). This latter topic, in accordance with DAVID HUME's masterful analysis, was also dealt with in §§ 76 and 77 of the present author's monograph '*Mind and Matter*' (K., 1961). Among 'sciences' covering an 'all-inclusive' or

among Men and Animals, and among Human 'races'. Cultural anthropology is concerned with the development of ethnic or racial traits, and thus likewise with the classification and distribution of races, moreover with 'primitive' and 'advanced' societies, with cultures,[54] and also with languages and allied topics. For a general orientation about the various basic aspects and the historical development of anthropology the following publications, containing relevant bibliographic data, may be pointed out: CALVERTON (1931, 1937), DARLINGTON (1969), FORD (1939), KROEBER (1953), PENDELL (1942), SALLER (1930, 1964), and WILSON (1975). Additional references, including some contemporary views and controversies, will be given further below in connection with the discussion of specific topics.

The problem of the *origin of Man* is of import to both above-mentioned subdivisions of anthropology as well as to the anatomical and biological sciences whose domains have remained autonomous and quite independent of elaborate anthropologic speculations.

Although one may be reasonably convinced that recent Man is a highly developed anthropoid Ape whose ancestors were particular Cenozoic, perhaps Miocene or Pliocene Primates (cf. Fig. 154B), nothing certain whatsoever, despite multitudinous publications and spec-

---

'catch-all' domain, comparable, in this respect to Anthropology, one could mention 'Psychology', Geography, and History. In contradistinction to the 'exact sciences' (Mathematics, Physics, Chemistry, Astronomy), they can be classified as '*soft sciences*'. There are, of course, also 'soft spots' in Mathematics, Physics, and Astronomy, but few if any in Chemistry, in which the technologic aspect predominates.

[54] One is at loss either to formulate for oneself, or to accept, as given by numerous authors, a reasonably overall valid definition of '*culture*', which sometimes is synonymous with '*civilization*', and sometimes differs in various respects from the meaning of said term. Thus one could say that '*civilization*' denotes an 'advanced' stage of 'material', architectural, technological, scientific accomplishments combined with a corresponding complex social organization, while '*culture*' stresses the artistic, literary and intellectual achievements with particular regard to 'enlightenment' and 'refinement'. 'Culture' is also said to subsume the mores, folkways, institutions and traditions which distinguish one ethnic group, nation, or 'race' from another. An amusing paperback volume by the anthropologist ASHLEY MONTAGU (1959) purports to teach, for 35 cents, how to become a 'cultured man' by memorizing a selection of sciolisms and absorbing, hook, line and sinker, that author's opinions and prejudices. Again, 'culture' is said to indicate the use of implements, utensils, fire, clothing, domestic animals, agriculture and the manner of housing, respectively the nomadic or hunting activities, as well as the attempts at artistic decoration and drawing by early Man. Various nomadic or settled cultures are also designated as 'primitive' cultures in contrast to the more 'advanced' ones, such as classical Greek, classical Chinese, ancient Hindu, or modern Western 'culture'.

ulations by numerous authors with divergent views, is known about the details of Man's origin and the presumably concomitant development of articulated speech resulting in the differentiation of languages. If, quite arbitrarily, the Simian phylogenetic 'radiation' representing the 'origin' of Man is assumed to have 'branched off' about one million years ago, then the extant species *Homo sapiens* with its presently obtaining diverse subspecies is, *qua* genetic combinations, recombinations, shufflings, mutations, and other aspects of heredity, the product of roughly about 30 000 to 40 000 generations with considerable inbreeding and domestication effects.

The earliest Primate stages in the evolution of Man are inferred from the interpretation of fossil bones. Human activities at the Pleistocene epoch (from one million to about 20 000 or more years ago), and at the prehistoric recent epoch, are evidenced by remnants of bones as well as of artefacts. The concepts of Palaeolithic, Neolithic, Bronze, and Iron cultural stages or of so-called 'horizons' have been elaborated. These stages merge, respectively partly overlap not only with the origin of more or less coherent historical tradition, but in some regions even the earlier ones have persisted to recent times. Thus, KAPPERS (1929) and others consider Australian tribes as representing a 'recent palaeolithic race', 'related to an unknown Homo sapiens fossilis, without any genetic relationship with Neanderthalmen'.

Following the prehistoric periods, in which the use of artefacts and of fire, the origin of articulated speech, early artistic activities, such as decorations and drawings etc., and various sorts of disposal of the dead (e.g. *burial*) represent important phases distinguishing Man from other animals, the historical stage can be said to begin with the invention of *writing* and the preservation of *permanent written records*. In addition to 'historical events', individually and collectively acquired experiences and skills became more efficiently transmitted, thereby accelerating the 'progress of civilization'.[55]

The earliest systems of writing were independently devised, in two differing manners, by the Mesopotamian *Sumerians*, a people of unknown racial provenance, and by the Mediterranean 'Hamitic' *Egyp-*

[55] In discussing the evolution of the Human brain with respect to the special status *(Sonderstellung)* of Man, SPATZ (1955) justly commented: '*Der Mensch besitzt ausser der biologischen Art der Weitergabe, die man Vererbung nennt, noch eine andere Art, die nur ihm eigen ist. Diese "neue Art der Weitergabe", wie wir sagen wollen, besteht in der Übermittlung von im individuellen Leben erworbenen Erfahrungen und Fähigkeiten auf kommende Generationen, wobei Bildnerei, Sprache und Schrift eine führende Rolle spielen.*'

tians. The former developed the *cuneiform script*,[55a] perhaps as early as 4000 B.C. and took the first steps toward the introduction of formal education (cf. the highly interesting treatise by KRAMER, 1963). The Egyptians, possibly about the same time or somewhat later, devised their *hieroglyphs* as a *pictographic script*. Subsequently, from both systems, simplified *syllabic* and *alphabetic notations representing sounds* were developed and adopted by diverse ethnic groups. The Greek and Roman alphabets are presumably derived from adaptations devised by the Phoenicians or by the Minoans. Other syllabic adaptations, perhaps from the same sources, are, *inter alia*, the diverse scripts of the Hindus, such, e.g. as the Sanskrit Devanâgarî.

Substantially later than the Egyptians, but doubtless earlier than 1000 B.C., the *Chinese* apparently independently developed their *ideographic script*, which was introduced into Japan about 600 A.D. and greatly modified the Japanese language, whose structure is entirely different from that of the Chinese. The Japanese, in turn, roughly about 800 A.D., devised two *syllabic alphabets* derived from the Chinese characters, but particularly adapted to the sounds of their own language.

A fourth independent invention of writing originated, much later than the previous ones, in *Middle America*, where, perhaps about 200 B.C. or somewhat earlier, a hieroglyphic writing and a calendar-system may have made their appearance. Subsequently, the principle of *place value for numerals* and the *zero* seem to have been introduced into said systems of notations.[56] The *Mayan* ruins in Yucatan are examples of vestiges of these pre-Columbian Middle-American civilizations, which could, except for the numerals and the zero, be evaluated as having lagged several thousand years behind the pioneering ones of the Old World, and flourished between 250 to 900 A.D. At that

[55a] This script was subsequently adopted by the Semitic Akkadians, Babylonians, and Assyrians, and by the Iranian Medians and Persians. Bilingual and trilingual cuneiform inscriptions such as those from Persepolis and Behistun provided the basis for the decipherment of cuneiform scripts by GROTEFEND, RAWLINSON and others in the course of the 19th century, just as the well-known *Rosetta Stone* provided the basis for the decipherment of the Egyptian hieroglyphs by CHAMPOLLION and others.

[56] In the Old World, the number system based on place value, and the *zero* were devised, perhaps before 800 A.D. by the Hindus, and then adapted by the Arabs, being therefore known as Arabic numerals. The Babylonians had also used a character denoting the absence of number and made desultory attempts at devising a crude place value in numeration, but apparently did not succed in creating a number system in which the zero played any such part as it does in the system elaborated by the Hindus (SMITH, D.E., *History of Mathematics*, vol. II, p. 69).

period, however, a highly accurate astronomic calendar, superior to the then contemporary Old World ones, had been developed. The *Mayan glyphs* are still not completely deciphered, but substantial progress in their interpretation has been made in recent years. The lingering Middle American civilizations, as is well known, were overthrown, after 1519, by the Spaniards.

Seen from the neurobiological viewpoint, these independent origins of writing and of numerals as well as many other resemblances of independently developed concepts and practices can be attributed to the roughly identical basic structures and mechanisms displayed by all normal or well developed Human *brains*.[57] The cerebrum, in this respect comparable to a computing mechanism, responds to the problems posed by the outside world with a repertory of performances in accordance with relatively few fixed function-rules, whose application nevertheless remains quite flexible.[58] Thus, discounting the numerous interfering variables, the Human brain tends to provide identical or comparable answers to identical problems.

In the *prehistoric periods*, the use and production of fire,[58a] the manufacture of tools and clothing, the beginning of metallurgy, artistic decoration and drawing, elementary knowledge of edible and poisonous plants, beginnings of agriculture, domestication of animals, diverse methods of housing, the use of canoes or boats, the change of the seasons, the observation of celestial bodies and the apparent rotation of the celestial sphere, methods of counting and of calendar notions, all involved a diversity of problems which, to a significant extent, were independently, and in a diversity of fundamentally similar manners, as well as with various degrees of success, dealt with by the brains of prehistoric and early historic Man in widely scattered region of this globe. On the other hand, there were also many instances of cultural interactions, whereby accomplishments of one ethnic group became adopted by another.

With respect to *independently developed answers* to particular problems,

---

[57] This, of course, does not exclude considerable individual differences in the degree of structural 'perfection' and functional 'efficiency'.

[58] Cf. K. (1957, pp. 131–132, with reference to some of ASHBY's concepts).

[58a] According to MONTAGU (1959), the only ethnic groups unable to make fire in the 20th century were the *Andaman Islanders* and the *Congo Pygmies*. The former, although ignorant of fire-making, are said to have obtained fire from others and to have kept it by never letting the fires go out. The *Pygmies* are said to have purchased their fire from their Bantu neighbors.

and proceeding to more recent times, it will also be recalled that the general concept of organic evolution was independently expressed by numerous authors since classical antiquity and, with regard to the present day formulations of 'natural selection', was independently and almost simultaneously propounded by DARWIN and WALLACE. Again, a concept perhaps comparable in importance to those of the heliocentric system,[58b] of the circulation of blood, and of the law of gravitation, namely the *fundamental Mendelian rules of inheritance*, discovered by the *Augustine* monk and abbot GREGOR MENDEL in *Brünn*, but ignored following their publication in 1865, could be mentioned as another example. These rules were almost simultaneously and independently re-discovered about 1900 by CORRENS, DE VRIES, and TSCHERMAK, and then properly credited to MENDEL (cf. sections 4, 5, and 7, chapter II, vol. 1).

Reverting to the available *recorded history*, it is opportune, as pointed out by such different authors as the philosopher SCHOPENHAUER and the historian BURY, to distinguish the political history (*Staatsgeschichte*, particularly emphasized by RANKE and many other historians) from the history of the diverse manifestations of Human activities, e.g.

[58b] This was propounded by ARISTARCH of *Samos* about 280 B.C., and apparently independently again re-introduced by COPERNICUS (1543). It should, however, be pointed out that the original *Copernican system* remained quite erroneous with respect to two important aspects and did not correspond to what is now generally referred to by that term. As regards the first aspect, COPERNICUS did not relinquish the old concept of circular orbits, and was thereby compelled to retain a large number of complex *epicycles* and other artificial features of the *Ptolemean system*. The present-day heliocentric concept became established by KEPLER (1609) who showed that the planetary orbits were *elliptic*, and who formulated his well-known three laws. On the basis of KEPLER's discovery and on that of the laws of falling bodies, established by GALILEO (1638), NEWTON was then enabled to formulate the *law of gravitation* (1687), Second, COPERNICUS did not contest the old cosmologic concept of an 8th sphere of fixed stars, as e.g. described in the medieval cosmology of DANTE (*Divina Commedia*, 1331) and in that of the great poet CAMÕES (*Os Lusiadas*, Canto X, 1572). It was the great philosopher and martyr GIORDANO BRUNO (1548–1600) who emphasized the true nature of the 'fixed stars'. That author's *Cena delle ceneri* (ca. 1585; *Das Aschermittwochsmahl*, transl. and commented by L. KUHLENBECK, 1906) brings a vivid account of the contemporary cosmologic controversies and of the unsavory traits displayed by the *Oxford* pandits of that time. Hindu astronomy, whose Greek origin seems to have been conclusively established, remained a curious mixture of old fantastic ideas and adopted foreign geometric methods of calculation. ÂRYABHATA, about 530 A.D., is quoted by Hindu authors as having (rather vaguely) assumed some kind of motion or revolution of the earth, producing the daily rising of stars and planets, the sphere of stars being stationary. His views are interpreted by some as propounding some sort of a 'heliocentric system'.

history of science, of philosophy, of religions, literature, art, industry, economics, and law.

To those who consider Human affairs from a 'philosophical' or 'detached' viewpoint, the history of states or nations, i.e. the political history, will disclose an immense as well as unending sequence of the most repulsive abominations, outrages, brutalities and deliberate cruelties, as well as of endless miseries afflicting Mankind. Thus, *miseria humana*, *nequitia humana* and *stultitia humana* are shown to play the predominant role, characterizing, as it were, the '*martyrdom of Man*' (cf. e.g. READE, 1872, 1925). Fiendishness, depravation, fraud and lying are here frequently, but by no means always or necessarily combined with stupidity. Although this history, recording events, includes a succession of presumably 'true' or actual facts, it is distorted by inherent mendaciousness. Historical documents not only have gaps but are mostly biassed, and very frequently calculated to deceive.[59] Still worse, in this respect, are the writings by most historians.[60]

Yet, although Man, not intrinsically but only in degree different from Apes and from animals in general, is, more often than not, wretched, ignorant, and vicious,[61] an undefined number of Human beings have led relatively untroubled lives and found satisfaction in their activities. Moreover, said degree in difference from animals provided for the emergence of a likewise undefined number of Human beings of substantial ethical or intellectual merits or of both. The accomplishments of these perhaps relatively uncommon individuals (*apparent rari nantes in gurgite vasto:* VIRGIL's *Aeneid* I, 118) provide the basis of the diverse other types of history mentioned above, which nevertheless, also disclose many unpleasant happenings. On the other hand, even

---

[59] *Per contra*, in the various historical natural sciences, such as geology, paleontology, phylogeny, and cosmogony, the data may deceive the observer, but they are not 'deliberately trying to deceive him'.

[60] SCHOPENHAUER (V, p. 473 ed. *Grisebach)*, in commenting on the imperfections of the historical records, emphasizes the additional mendaciousness and states '*dass die Geschichtsmuse* KLIO *mit der Lüge so durch und durch infiziert ist, wie eine Gassenhure mit der Syphilis*'. There is an old German quip: *Geschichte ist ein Roman, den man glaubt, und ein Roman ist eine Geschichte, die man nicht glaubt.* FONTENELLE (1659–1757) called history '*une fable convenue*'.

[61] SCHOPENHAUER (V, p. 219) quotes with approval GOBINEAU's statement: '*l'homme est l'animal méchant par excellence*'. In his tale '*Micromégas*' VOLTAIRE remarks: '*Si l'on en excepte un petit nombre d'habitants fort peu considérés, tout le reste est un assemblage de fous, de méchants, et de malheureux.*' Elsewhere VOLTAIRE also commented: '*Nous laisserons ce monde-ci aussi sot et aussi méchant que nous l'avons trouvé en y arrivant.*'

among the abominations of sociopolitical history, a few phenomena displaying positive ethical values become manifested.

On the whole, that what may arbitrarily be called '*progress*' is particularly evident in the history of the 'intellectual' pursuits. Thus, SCHOPENHAUER states: '*Dieses intellektuelle Leben schwebt, wie eine ätherische Zugabe, ein sich aus der Gährung entwickelnder wohlriechender Duft über dem weltlichen Treiben, dem eigentlich realen, vom Willen geführten Leben der Völker, und neben der Weltgeschichte geht schuldlos und nicht blutbefleckt die Geschichte der Philosophie, der Wissenschaften und der Künste.*'

The concept of a so-called noösphere, propounded by the anthropologist TEILHARD DE CHARDIN (1959), can be said to represent a rather naive reification[61a] or hypostatization of the mental domain described by SCHOPENHAUER's metaphor.

EINSTEIN (1954, p. 65) comments on this topic as follows: 'There are only a few enlightened people with a lucid mind and style and with good taste within a century. What has been preserved of their work belongs to the most precious possessions of mankind. We owe it to a few writers of antiquity that the people in the Middle Ages could slowly extricate themselves from the superstition and the ignorance that had darkened life for more than half a millenium.' He could well have said: for more than a whole millenium.

[61a] Cf. on pp. 441–444 of the monograph '*Mind and Matter*' (K., 1961) the comments on a comparable reification by another anthropologist (L. A. WHITE) concerning the so-called 'locus of mathematical reality'. Conspicuously naive reifications are likewise POPPER's '*Three-World concept*' and its further 'neurobiologic' elaboration, with quaint diagrams of Worlds 1, 1b, 2, 3a, and 3b, by ECCLES (1970). POPPER (1959) previously had also propounded some peculiar views on 'the logic of scientific discovery'. A far more reasonable and sober concept, comparable to SCHOPENHAUER's metaphor of a particular mental domain, and stressing '*die Einheit des Geisteslebens in Bewusstsein und Tat der Menschheit*', was presented by my teacher R. EUCKEN (1888), under whose sponsorship I obtained the degree of *Doctor philosophiae* in 1920 at the University of Jena. EUCKEN stressed the axiological values of the Human mind, which he defended against their corrosion by unwarranted encroachments or by expansions into the axiologic domain, of materialism, mechanism, naturalism and positivism. Both EUCKEN and my father, LUDWIG KUHLENBECK (1857–1920), who was particularly concerned with the interpretation of GIORDANO BRUNO's philosophy, can be regarded as conservative axiologists combining a profound understanding of epistemologic problems with a comprehensive consideration of historical and cultural developments. EUCKEN's philosophy justly has been characterized as being essentially 'a call to arms against the deadening influences of modern life'. EUCKEN's counterpart at the University of Jena was ERNST HAECKEL, whose frame of mind nevertheless remained deeply religious (in the wider sense) and who attempted, with debatable success, to integrate his penetrating biological insight into a system of religious monistic philosophy.

The concept of '*progress*', ably discussed by BURY (1955), is a rather ambiguous one, including 'objective' as well as 'subjective' (emotional) aspects. Quite evidently, 'scientific' and medical knowledge, and technological accomplishments have greatly increased, particularly since the Western Renaissance in the 15th century, initiating present-day Western civilization, and many savageries of early 'laws' and customs gradually became somewhat attenuated. Yet, it is questionable whether all aspects of the complex manifestations of 'progress' are either 'desirable', 'good', or 'beneficial'. Such axiologic evaluations are, of course, subjective, emotional, and irrational'.[62] 'Favorable' aspects of 'progress' seem to be combined with 'unfavorable' 'side-effects'. Another moot question is that of 'indefinite progress'.

DAVID HUME, in one of his essays, expressed a sceptical opinion. He argued that there is little ground to suppose that the world is eternal. 'It is probably mortal, and must therefore, with all things in it, have its infancy, youth, manhood, and old age. Man will share in these changes of state. We must then expect that the Human species should, when the world is in the age of manhood, possess greater bodily and mental vigor, longer life, and a stronger inclination and power of generation. Physically and in mental powers men have been pretty much the same in all known ages. The sciences and arts have flourished now and have again decayed, but when they reached the highest perfection among one people, others were perhaps wholly unacquainted with them. We are therefore uncertain whether at present' (i.e. at HUME's time) 'Man is advancing to this point of perfection or declining from it.'

Yet, HUME insisted on the improvements in art and industry and on the greater liberty and security enjoyed in contemporary Europe. With special emphasis on the abolition of slavery in Europe he remarked: 'to one who considers coolly on the subject, it will appear that human nature in general really enjoys more liberty at present in the most arbitrary governement of Europe than it did during the most flourishing period of ancient times'.

Reverting to the neurobiologic aspect, there is no evidence whatsoever that brain capacity or 'intelligence' have increased or improved in the course of the roughly 6 000 to 4 000 years of recorded history. The actually achieved progress seems entirely due to the accumulated

---

[62] Cf. again, as mentioned in footnote 53, the comments on 'reason' and 'emotion' in § 76, pp. 452ff. of the monograph '*Mind and Matter*' (K., 1961).

transmission of acquired experiences and skills, as my esteemed old friend SPATZ (cf. above footnote 55) justly pointed out.

According to the dictum of THOMAS JEFFERSON, incorporated in the American Declaration of Independence, 'All men are created equal'. This statement is obviously belied by the evident discrepancies in physical and mental endowment, as well as meaningless in view of the obtaining social inequalities.[63] Yet, because of the vagueness and ambiguities of language, some equality at several levels indeed obtains for Man. Thus, paraphrasing a well-known dictum,[64] one could say that 'what is true for *Escherichia coli* is true for Man' *(qua* macromolecular organic being). As a sentient being experiencing pain and suffering all men and comparable sentient animals are likewise essentially equal.[65] Finally, despite its tinge of hypocrisy, JEFFERSON's statement was not without influence in working for the liberalization of institutions, the theoretical 'equality before the law', and the more generalized abolishment of slavery, in agreement with the above-mentioned views of DAVID HUME.

In the *Judeo-Christian mythology*, various aspects of which seem derived from Sumerian and subsequent Babylonian sources (cf. e.g. DELITZSCH, 1901; KRAMER, 1963), Man is said to have been created by God in his own image and after his likeness. Although God saw that everything he had made was very good, subsequent events showed the mistake of his premature appraisal. The wickedness of Man became great, all his imaginations and the thoughts of his heart being only evil continually. Thus, God repented to have made Man, and resolved to destroy him, saving only *Noah* and his kin, together with a selection of animals. Strangely enough, while most terrestrial animals were thus liquidated together with almost all humanity, aquatic animals evidently remained unaffected by God's punitive measure.

Nevertheless, after this large-scale elimination, wickedness continued to spread and God let events go on despite the fact that Man's heart remained evil from his youth.[66] God merely chose to destroy

---

[63] It is, in this respect, of interest to note that JEFFERSON and other signers of the Declaration of Independence were slave holders. Moreover, JEFFERSON, WASHINGTON and other '*founding Fathers*', although with a benevolent attitude, believed in the inferiority of Negroes.

[64] Cf. p.11, footnote 8 of volume 3/II.

[65] In the Hindu thought, this is expressed by the famed *Mahâvâkya* (Great Saying) many times repeated in the *Châdogya Upanishad* (VI, 8, 9, 10, 11, 12, 13): *tat twam asi (this art thou)*.

[66] Genesis 8, 21.

Sodom and Gomorrah[67] by a cataclysm comparable to the dropping of an atomic bomb. As far as the wickedness of Humanity in general was concerned, he merely threatened to visit the iniquity of the fathers upon the children, upon the third and upon the fourth generation.[68]

The *Mohammedan Koran*,[68a] derived from Judeo-Christian sources, is still more explicitly revengeful, breathing hell-fire with tiresome repetitions and always in the name of God, 'the Compassionate, the Merciful'. Except for the predestinated chosen believers, eternal damnation with terrible retribution and suffering, is, according to the *Koran*, decreed to be the inevitable fate of Humanity.

The willingness to accept and to reconcile, respectively the inability to realize, the numerous contradictions, incongruities and absurdities of these allegedly divinely inspired mythologies can be evaluated as a manifestation of schizophrenic logic and thought, which to some extent are a general characteristic of the Human mind.[69] Said contradictions and absurdities were already concisely exposed by the *Sceptic* SEXTUS EMPIRICUS, who lived circa 200 A.D., and presumably was a contemporary of the Roman emperor SEPTIMIUS SEVERUS (193–211 A.D.), in Book III, chapter 3 (περὶ θεοῦ) of his '*Outlines of Pyrrhonism*'. SAINT AUGUSTINE (354–430) and MARTIN LUTHER (1483–1546) squarely faced the resulting dilemmas, but resolved them by conspicuously schizophrenic reasoning. VANINI (1586–1619) clearly disclosed said incongruities in a dialogue, in which, *pro forma*, they were upheld by himself as one of the interlocutors using the orthodox arguments, whose fallacy and futility became self-evident to the intelligent reader. It will be recalled that VANINI was later atrociously burned alive at the stake, after his tongue was torn out, by order of the *Roman Catholic hierarchy*, who, 19 years earlier, had officially murdered GIORDANO BRUNO in the same abominable manner.

---

[67] It is here of interest to note the haggling of ABRAHAM with God about the sparing of these cities if 50, 45, 40, 30, 20, 10 righteous men could be found therein (Genesis 18).

[68] Exodus 20, 5.

[68a] With regard to this scripture, SCHOPENHAUER (W. a. W. u. V., vol. 2, chapter 17) states: '*Man betrachte z.B. den Koran: dieses schlechte Buch war hinreichend, eine Weltreligion zu begründen, das metaphysische Bedürfnis zahlreicher Millionen Menschen seit 1200 Jahren zu befriedigen, die Grundlage ihrer Moral und einer bedeutenden Verachtung des Todes zu werden, wie sie zu blutigen Kriegen und den ausgedehntesten Eroberungen zu begeistern. Wir finden in ihm die traurigste und ärmlichste Gestalt des Theismus. Viel mag durch die Übersetzungen verloren gehen; aber ich habe keinen einzigen wertvollen Gedanken darin entdecken können.*'

[69] Cf. § 78, p. 500f. in the monograph '*Mind and Matter*' (K., 1961), and pp. 749–751, section 8, chapter VI of volume 3/II.

Numerous sophisticated theologians, apologetes and exegetes have attempted, by intricate and devious lawyers' arguments, to mitigate the crudity of the Judeo-Christian mythology and to explain it away in terms of diverse, highly unconvincing, conflicting and controversial elaborations, including, *inter alia*, also interpretations purporting to suggest a 'higher meaning' hidden by a necessary symbolism adapted to ignorant Man. Despite all these attempts at vindication it can be replied that what is written in the Scriptures stands, and exactly says what it says.

It should, however, be conceded that the Jewish *Torah* and related religious writings, respectively their adoption as the Christian Old Testament, and, perhaps to a lesser extent, at least parts of the Christian New Testament, are very remarkable and valuable collections of literature with substantial cultural import.

On the other hand, even in the purportedly 'monotheistic' Christian religion with its 'Trinity', a whole host of good and evil supernatural beings, such as archangels, angels, miracle working saints, and a satanic devil with his subordinate demons, witches, sorcerers, etc., was quite generally, and still is, assumed to exist.[69a] There is hardly much difference between the throngs of these various beings and the pantheons of polytheistic religions.

Again, the Oriental religious and philosophical systems, such as the various forms of *Hinduism* and of *Buddhism*, moreover the Chinese *Taoism* and *Confucianism*, are likewise characterized by diverse schizophrenic tendencies. Nevertheless, some aspects of these systems could, depending on the viewpoint, be evaluated as fully compatible with the 'orderliness of nature' and as far superior to the doctrines of the Judeo-Christian mythology.

Buddhism conceives Man as a sentient being without an essential or permanent self,[70] i.e. without a soul, not intrinsically differing from

[69a] As regards *Satanism*, cf. e.g. SELIGMANN (1948), and particularly 'The history of witchcraft' by the *Reverend Father* MONTAGUE SUMMER (1926, 1956). This latter, a Roman Catholic priest in good standing until his death in 1948, firmly believed in Satanism and exorcism. He vigorously defended as fully commendable all the abominations ever perpetrated by his Church in order to extirpate heresy and witchcraft. Some of his writings conspicuously display the official '*imprimi potest*' and '*imprimatur*' of his high-ranking ecclesiastic superiors.

[70] The three fundamental tenets of *Buddhism* (Sanskrit: *Trividyâ*, i.e. *threefold knowledge)* are: *Anâtman (non-self)*, *Anitya (intrinsic non-permanence)*, and *Duhkha (immanent suffering)*. The ancient Nordic mythology, with its Gods who must die *(Götterdämmerung)* is likewise full of the sense of impermanence and suffering. But the presumably genetically 'racial'

animals, and, like these latter, arisen from 'causes' in accordance with an impersonal orderliness.[71] Yet Buddhism, especially in the grandiose cosmologic phantasmagories of the Mahâyâna, confers upon the potentially enlightened Human being, rising through his own efforts, an almost cosmic dignity, perhaps unequalled in any other religious system.

In the *Confucian* philosophy of ancient China (ca. 500 B.C.) the ideal of the Superior Man (*Chün-tzu*, Sino-Japan.: *Kun-shi*) is stressed. The *Doctrine of the Mean* (*Chung-yung*, Sino-Japan.: *Chû-yô*) states (XI, 3): 'The Superior Man accords with the course of the Mean; though he may be all unknown, unregarded by the world, he feels no regret. It is only the sage who is able for this.' Another saying, in chapter XIII, 3, reads: 'when one cultivates to the utmost the principles of one's nature, and exercises them on the principle of reciprocity, one is not far from the path. *What you do not like when done to yourself, do not do to others.*'

As regards various other concepts of desirable perfection, which imply the mediocrity, viciousness and wretchedness of Humanity in general, it will be sufficient to mention the Hindu *Sâdhus*, the unattainable, rather eccentric as well as contradictory goal of the classical *Stoic* '*Sage*', the Christian ideal of the '*Saint*', and the frenzied '*Superman*' of the erratic genius NIETZSCHE, who could perhaps be characterized as an 'enraged mystic'.

From the *neurobiological viewpoint*, ethical concepts may be evaluated as resulting from interactions between limbic system and isocortical activities, that is, as a combination of affectivity, emotion, and reason (logical thought). Ethical precepts could roughly be classified into four categories concerning: (1) general ethical behavior, (2) family relations, (3) other aspects of social structure, including duties to nation and community, Human, respectively individual, 'rights', and (4) miscellaneous ritualistic or religious ordinances.

Nordic temperament bade a man fight and face his fate. *Per contra*, Hinduism (except for *Karma Yoga)*, Buddhism, and some forms of Christianity sought deliverance or escape by means of ascetic monasticism.

[71] The substance of Buddhist doctrine is expressed by a famed verse in both *Pâli* and *Sanskrit* version, of which the latter reads:

Ye dharmâ hetuprabhavâ / hetum tesâm Tathâgata
hyavadat tesâm ca yo nirodha / evamvâdî mahâs'ramanah.

The norms originating from causation,
their cause the *Tathâgata* has explained,
and also their extinction. This the Great *Shramana* teaches.

In the aspect here under consideration it will suffice to state that the precepts under (1) are almost identical in all systems and directed (a) against killing respectively against cruelty or inflicting pain,[71a] (b) against stealing or transgressions upon the right of others, and (c) against lying and deceit, i.e. against dishonesty.

The precepts under (2), bidding to honor one's parents and dealing with paternal, maternal, brotherly and sisterly love, are, on the whole, likewise rather similar in all systems. Those under (3) on the other hand, greatly differ, and, *qua* Human 'rights' may permit or justify slavery, which one could regard as one of the most vicious institutions devised by Man.[72] The precepts under (4) differ still more widely and include laws concerning the relation of the individual to deity or supernatural powers, moreover ordinances concerning sacrifices, rites, ceremonials, taboos, sexual behavior, use of intoxicants, and other matters of this type.[72a]

Although presumably of multifactorial origin, e.g. *inter alia* concerning questions of health preservation, the prescriptions under (4) are essentially expressions of the above-mentioned schizophrenic aspect of the Human mind, which can also become evident in some of the features that are manifested by the diverse formulations of precepts pertaining to the other three categories.

Concerning the ethics of Human behavior it is perhaps appropriate to qote a remark by A.N. WHITEHEAD (1925): 'It does not matter what men say in words, so long as their activities are controlled by settled instincts. The words may ultimately destroy the instincts, but until this has occurred, the words do not count.'

[71a] There obtains, however, a wide divergence in their interpretation with regard to self-defense, justifiable war activities, and punishment of malefactors. Likewise, no agreement obtains concerning the application of these precepts with regard to animals.

[72] It is here of interest to note that, in an evidently modified form, the institution of 'slavery', and the keeping of 'domestic animals' also obtain among some Invertebrate social Insects, namely as displayed by certain species of Ants.

[72a] JOHN LYDGATE (1370–1450), a younger contemporary of CHAUCER (1340–1400), remarked in this respect:

'*But man alone, alas the hard stond,*
Full cruelly by kindes ordinance
Constrained is, and by statutes bound,
And debarred from all suche pleasance;
What meaneth this, what is this pretence
Of laws, I wis, against all right of Kinde,
Without a cause, so narrow men to binde?'

Reverting to the topic of inequality among Men, which obtains despite the above-mentioned statement by THOMAS JEFFERSON, people of above-average or outstanding achievements may become more or less widely recognized[73] as prominent artists, writers, scholars, scientists, inventors, statesmen, or 'leaders' in diverse fields, and as representing a Human '*elite*', of various ranks, the highest being that of '*genius*'. The eminent psychiatrist KRETSCHMER (1931) uses the term genius for those men or women who are able to arouse permanently, and in the highest degree, that positive, 'scientifically grounded' feeling of worth and value in a wide group of Human beings. KRETSCHMER restricts this only to those cases where the value arises with 'psychological necessity', out of the special mental structure of the bringer of value, not where merely a stroke of luck or some coincidences of factors had played the decisive role.

High achievement and genius are evidently the results of two main factors, namely the mental faculties provided by the *brain* and by the *conditions of the environment*. The development of the brain obviously depends on genetic factors, but the perfection of its capacities is also determined by the environmental factors of '*nurture*'. It is exceedingly difficult if not impossible to determine or 'measure', that is, to give quantitative values to either sort of factors involved. This difficulty results in the opposite views of the '*egalitarians*' or '*environmentarians*'[74] stressing 'nurture' and of the '*hereditarians*' emphasizing the genetic factors.

Although '*intelligence*' is a very poorly definable 'capacity' manifested by aspects of behaviour (cf. K., 1965a), it can, to some extent, be 'measured' by 'intelligence tests' of diverse sorts which yield rather controversial, arbitrary values. Yet, *faute de mieux*, this is still perhaps the best available method of very roughly measuring certain 'mental faculties'.

Despite the alluded controversies (cf. e.g. LAYZER, 1974, 1975; KAMIN, 1974, and others), the findings of HERRNSTEIN (1973a, b and

[73] To some extent, such recognition is both a question of 'chance' and of arbitrary judgement. Thus, in commenting on a selection of 'outstanding neurologists' my esteemed colleague and friend WEBB HAYMAKER (1953) justly remarked: 'who would expect any two persons to see exactly eye-to-eye in judging greatness ?'.

[74] Thus, as vividly described by ZIRKLE (1949) the 'geneticist' LYSENKO, an apostle of 'nurture', through his influence on the late and unlamented STALIN, nearly wrecked the science of genetics in the Soviet Union, until *Lysenkoism* finally became discredited. ZIRKLE's 'Death of a science in Russia' was apparently a deep coma rather than actual death.

other publications), and JENSEN (1972) strongly support the views of the 'hereditarians'. Yet, since, on biological grounds, relevant roles can be assumed for both genetic and environmental factors, much still unresolvable uncertainty obtains. The cerebral basis of 'mentality' or 'behavior', moreover, includes mechanisms pertaining not only to 'intelligence' but to '*affectivity*'. Thus, the genetics of behavior[74a] must consider factors, respectively components whose definition and 'measurement' are much more difficult and arbitrary than the evaluation of intelligence. This is likewise the case as regards the environmental factors. Among the manifestation of behavior pertaining to affectivity and presumably related to the 'limbic system' are, *inter alia*, such different components as artistic ability, perseverance, stability, application to hard work, honesty or dishonesty, selfishness or altruism, submissiveness or aggression. This latter behavioral manifestation has been particularly considered by ethologists (e.g. LORENZ, 1966; TINBERGEN, 1968; GILULA and DANIELS, 1969; JOHNSON, 1972). Elaborations on the inconclusive and controversial views concerning the aspects of emotion can also be found in the publications by KNAPP (1963), IZARD (1971), and SCHWARTZ *et al.* (1975). An older attempt is that by SPINOZA in his posthumous *Ethica* (1677), *Pars III*, *De origine et natura affectuum*, and *Pars IV*, *De servitute humana seu de affectuum viribus*.

GALTON (1869, 1892) emphasized the inheritance of exceptional ability, KRETSCHMER (1931) was concerned with the particular psychological qualities which distinguish the genius from persons who merely possess exceptional ability, and BRAIN (1960) discussed various topics related to genius in general. Again, GALTON assumed that, in surveying the entire course of recorded Human history, the appearance of not much more than about four hundred extraordinary geniuses could be recognized.

SPÄTH (1860) and GALTON (1875) stressed the method of comparing incidents of traits in monozygotic and dizygotic twins for distinguishing the relevant roles played by genes or by environmental factors. Yet, a variable 'penetrability' of genes may occur in discordant monozygotic twins.

[74a] For contemporary views on the 'genetics of behavior' cf. the publication edited by VAN ABEELEN (1974). Although a multitude of new genetic data has accumulated within the last 50 years (cf. e.g. the notable text by LEVITAN and MONTAGU, 1971), the popular publication by the eminent geneticist GOLDSCHMIDT (1927) still gives a masterful overall presentation of the general orderliness displayed by the mechanisms of heredity.

Another classic method of Human genetics is the *analysis of pedigrees* which, despite its limitations, has yielded significant genetic references. As regards exceptional abilities, there are some families of musicians, (e.g. *Bach*, *Haydn*, *Mozart families*), and of scientists, poets and literary personalities. Some forms of hereditary ability manifest themselves in various different spheres of activity, but others do not. The *Bernoulli family* excelled in other fields besides mathematics (cf. BELL, 1937). Statesmen, poets and literary men appear fairly frequently in each other's families, but eminent relations of artists often tend to be artists, and those of musicians to be musicians. The well-endowed *Edinger family* included, among its relatives, the eminent geneticist RICHARD GOLDSCHMIDT. The outstanding neurobiologist LUDWIG EDINGER had a likewise outstanding daughter, TILLY EDINGER, one of the founders of paleoneurology, and a very gifted son, the physician FRITZ EDINGER (cf. chapter XIII, vol. 5/I, p. 620, footnote 121). BRAIN (1960) points out that some allowance must be made for environment in the shape of family tradition, but remarks that this does not explain the variability of the manifestations of hereditary ability in some families and its comparative fixity in others. Paraphrasing a comment by MARTIN (1956), one could here say that having shuffled the pack and dealt an excellent hand, the hereditary mechanisms reshuffle the pack and distribute the cards again with or without tendency toward repetitive patterns.

Again, while descendants of highly gifted persons or families may be rather mediocre individuals, geniuses or unusually gifted persons can suddenly occur in obscure, but presumably well endowed families, and amidst quite unfavorable environments as e.g. in the case of GAUSS and of FARADAY. BRAIN (1960) also alludes to the psychological characteristics that distinguish the great military leader, statesman and administrator. These characteristics are, as various genealogies seem to show, often inherited. BRAIN assumes that the genius in these fields must possess an outstanding intelligence which operates on the minds of men as well as on their material circumstances. BRAIN admits, however, that a 'special blend of feeling' i.e. a particular component of affectivity respectively 'will' is required to direct the 'cerebral schemas' of such geniuses, which he considers 'artists in action'.

It is here of particular interest that General VON CLAUSEWITZ (1780–1831), the eminent theoretician and philosopher of war, with diversified wide experience both in actual combat and as Chief of Staff, and as instructor in the War Academy, did not consider outstanding intelligence to be a requirement for great military leaders.

In the private course of military instruction given 1810–1812 to the Prussian Crown-prince (later King FRIEDRICH WILHELM IV) CLAUSEWITZ (Appendix to '*Vom Kriege*' p. 685–800, 1915 edition) stated:

'*Die Grundsätze der Kriegskunst sind an sich höchst einfach, liegen dem gesunden Menschenverstande ganz nahe, und wenn sie in der Taktik etwas mehr als in der Strategie auf einem besonderen Wissen beruhen, so ist doch dieses Wissen von so geringem Umfange, dass es sich kaum mit einer anderen Wissenschaft an Mannigfaltigkeit und Ausdehnung vergleichen lässt. Gelehrsamkeit und tiefe Wissenschaft sind also hier nicht erforderlich, selbst nicht einmal grosse Eigenschaften des Verstandes. Würde ausser einer geübten Urteilskraft eine besondere Eigenschaft des Verstandes erfordert, so geht aus allem hervor, dass es List oder Schlauheit wäre.*'

'*Noch in dem Revolutionskriege haben sich gar viele Leute als geschickte Feldherren, oft als Feldherren der ersten Grösse gezeigt, die keine militärische Bildung genossen hatten. Von* CONDÉ, WALLENSTEIN, SUWAROW *und vielen anderen ist es wenigstens sehr zweifelhaft.*'[74b]

'*Das Kriegführen selbst ist sehr schwer, das leidet keinen Zweifel; allein die Schwierigkeit liegt nicht darin, dass besondere Gelehrsamkeit oder grosses Genie erfordert würde, die wahren Grundsätze des Kriegführens einzusehen; dies vermag jeder gut organisierte Kopf, der ohne Vorurteil und mit der Sache nicht durchaus unbekannt ist. Sogar die Anwendung dieser Grundsätze auf der Karte und dem Papier hat keine Schwierigkeit, und einen guten Operationsplan entworfen zu haben, ist kein grosses Meisterstück. Die grosse Schwierigkeit besteht aber darin:*

den Grundsätzen, welche man sich gemacht hat, in der Ausführung treu zu bleiben.'

'*Das ganze Kriegführen gleicht der Wirkung einer zusammengesetzten Maschine mit ungeheurer Friktion, so dass Kombinationen, die man mit Leichtigkeit auf dem Papier entwirft, sich nur mit grossen Anstrengungen ausführen lassen.*'

[74b] As a more recent instance, corroborating CLAUSEWITZ's conclusion, LEON TROTSKY (LEIB DAVIDOVITCH BRAUNSTEIN) could be mentioned, who, without previous military training, became a '*Feldherr der ersten Grösse*'. He organized and led the *Red Army*, and, by his superior strategy, saved the *Bolshevik regime* from collapse in the campaigns of 1918–1920. He fought on numerous fronts against the *White Armies* of KORNILOV, DENIKIN, JUDENITCH, PETLJURA, and KOLCHAK, which had foreign support, as well as against the British-American interventions at *Murmansk* and *Archangel*, the British Intervention in *Georgia*, *Armenia* and *Turkestan*, and the British-American-Japanese intervention in *Vladivostok*. Likewise in the *American Civil War*, numerous civilians became very competent generals, although the superior military leader, general *Robert E. Lee*. was a professional soldier. Again, in the recent *Israeli-Arab wars*, Jewish military leaders with civilian background displayed superior generalship of the first rank.

Mutatis mutandis, these latter comments, concerning planning, frictions and perseverance, also apply to the activities of purposeful Human life in general, such life being not unlike to continuous warfare. Thus, VOLTAIRE justly remarked: '*On ne réussit dans ce monde qu'à la pointe de l'épée, et on meurt les armes à la main.*'

Another interesting and controversial question, not considered by BRAIN in his '*Reflections*' (1960), but taken up by KRETSCHMER (1931) is the relation between *genius* and *sex*. The well-known psychiatrist MÖBIUS (1911), noted as a founder of '*Individualpsychologie*' and for his numerous 'pathographies' on geniuses and eminent personalities, published an essay entitled '*Über den physiologischen Schwachsinn des Weibes*'. His pronounced misogyny presumably was the result of his unhappy marriage, as FOREL (1935) pointed out. KRETSCHMER, however, admits that there is no record of a woman genius in the strictest sense, although, of course, numerous women have reached a high degree of eminence. To the objection that a lack of the female geniuses is due to the fact that women were traditionally prevented from participating in intellectual or public life, KRETSCHMER replies that, although they were not prevented from musical activities, no truly great woman composer can be named, notwithstanding the creditable contributions by KLARA SCHUMANN, FANNY MENDELSOHN, and CORONA SCHRÖTER, who were inspired by their personal environment.

There are, however, below the level of genius, numerous records of eminent women poets, writers, scientists, physicians, administrators and statesmen, as well as of some notable queens. It will here be sufficient to mention the following ones among many other instances: the lyric poetess SAPPHO of *Lesbos* (*floruit* ca. 600 B.C. ?), the Neoplatonic philosopher HYPATIA of *Alexandria*, murdered 415 A.D. by a Christian mob, and the Japanese court ladies MURASAKI SHIKIBU and SEI SHÔNAGON, who both flourished about A.D. 1000. The former is the author of the epic romance *Genji Monogatari*, one of the masterworks of Japanese literature, and of a famed diary. SEI SHÔNAGON wrote a likewise celebrated volume of miscellanies, entitled *Makura no Sôshi (Pillow Sketches)*. In the West, and about the same time, the gifted Benedictine Nun ROSWITHA (HROSVITHA) VON GANDERSHEIM (ca. 980) composed creditable poetic and dramatic works in Latin. MARIE DE FRANCE (ca. 1165) was the author of courtly tales; with ROSWITHA and HÉLOÏSE (ca. 1130, the beloved of PIERRE ABÉLARD) she is considered one of the three noteworthy medieval women authors.

Again, in later times, there are the obstetrician-midwives LOUISE

BOURGEOIS (1563–1636) and MARGUERITE DU TERTRE (ca. 1660), both in France, and JUSTINE SIEGEMUND (ca. 1680) in Germany, all three authors of creditable treatises on obstetrics, and, in England, the novelists JANE AUSTEN (1775–1817), MARY WOLLSTONECRAFT SHELLEY (1797–1851, author of *Frankenstein*), the BRONTË *Sisters* (CHARLOTTE, 1806–1855; EMILY, 1818–1848; ANNE, 1820–1849), writers of noted novels and poems, likewise GEORGE ELIOT (MARY ANN EVANS, 1819–1880, whose mother was also an author, championing the rights of woman). In Germany, the poetess ANETTE VON DROSTE-HÜLSHOFF (1799–1848), and the talented writer MALVIDA VON MEYSENBUG (1816–1903, a friend of NIETZSCHE) could be named among a few others. In France there was GEORGE SAND (AURORE DUPIN, baronne DUDEVANT, 1804–1876, the beloved of CHOPIN). In this short list of outstanding women should also be included the eminent Russian mathematician SONJA KOVALEVSKY (1850–1891, pupil of WEIERSTRASS), and, in more recent times, the scientists *Madame* CURIE, *Madame* DEJERINE, *Madame* VOGT, and the woman statesman GOLDA MEIR.

Having been, for many years, an educator of women in medicine, I can testify to their overall excellence as physicians, many of whom are far superior to numerous male colleagues. My students at the former *Woman's Medical College of Pennsylvania* included, besides Caucasians, representatives of the Mongolian and the Negro races, but evidently represented, *in toto*, an already highly selected group.

In ancient Germany, as reported by TACITUS (*Germania*, chapt. 8), women administered treatment to the sick and wounded: '*ad matres, ad coniuges vulnera ferunt; nec illae numerare aut exigere plagas pavent*'. Concerning women in medicine, there is also a detailed publication by HURD-MEAD (1938).

It seems obvious that not only woman suffrage but also free access of qualified women to prominent positions traditionally held by men are fully justified. On the other hand, and from a statistical viewpoint, relevant biologic differences, including behavioral aspects, doubtless obtain between the two sexes and have resulted in the thereto related sociologic pattern which prevailed in Western and other civilizations with *apparent* male dominance. The now emerging so-called *Women's Liberation Movement ('Women's Lib', 'Equal Rights Amendment'*, ERA), propounded by vociferous viragos and termagants, can be evaluated as one of the many symptoms characterizing the disintegration *(paracme)* of Western and perhaps, generally speaking, Human civilization.

not play dice *('der liebe Gott würfelt nicht')*. EINSTEIN meant thereby to express his belief in universal *causation* as opposed to the denial of causality by quantum physicists. Yet, although the mathematical formalism of quantum physics evidently precludes causality, the outcome of a throw of dice or that of any game of chance *does not*, and can be conceived as a *strictly causal event*. Thus, in *apparent* contradiction to EINSTEIN's dictum, and in accordance with the probabilistic aspect of natural biologic and genetic events, one could here emphatically reply: the good Lord indeed plays a cosmic game of dice, involving phantastic odds.

This seems to be corroborated by the apparent scarcity of organic life produced by said cosmic game of chance in our solar system as well as presumably in the Universe at large with its innumerable galaxies. One might here say: what a waste of time and space! Yet, in view of the perhaps still very large number of thinly scattered stars with planetary systems of which some members might have originated organic life and sentient beings, one could also say: what a scope for suffering, misery, and madness!

A well known Buddhist *Hînayâna sûtra* stresses, in this cosmic perspective, the exceptional rareness of the chance of Human birth.[75c] At present, and on our planet, there obtains, of course, the apparently opposite condition of overpopulation with still unforeseeable but rather ominous prospects about whose implications and perhaps belated or futile remedies controversial opinions are expressed (cf. e.g. CALHOUN, 1962).

Reverting now to persons with exceptional ability, respectively to *geniuses*, it can evidently be asked how the brains of such prominent

---

Caché sous le manteau de Descartes son maître,
Marchant à pas comptés s'approcha du Grand Etre,
– Pardonnez-moi, dit-il, en lui parlant tout bas
Mais je pense, entre nous, que vous n'existez pas.'

[75c] This sermon from the Pâli *Sutta Nipâta* can approximately be rendered as follows: The Lord Buddha said: 'Just as if, brethren, a man should throw into the mighty ocean a yoke with one hole, and then a one-eyed turtle should pop up to the surface only once at the end of every hundred years.

Now what think ye brethren? would that turtle push his neck through that yoke's hole each time when it popped up to the surface?'

– 'It might be so, Lord, now and again, after the lapse of a long time.'

– 'Well, brethren, sooner, do I declare, would that turtle push his neck through the hole of that drifting yoke than would come to pass the birth of man. So rare as this is the chance of human birth.'

people differ from those of 'ordinary' men. Numerous studies on these so-called *elite brains* have been undertaken by diverse authors. It will here suffice to mention, among others, the following contributions to this topic. RETZIUS, in his '*Biologische Untersuchungen*' (1898, 1900, 1904, 1905) examined the brains of the astronomer GYLDÉN, of the eminent woman mathematician SONJA KOVALEVSKI, and of other prominent persons (an educator, a statesman, and a physiologist). BECHTEREW and WEINBERG (1909) reported on the brain of the eminent chemist D.J. MENDELEJEW. The brains of the noted painter v. MENZEL, the chemist BUNSEN, the physiologist HELMHOLTZ, and the historian MOMMSEN were examined by HANSEMANN (1907). DONALDSON and CANARAN (1929) reported on the brains of three scholars, SPITZKA (1907) published a study of the brain of six eminent scientists, MAURER (1924) studied the brain of ERNST HAECKEL, and RIESE and GOLDSTEIN (1950) reported on the brain of LUDWIG EDINGER (Figs. 66–68). RIESE (1953) also examined the brain of the gifted physician and scientist T. BURROW. SPERINO's report (1901) on the brain of the anatomist GIACOMINI was mentioned above in connection with the duplication of sulcus centralis.

LANDAU (1923), in a general discussion of this topic, refers to his own observations concerning the brain of the orator and statesman GAMBETTA, which had also been previously studied by CHUDZINSKI and DUVAL (1886). General comments on the brains of prominent people were expressed by RIESE (1954, 1966). A detailed description of the most suitable procedures for the investigation of 'elite brains' was published by v. ECONOMO (1929).

Although I made an effort to familiarize myself thoroughly with the literature on this topic, my first-hand acquaintance with 'elite brains' has remained exceedingly sketchy and incidental. I merely had the opportunity, through the courtesy of Prof. HOPF, to inspect the unprocessed brain of OSCAR VOGT, and the sections of AUGUST FOREL's brain, which latter is characterized by an extensive left-sided hemorrhage lesion resulting in a right hemiparesis at the age of 64. FOREL, nevertheless, overcame this condition, learned to write with his left hand, and remained mentally alert and active until his death at the age of 83 (cf. my biographic sketch of FOREL in WEBB HAYMAKER's '*Founders of Neurology*', 1953, 1970). Through the courtesy of Dr. TH.S. HARVEY, then pathologist at *Princeton Hospital*, I also had the opportunity to inspect slides from EINSTEIN's brain (weight: 1 230 g) and illustrations of that brain's convolutional pattern. I understand that a de-

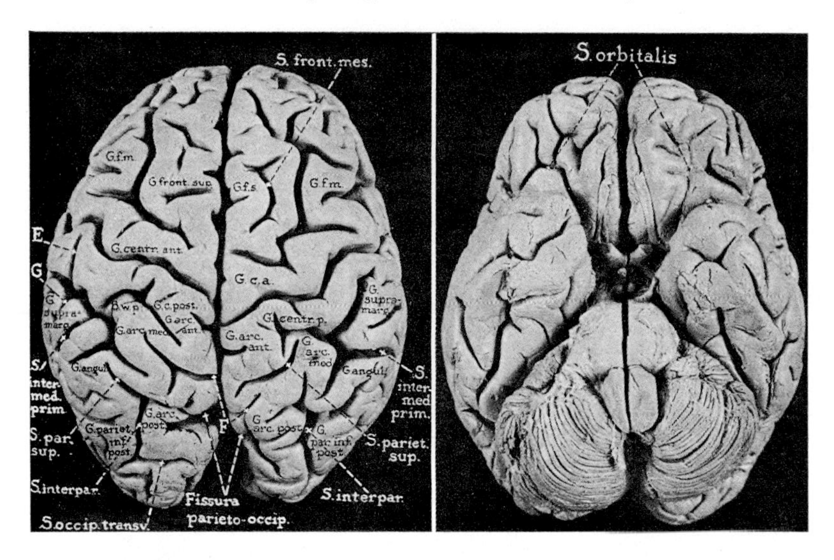

Figure 66. Dorsal and basal aspect of LUDWIG EDINGER's brain (from RIESE and GOLDSTEIN, 1950). E: sulcus incompletely subdividing dorsal and basal part of left postcentral gyrus; F: bifurcation of left postcentral sulcus; G: caudal spur of left postcentral sulcus. Other designations in Figures 66–68 self-explanatory.

tailed study of EINSTEIN's brain by several neurobiologists is projected or already in progress.

With few recent exceptions, the studies on 'elite brains' concern their particular fissuration patterns, or, in other words, a very gross morphologic aspect. The variations as well as the asymmetries[76] between right and left hemisphere obtaining in these elite brains are in no way different from those that can be observed in the Human population at large. However, being acquainted with the particular personal characteristics, abilities, and accomplishments of the prominent individuals whose brain they studied, the investigators generally yielded to the temptation of interpreting said variations in accordance with their views on the 'localization of faculties'.

Thus, the brain of GAMBETTA, who was a noted orator, displayed a peculiar duplication of the left inferior frontal gyrus *(Broca's convolution)*, supposedly correlated with GAMBETTA's abilities. Yet, LANDAU (1923) found exactly the same variation in the brain of a young French

[76] Such asymmetries may likewise be quite pronounced not only in Anthropoid Apes but also in all gyrencephalic Mammals (cf. e.g. Fig. 15A of the Dog's hemisphere).

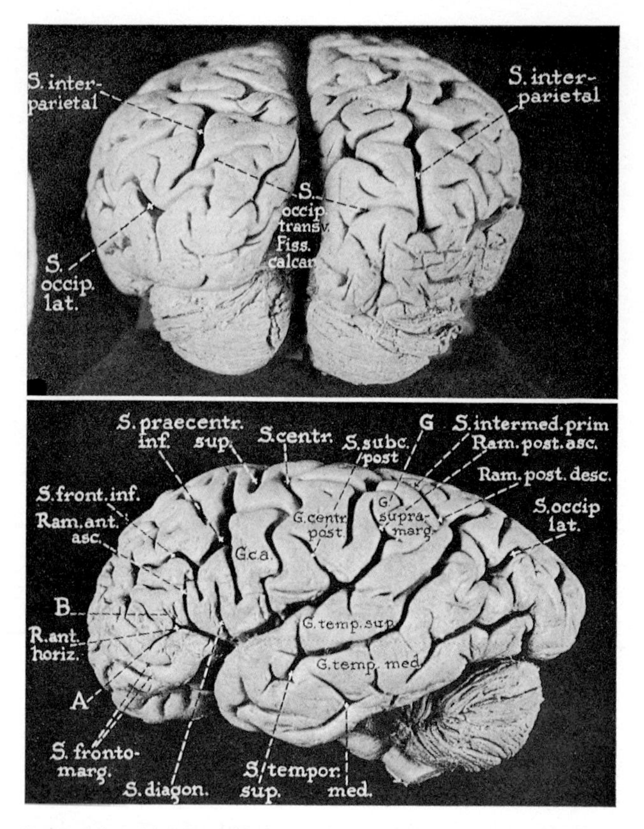

Figure 67. Caudal and left lateral aspects of LUDWIG EDINGER's brain (from RIESE and GOLDSTEIN, 1950). A, B: rostrobasal and rostrodorsal branch of left anterior horizontal ramus fissurae Sylvii. I would interpret s. occipitalis lat. as sulcus praelunatus and its occipital curved expansion as sulcus lunatus.

soldier, who was known in life as an average decent person, but with less than mediocre oratory aptitude.[77]

RIESE (1950, 1954, 1966) believes that in the brain of outstanding right-handed people, the more complicated convolutional pattern is found in the left hemisphere and vice-versa. Yet, the interpretation of what is 'more complicated' seems, quite frequently, to be rather questionable.

---

[77] LANDAU (1923) quotes a very pertinent remark by DEJERINE (1895, p.261): '*Il y a lieu de faire à cet égard de nombreuses réserves car nous avons rencontré bien des fois dans des autopsies quelquonques un remarquable développement du pied de la circonvolution de* BROCA *sans que pour cela l'individu qui en était porteur eût été pendant sa vie doué de facultés oratoires particulières. Ici, comme pour d'autres circonvolutions, ce n'est pas le volume de l'organe qui fait la fonction.*'

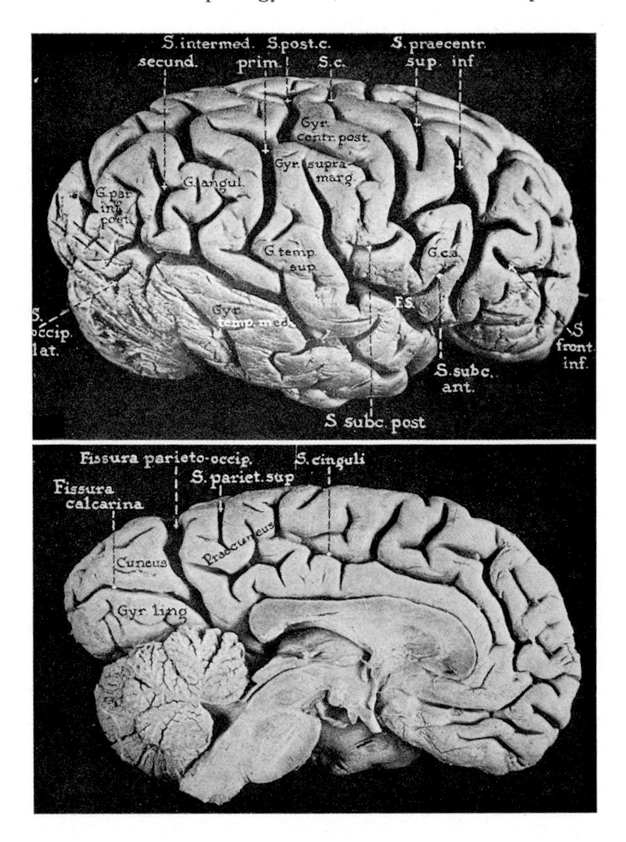

Figure 68. Right lateral and left medial aspects of LUDWIG EDINGER's brain (from RIESE and GOLDSTEIN, 1950).

LUDWIG EDINGER was left-handed, and RIESE and GOLDSTEIN (1950) evaluate his right hemisphere as exhibiting an increased development with more complicated fissural pattern (cf. Figs. 66–68). The cited authors also stress the right–left asymmetry of that brain's fissural pattern. However, as mentioned above (cf. footnote 76), such asymmetries are very common not only in Man, but in other Primates and gyrencephalic Subprimates. As regards EDINGER's more complicated right pattern, the evaluation must remain rather subjective, and, even if accepted, the greater complication is hardly very pronounced. The reader may, by inspecting the Figures, attempt to judge for himself. Another 'elite brain', that of HELMHOLTZ, as studied by EDINGER (1912), is depicted further below in Figure 73C.

Again, the extension of the visual area striata toward the lateral side, with or without sulcus lunatus, is taken to indicate 'visual gift'. This extension is fairly often seen in average individuals and, moreover, is not necessarily correlated with the actual size of the area striata, whose boundary lines may merely be shifted. Finally, the fissuration patterns of average individuals may be as complex or even more so than those of some elite brains.[78]

In addition to the fissuration pattern of these brains, cytoarchitectural features have been investigated in some of the recent studies (e.g. RIESE, 1954, and in other studies quoted by that author). It cannot be said these rather ambiguous cytoarchitectural data have yielded convincing or relevant results. Although it can be reasonably assumed that the brains of gifted people or of geniuses have higher capabilities of performance than those of 'average persons', and that these capabilities are based on 'physico-chemical' differences to an essential extent determined by the 'genome', it seems most likely that said differences are related to details of circuitry and biochemistry which still remain undetectable by the available methods of investigation. Thus, the opinion can be maintained that the studies of 'elite brains' have, so far, not yielded any conclusive or significant results. The complexity of the relevant multiple variables and parameters can be inferred from the fact that the cerebral cortex of prominent scientists may even display conspicuous senile changes although the mental capacity of these persons had remained unimpaired up to their death (cf. vol. 3/I, p. 722, and RIESE, 1953).

Another attempt at anthropological interpretation of fissuration pattern involves the investigation of brains pertaining to different Human *races*.[79] An extensive literature on that topic, which was also particularly considered by KAPPERS (1929, 1947) is extant. It may here be sufficient to point out, as further contributions, those by BEAN (1906, 1914), VAN BORK-FELTKAMP (1930, 1933), BUSHMAKIN (1928), CHI and CHANG (1941), HAYASHI and NAKAMURA (1914), KEEGAN (1926), KOHLBRUGGE (1906), KOOY (1921), KUHLENBECK (1928a, b), KURZ (1924), SHELLSHEAR (1937), ELLIOT SMITH (1903/04a), SPITZKA (1902), and WOOLLARD (1929). Comments on this subject matter are also included in the publications by CONNOLLY (1950) and by LANDAU (1923).

[78] Cf. e.g. Figure 66A with Figures 73A, B.

[79] Comments on the inconclusive results of studies concerning brain weight in different races were given in section 8, chapter VI of volume 3/II.

Concerning the highly controversial overall topic here under consideration, one could perhaps formulate the following five questions:

1. What is a Human *race?*
2. What are reasonably well measurable or definable anatomic respectively physical and additional 'objective' racial characteristics?
3. Does the fissuration pattern respectively the morphology of the Human brain display recognizable racial characteristics?
4. Do racial differences of behavioral manifestations exist, and if so, to which extent are they a result of genetic or of environmental factors?
5. Are axiologic evaluations of races or of racial characteristics justified?

As regards the concept of '*race*' it must be emphasized that all present-day Human beings apparently pertain to one single, interbreeding '*species*' designated as '*Homo sapiens*'. The difficulties in formulating a suitable definition of 'species' were pointed out and discussed in section 3, chapter II of volume 1. The difficulties in defining 'race', i.e. 'subspecies', and in classifying different races are by far much greater and have not been solved in a satisfactory manner.

Quite evidently, with respect to what could be called Human features, i.e. to the general characteristics of the 'species', all 'normal' Human beings can be assumed to have a common complex of the genome, and may, in this rapport, be considered homozygotic (cf. e.g. SALLER, 1930).

On the other hand numerous inheritable differences obtain with respect to physical characteristics such as skin pigmentation, type of hair, form of head, of nose, and additional anatomical traits, whereby the Human species displays a variety of 'subspecies'. These latter, moreover, are by no means homogeneous, but manifest a diversity of further subsets which can be arranged in descending taxonomic relationships, respectively types, from 'subraces' down to family lines. Family types, although pertaining to one and the same 'race' or 'subrace' may differ considerably from each other, thereby displaying heterozygotic characteristics within a common category of subspecies. The vagueness of the term 'race' thus seems obvious, and the concept of '*pure races*' cannot be upheld. Even at the undetermined period of its origin from an unknown 'genus' or 'species' of Anthropoid apes, the evolving Hominid species from which the present-day Human races presumably derived, may already have represented, *ab initio*, a highly 'mixed' heterozygotic gene pool.

The uncertainty of all speculations concerning the relevant details of Man's origin and subsequent distribution over our planet should again be emphasized. Because recent Man represents a common, single species, its monophyletic origin is, however, strongly suggested, and polyphyletic views, such e.g. as propounded by KURZ (1924), who assumed orangoid, gorilloid, and chimpanzoid Human races, arising from three different 'radiations' to each of which one of the above-named Anthropoids is supposedly related, can hardly be taken seriously.[80] Although presented with numerous 'palaeontological arguments', the various theories on the differentiation of Human subspecies elaborated by COON (1962) and others likewise remain unconvincing.

As regards general aspects of Human evolution from Anthropoids, the principle of fetalization, retardation or '*neoteny*' propounded by BOLK (1926) does not seem implausible and may have been one of the relevant factors. The cited author pointed out that the Anthropoid and the Human *newborns* respectively infants or juveniles are much more similar to each other than adults (cf. Figs. 69 A–D). The greater differences concomitantly arising with further aging can be described as an increasing deviation of Anthropoid morphological features from the late fetal or newborn stage, while adult Man retains a greater resemblance to juvenile characteristics. This is particularly the case with regard to the relationship between neurocranium and facial cranium.[81]

[80] Of the three large Anthropoid apes, the Chimpanzee might perhaps stand closest to the Hominid evolutionary line, being thus, among our three or four distant Anthropoid cousins, the nearest relative. The results of serologic tests can be interpreted in agreement with this view (cf. SALLER, 1930).

[81] Other characteristics, pointed out by BOLK, are e.g. shape of the sacrum, relative lack of skin pigmentation, sparse body hair, shape of ear. It should also here be added that the characteristic aspects of the juvenile head in the Orang and in other Primates, contrasting with the subsequent adult shape, were already pointed out by CUVIER (1769–1832) and FLOURENS (1794–1867). SCHOPENHAUER, in chapter 2, volume 2 of his '*Welt als Wille und Vorstellung*' extensively quotes both above-mentioned authors with regard to this topic, adding further elaborations of his own. However, neither SCHOPENHAUER's nor CUVIER's and FLOUREN's comments suggested a phylogenetic implication of said juvenile features as expounded by BOLK.

Figure 69 A. Contrast between a very young Chimpanzee infant and a fully mature adult individual (from KATZ, 1948).

Figure 69 B. Skull of a young Chimpanzee (from MAURER, 1928). Note the greater relative size of neurocranium respectively cranial capacity, the lesser size of the supraorbital ridges, and the greater facial angle in comparison with Figure 69 C.

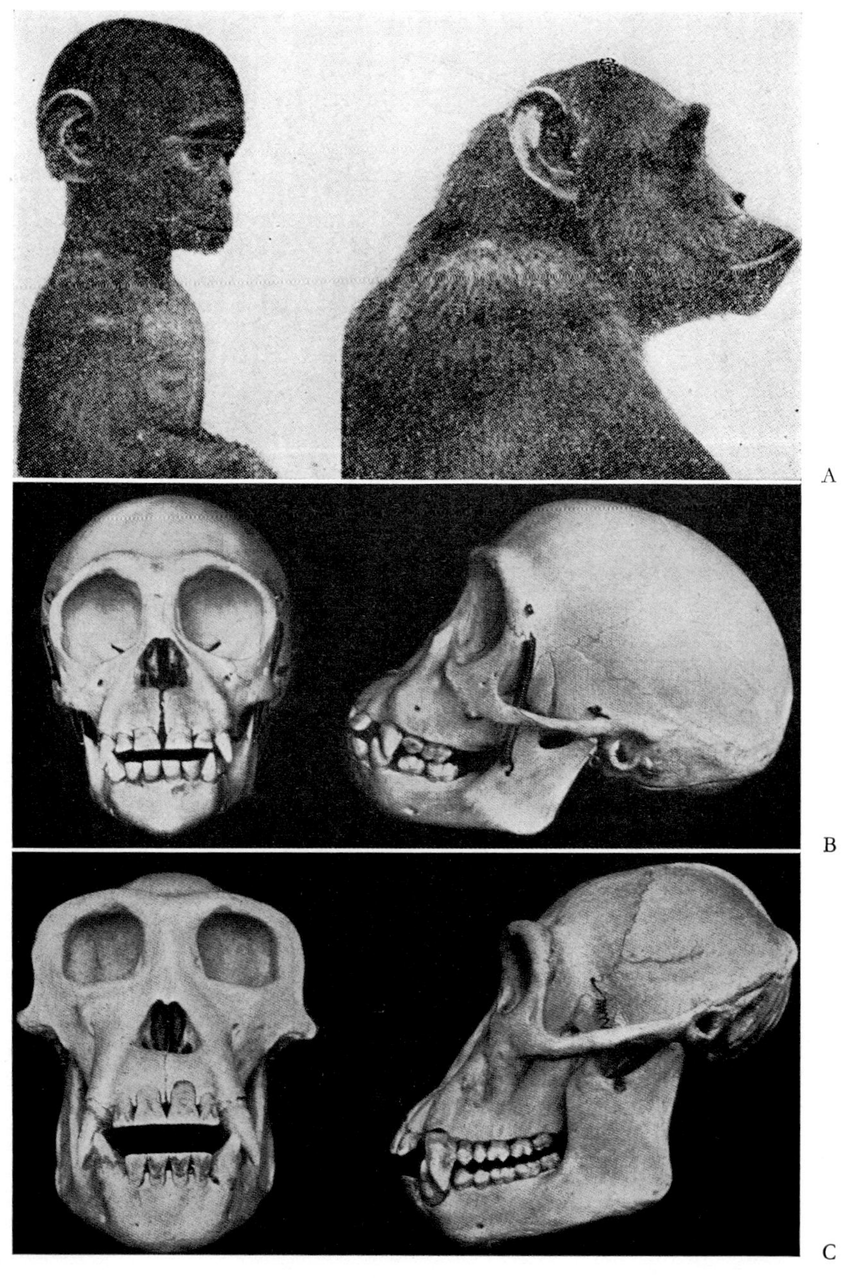

Figure 69C. Skull of an adult Chimpanzee (from MAURER, 1928). Note relative sizes of neurocranium and visceral cranium, as well as greater prognathism in comparison with Figure 69B.

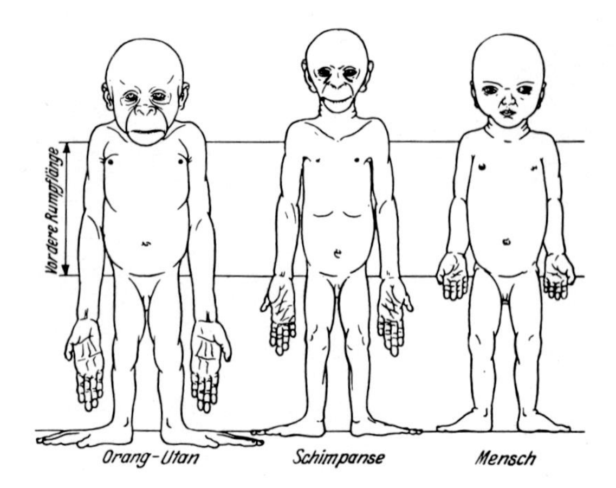

Figure 69 D. Body proportions in newborn Orang, Chimpanzee, and Man (after A. H. Schultz, from Saller, 1930).

Anthropoids, *relatively* orthognathous (cf. below, p. 113) at birth, become gradually pronouncedly prognathous (cf. Figs. 69 A–C) in contradistinction to the permanent more or less orthognathous character of the Human skull. The large *relative* endocranial capacity in both Anthropoid and Human newborns becomes subsequently much less reduced in Man. Despite various perhaps more or less justified criticisms directed at details of Bolk's theory, the facts emphasized by this latter author are incontrovertible, and can be interpreted to indicate that some hormonal effects,[82] very roughly comparable to those obtaining in neoteny, may have played a role in the multifactorial events leading to the origin of Man.

The shape of the cranial cavity and its relation to cranial capacity at various stages of Anthropoid respectively Hominid evolution is illustrated by the diagram of Figure 69 E.

More recent speculations on Human evolution, Hominid lineage, and evolutionary molecular population genetics are, *inter alia*, those by Mc Henry (1975), Nei (1975), and Oxnard (1975), none of which can be considered sufficiently convincing or conclusive.

Although quite definite distinctions between ethnic groups were made in the literature of antiquity, in which, e.g. reference to characteristics of the black '*Ethiopians*' and to the characteristics of *Scythians*,

[82] Hormonal effects were also taken into consideration by Keith (1947, 1950). Endocrine activities can be assumed to depend on both genetic and non-genetic factors.

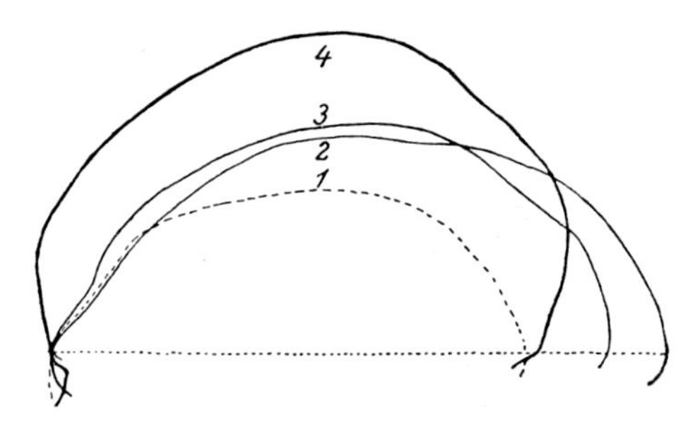

Figure 69 E. Outlines depicting profiles of neurocranium in Chimpanzee, Pithecanthropus, Neandertal Man and recent Man (after KLAATSCH, from MAURER, 1928). 1: (adult) Chimpanzee; 2: Pithecanthropus; 3: Neandertal Man; 4: Recent Man. Similar outlines drawn by other authors show slight differences of detail but agree as regards the overall relationships.

Britons, *Gauls* and *Germans* were made,[83] the biologic concept of '*race*' is comparatively recent, and said term was perhaps, in this sense, first used by BUFFON (1707–1788), who did not, however, elaborate a systematic racial classification.

The Judeo-Christian Old Testament derived the Human ethnic groups from the three sons of *Noah*, namely *Shem (Semites)*, *Ham (Hamites*, *non-Semitic Mediterranean*, later assumed by theologians to include Negroes), and *Japheth (northern people)*.[84]

The first taxonomic classification of races seems to have been that by LINNÉ (1707–1778) who subdivided his species '*homo sapiens*' into four varieties, based on skin color (and type of hair), namely *Homo Americanus* (red), *Europaeus* (white), *Asiaticus* (yellow) and *Africanus* (black). LINNÉ also added differences in temperament (choleric, sanguinic, melancholic, phlegmatic, in the above given order) and other mental traits, but apparently did not realize that these general characteristics were by no means specifically distributed upon, or restricted to, any one of his four groups.

J. F. BLUMENBACH (1752–1840), who is generally considered to be the founder of scientific anthropology, subdivided the Human species into five variations or 'races', mainly on the basis of skin color, namely

[83] Cf. the Βιβλιοθήκη ἱστορική (Book III, 8) of DIODORUS SICULUS (about 30 B.C.) and the *Germania* of TACITUS (about 98 A.D.).

[84] CUVIER (1707–1788) and GOBINEAU (1816–1882), although with some modifications, still essentially based their concepts on the origin of races upon this biblical myth.

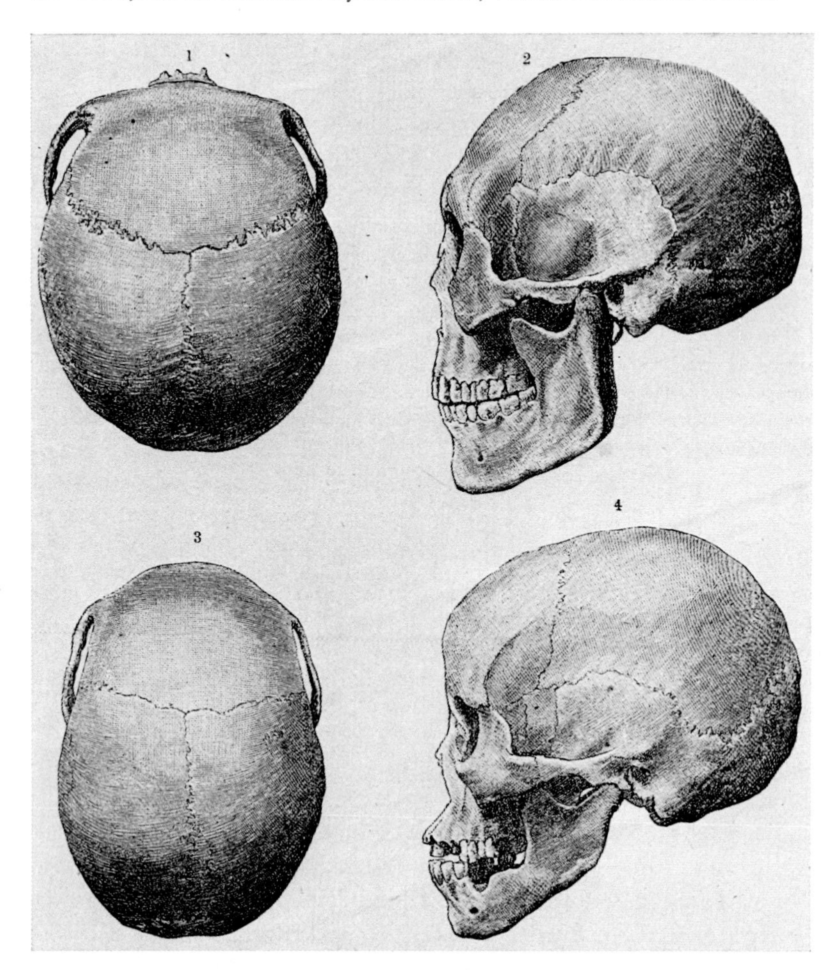

Figure 70. Conspicuous brachycephaly and dolichocephaly as displayed by the cranium in a Mongoloid and in a Negroid (from RANKE, 1923). 1, 2: cranium of a Kalmuck (Mongoloid); 3, 4: cranium of a Negro.

into *Caucasian* or white, *Mongolian* or yellow, *Ethiopian* or black, *American* or red, and *Malayan* or brown. Subsequently, anthropologists stressed the significance of cranial shape, of body proportions, of hair characteristics and of language. With respect to particular procedures of measurement, craniometry has been much emphasized, and so-called cephalic indices were established.[85]

---

[85] Denoted by 100 times the maximal head breadth divided by the maximal head length. Thus: *brachycephaly:* index of above 80; *dolichocephaly:* index of below 80.

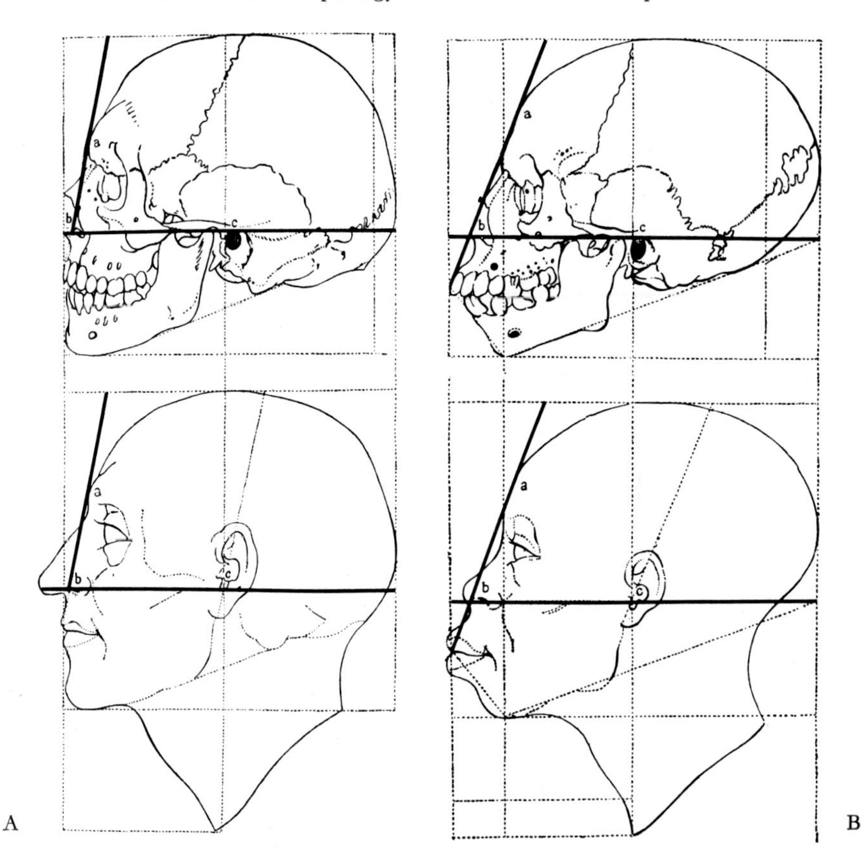

Figure 71 A, B. Cranium and facial outline in a Caucasian and in a Negro (after R. CAMPER, from RANKE, 1923). A: 'European'; B: Negro; abc: facial angle.

Figure 70 shows extreme brachycephaly in a Mongoloid, and extreme dolichocephaly in a Negro. The significance of these cranial configurations has been exaggerated: both types of shapes occur, to some degree, in all races. Of slightly greater import is perhaps the *facial angle* pointed out by PETRUS CAMPER (1722–1789), and indicated by a 'horizontal line' drawn from upper border of external acoustic meatus to lowest level of orbit, intersecting with a line connecting *nasion* (middle point of frontonasal suture) with *prosthion* (central point of the lower edge of upper alveolar arch). *Orthognathous* skulls have a facial angle above 85°, *mesognathous* ones between 80° and 85°, and *prognathous* of below 80°. *Prognathism* is quite frequently but not exclusively displayed

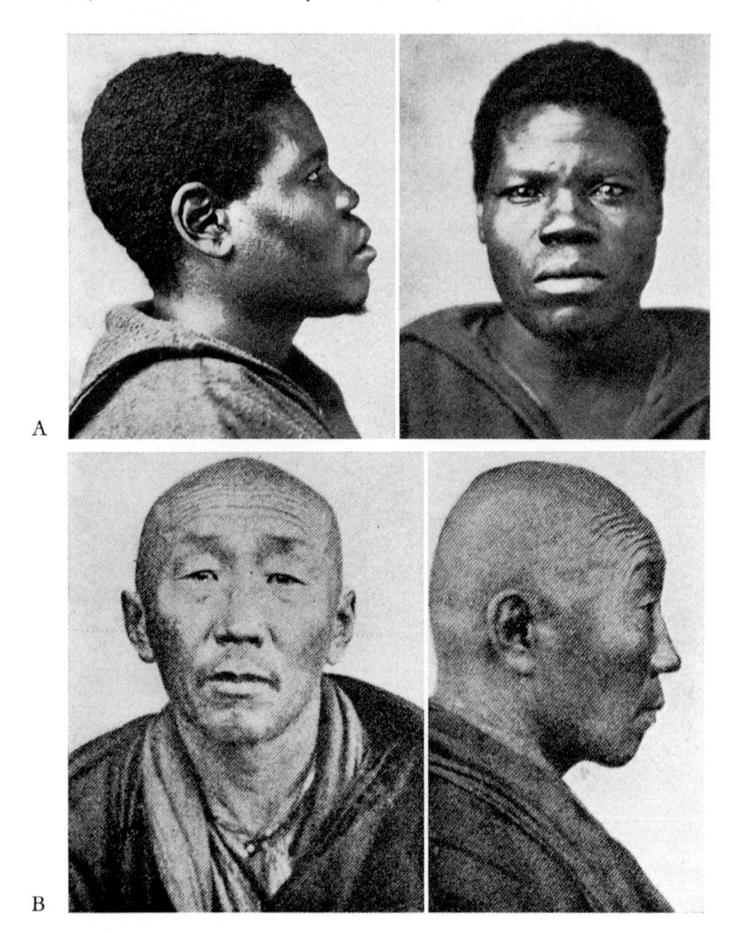

Figure 72A. East African Negro (photogr. by G. FRITSCH, from MAURER, 1928). Compare with Figure 71B for facial angle.

Figure 72B. Mongoloid Buriat (after MONTANDON, from SALLER, 1930).

Figure 72C. Caucasian, Norwegian Nordic subvariety (after SCHREINER, from SALLER, 1930).

Figure 72D. Caucasian, North-German Nordic subvariety (after SALLER, 1930).

Figure 72E. Sephardic Jew from Saloniki: Caucasian Mediterranean of Semitic subvariety (after PASSARGE, from SALLER, 1930).

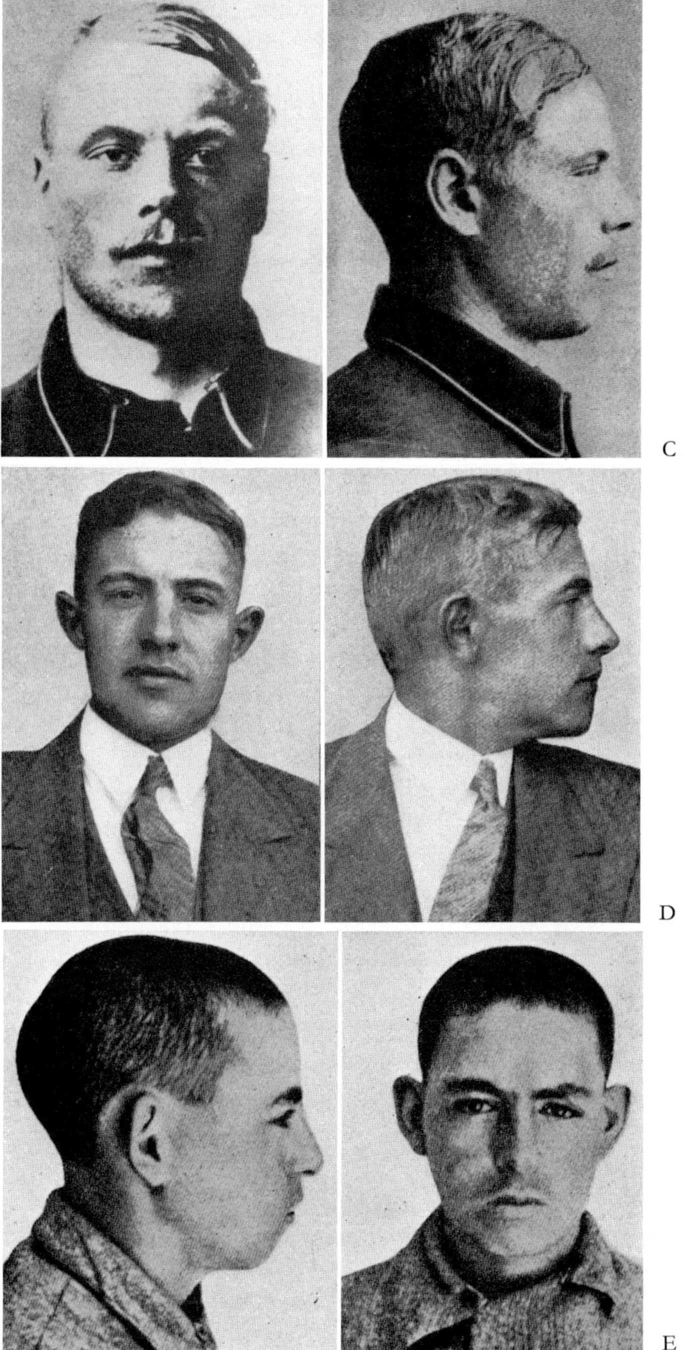
C
D
E

by Negroes[86], in contradistinction to the generally *orthognathous* Mongoloids and Caucasians (cf. Figs. 71 A, B; 72 A–E).

As regards other facial features, the ancestral Apes presumably had a flat nose. Negroes and Australians likewise generally have flat, broad noses, while Caucasians tend to have narrow and prominent noses. On the other hand, Apes have narrow, thin lips, while in Negroes the thickest, most 'human' lips predominate, those of Caucasians being, on the whole, thinner.

Yet it should be mentioned that, according to BOAS and some others, not only size of the body but also forms of the head and face may, in some way, be influenced by the environment in which the people live. Whether such changes merely effect the phenotype or, in an unknown manner (cf. above in section 1 the comments concerning the domestication problem) become 'hereditary', remains an open question. The hair coat of Apes is fairly strong. Among Human races, Caucasians, Australians and a few scattered other groups have the amplest body hairs. Mongoloids have the least. Concerning the hair types, *lissotrich* (straight), *kymotrich* (wavy), and *ulotrich* (woolly) forms are the main sorts distinguished, which can again be subdivided. The hair color varies from blond over black-brown and black to red. The black color predominates in the ulotrich Negroes, the generally kymotrich Australians, the lissotrich Mongolians and is common among the generally kymotrich Caucasians, among which the so-called Nordic subgroup frequently displays blond hair (and blue eye color), while other Caucasian groups may also display numerous mixtures of hair color.

As regards some other characteristics, the distribution of the *blood groups* O, A, B, and AB is likewise of interest. Group O is found in all races and seems to be the commonest type. Group A is also common, while group B and especially AB are less common. AB seems not to be present in Australian aborigines and in some American Indian tribes. Nevertheless, with few exceptions as e.g. just mentioned, in nearly every Human racial group examined, a mixture of all four blood groups was found, but racial differences obtain in the ratio of said blood groups' distribution.

With respect to other 'objective' characteristics, that of *language*, which, except for the required brain mechanisms, essentially depends on 'nurture', has been considered. FRANZ BOPP (1791–1867), one of the

[86] Cf. also the prognathism in Anthropoid apes (Figs. 69 A–C).

founders of comparative linguistics, pointed out the common features of Sanskrit, other Indian languages, Persian, Greek, Latin, Celtic tongues, French, English, German, Scandinavian tongues, Lithuanian, Russian and other Slavonic tongues, Armenian etc., now generally subsumed under the *Indo-Germanic* or *Indo-European language family*. The linguist and ethnologist FRIEDRICH MÜLLER (1834–1894) based his classification of Human races on the two criteria of hair type and of language, which he considered to be more constant than cranial characteristics. He was, however, fully aware that language, by itself, can not be taken as a valid criterion of physical race. It is rather obvious that language and race are not necessarily correlated, but linguistic 'lines of cleavage' may nevertheless, to a certain and variable extent, be correlated with the ill defined racial ones.

At present, no agreement obtains as to classification and subdivision of Human races. Thus, DUNN and DOBZHANSKY (1957), in discussing this topic, point out, among the multitudinous proposed systems of classification, the following two fairly recent ones. In 1933 VON EICKSTÄDT recognized three basic races, '*Europid*', '*Negrid*', and '*Mongolid*', with 18 'subraces', 11 'collateral subraces' and 3 'intermediate forms'. In 1950, COON, GARN, and BIRDSELL distinguished 6 'putative stocks', namely *Negroid*, *Mongoloid*, *White*, *Australoid*, *American Indian*, and *Polynesian*, and 30 different races.

In order to formulate a reasonably valid first approximation I am inclined to distinguish three main racial groups, namely *Caucasoid*, *Mongoloid*, and *Negroid*. The *Caucasoid*, again, could be subdivided into a number of subgroups, but it might here be sufficient to distinguish the western Nordic, Alpine, Mediterranean subvarieties and the eastern Indo-Iranian group, all of which are highly 'mixed'. The Mediterraneans would include, *inter alia*, the Semitic and 'Hamitic' subgroups.

The *Mongoloids* could be said to include various populations adhering to the Turanian or Ural-Altaic language groups; moreover the Chinese, Indo-Chinese, Japanese, Koreans, Tibetans, Malays, Polynesians, Eskimos, and the Amerindians.

The *Negroids* would include the African Bantu, Hottentot, Bushmen and others. There are, moreover, a number of smaller racial groups whose inclusion in any general classification scheme is exceedingly difficult. The *Australian aborigines* certainly differ considerably from the African Negroids and represent a special subgroup; much the same could be said about *Melanesian* and *Micronesian* populations, the various African and Asiatic *Pygmies*, the dark *Dravidians* of India and Ceylon, and the *Weddas* and related groups of Ceylon and Celebes, also the indigenous *Ainus* of Japan, which latter are perhaps distantly related to the Caucasoids.[87]

Generally speaking, it could be said that the racial subdivisions of the species Homo sapiens are quite analogous to the races or subspecies of animals or plants in nature, and to the breeds or varieties of domesticated animals and cultivated plants. All of these seem to arise in the course of mutations, gene recombinations, and natural or artificial selection of those collections of genes which are suited to certain environments. Human racial groups, moreover, are subject to cultural and social interactions differing from the 'biological' factors affecting animals and plants, and these two sorts of influences are in continual interaction with each other (DUNN and DOBZHANSKY, 1957). Races and their subdivisions have been defined as populations which differ in the frequencies of some gene combinations, but this is a rather vague and inadequate definition. One could also speak of races as '*genetic pools*', but *ethnic groups*, consisting of 'racial mixtures' can likewise be considered 'genetic pools', particularly if insular, as in the cases of e.g. Great Britain or Japan. Again, it becomes rather arbitrary to select the 'essential racial features', resulting from a common inheritance, which characterize a 'race'.

'Race' and 'subraces' are obviously abstractions without physical existence and not observable as such. Only the individuals are observable, and arrays of individuals can be classified in accordance with a diversity of rules, defining the extension and the intension of the resulting sets.

Again, 'races' as characterized by physical traits must be distinguished from *ethnic groups* with common cultural traditions and language (*'people'*, *'Volk'*), as well as from nations, which are political units, and have certain historical traditions. The arbitrarily defined conceptual 'units', race, ethnic group, people and nation may overlap and intersect in the most diversified patterns. Thus, in accordance with said terminology, we have the Swiss nation with its historical tradition, includ-

---

[87] With regard to the obtaining uncertainties, it should again be emphasized that almost every anthropologist proposes his own system of classification. Quite sober and still useful discussions on that topic can be found in the publications by VATTER (1927) and SALLER (1930), both of which, with references to the relevant preceding authors, present their own classification. The three main groups (Caucasoid, Mongoloid, Negroid) are, however, rather generally recognized.

ing French, German, Italian and Romanisch people with their own language and cultural traditions, and with a mixed but entirely Caucasoid racial population. The United States of America, again, are a nation with historical tradition, including a diversity of people essentially adopting the English language, and with a diversity of Caucasoid, Mongoloid, and Negroid racial mixtures. In both cases, as well as in other nations a variety of 'genetic pools' can be delimited in arbitrary fashion.

The designation '*ethnic group*' is occasionally and not without some justification used as a synonym for 'race' or 'subrace', but emphasizes the common traditions rather than the genetic aspects implied by the term 'race'.

Following the foregoing tentative and of necessity rather unsatisfactory answers to the first two questions enumerated above on p. 107, we may now turn to the third one, concerning racial characteristics of the *cerebral hemisphere's fissuration pattern.* KAPPERS (1929, 1947) who also discussed the highly ambiguous and inconclusive hypothetical features of the Pithecanthropus and Neanderthal brains inferred from endocranial casts, reviews, with numerous illustrations and bibliographic references, the various aspects of the brain in Australian aboriginals, Bushmen, Hottentots, Negroes, Eskimos, American Indians, Mongols, Chinese, Indonesians, and Caucasoids. Reference may here also be made to the publications cited in section 4, dealing with the sulcus lunatus in diverse races. As regards the brain of Eskimos, there are papers by SPITZKA (1902) and VAN BORK-FELTKAMP (1935). The brain of Australian aboriginals was investigated by WOOLLARD (1929, 1931).

KAPPERS (1929, 1947), who admits that the anthropology of the brain is still far behind the available physical anthropological knowledge concerning other parts of the body, nevertheless concludes that the available data '*ne sont certainement pas dénuées de faits intéressants*'.

On the basis of my own experience,[88] I believe, however, that it is impossible to identify, according to fissuration pattern, form and size, a Human brain of unknown provenance as pertaining to an individual

[88] Since 1919, starting as a medical student admitted to MAURER's *privatissimum* with research laboratory privilege at the Anatomical Institute in Jena, I have studied gross morphological as well as cyto- and myeloarchitectural features of the Human brain, continuing with these studies in Japan, in Breslau, and in Philadelphia, both in connection with my own investigations and with the anatomical and neuroanatomical courses under my direction. With Dr. GLOBUS at the *Mount Sinai Hospital* in New York, and with Dr. WEBB HAYMAKER at the *Army Institute of Pathology* in Washington, D.C., I also examined

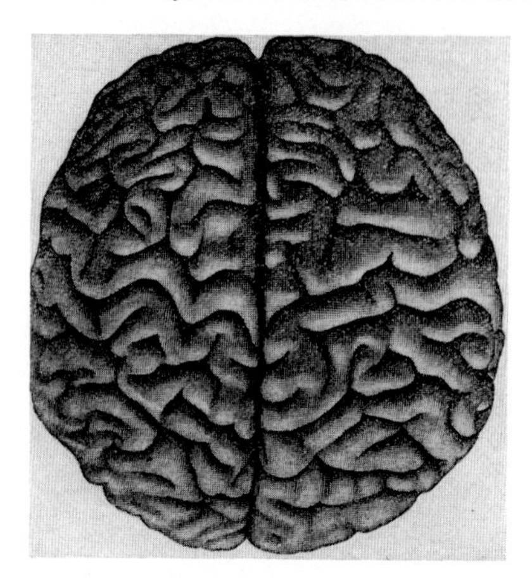

Figure 73 A. Brachycephalic brain (Caucasian Alpine) of a South German (Munich) laborer (from RANKE, 1923).

of any particular race, or for that matter, as that of a male or a female. This statement applies also to microscopic features,[89] including cytoarchitecture, and, as pointed out above, to a distinction between 'elite brains' and brains of average individuals.

LE GROS CLARK (1949) likewise concluded that 'there is no macroscopic or microscopic difference by which it is possible for the anatomist to distinguish the brain in single individuals of different races'. It will here be sufficient to illustrate, with Figures 73 A and B, a brachycephalic Caucasoid brain for comparison with a dolichocephalic Negroid one, moreover, with Figure 73 C, a dolichocephalic Caucasoid 'elite brain' (HELMHOLTZ) and a likewise dolichocephalic Papuan brain. The former does indeed display a more complex fissuration pattern than the latter. Nevertheless, complex fissuration patterns can be seen in all Human races, and in some 'elite brains', of doubtless eminent personalities, the fissuration pattern displays only 'average' com-

---

numerous pathologic and additional normal Human brains. Thus, over more than 50 years, I had ample opportunity to become acquainted with the relevant features of a large number of Caucasoid, Mongoloid, and (American) Negroid brains.

[89] Here, of course, by looking for the so-called *nucleolar satellites* of BARR and BERTRAM, it might be possible, with some degree of probability, to determine the sex (cf. vol. 3/I, chapter V, section 2, p. 97). As regards skeletal features, the pelvis generally displays a clearly recognizable sex-difference, although, even here, some rather rare intermediate forms, e.g. *qua* angulus pubis, can occur.

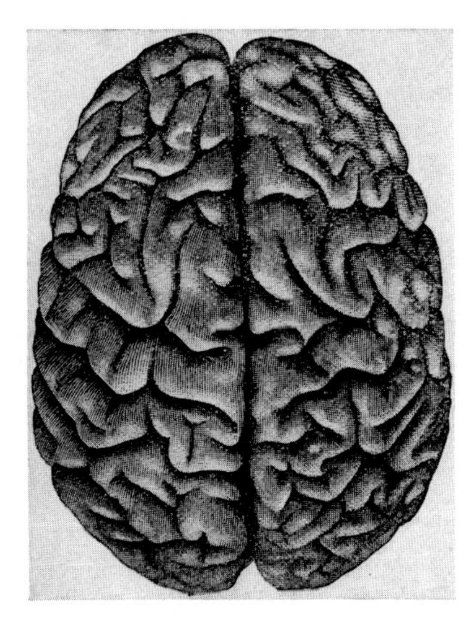

Figure 73 B. Dolichocephalic brain of a Negro (from RANKE, 1923).

plexity. A critical and competent paper on '*Rassengehirne*' by NGO-WYANG (1936), then Professor of Brain Anatomy at the University of Nanking, ends with the conclusion: '*Es scheint mir daher verfrüht, etwas Sicheres über Verschiedenheiten von Rassengehirnen auszusagen.*'

Thus, answering the third of the above-mentioned five questions, it could be stated that none of the described fissuration patterns of the Human brain are characteristic for any particular race. This, however, as e.g. in the case of blood groups, does not exclude the possibility that various patterns of fissuration display different statistical frequencies in the diverse races. With regard to the sulcus lunatus, such different frequencies are indeed reported, as mentioned above in section 4. Concerning the statistical distribution of other sorts of fissuration patterns, the available data are not very conclusive, particularly also because of the diverse interpretations and classifications proposed by the different authors.

Turning now to question 4, one is justified to state that different ethnic groups, people, or nations, characterized by different racial compositions, have developed, displayed or still display sorts of cultures, which are reducible to 'behavorial manifestations', respectively 'psychologic traits' depending on either genetic or environmental factors or both. Whether or not Human 'races' differ in hereditary 'psy-

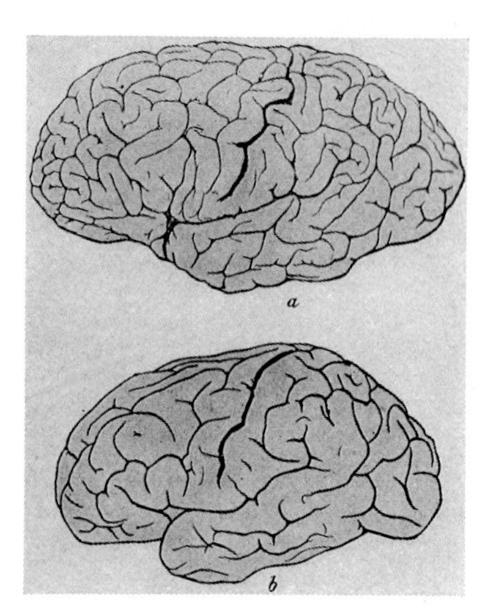

Figure 73C. Lateral aspect of a dolichocephalic Caucasoid elite brain (a: HELMHOLTZ) and (b) brain of unknown Papuan (a after EDINGER, b after SPITZKA, both from EDINGER, 1912). The brain of HELMHOLTZ was also studied by HANSEMANN about 1899.

chologic traits' is a hotly disputed issue between the '*hereditarians*' and the '*egalitarians*'. The lack of racial specificity displayed by the brain's fissuration pattern neither supports the latter, nor contradicts the former viewpoint.

Prima facie, the assumption of different fundamental hereditary traits in diverse Human races is clearly suggested by the hereditary different 'mental' (behavioral, temperamental) traits obtaining in races of domesticated animals, e.g. breeds of Horses or Dogs. These obvious hereditary traits are even conceded by avowed '*egalitarians*' and '*environmentalists*' such, among others, as the anthropologist FRANZ BOAS. The '*environmentalists*', however, argue that the races of domesticated animals are developed by carefully controlled inbreeding, and resemble Human family lines rather than Human races, which latter should be compared to wild animal races. Another *(environmentalist)* argument claims that, although the amount as well as the kind of variation which a trait shows in different environments is decided by the hereditary make-up of the organism, the traits themselves depend upon environmental factors, the temperamental and behavioral Human manifestations being thereby much more plastic and not comparable to those displayed by domestic animal races.

It could be replied that these '*environmentalist*' arguments are unconvincing. First of all, the heritability of 'mental traits' in animals appears well established, and Man, in his biologic aspects, is an animal. Second, considerable *self-domestication* (cf. above, p. 13) and inbreeding obtains in Human ethnic groups.

The noted geneticist RICHARD GOLDSCHMIDT (1927) justly comments as follows on Human '*Erbeigenschaften, auf denen seine Kulturfähigkeit beruht*': '*Nehmen wir willkürlich ein paar Eigenschaften (ohne uns darum zu bekümmern, ob sie einfache Erbeigenschaften vom Standpunkt der Vererbungslehre sind und im Bewusstsein, dass sie der Psychologe nicht als einfache Qualitäten anerkennen würde), die da in Betracht kommen könnten: schnelle Auffassungsgabe, logischer Sinn, Formensinn, Fähigkeit zum Nachahmen, Handgeschicklichkeit, gutes Tastgefühl, Eigenbrödelei, Herdentrieb, Brutalität, usw. Es ist klar, dass es als Grundlage für eine Kulturentwicklung nicht gleichgültig ist, wieviele und welche von diesen und Hunderten anderer Eigenschaften miteinander vereinigt sind und fernerhin, dass viele Kulturzustände überhaupt nur verwirklicht werden können, wenn als Grundlage bestimmte Kombinationen solcher Eigenschaften vorhanden sind. Betrachten wir nun die Geschichte jener Teile der Menschheit, die eine höhere Kultur entwickelt haben, von dem Anfang des Haustierdaseins des Menschengeschlechts an. Es braucht nicht vieler Beispiele, um zu zeigen, dass hier das verwickeltste Kreuzungsexperiment vorliegt, das die Natur kennt. Ständig schoben sich Gruppen von Menschen von bestimmtem Rassencharakter, die selbst bereits das Produkt vorausgegangener Kreuzungen waren, nach neuen Wohnsitzen hin und kreuzten sich mit der vorgefundenen ansässigen Bevölkerung, und dann folgte eine ungeheure Umkombination von Erbcharakteren. Kaum ist eine gewisse Ruhe eingetreten, so folgt eine neue Welle, neue Kreuzungen, neue Kombinationen. Jede Kombination aber eröffnet, je nach ihrer Art, neue Möglichkeiten der Kulturentwicklung, sei es zum Guten, sei es zum Bösen.*'

'*In einer abgeschlossenen Gruppe von Menschen, also sagen wir etwa auf einer Insel, auf der ein Menschenstamm nach stattgefundenen Kreuzungen isoliert würde, liegen die Verhältnisse nun so: Von einem Menschenpaar sollten bei der jetzt herrschenden durchschnittlichen Vermehrungsrate in etwa 1700 Jahren soviel Nachkommen abstammen, als die Gesamtzahl der Menschen auf der Erde beträgt. Wenn also, wie es nun in Wirklichkeit der Fall ist, in einem isolierten Land die Zahl der Bewohner in der gleichen Zeit nur wenig (im Vergleich zur möglichen Vermehrung) wächst, so müssen alle Bewohner schliesslich eine Menge gemeinsamer Ahnen haben, mehr oder weniger miteinander verwandt sein. Mit anderen Worten, es hat ein beträchtliches Mass von Inzucht stattgefunden. Inzucht nach Kreuzung ist nun, wie wir wissen, ein sicheres Mittel, um Einheitlichkeit, weitgehende Homozygotie zu erzielen. Daraus folgt, dass, nachdem Kreuzung eine Vorbereitung der Kulturfähigkeit bewirkt hat, Isolierung und Inzucht eine Vereinheitlichung und Festigung dieser Kultur bedingt. Als Beispiel liesse sich die Entwicklung Englands und Japans anführen.'*

As regards inbreeding, it could be pointed out that a given generation of siblings has 2^1 parents, and, *if no inbreeding would obtain*, 2^2 grandparents, 2^3 great grandparents, etc., such that, assuming roughly 30 years for a generation, $2^{100} \cong 10^{30}$ ancestors would have existed 3000 years ago, and $2^{1000} \cong 10^{300}$ ancestors[90] 30 000 years ago, at the postglacial, late Pleistocene epoch. This is obviously impossible. At present the entire living Human population is of a numerical order of magnitude corresponding to roughly 4×10^9. *Per contra*, although there are no substantial data for estimates of Human population figures before the 17th century, demographers, on the basis of reasonable extrapolations, assume a total population of roughly five million about 8000 B.C., before the dawn of history. An increase to a total between 200 and 300 million at the beginning of the Christian area is believed to have occurred. In 16 centuries a further increase to 500 million may have taken place about 1650. A substantial growth presumably began after 1650, such that the population doubled in the next 200 years to one billion around 1850, doubled again to two billion by 1930, and perhaps some time before 1976 will have doubled again (STEELE, 1975).

As a third reply to the 'environmental' argumentation it could be stated that *wild* as well as *domestic* animal races display hereditary 'behavioral traits'.

As a fourth reply to the 'environmentalists', concerning their argument that specific 'traits' are not inherited, but merely 'the manner, in which an organism responds to its environment' the logical weakness of this formulation could be pointed out. Clearly, if the gene pattern provides the propensity toward, i.e. the *anlagen*[91] of certain traits, these latter still depend upon inheritance, although developed by nurture,

[90] It will be recalled that this number by far exceeds the estimated number of atoms in the entire astronomically visible universe (about 10^{70} to 10^{75} as calculated by reasonably valid extrapolations and first approximations).

[91] VATTER (1927) concludes '*dass sich ebenso wie die körperlichen Rassenmerkmale auch bestimmte seelische und geistige Anlagen nach den Mendelschen Gesetzen vererben, aber wir sind noch nicht in der Lage, in dem Komplex der seelischen und geistigen Äusserungen, den wir bei irgendeiner Gruppe von Menschen feststellen, zu unterscheiden, ob es sich im einzelnen Falle wie im allgemeinen um das Resultat einer bestimmten rassenhaften Veranlagung oder gemeinsamen historischen oder kulturellen Erlebens handelt*'.

whose substantial significance no reasonable *'hereditarian'* will question. The argument resolves itself to an arbitrary point of emphasis, respectively of semantics.

The difficulties in accurately defining, and the still greater difficulties in assigning numerical values to, and thereby 'measuring' mental traits were pointed out above in dealing with the evaluation of individuals. As regards racial groups these difficulties become again far greater. The formulation of 'general laws' and valid concepts in this entire field is severely impeded by the fact that the observed phenomena are poorly definable and, moreover, affected by many factors which are hard to evaluate separately.

Yet, it can be maintained that hereditary 'mental differences' do exist not only between individuals, but also, statistically speaking, between groups of individuals, that is between populations which may differ in their *'stocks of genes'*. Instead of 'genes' one might also say *'genetic factors'*, since mental traits are presumably not based on single 'genes', but on the cooperation of many factors which can be conceived as 'units'.

Detailed studies concerning the IQ in two racial groups, namely in American Negroes and Caucasians have been undertaken by JENSEN (1973),[92] and seem to demonstrate that genetic (as well as environmental) differences are involved in the average disparity between these Negroes and Caucasians in intelligence and educability (cf. also SCARR-SALAPATEK, 1971; LOEHLIN *et al.*, 1975). Similar views were expressed by SHOCKLEY (cf. ANONYMOUS, 1967), who also pointed out the high IQ averages of Mongoloids ('Orientals') which to some extent even surpass those of Caucasians. JENSEN, SHOCKLEY, as well as HERRNSTEIN (who is essentially concerned with the relationship of IQ to social status respectively success) have been strongly attacked, and even prevented from presenting their views, by militant radical *'egalitarians'* and *'environmentalists'* who advocate freedom of research and speech for themselves while denying it to those of other opinion (cf. e.g. HOLDEN, 1973).

---

[92] One might here also mention the ambitious and highly vocal British psychologist EYSENCK (The IQ Argument; Library Press, Freeport 1971). However, the candor of this author is open to question. In a previous publication (Sense and Nonsense in Psychology; Penguin-Pelican Books, 1962) he made mendacious and preposterous reference to some views of SCHOPENHAUER, which he maliciously distorted. EYSENCK was therefore severely and justly castigated on pp. 192–193 of the *Schopenhauer Jahrbuch 54* (1973). Another comment on EYSENCK can be found on p. 450, footnote 128 of the monograph *'Mind and Matter'* (K., 1961).

Assuming, as seems most likely, that JENSEN, SHOKLEY and others of similar opinion are justified in their conclusions, which refer to statistical data, it appears nevertheless evident that an undefined but large number of 'Negroes' would be more 'intelligent' than a likewise rather large undefined number of 'Caucasians' or 'Mongoloids'.

Generally speaking, one could also say that numerous ethnic groups pertaining to the Caucasoid and Mongoloid main 'races' have produced considerable 'cultural' or 'civilisatory' accomplishments, while others, likewise pertaining to these two main groups, or to the poorly classifiable subvarieties mentioned above on p. 117, and the Negroid main group as a whole, have not. This appraisal leads to the fifth question concerning race, namely to the disputed topic of axiologic evaluation.

In attempting to deal with such evaluation of racial characteristics it must again be emphasized that, from a rigorous logical and semantic viewpoint, no axiologic evaluation whatsoever is based on reason. Any evaluation implying a 'higher' or 'lower' (i.e. better or worse) status or pattern of events ('condition', 'progress', 'decline', 'decadence', 'degeneration', etc.) involves affectivity (emotion), respectively teleologic concepts based upon affectivity (choice, preference, decision, like or dislike, purpose) and cannot be, in this respect, purely 'scientific'. This was clearly recognized and elaborated by DAVID HUME (cf. vol. 3/II, p. 749, footnote 335, and K., 1961, p. 454f.).[93]

From the beginning of recorded history, ethnic groups, through their spokesmen, have tended to stress their own superiority. The Jews considered themselves God's chosen people. The ancient Greeks held themselves superior to the non-Hellenic nations which they designated as βάρβαροι. The designation *Arian* (or *Aryan)*, from the Sanskrit *ârya* (Pâli: *ariya)*, i.e. noble[93a], was used for the three upper Hindu

[93] The eminent statesman OTTO FÜRST von BISMARCK, in volume II, p.155 of his '*Gedanken und Erinnerungen*' (1898) comments as follows: '*dass in der Politik und in der Religion keiner dem Andersgläubigen die Richtigkeit der eigenen Überzeugung des eigenen Glaubens concludent nachweisen kann, und dass kein Gerichtshof vorhanden ist, der die Meinungsverschiedenheiten durch Erkenntnis zur Ruhe verweisen könnte.*'

'*In der Politik wie auf dem Gebiete des religiösen Glaubens kann der Konservative dem Liberalen, der Gläubige dem Ungläubigen niemals ein anderes Argument entgegenhalten, als das in tausend Variationen der Beredsamkeit breitgetretene Thema: meine politischen Überzeugungen sind richtig und die deinigen falsch; mein Glaube ist Gott gefällig, dein Unglaube führt zur Verdammnis.*' BISMARCK's comment, of course, applies, *mutatis mutandis*, to all axiologic evaluations.

[93a] E.g. the *four noble truths* or the *noble eightfold path* of Buddhism (Pâli: *cattâri ariyasaccâni; ariyo attangiko maggo)*.

castes, and more generally for the Indo-Iranian people. It was also applied to the Indo-European group of languages. The abuse of that term by the Nazis is notorious.

The Chinese believed that all other nations had to obey their emperor,[94] the Son of Heaven, ruling the Middle Kingdom; they called other people, considered of lower culture, *man-tzu* (barbarians), using an ideogram with the radical 142 *(ch'ung)*, designating insects respectively vermin. Subsequently, upon contact with the Western world, the long-nosed Caucasians were called *yang-kuei-tzu* (foreign devils). The Japanese, likewise using the original Chinese ideogram, designated uncivilized ethnic groups as *ban-jin*, but adapting to the Western ways, more politely called the Caucasians *sei-yô-jin*, i.e. Western Ocean People. Diverse other examples, such as KIPLING's 'lesser breed without the Law', etc., could be cited.

Some measure of *esprit de corps*, of common spirit pervading the members of an ethnic, cultural, national or any other group, is, of course, an important factor for the stability, success or progress of such group, and implies devotion, regard for the group's 'honor', and a variable degree of pride. This attitude easily becomes exaggerated, but no agreement on its proper measure, merit, value, commendability, respectively its evil effects and reprehensibility can be reached, since such judgements depend on personal attitudes presumably conditioned by 'heredity' *and* 'nurture'. SCHOPENHAUER, an individualist, considered national pride to be the cheapest sort of pride.[95] He also emphasized that *'jede Nation spottet über die andere und alle haben Recht'*. Because of the aggressiveness and viciousness of Man, most or all nations and ethnic groups have reciprocally suffered abuses from each other, resulting in distrust and hatred, which engenders more hatred as well as reprisals.[96]

---

[94] Cf. the arrogant exhortation included in the answer given by the Chinese *Manchu emperor* CHIEN-LUNG in 1793 to the envoy of England's *King* GEORGE III: *'It behooves you, o King, to respect my sentiments and to display even greater devotion and loyalty in future, so that by perpetual submission to our Throne, you may secure peace and prosperity for your country hereafter.'*

[95] SCHOPENHAUER (ed. *Grisebach*, vol. V, pp. 404–405) remarks: *'Die wohlfeilste Art des Stolzes hingegen ist der Nationalstolz. Denn er verrät in dem damit Behafteten den Mangel an individuellen Eigenschaften, auf die er stolz sein könnte'*. *Mutatis mutandis*, this could also be said with regard to racial pride.

[96] The *Dhammapada (Path of the Law)* contained in the *Khuddakanikâyo* of the *Pâli Sutta Pitakam* of *Hînayâna Buddhism*, proclaims (I, 5):

'Never by hatred is hatred appeased
Hatred ceases by kindness; this is an eternal law.' (Continuation on page 128)

Reverting to the highly arbitrary *criteria* for an axiologic evaluation of 'higher' or 'lower' cultural status of ethnic groups, nations, or 'races', the following could roughly be distinguished: achievements in literature, in philosophy, in religion, in art and architecture, in science, in technology, in legislation and social organization, trade, communication and navigation, and military organization.

Western civilization, slowly beginning to emerge in the 15th century on the basis of re-discovered values from *Graeco-Roman antiquity*, and with various 'nuclei' of origin in Italy, in the Iberian Penninsula and the more northern countries, reached perhaps, after the totalitarian power of the Roman Pontifical Imperium was definitely broken, its flowering, very roughly comparable to the *Periclean* or *Augustean* epochs, in the 19th century.

Prima facie, and assuming racial differences in mental traits, it does not appear unreasonable to consider Western civilization as produced by the mentality of the Caucasian 'race', and to ask which subraces or ethnic groups predominantly contributed to said civilization's development. The *Comte* DE GOBINEAU, a learned and gifted writer[97] published an extensive work in four volumes (1853–1855) entitled '*Essai sur l'inégalité des races humaines*'. He was, however, highly biassed, extolling the blond Nordics and disparaging the Mediterraneans respectively the Celtics. As an aristocrat, he was embittered about the

---

By a strange coincidence, SPINOZA, who could not possibly have been acquainted with that scripture, which at this time was still entirely unknown in the West, states in *Ethices Pars tertia, Propositio XLIII:*

'*Odium reciproco odio augetur, et amori contra deleri potest.*'

In the Christian New Testament, *Matthew* V, 44 likewise reads: 'love your enemies, bless them that curse you, do good to them that hate you'. The Greek original reads: 'ἀγαπᾶτε τοὺς ἐχθροὺς ὑμῶν' *(diligete inimicos vestros)*. 'Αγαπᾶν is, of course, not φιλεῖν nor ἐρᾶν, and *diligere* is not *amare*, but rather *carum habere*. '*Love*' has a great extension correlated with very ambiguous intension. What can here be reasonably meant, is ἀγάπη, *caritas*, namely kindness or goodwill (*Wohlwollen*) respectively understanding. In a more 'Nordic' version, adapted to the spirit of a fighting man (cf. above footnote 70) the gist of the cited axiom could be modified to the well-known exhortation: in war fight without hatred but with firm resolution, in defeat retain defiance, in victory show magnanimity, and in peace promote good will.

[97] *Count* GOBINEAU joined the French diplomatic service in 1849 at the invitation of ALEXIS DE TOCQUEVILLE and resigned in 1877, having successively served, in the course of his career, as ambassador to Persia, Greece, Brazil and Sweden–Norway. He retired to Italy and, two years before his death in 1882, became a close personal friend of RICHARD WAGNER. GOBINEAU's numerous writings of literary value include '*Nouvelles Asiatiques*' (1876) and '*La Renaissance. Scènes historiques*' (1877).

French Revolution, which he considered as a revolt of the 'plebeian' Gauls against the 'noble' Franks. His notion of 'race' was, moreover, quite rudimentary, and based on the old myth of *Shem*, *Ham*, and *Japheth*. Despite its evident imperfections, GOBINEAU's adroitly written work was widely read and understandingly well received by those favoring his thesis. It made a great impression on the composer RICHARD WAGNER, who had a predilection for Nordic mythology.

HOUSTON STEWART CHAMBERLAIN, a Germanized Englishman, who married RICHARD WAGNER's stepdaughter, attempted to trace the evolution of Western civilization in his treatise '*Die Grundlagen des neunzehnten Jahrhunderts*'. Although doubtless a *crank*, he was an erudite writer with some biological background and, with a better understanding of the concept 'race', he pointed out the mythologic nature of GOBINEAU's views on racial origin. CHAMBERLAIN was, nevertheless, no less biassed in overestimating the accomplishments of the Teutonic strain, which he correctly considered a 'racial mixture'. He obstinately claimed Germanic origin for almost all the great men in history, including *Jesus Christ*. He was, moreover, carried away by his irrational antagonism towards the Jews, and, by numerous learned but highly specious and hollow arguments, made a sustained effort to cast aspersion upon and to detract this ethnic group. Although appealing to a prejudiced and naive coterie, CHAMBERLAIN will be considered an unsympathetic and pretentious *crackpot* by impartial and equitable readers.

It is hardly necessary, merely mentioning the Old Testament, moreover JOSEPHUS FLAVIUS, PHILO JUDAEUS, the various noted Rabbis and Jewish physicians tolerated by the Islamic culture, furthermore SPINOZA, MENDELSSOHN, and the numerous more recent eminent scientists such as EDINGER, EINSTEIN, and many others, to point out the outstanding contributions to Western culture by that ethnic group, including its *Askenazic* and *Sephardic* subdivisions.

As regards the multifactorial origins of *antisemitism*, Christian religious prejudice related to the crucifixion of CHRIST (cf. MATTHEW 27: 24–25) doubtless played a significant role, while the strong ethnic emphasis and cohesion of the Jews both in their ancient body politic and in their subsequent Diaspora represented another important factor. The medieval segregation of the Jews and the concomitant mutual antagonism provided a *circulus vitiosus* increasing hostile feelings, and leading to notorious persecutions such as in Spain and later to the pogroms in Czaristic Russia and Russian dominated Poland. Political, economical and social crises provided a fertile ground for antisemitism, motivated by the general Human tendency of looking for a particular scapegoat bearing blame for adversities and calamities.

England expelled Jews in 1290, France in 1306, Spain in 1492, Portugal about 1497 (cf. ADLER and SINGER, *The Jewish Encyclopedia*, 1904). The *Sephardim*, to which SPINOZA belonged, found sanctuary in the Netherlands. In Central Europe, Jews were confined to ghettos. Under CROMWELL (about 1655) Jews returned to England, and in 1858 the first British Jew was elected to Parliament. Ten years later, DISRAELI became Great Britain's Prime Minister.

About 1791 the French Assembly granted French Jews full citizenship, and similar developments followed in various other European countries. Yet, an undercurrent of antisemitism remained, as, *inter alia*, exemplified in France by the *Dreyfus-affair* and by the endless diatribes in the widely read '*Libre Parole*'. Antisemitism was likewise to a certain extent, as in other countries, present in Germany before the advent of the Third Reich, but did not, despite some evident discrimination, prevent Jews from attaining adequate status in elevated social, academic and high official positions, nor from obtaining appointment as Army officers. It may be recalled that numerous Jews were full Professors and Directors of Institutes at Universities in Imperial Germany. Thus, EINSTEIN was made a member of the Royal Prussian Academy of Sciences and of the *Kaiser Wilhelm* (now *Max Planck) Society for the Advancement of Sciences*, of which RICHARD GOLDSCHMIDT was also a member of high standing in the Second Reich and in the Weimar Republic. Under the conditions resulting from the infamous *Versailles Treaty*, virulent antisemitism was revived by the ideology of the fiendish *Pied Piper (Rattenfänger)* HITLER.

My esteemed old friend GEORGE JAFFÉ (1880–1965) who was Ordinarius of Physics at Giessen University and had served with distinction as a Signal Corps officer in World War I, rightly regarded such antisemitism as a bitter injustice *(bitteres Unrecht)*. He considered himself, like R. SEMON and others, an 'assimilated Jew'. Following his dismissal by the *Nazis*, he became Professor of Theoretical Physics at Louisiana State University. At the beginning of his career he had enjoyed the privilege of working in the famed laboratories of OSTWALD in Leipzig, of J. J. THOMSON in Cambridge, and of the CURIES in Paris, about which three laboratories he much later published highly interesting recollections (JAFFÉ, 1952).

Thoughtful comments on the Jewish people and on antisemitism, from the viewpoint of a sensitive and highly aggrieved Jew, can be found in EINSTEIN's '*Ideas and Opinions*' (1959). The eminent Jewish geneticist R. GOLDSCHMIDT, who maintained the genetic aspects of race and the significance of racial traits (cf. above, p. 123), ably vindicated the Jewish cause in his autobiography (1960).

On the other hand, and as an example of the *coincidentia oppositorum*, ASHLEY MONTAGU's publication '*Man's most dangerous myth: the fallacy of race*' (1952) may be pointed out as the counterpart to CHAMBERLAIN's above-mentioned treatise (cf. p. 129). The erudite anthropologist MONTAGU represents, as it were, an '*egalitarian*' *Chamberlain*, comparable to this latter in wishful respectively willful thinking and speciousness of argumentation.[98] Both authors proffer mere assertions as proven facts and can be regarded as pertaining to diametrically opposite but nevertheless comparable domains of the *lunatic fringe*.

The eminent psychiatrist E. KRETSCHMER, in chapter 5 ('Genie und Rasse') of his treatise '*Geniale Menschen*' (1931) justly remarks: '*Das Kapitel der Rassenpsychologie ist bis heute ein besonders unerfreuliches. Nicht deshalb, weil die Angaben der Rassentheoretiker über einzelne seelische Eigenschaften von menschlichen Rassen durchweg unrichtig wären; sondern deshalb, weil sie fast stets so einseitig und tendenziös ausgewählt und beleuchtet werden, dass der entstehende Gesamteindruck ein völliges Zerrbild ergibt. Fast stets sind diese psychologischen Rassebücher so geschrieben, dass der Autor die Verherrlichung seiner eigenen Rasse oder mindestens seiner eigenen politischen Tendenzen oder idealistischen Schwärmereien mit scheinbar wissenschaftlichen Mitteln anstrebt.*'

The European (Caucasian) races which KRETSCHMER recognizes and deals with in his study of geniuses are the 'Nordic', 'Alpine', 'Dinaric', and 'Mediterranean' subvarieties. He is particularly interested in the distribution of schizothymic and cyclothymic traits upon theses races, but concludes that the relevant constitutional types are not, as such, characteristic for any racial subgroup. His iconography of

---

[98] MONTAGU's treatise is somewhat more abundant in invectives than that of CHAMBERLAIN, whom the former characterizes as 'an apostate Englishman glorifying the Teutonic spirit, the German brand in particular, at the expense, among others, of his ancestral land and heritage'. In contradiction to the last part of MONTAGU's statement it will be sufficient to quote CHAMBERLAIN: '*England ist durch seine Insellage so gut wie abgeschnitten; die letzte (nicht sehr zahlreiche) Invasion fand vor 800 Jahren statt, seitdem sind nur einige Tausend Niederländer und einige Tausend Hugenotten hinübergesiedelt (alles Stammverwandte), und so ist die augenblicklich stärkste Rasse Europas gezüchtet worden.*' (vol.I, p.274). '*Germanen im engeren Sinne des Wortes — die Deutschen, die Angelsachsen, die Holländer, die Skandinavier*' (vol.II, p.854).

geniuses and eminent people comprises about 78 portraits, including at least 3 persons of unquestionable Jewish origin (SPINOZA, MENDELSSOHN, HEINE), and 4 women.[99]

Reverting to the valid aspect of GOBINEAU's thesis, namely to the question which 'races' respectively ethnic groups have contributed to the development of Western civilization and, again, to which extent, one could state that said civilization is essentially of Caucasian origin with contributions from all Caucasian subraces. To a substantial degree it is based on Judeo-Christian tradition and Graeco-Roman civilization. This foundation, in turn, owes much to Egyptian, Mesopotamian, Minoan, and other Mediterranean ethnic groups. Moreover, non-Caucasian populations related to the Ural-Altaic language group, such as Magyars and Finns, made additional contributions.

TOYNBEE (1947, 1957) has made a very interesting although controversial attempt at a comparative study of the diverse Human civilizations, their genesis, growth, breakdowns, and disintegrations. Of the 21 civilizations which this author distinguishes, and disregarding details of his concepts, only about seven may be said to be independently original (*'unrelated'*), i.e. aboriginal or *autochthonous*, namely the *Sumeric*, the *Egyptian*, the *Chinese*, apparently the *Minoan*, apparently the *Indic*[100] (precursor of the *Hindu*), the *Andean (Inca)*, and the *Central American (Mayan, Yucatec, Mexic)*. Discounting the two American ones, all subsequent civilizations were 'apparented' to one or more of the five first enumerated 'cultures'.

Some ethnic groups pertaining to the Mongoloid race, e.g. Polynesians, Eskimos, North-American Indians and various South-American Indians remained unrelated, developing 'arrested societies' or 'abortive civilizations' adapted to Oceanic insularity and navigation (Polynesians), Arctic conditions (Eskimos), respectively to nomadic or primitive settled life. Much the same could be said about the Australians, Melanesians, and other ill-defined groups mentioned on p. 117–118).

TOYNBEE who is sceptical about the significance of racial factors for civilization, nevertheless lists the contributions of Caucasians, Mongoloids (including Malays in Indonesia) and Dravidians to the various

---

[99] Cf. his comments concerning women cited on p. 97.

[100] The relationship of the apparently non-Caucasian but not Negroid dark *Dravidian (Tamil)* ethnic group to 'Indic' or Hindu civilization is poorly understood. The Dravidian literature, influenced by Sanskrit, includes the notable poetic and philosophic *Kural* by TIRUVALLUVAR (2nd–3rd century A.D.).

civilizations, and states that the Negroids have not contributed to any.[101]

This would, in some way, agree with the conclusions of JENSEN, SHOCKLEY and others mentioned above on p. 125. ALBERT SCHWEITZER, who devoted himself as a physician to treating African Negroes, had a poor opinion of their mentality, and is reported to have thought 'they should be kept in their place' (cf. BRAIN, 1960, p. 149). Similar, even much stronger views were repeatedly expressed by AUGUST FOREL, who was a liberal humanitarian, pacifist, socialist, an advocate of sane attitudes toward sex, and a champion of Woman's emancipation. On the basis of his experiences in America and North Africa he considered Negroes as '*kulturunfähig*' and '*minderwertig*', emphasizing that '*das muss einmal deutlich und ohne Scheu erklärt werden*'.[102]

Militant Negroes, on the other hand, claim that the impressive stone ruins of *Zimbabwe (Symbahje)* in Rhodesia are proof of an ancient advanced Negro civilization, but it appears most likely that *Zimbabwe* was built by Arab traders, perhaps in the 11th century, and later abandoned. However, reports on these diverse historical events, and on the subsequent various native, e.g. Yoruba, Kwa and other kingdoms which persisted before the full impact of European colonization, seem confused and highly unreliable.

Islamic Arab penetration into East Africa, beginning in the 7th century, was very extensive, and Arabic writing was adopted by various African populations under Arab influence. Thus a Suaheli translation, in Arabic script, of parts of the '*Book of a Thousand Nights and a Night*' is extant, and a few medieval manuscripts, of Suaheli tales and poems, likewise in Arabic script, have been collected by missionaries.

---

[101] 'The Black races alone have not contributed positively to any civilization – as yet' (TOYNBEE, 1947, vol. I, p. 54).

[102] In his autobiography (1935, p. 158) FOREL asks: '*Welche Rassen sind für die Weiterentwicklung der Menschheit brauchbar, und welche nicht? Und wenn die niedrigsten Rassen unbrauchbar sind, wie soll man sie allmählich ausmerzen?*' It is difficult to imagine how this could be humanely accomplished. In the monograph '*Mind and Matter*' (1961, p. 511) I have unequivocally expressed my opposition to any planned 'genetic' or 'eugenic' control. At most, within a given nation, a humane '*genocide*' might be attempted by encouraging intermarriage (hybridization) for the assimilable individuals, and remunerated voluntary infertility for the unassimilable ones, of an 'undesirable race'. Even such very hypothetical special case remains highly fanciful. Who is to draw a clear-cut distinction between desirable and undesirable respectively assimilable and non-assimilable groups? These latter, at most, are only clearly represented by definitely criminal or deviant individuals.

Much the same can be said about the penetration of Mediterranean cultural influence into West Africa, where the Islamitic kingdoms of *Ghana*, *Songhai*, *Mali* or *Mandingo* and others were formed by, or arose from, the impact of Semitic or 'Hamitic' people whose religion and civilization had been imposed on native groups. In the 18th century, about 1730, a native, of Ghana, ANTON WILHELM AMO, studied medicine and philosophy in Germany at the universities Halle, Wittenberg and Jena, and even became a *Privatdozent* (SUCHIER, 1915/16). He is considered by some to be 'the most famous African scholar' of that century.

At *Timbuktu* and some other places Moslem schools or universities with numerous teachers flourished and large libraries contained, *inter alia*, Arabic works of Moslem and of translated Greek philosophers. *Timbuktu* was visited about 1352 by the great Moroccan traveller IBN BATUTA, who wrote a detailed description of that region. Another description was later given, about 1500, by the originally Moslem traveller LEO AFRICANUS *(Alhassan ibn Mohammed Alwazzan)* of Granada, who later became a Christian convert. Following the decline of Islamic power after the expulsion of the Moslems from the Iberian peninsula, their cultural expansion into West Africa likewise gradually decayed. Having been irritated as well as to some extent also abused by European colonization, and, particularly in America severely aggrieved by slavery[103] and, at least in the South, by wrong treatment following emancipation, a strong and militant reaction of the Negroid ethnic groups, demanding equality and even preferred status, has developed, favored by the present decay of Western civilization following the two recent World Wars (in this respect comparable to the Peloponesian Wars).

It should, however, also be emphasized that, discounting minor and abortive Negro-slave insurrections in Latin America as well as some aspects of the turbulent history of Haiti, the emancipation of Negroes from slavery was entirely accomplished by the efforts of 19th century white abolitionists (mainly British, American and French) and not by any efforts of the blacks. American black 'racists' seem to

[103] It should be recalled that African tribal chiefs and Arabic slave traders were as much responsible for the institution of Negro slavery as the Westerners. This, of course, does not excuse the misdeeds of the latter. One could here say, with SCHILLER:

'*Das eben ist der Fluch der bösen Tat,*
Dass sie, fortzeugend, immer Böses muss gebären.'
(*Wallenstein, Die Piccolomini* V, 1).

forget this fact and blame the subsequent difficulties of cultural assimilation exclusively on the whites. Black antagonism to Western culture is, *inter alia*, expressed by assuming spurious Moslem names such as *Muhammad Ali* etc., and by the foundation of mutually conflicting '*Black Muslim*', '*Hanafi Moslem*' and similar groups, disregarding the fact that the Moslem Arabs were original and predominant slave dealers and that Negro slavery persisted in diverse Arab countries almost to the present time.

Concerning the United States, where the census of 1970 recorded about 22 500 000 'Negroes', i.e. somewhat more than 10 per cent of the population, relevant publications on this problem are those by HERSKOVITS (1928), DRAKE and CAYTON (1945), and MYRDAL (1944, 1962). DRAKE and CAYTON, who are educated Negroes, ably present their cause, which is likewise championed by MYRDAL, a learned but somewhat biassed Swedish sociologist. As regards the overall biology of the Negro, with particular emphasis on conditions in the United States, a scholarly publication has been prepared by LEWIS (1942).

Among other racial minorities in the United States are approximately about 790 000 Indians, and about one million Orientals (mainly Chinese and Japanese). The Mongoloid Orientals, presumably better endowed than the Negroids, have overcome discrimination and satisfactorily established themselves.[104] The Indian problem remains, but, because of the lesser size of this ethnic group, its impact ist not comparable to that of the Negro problem.

As regards this latter, and in contradistinction to the view of FOREL and similar others, it seems evident that the racially very mixed Negroid ethnic group includes numerous well educated, intelligent, capable, respectable and honorable persons, far preferable to some Caucasian ones, and regardless of the degree of admixture of 'Caucasian genes', which widely obtains in this population. Quite arbitrarily assuming for such non-measurable traits *Gaussian curves*, one could say that those for Negroids and Caucasians overlap but do not coincide, such that the right end of the left former would not reach that of the right latter, being thereby in agreement with the findings of JENSEN or SHOCKLEY and with the fact that even the most creditable accomplishments of the Negroids who have reached eminence do not reach the level of 'genius' and are achieved by relatively less individuals than in

---

[104] This can also be said about disfavoured Caucasian minorities pertaining to various ethnic groups including, in particular, the Jewish one.

the Caucasoid and Mongoloid groups. Except for the writings of educated Negroes more or less assimilated to Western civilization and for the above-mentioned Arabian-inspired haphazard African writings, it can be maintained that there is no Negro literature. There is, of course, a diversified autochthonous African Negro art, admired by some, while others consider it 'primitive' or 'barbarous' (cf. the comments in chapter VI, section 8, pp. 749–750, footnotes 335, 336, vol. 3/II).

Again, contrary to the claims of those favoring extreme 'racial' views, one could maintain that no 'race', as such, is 'unassimilable' to Western civilization. The personal qualities of an individual are far more important than the race to which that person pertains, and educable individuals[105] can be assumed to obtain in all racial subdivisions of the Human species. *Per contra*, 'unassimilable' individuals can also be found in all races. Thus, the numerous ones pertaining to the *Juke* and *Kallikak* families, mentioned above (p. 98) were of native-born Caucasian American stock.

Likewise, as regards hybridization, and as pointed out by GOLDSCHMIDT (1927) and others, there is no foundation for the belief that, in the case of racial mixtures always only the 'undesirable' racial traits of both parents are transmitted to their offspring, respectively that human race hybrids are 'inferior to both of their parents and somehow constitutionally unbalanced. The outcome of hybridization, 'desirable' or 'undesirable' as the case may be, must be assumed to depend upon the individual genomes of the parents and their chance combinations.

Two conspicuous examples of Caucasian-Negroid hybrids attaining eminence are the DUMAS family and PUSHKIN. ALEXANDRE DAVY DUMAS (1762–1806), who became a competent French general, was the son of a French nobleman and a Dominican Negress. The general's son, in turn, was ALEXANDRE DUMAS *Père* (1802–1870) the famed novelist, author of the '*Comte de Monte Christo*' and many other works. ALEXANDRE DUMAS *Fils* (1824–1895), son of the novelist, also became a noted dramatist, author, among other works, of '*La Dame aux Camélias*' (adapted by GIUSEPPE VERDI as '*La Traviata*').

ALEXANDER PUSHKIN (1799–1837) was the great grandson of a Negro boy presented to the Czar PETER I (1682–1725) and later allow-

[105] The *Analects* of CONFUCIUS contain, in Book XV, chapter 38, a terse saying consisting of 6 ideograms:

Tzu-yüeh yu-chiao wu-lei.

It can be translated to mean:

The Master said: 'where there is education, there are no races.'

ed to marry into Russian aristocracy. PUSHKIN, author of '*Eugène Onegin*' and other works, is considered one of the greatest Russian poets, also ranking high among the eminent personalities of world literature. He is apparently the only Negro hybrid reaching the rank of '*genius*'.

Studies on Caucasian-Negroid (Hottentot) hybrids, the so-called *Rehoboth bastards* in Southwest Africa (FISCHER, 1913; cf. also SALLER, 1930) likewise may be interpreted to indicate that such hybridization can produce offspring quite devoid of 'unfavorable' traits.[106]

It seems, moreover, apparent that a very substantial part of the North-American so-called 'Negro' or 'black' population consists of Negroid-Caucasoid hybrids in diverse proportions. Many such hybrids provide the black 'intelligentsia' or leadership. In the United States all Negroid hybrids commonly are considered to be 'Negroes' or 'blacks'. Yet, among the most vocal black leaders there are not few individuals who, even by 'racial standards' comparable to those postulated by the notorious '*Nuremberg laws*', could evidently be classified as 'Caucasoids' respectively 'whites'.

Statistically, however, there may be more 'unassimilable' individuals among the Negroids than among some other 'racial groups' or 'genetic pools'. The conspicuous foci of decay displayed by the '*Negro-Jungles*' (wrongly called '*Ghettos*') of large American cities, combined with extreme criminality, although explained by '*egalitarians*' as exclusively due to '*discrimination*' and poverty, may well be, to a very substantial extent, the result of 'genetic factors'. In addition to excessive criminality, a lesser sense of responsibility is attributed to Negroes as a generalized 'group' by many Americans on the basis of their unfavorable experiences in living side by side with said ethnic group. The effect of the resulting attitudes, partly based on undeniable facts, and partly on 'social prejudice', is a 'vicious circle' intensifying mutual antagonism.

TOYNBEE, as cited above on p. 132 in dealing with genesis and disintegration of Human civilizations or societies, made an attempt at their systematic classification. SPENGLER (1919, 1927), who likewise distin-

---

[106] Similar studies, concerning, however, Caucasian-Mongoloid (Malayan) hybrids are those by RODENWALDT (1927) on the so-called *mestizos of Kisar*, also quoted by SALLER (1930). The cited studies dealt primarily with the multifactorial aspect of the inheritance of physical racial traits, but support the view that such hybridization, *per se*, seems quite unobjectionable with respect to 'mental traits' whose inheritance, despite some general 'racial predispositions' as e.g. considered by KRETSCHMER (1931) in Caucasians, essentially depends on the individual 'genomes' of the parents.

guished diverse cultures, propounded a '*morphology*' of history, endeavoring to define '*homologies*' in the various stages of developing and evolving civilizations. Like HUME (cf. above, p. 85) and others he assumed that civilizations passed through the same succession of stages as organisms like plants or human beings, manifesting, as it were, infancy, youth, manhood, old age, and decay.

TOYNBEE (1947, 1957), takes exception to this concept, insisting, with various arguments, that societies are not in any sense living organisms, and that the course of history is not predetermined. He detracts SPENGLER's 'dogmatic and undocumented determinism', and, with some justification, characterizes the said author as 'pontifical', but conveniently forgets his own pontifical attitude. One is here reminded of ST. MATTHEW 7, 3: 'Why beholdest thou the mote that is in thy brother's eye, but considerest not the beam that is in thine own eye'. A contemporary Oxford historian remarked that 'God had found a new prophet, and his name was ARNOLD TOYNBEE'.

Although societies or civilizations are indeed abstractions, namely semantic, logical entities without 'material' ('physical') existence, but fictions with logical existence, they nevertheless represent, as interacting arrays of individuals, biological systems which essentially depend upon *brain functions*.

From a mathematical viewpoint, all systems, be they non-living mechanical or biologic, can be analyzed in terms of their observable properties, and invariants respectively analogies or even homologies may be defined.[107] Despite its numerous weaknesses, the vagueness and ambiguities of its emotional mystical concepts of 'form' and various rather questionable interpretations, SPENGLER's scholarly work contains many highly interesting opinions which are perhaps better supported than some of TOYNBEE's views. Diverse aspects of present-day developments appear to be in full conformity with SPENGLER's main thesis, the *decline of the West*.

Both TOYNBEE and SPENGLER lack an adequate biologic background, and discount the genetic significance of 'race'. SPENGLER cryptically elaborates on soul and landscape *(Seele und Landschaft)*, claiming that the relevant 'fundamental experiences' *(grundlegende Bestimmungen)* '*nicht mehr im Bereiche der Mitteilbarkeit durch Begriff, Definition und Be-*

---

[107] Thus, RASHEVSKY (1968) and others have shown that, to some extent, it is possible to 'mathematize' history (cf. also the critique, by FOGEL, of RASHEVSKY's treatise in Science *163:* 666, 1969).

weis liegen, dass sie vielmehr ihrer tiefsten Bedeutung nach gefühlt, erlebt, erschaut werden müssen'.

TOYNBEE's logic is at times rather weak. He may be right in claiming that the study of the course of history does not succeed in formulating rigid 'laws'[107a] and exact predictions as to future developments, but his arguments against fate and determinism are quite unconvincing. He confuses uncertainty or unpredictability with indeterminism. Of the cited two authors, SPENGLER can perhaps be considered the more brillant one.

Although Human societies, civilizations, and ethnic groups are less numerous than sets of biologic organisms such as species, genera, orders, and classes,[108] their diversity is very substantial, and their highly complex characteristics are less well definable than those in the taxonomy of organisms.

As regards comparisons, such as attempted by SPENGLER, it becomes necessary to be aware of the considerable differences in life durations of individual organisms, from a few days in certain Insect imagines to about a millenium in Sequoia trees. Again, some species respectively orders persisted relatively unchanged for long periods, while others became extinct (e.g. many Reptiles, Birds, etc.). As in the case of Human 'cultures', new plant and animal forms emerged in the course of organic evolution, whose future progression seems quite unpredictable.

Per analogiam, some Human societies displayed stability of long duration, others decayed or disintegrated, and new civilizations evolved through changes in persisting ethnic groups. In these diversified events, 'invariants' become detectable,[109] some of which SPENGLER

[107a] It will be recalled that HEGEL (1770–1831), in his posthumous '*Vorlesungen über die Philosophie der Geschichte*' claimed that a divine 'Idea' manifested itself in history: '*Dass die Weltgeschichte dieser Entwicklungsgang und das wirkliche Werden des Geistes ist, unter dem wechselnden Schauspiele ihrer Geschichten, dies ist die wahrhafte Theodizee, die Rechtfertigung Gottes in der Geschichte.*'

[108] In accordance with TOYNBEE's enumeration (cf. above, p. 132) the number of civilizations would, however, roughly correspond to the number of animal phyla pointed out in dealing with taxonomy (chapter II, section 2, pp. 52f., vol. 1).

[109] This may be interpreted as the meaning of the old dictum: '*non nova sed nove*'. In a French version it is expressed as: '*plus ça change, plus c'est la même chose*'. GIORDANO BRUNO, about 1588 in *Wittenberg*, stated in a comment:

'*Salomon et Pythagoras:*
Quid est quod est?
Ipsum quod fuit
Quid est quod fuit?
Ipsum quod est
Nil sub sole novum.

conceives as '*homologies*', but which, from a more rigorous viewpoint, mostly represent mere *analogies*. Many far fetched 'homologies' of this author could be likened to an attempt at comparing e.g. the morphology of Vertebrates with that of quite different Invertebrates in terms of 'homologies'.

Roughly speaking, accomplishments of a civilization are displayed by literature, art, religion, social and military organization, science and technology. Taking as criteria several or all of these sorts of accomplishments in diverse combinations, one could, by arbitrary axiologic evaluation, rate some civilizations as superior, others as high, a few as pioneering, and still others as inferior.

Pioneering civilizations are doubtless the Sumerian, the Egyptian, the Chinese, and the Central American, which latter displayed pronounced schizophrenic features and was particularly obsessed with bloody Human sacrifices. Mayan and Aztec religious ceremonies required yearly thousands of victims, whose still beating hearts were torn out after cutting open the chest with an obsidian knife. H. G. WELLS (1949)[110] designates these American civilizations as aberrant, and justly points out that many Mayan inscriptions resemble a sort of elaborate drawings made by lunatics in European asylums.[111]

Among the *superior civilizations* one could include the classical Greek, the Roman, the Chinese and its Japanese derivative from about 600 A.D. to 1867, the Hindu, and the Western civilization subsequently to the Renaissance, a civilization also adopted by Japan after 1867.

The Greek civilization flourished in city-states such as Athens, Thebes, and the Ionian cities of Asia Minor. Roman civilization, influenced by the Greek one, evolved as that of a dominant single Imperium.[112] The superior Chinese civilisation evolved directly from a pioneer-

---

[110] This work, by a noted novelist and non-professional historian, displays the particular prejudices and emotional valuations of English lower middle class mentality as obtaining during the period before midcentury. It contains a strange mixture of shrewd, well-balanced judgements and crudely biassed, unduly warped interpretations. Yet, it is a remarkable book by a gifted author, and presents a useful summary of main events in history, in some respects even superior to the writings of professional, academic historians.

[111] Concerning the art of the insane, cf. the comments on pp. 750–751, footnote 336 in chapter VI, section 8 of volume 3/II.

[112] Interesting and diversified, but to some extent unconvincing or inconclusive elaborations on Graeco-Roman civilization respectively its decay are expressed by FUSTEL DE COULANGES (1830–1889; *La cité antique)*, by MONTESQUIEU (1689–1753; *Considérations*

ing one, and maintained itself, through various political vicissitudes and fluctuating periods of decline until its final decay under the Manchu.[112a] It displayed several flowering periods, in the *Chou* era around 500–300 B.C. (LAOTSE, CONFUCIUS, MENCIUS), in the *Han* era (206 B.C. to 220 A.D.), which toward its end was perhaps the highest civilization existing on our planet at that particular time, and became, through the introduction of *Buddhism*, about 67 A.D., substantially influenced by Hindu civilization. Another flourishing period of Chinese civilization occurred in the *Tang* era (618 to 907 A.D.). The Japanese, a barbarian island population, adopted Chinese culture since about 600 A.D. and evolved, upon this basis, their own civilization which reached its final stage in the *Tokugawa* period (1600–1867 A.D.). There is an evident analogy in the relationship of Chinese to Japanese civilization and that of Graeco-Roman to Western civilization, which latter was likewise developed, although by several ethnic subgroups, among barbarians.

Hindu civilization, with numerous vicissitudes, developed among diverse frequently warring kingdoms and oligarchic states, as e.g. at the time of the BUDDHA (about 500 B.C.), became influenced by Greece following ALEXANDER's expedition of conquest, and was partly united in ASOKA's empire (274–236 B.C.). Its further development was recorded by the great Chinese Buddhist pilgrims FA-HIAN (FA-HSIEN; travelled 399–414 A.D.), HUI-SHENG and SUNG-YÜN (travelled 518–522), HSÜAN-CHUANG (HSÜAN-TSANG; travelled 629–645), and I-CHING (travelled 671–690). At FA-HIAN's time, there were numerous independent Buddhist kingdoms, and, upon his return, FA-HIAN arose some antagonism at home by extolling the conditions prevailing in Indian countries, which he called *Chung-Kuo*, a term conventionally restricted to China (the 'Central Kingdom'), and which he described as comparing advantageously to his less favored homeland.[113] I-CHING's report

sur les causes de la grandeur des Romains et de leur décadence), by GIBBON (1737–1794; *The decline and fall of the Roman empire)*, and by the more recent historian O. SEECK *(Geschichte des Untergangs der antiken Welt*, in 6 volumes 1895–1921).

[112a] It is of interest that the Mongoloid *Manchus*, who conquered China about 1644 and imposed upon the Chinese the wearing of the Manchu *queue* as a sign of submission, nevertheless completely adopted Chinese civilization. Previously, this latter was also completely adopted by the Mongols who overthrew the *Sung* dynasty under KUBLAI KHAN in 1280. The *Yüan* dynasty lasted until replaced, in 1368, by the cultured indigenous Chinese *Ming* dynasty.

[113] FA-HIAN lived later than the *Han*, and before the *Tang* period, at a time of division between north and south China.

over 250 years later, disclosed, on the other hand, evidence of decay in India.

In the 2nd century of the Christian era, the influence of Hindu civilization had also extended eastward to Cambodia, Champa (Annam) and other regions of the Indo-Chinese peninsula, and to the Indonesian Archipelago (e.g. Sumatra and Java). The ruins of *Angkor Vat* in Cambodia, and of *Borobodur* in Java, as well as the Hinduism of Bali, are relics pertaining to various subsequent epochs of that cultural expansion. The decline of Hindu civilization was accelerated by the Islamic conquests.[114]

Western civilization, perhaps substantially influenced, following the fall of the Byzantine empire in 1453, by an influx of refugees bringing Greek learning, was developed, slowly beginning approximatly in the 15th century, by a diversity of national groups, extending, as it were, from Italy and the Iberian peninsula to Western, Central, Northern, and Eastern Europe.

Portugal, a small country whose people numbered little more than one million, and whose literary accomplishments culminated with CAMÕES (1524–1580), spread its power over the oceans, opening, since 1487, the way to South and East Africa as well as to India and the Far East. Since 1492, the Spaniards expanded their power to the American continent. In 1493, the unsavory Pope ALEXANDER VI (BORGIA) decreed his '*Line of Demarcation*', readjusted in 1494 by the *Treaty of Tordesillas*, as a transpolar great circle passing 370 leagues west of the Cape Verde Islands, and dividing up the non-Christian world between the Spanish to the west, and the Portuguese to the east of said demarcation line.

In 1580 Portugal fell to Spain, and did not recover all of its possessions with the recovery of independence in 1640. The sea-power of Spain was challenged by Holland and by England, which latter, overcoming both Spain and Holland, finally became, as Great Britain, and despite the successful emancipation of its North American colonies, the greatest imperial power of Western civilization before the present decline of the West.

Before World War I there were, in addition to the British Empire, seven great powers *(Grossmächte)*, namely the United States, Germany, France, Russia, Japan, and, rather more nominally than actually,

---

[114] Although the diverse sorts of *Islamic* civilization may be considered 'high', one could, rightly or wrongly, rate them as below the grade of 'superior'.

the Danube Monarchy and Italy. All of these, including, despite its Oriental tradition, Westernized Japan, pertained to Western civilization.

Concerning the decline of Spain, it is of interest that CAJAL, a patriot who was keenly aware of his country's plight and of the concomitant '*atraso cientifico*', devotes an entire chapter of his '*Reglas y consejos sobre investigación cientifica* (6th ed., 1923) to this topic. Since the 16th century, '*la cima de nuestra intelectualidad*', Spain is said to have become '*un pais intelectualmente atrasade, no decadente*'.

CAJAL elaborates on six different sorts of environmental and sociologic factors generally assumed to have played a role, and designated as '*explicaciones fisicas, historicas y morales:* (1) *hipotesis termica*, (2) *teoria oligohidrica*, (3) *teoria economico-politica*, (4) *teoria del fanatismo religioso*, (5) *hipotesis del orgullo y arrogancia españoles*, (6) *teoria de la segregación intelectual*.' Strangely enough, CAJAL, himself a biologist, does not take into consideration a possible genetic deterioration, caused by reshuffling of the genes, mutations, and additional unknown genetic mechanisms as well as by the numerous negative aspects of selection within the given genetic pool.

With diversified and different manifestations, the multifactorial decline of Spain[115] deplored by CAJAL, who wishfully assumed backwardness rather than decadence, has been followed by a decline gradually involving all other Western national or ethnic groups, partly concomitant, however, with an outburst of scientific and technologic activity whose chaotic results recall the Biblical parable of the Tower of Babel.

According to SPENGLER, this stage is characterized by the '*miracle of the Cosmopolis*': '*das Wunder der Weltstadt, das grosse steinerne Sinnbild des Formlosen, ungeheuer, prachtvoll, im Übermut sich dehnend. Sie zieht die Daseinsströme des ohnmächtigen Landes in sich hinein, Menschenmassen, die wie Dünen aus einer in die andere verweht werden, wie loser Sand zwischen den Steinen verrieseln. Hier feiern Geist und Geld ihre höchsten und letzten Siege. Es ist das Künstlichste und Feinste, was in der Lichtwelt des menschlichen*

[115] This decline was likewise, as it were *ab initio*, manifested by Latin America. It will here be sufficient to point out CAJAL's relevant remark (1923, p. 249): '*continuamos a la zaga de las pequeñas nacionalidades del Norte de Europa. Pueblos hermanos como Portugal y las Repúblicas sudamericanas, donde la despreocupación dogmática es acaso mayor que entre nosotros, viven, sobre poco mas o menos, en el mismo plano cultural*'. CAJAL's mention of '*despreocupación dogmática*' refers to his evaluation of factor (4), namely religious fanaticism.

Auges erscheint, etwas Unheimliches und Unwahrscheinliches, das fast schon jenseits der Möglichkeiten kosmischer Gestaltung steht.'[116]

The democratic ideals of equality for all, natural rights, freedom of opinion, and universal suffrage are considered illusions by the cited author. Freedom of public opinion involves manipulation ot that opinion, requiring money. Freedom of the press depends on possession of the press, again requiring money. Universal franchise involves electioneering with its propaganda, depending on the wishes of those who provide the required exorbitant financial backing.

Thus, in the form of so-called democracy, money has won. There is a period in which politics are almost its preserve. But when the old orders of the culture have been destroyed, then, according to SPENGLER, the chaos gives forth a new and overpowering factor in the form of '*Caesarism*' which brings to an end the politics of money and intellect. By '*Caesarism*' SPENGLER means any kind of totalitarian government which, irrespective of any constitutional formulations that it may have, is in its essential nature what he calls 'thoroughly formless'.

The coming of *Caesarism* breaks the dictature of money and its political weapon democracy. The machine with its human retinue, the real queen of this century, succumbs to a higher power. But with this, money, too, is at the end of its success, and the last conflict is at hand, through which civilization receives a new form, the conflict between money and blood. In *Spengler's* words (1927): '*So schliesst das Schauspiel einer hohen Kultur, diese ganze wundervolle Welt von Gottheiten, Künsten, Gedanken, Schlachten, Städten wieder mit den Urtatsachen des ewigen Blutes.*'

'*Die Zeit ist es, deren unerbittlicher Gang den flüchtigen Zufall Kultur auf diesem Planeten in den Zufall Mensch einbettet, eine Form, in welcher der Zufall Leben eine Zeitlang dahinströmt, während in der Lichtwelt unserer Augen sich dahinter die strömenden Horizonte der Weltgeschichte und der Sterngeschichte auftun.*'

Be that as it may, SPENGLER, TOYNBEE and other interpreters of history fail to take due cognizance of the predominant role played by the *mechanisms of the Human brain.* Since my early contacts with psychiatry under HANS BERGER, in association with my fellow student, colleague and friend, the late E. VON DOMARUS (1893–1958), who was engaged in a penetrating study of *schizophrenic logic and thought*, I be-

[116] One may wonder how SPENGLER would have been impressed by the shoddiness, filth, litter, grafitti and decay of diverse American metropolitan city districts, and with the foul stench of their subways.

came impressed with what, for lack of a better word, I would call the schizophrenic trends or aspects representing a presumably intrinsic, and to a substantial degree genetically determined component of the 'normal' Human mind.

These trends, I believe, have been an important and much underestimated factor in historical events and, particularly, in the fluctuations displayed by evolution and decay of civilizations. Said factor may perhaps, together with other relevant ones such as overpopulation and the numerous negative aspects of selection, lead to decline or even final extinction of Humanity.[117] In his '*Generelle Morphologie der Organismen*' HAECKEL (1866) emphasized that both in ontogeny and in phylogeny of the organisms, three stages can be recognized, namely, *epacme*, *acme*, and *paracme*, corresponding to development, modification and degeneration *(anaplasia, metaplasia, cataplasia; Aufbildung, Umbildung, Rückbildung)*. This, as HUME (cf. above, p. 85) already anticipated, presumably applies not only to individual organisms and botanical or zoological taxonomic groups, but also to the peculiar Human species, which seems to have entered its paracme, and indeed faces an ominous future.[118] Paraphrasing an old quip, one could either grimly say that the situation is very serious but not absolutely hopeless, or, with resigned insouciance, that it is quite hopeless but not serious.

Apparently confirming SPENGLER's anticipation of what he called the late '*Caesarism*', the emergence of a new sort of imperialistic societies, based on the *Marxist religion*, seems now to become manifested. Of the above-mentioned eight great powers previously to World War I, only two have retained that status, and a new third one emerges. These three imperialistic powers are (1) the United States, weakened by internal dissensions and widespread corruption, but nevertheless a last although shaky and corroded stronghold of the decaying West, (2)

---

[117] The tentative hypothesis of a *progressive schizophrenic degeneration of the human brain* was pointed out in volume 3/II, chapter VI, section 8, pp. 749–751. It was also suggested in previous publications of the author (K., 1961, p. 511; 1966, p. 323; 1973, p. 366).

[118] Some of the sociologic changes concomitant with the decay of Western civilization and the accelerating overpopulation have been dealt with, in a sensational 'best-seller' type of publication, and, as it were, in '*Hollywood-style*' by a journalistic writer (TOFFLER, A: *Future shock*. Random House, New York 1970, 1971). Said publication, although concerned with some of the obtaining actual topics, is characterized by exaggerations, half-truths, pretentious superficial sciolism, and fanciful interpretations. It cannot be taken seriously but is well adapted to the uneducated public at large, whose mentality is fed by the so-called '*mass-media*'.

the seemingly potent Soviet Union, and (3) the apparently consolidating, potentially highly powerful People's Republic of China.

The mutually antagonistic aims, ambitions or policies of these three great powers are restrained by fear of each other, and, while the threat of nuclear war hangs like a *Sword of Damocles* over the world, the lesser nations[118a] or ethnic groups variously aligned with or at least somehow dependent upon the one or the other of the three present great powers, retain an uncertain measure of free play or illusory independence, except for those already directly incorporated into the two Marxist Imperia. These new forms of society, respectively of civilization, will be welcomed by some, and deprecated by others. Increasing regimentation is not only characteristic for the totalitarian societies but likewise manifests itself in the still surviving so-called democratic ones.

In Western 'democratic' societies, nevertheless, the individual generally still retains a substantial degree of personal liberty, including freedom of speech, of residence, and of foreign travel. Individualists will therefore prefer such forms of government, representative of the so-called *'free world'*, despite their manifold increasingly conspicuous shortcomings. Totalitarian societies of the late '*Caesarism*' on the other hand, prevent, *inter alia*, free exit, thereby clearly indicating that their subjected citizens are *de facto* enslaved captives.

The stability or durability of these totalitarian societies depends, of course, on the unpredictable outcome of extrinsic and intrinsic events, which latter include internal strife within the ruling power elites. An essential intrinsic requirement for stability is the thorough indoctrination of the subjected citizens. Non-converted ones, dissidents, or undesirables are restrained, terrorized, imprisoned and abused (cf. e.g. the well-known writings of MILOVAN DJILAS and of SOLZHENITSYN), as well as confined in slave-labor camps or exterminated. The number of Human lives destroyed by the internal measures of totalitarian

---

[118a] These lesser nations include, among others, the fallen former great powers and the multitudinous barbarous new nations hatched from the disintegration of said powers' colonial empires. One could, arbitrarily, distinguish second, third, fourth class (etc.) nations. Barbarous or decayed nations are now euphemistically or hypocritically called 'developing countries' *('Entwicklungsländer')*. Most of them do indeed 'develop' by uninhibited overpopulation. Some of the lesser *'powers'* of Western civilizationn nevertheless remain industrially rather strong, notably Japan, West Germany, France, and the United Kingdom. Yet, this industrial 'power' remains very shaky, and, discounting various additional unpredictable circumstances, entirely depends on the protection still provided by the United States.

governments is estimated to total many millions and presumably exceeds the unknown but very large total number of live victims sacrificed in the religious rites of ancient Middle American societies throughout the duration of their 'flowering'.

As regards stability, it will be recalled that, because of extrinsic events, HITLER's '*Tausendjähriges Reich*' collapsed within 12 years, and MUSSOLINI's fascist state lasted from 1922 to 1943. The Soviet Union, established in 1917, appears, after more than 50 years, unshaken and is perhaps the strongest of the now extant three great powers. The United States not only saved it from destruction in World War II, but, until a reversal of policy took place, significantly abetted and furthered Soviet expansion at that war's termination. Russia's design seems to be an ultimate imperial *pax Sovietica* encompassing the entire globe, comparable, in some respects, to the imperial *pax Romana*.[118b]

China, in contradistinction to Japan, was never culturally westernized to a relevant degree. As one of the original pioneering civilizations, it developed a fluctuating, interrupted sequence of superior cultures, with final decay in the *Manchu* era. This was followed by the present *Spenglerian* '*late Caesarism*' of the People's Republic of China, which, consolidated since 1949, seems to show signs of growing strength.

Because the *brain* and its functions play a highly relevant role in the manifestations of organic life pertaining to the purview of 'anthropology' and history, the foregoing lengthy discussion of some apparently remote, but definitely apposite topics was deemed appropriate.

The eminent neurobiologist AUGUST FOREL emphasizes this fact in his autobiography (1935, p. 155). He states: '*ja, das Gehirn ist der Mensch im Menschen, und seine Krankheiten bedeuten demnach alle Störungen des inneren Menschen in seinem Denken, Fühlen und Handeln. Was sind Theologie, Recht, Wissenschaft, Kunst, Phantasie, Krieg, Lust, Unlust, Hass, Liebe in unserer ganzen Kulturgeschichte, wenn nicht individuelle und soziale*

[118b] '*Raptores orbis, postquam cuncta vastantibus defuere terrae, jam et mare scrutantur; si locuples hostis est, avari, si pauper, ambitiosi, quos non Oriens, non Occidens satiaverit: soli omnium opes atque inopiam pari adfectu concupiscunt. Auferre trucidare rapere falsis nominibus imperium, atque ubi solutudinem faciunt, pacem appellant.*' (TACITUS, *Agricola* 30).

KANT's commendable views, ably propounded in his treatise '*Zum ewigen Frieden*' (1795) are, *rebus sic stantibus*, obviously utopian. It is hardly necessary to add that the defunct '*League of Nations*' and the contemporary '*United Nations*' represent hypocritical and preposterous organizations *ab initio* destined to complete failure.

Vorgänge unseres Gehirnlebens? Und wie viele dieser Vorgänge erweisen sich nicht als pathologisch?'

LASSEK (1957), an anatomist and neurobiologist, noted for his studies on the pyramidal tract (cf. e.g. vol. 4, chapter VIII), dealt in an entire volume, '*The human brain from primitive to modern*', with diverse aspects of the overall topic discussed in section 5 of the present chapter.[119]

It should be understood that, within this context, only the general outlines of the multitudinous obtaining problems could be intimated. From a viewpoint of scepticism, and paraphrasing DAVID HUME, one might say that the alluded questions admit of no conclusive or convincing answers. The only effect of the thereto pertaining arguments is 'to cause that momentary amazement and irresolution and confusion, which is the result of scepticism'.

6. Concluding Remarks on the General Evaluation of Mammalian Telencephalic Surface Patterns

Reverting now to the relevant gross anatomical features of the Mammalian telencephalon, manifested by the *fissuration pattern*, it seems evident that four *primitive grooves*, namely sulcus rhinalis lateralis with its anterior and posterior portion, sulcus endorhinalis, likewise with anterior and posterior portion, sulcus hippocampi, and sulcus limitans hippocampi are definitely related to longitudinal cortical zones of *grundbestandteil* value. Of these grooves, the first indicates the boundary between neocortex and piriform lobe cortex, the second runs along the boundary between that latter cortex and the basal cortex. The third groove corresponds to the transition of hippocampal formation into parahippocampal cortex. In the paraterminal region,

[119] The interested reader, comparing LASSEK's treatise with the foregoing discussion by the present author will note various differences in our selection, presentation and interpretation of problems pertaining to this extensive subject matter, which is also considered in my monographs '*Brain and Consciousness*' (1957), '*Gehirn und Bewusstsein*' (1973), '*Mind and Matter*' (1961), and in several papers (K., 1965a, b, 1966, 1968, 1972). Other publications by neurobiologists dealing with some aspects of this topic are those by SHERRINGTON (1951), HERRICK (1956), BURR (1960, 1962), and ECCLES (1970). In his reflexions of an octogenarian (*El mundo visto a los ochenta años*, 1934) and in his coffee-house aphorisms (*Charlas de café*, 1944), CAJAL likewise included a variety of comments on the themes under discussion in the present section.

the fourth groove demarcates hippocampal formation from the basimedial grisea.

A fifth groove, which can be included into the group of primitive sulci, indicates the transition of parahippocampal cortex into neocortex *sensu strictiori*. In the brains of various lissencephalic Mammals, said groove may be rather faintly developed, but in gyrencephalic 'intermediate' and 'higher' Mammals, becomes the conspicuous *genualis-splenialis system*,[120] respectively the *sulcus calloso-marginalis* approximately delim iting the limbic lobe from the neopallium *sensu strictiori*.

The strictly constant and close relationship of the first four abovementioned sulci to well-defined major cortical regions is obvious, and there can be no doubt about their homologization in the diverse Mammalian forms.

Concerning the fifth sulcus, its overall interpretation as a limiting groove of the limbic lobe can be upheld, but the relationship of this sulcus and of its further expansions into a system seems somewhat less constant, respectively easily definable throughout the Mammalian series because of some uncertainties in clearly delimiting neocortex *sensu strictiori* from parahippocampal cortex in the diverse taxonomic groups.[121]

A groove *sui generis* is the *Sylvian sulcus* and its further expansion, related to the hemispheric bend, whereby the insula becomes opercularized. The primitive Sylvian vallecula, the pseudosylvian sulcus and the deep Sylvian fissure (fissura cerebri lateralis) of large Anthropoids and Man can be evaluated as being kathomologous.[122]

A standardized procedure of distinction between the terms 'sulcus' and 'fissure', despite some attempts, has not been consistently applied in the pertinent literature. Yet, it seems appropriate to use the term 'fissure' for the complex Sylvian system of 'intermediate' Mammals and of the Anthropoid Primates. There are, moreover, two rather deep fissures of a different type, namely the *fissura longitudinalis cerebri*, separating the two hemispheres, and the *fissura transversa cerebri*. Part of the 'bottom' of the former is provided by corpus callosum respectively, in

[120] The sulcus splenialis is occasionally also called sulcus marginalis (cf. Figs. 44, 46).

[121] As regards the extent of the parahippocampal cortex respectively the distinction of neocortex (isocortex) and allocortex, cf. the treatise by STEPHAN (1975). Relevant questions of interpretation and terminology will be dealt with further below in chapter XV.

[122] The topologic invariance of the *Sylvian system* is self-evident upon topologic mapping, despite the shifting relationships of its boundary neighborhoods to cortical areas.

various Aplacentalia, by the commissura (dorsalis) pallii ('hippocampi'). The fissura transversa cerebri is formed by the caudal extension of the hemisphere dorsally to diencephalon and basally to corpus callosum with adjacent fornix. In forms with large hemispheres this fissure extends above mesencephalic tectum and cerebellum. Between corpus callosum complex and diencephalic roof, it contains the tela chorioidea.

In addition to the primitive grooves, which may be said to represent telencephalic sulci of the first order or *primary sulci*, and to the Sylvian system, the gyrencephalic Mammalian brains display rather constant neopallial[123] gyri and *sulci of the second order (secondary sulci)* as e.g. characteristic for Carnivores. The position of these sulci and their topographical[124] relations are here quite constant, and some relationships to cortical areas likewise obtain, although the precise and strict conformity as well as permanence, fixity or stability of said relationship, claimed by WOOLSEY (1960) and others on the basis of unconvincing interpretations (cf. Fig. 47), cannot be said to hold.

Generally speaking, the secondary sulci of the Carnivore fissuration pattern can be homologized not only with those of the Ungulate pattern and that of other gyrencephalic Subprimate groups, but also with those of the Primate fissuration pattern. Nevertheless these homologizations are subject to some reservations concerning certain sulci and diverse controversial aspects dealt with in section 2 and 3.

As regards the Primate fissuration pattern, the constancy in general position and relations of the chief sulci, however different they may be in size and form, and however complex their association with new sulci in higher groups, is a most striking feature. This basic uniformity exists notwithstanding the fact that there is considerable variation throughout the Primate series in the immediate environment, namely the leptomeninges with their blood vessels, and the form of the cranium (CONNOLLY, 1950).

The significance of sulcus centralis as a boundary between motor and sensory areas is quite constant. Much the same could be said about the cingular sulcus system and the sulcus collateralis indicating the approximate boundary of the parahippocampal cortex. Likewise, the

---

[123] Here neopallium *sensu lateriori*, i.e. including the parahippocampal cortex.

[124] These topographic relations are likewise topologic in a relaxed, non-rigorous topology characterized by certain restraints as regards the permissible displacement of linear neighborhoods, but allowing for their junctions, interruptions and duplications.

calcarine sulcus as an axial sulcus of area striata and the sulcus parieto-occipitalis partly delimiting that area display rather definite relationships to distinctive cortical regions.

On the other hand, the cortical areas[125] shift in position with respect to numerous sulci when Primate fissuration patterns are compared in ascending taxonomic sequence. As the hemispheres become more differentiated, cortical areas are 'limited' by, respectively extend over, different sulci. Thus, as justly pointed out by CONNOLLY (1950), the same gyri in different groups may not always include the same area. In some genera there are cases where a sulcus limits an area only approximately, or only in a part of his course. Yet, because of the remarkable constancy of the pattern of the neopallial secondary sulci one is justified in establishing their homology, irrespective of their relations to cortical areas. One can therefore, as CONNOLLY states, assume 'a predetermined morphogenesis' of these sulci, which 'adapt themselves to new cortical areas'. One might also say that areas, as they expand or become displaced, shift beneath the fissuration pattern.

A particular case in point is here the sulcus lunatus, about whose homology there can be little doubt. KAPPERS (1929) already pointed out that the area striata of 'Prosimiae' is rostrally limited by this sulcus, which does not shift *pari passu* backward with said area, and consequently no longer delimits it in Anthropoids and Man, but lies within peri- or parastriate areas (cf. Figs. 62A, 63A, B). The area striata has here become displaced caudalward because of the visual association area's expansion.

It should again be emphasized that the presence of a sulcus lunatus does not *per se* imply 'lower mentality' respectively 'inferiority', but rather a still poorly understood 'random feature', quite generally related to Man's Simian ancestry, and not incompatible with the fissuration pattern of an 'elite brain' pertaining to an eminent personality. I would, for instance, interpret the configuration of EDINGER's left occipital lobe as displaying a sulcus lunatus with sulcus praelunatus (cf. Fig. 67).

In addition to the secondary or second order sulci, there appear both in the Carnivore-Ungulate pattern and its manifestations in other subprimate gyrencephalic Mammalian forms, as well as in the Primate

[125] Except for a few well-defined cytoarchitectural or myeloarchitectural areas, the mappings of most so-called cortical areas outlined by adherents of the parcellation concept, can be considered highly dubious (cf. the discussion of this topic in chapter XV).

fissuration pattern, numerous and diversified variable *tertiary sulci* whose homologization or descriptive assignment to a conventional scheme of fissuration respectively of gyri becomes rather uncertain. Because of various degrees of complexity, one could even distinguish still more variable and nondescript sulci and gyri of the *fourth order*, particularly in Man, Proboscidea and Cetaceans. Conspicuous asymmetries between right and left hemispheres, involving sulci of second, third, and fourth order, are known to occur in all gyrencephalic Mammals. LASSEK (1957) justly points out that, *qua* complexity of convolutions, Man does not possess, among gyrencephalic Mammals, the most furrowed brain, which latter is displayed by at least some Cetaceans.

Again, while some correlation of fissuration complexity with 'intelligence' is suggested by the recorded data, numerous unknown parameters seem to interact with, and apparently to counteract this tendency toward correlation. One may e.g. compare the brain of EDINGER (Figs. 66–68) with that of the Munich laborer or the East-African Negro (Figs. 73A, B).[126]

Thus, with respect to both the so-called 'elite brains' and the racial brains *(Rassengehirne)* the numerous data recorded in the extensive literature on this topic must be evaluated as inconclusive, and the various as well as differing specific interpretations propounded by the authors dealing with that matter, as unconvincing. At most, and comparably to the racial distribution of blood groups, certain fissuration patterns (including e.g. the sulcus lunatus) may manifest statistically relevant frequencies of occurrence in various races, without, *per se*, necessarily implying definable correlations with 'intelligence' or 'temperament'.

EDINGER (1911, 1912) cautiously assumed '*im allgemeinen*' and '*vornehmlich*' a complication of frontal lobe fissurization '*bei besonders hervorragenden Menschen*', and '*einseitige Vergrösserungen bestimmter Regionen*' in specialized '*einseitig besonders Begabten*'. He stressed here the brain of GAMBETTA, but as LANDAU (1923) pointed out, the particular finding in that eminent orator's brain is by no means of relevant significance.

Be that as it may, EDINGER nevertheless retained a sober scepticism

[126] Although perhaps not a 'genius' *sensu strictissimo*, EDINGER doubtless ranks high as an eminently gifted superior person of outstanding accomplishments, close to the undefined range of the 'genius'. One could, of course, claim that the 'obscure' persons whose brains are depicted in Figures 73A and B were of considerable potential ability inhibited by environmental conditions, but this neither provable nor disprovable assumption can be considered far-fetched and rather unlikely.

and remarked (1911, p. 426): '*so werden die Fragen, die hier interessieren, nicht durch Vergleichung etwa der Furchen zweier Gehirne zu lösen sein. Dieser Vergleichung hat man viel, und, wie mir scheinen möchte, fast überflüssige Arbeit gewidmet.*'

Fissuration, which increases the surface area, is of particular import for the expansion of the neocortex. LASSEK (1957), citing investigations by HENNEBERG (1910), KRAUS *et al.* (1928), MICHAELS and DAVISON (1930), asumes that nearly 73 per cent of the Human cerebral cortex is 'buried' within the relatively deep sulci. Detailed quantitative studies in the ascending taxonomic series Insectivores–Tupaia–Galago–Perodicticus–Aotes have been undertaken by STEPHAN (1975) and compared with findings in Homo. The relationship of neocortical surface to total telencephalic cortical surface is given, in per cents, as roughly 25, 55, 69 +, 83, 96 for the cited series Insectivores–Homo (STEPHAN, 1975).

7. References to Chapter XIV

ABEELEN, J.H.F. VAN (ed.): The genetics of behavior (Elsevier, New York 1974).

ADLER, D. and SINGER, I. (eds.): The Jewish Encyclopedia (Ktav Publishing, New York 1904).

ANONYMOUS: Racial studies: Academy states position on call for new research. Science *158:* 892–893 (1967).

ANTHONY, R.: Leçons sur le cerveau (Doin, Paris 1928).

ANTONI, N.R.E.: Ausbreitung und Flächenbeziehung der Area striata im menschlichen Gehirn. Folia neurobiol. *8:* 265–279 (1914).

BACHOFEN, J.: Das Mutterrecht; in CALVERTON The making of man, chapt. III, pp. 157–167 (Modern Library, New York 1931).

BAILEY, P. and BONIN, G. v.: The isocortex of man (University of Illinois Press, Urbana 1951).

BAILLARGER, J.G.F.: De l'étendue de la surface du cerveau et de ses rapports avec l'intelligence. Gaz. Hôp. *18:* 179 (1845).

BARRON, D.H.: An experimental analysis of some factors involved in the development of the fissure pattern of the cerebral cortex. J. exp. Zool. *113:* 553–581 (1950).

BEAN, R.B.: Some racial peculiarities of the Negro brain. Amer. J. Anat. *5:* 353–432 (1906).

BEAN, R.B.: A racial peculiarity in the pole of the temporal lobe of the Negro brain. Anat. Rec. *8:* 479–493 (1914).

BECCARI, N.: Neurologia comparata anatomo-funzionale dei vertebrati compreso l'uomo (Sansoni, Firenze 1943).

BECHTEREW, W. v. and WEINBERG, R.: Das Gehirn des Chemikers D. J. Mendelejew (Engelmann, Leipzig 1909).

Beck, E.: Die myeloarchitektonische Felderung des in der Sylvischen Furche gelegenen Teiles des menschlichen Schläfenlappens. J. Psychol. Neurol. *36:* 1–21 (1928).

Beck, E.: Die Myeloarchitektonik der dorsalen Schläfenlappenrinde beim Menschen. J. Psychol. Neurol. *41:* 129–262 (1930/31).

Bell, E.T.: Men of mathematics (Simon & Schuster, New York 1937).

Bismarck, O. Fürst von: Gedanken und Erinnerungen, 3 vols. (Cotta, Stuttgart 1898, 1922).

Bolk, L.: Das Problem der Menschwerdung (Fischer, Jena 1926).

Bork-Feltkamp, A.J. van: Uitkomsten van een onderzoek van een 60-tal hersenen van Chineezen; Dissert. (Versluys, Amsterdam 1930).

Bork-Feltkamp, A.J. van: Recherches sur 88 cerveaux de Chinois. Anthropologie *43:* 503–539 (1933).

Bork-Feltkamp, A.J. van: Fissuration of an Eskimo brain. Psychiat. neurol. Bladen 1935*:* 36–44 (1935).

Brain, R.: Some reflections on genius and other essays (Lippincott, Philadelphia 1960).

Brain, R.: Body, brain, mind and soul; in Huxley The humanist frame, pp. 11–63 (Harper & Brothers, New York 1961).

Brain, R.: Science and antiscience. Science *148:* 192–198 (1965).

Brauer, K. und Schober, W.: Katalog der Säugetiergehirne (Fischer, Jena 1970).

Broca, P.: Anatomie comparée des circonvolutions cérébrales. Le grand lobe limbique et la scissure limbique dans la série des mammifères. Rev. Anthropol. Ser. 2, *1:* 385–498 (1878).

Breathnach, A.S.: The olfactory tubercle, prepyriform cortex and precommissural region of the porpoise (Phocaena phocaena). J. Anat. *87:* 96–113 (1953).

Breathnach, A.S. and Goldby, F.: The amygdaloid nuclei, hippocampus and other parts of the rhinencephalon in the porpoise (Phocaena phocaena). J. Anat. *88:* 267–291 (1954).

Burr, H.S.: The neural basis of human behavior (Thomas, Springfield 1960).

Burr, H.S.: The nature of man and the meaning of existence (Thomas, Springfield 1962).

Bury, J.B.: The idea of progress. An inquiry into its origin and growth (Macmillan, New York 1932/Dover, New York 1955).

Bushmakin, N.: Characteristics of the brain in the Mongol race. Amer. J. phys. Anthrop. *12:* 221–243 (1928).

Cajal, S.R. y: Reglas y consejos sobre la investigación cientifica (Los tonicos de la voluntad); 6th ed. (Pueyo, Madrid 1923).

Cajal, S. R. y: El mundo visto a los ochenta años. Impresiones de un arteriosclerótico (Madrid 1934; Espasa-Calpe, Buenos Aires 1941).

Cajal, S. R. y: Charlas de Café. Pensamientos, anécdotas y confidencias. 3rd ed. (Espasa-Calpe, Buenos Aires 1944).

Calhoun, J.B.: Population density and social pathology. Scient. Amer. *206:* No. 2, 139–148 (1962).

Calverton, V.F. (ed.): The making of man. An outline of anthropology (Modern Library, New York 1931).

Calverton, V.F. (ed.): The making of society. An outline of sociology (Modern Library, New York 1937).

Chamberlain, H.S.: Die Grundlagen des neunzehnten Jahrhunderts. 2 vols.; 3. Aufl. (Bruckmann, München 1901).

Chi, T.K. and Chang, C.: The sulcal pattern of the Chinese brain. Amer. J. phys. Anthrop. *28:* 167–211 (1941).

CHUDZINSKI et DUVAL, M.: Description morphologique du cerveau de Gambetta. Bull. Soc. Anthrop. Paris, 3e s. *9:* 129–152 (1886).

CLARK, W.E. LE GROS: Deformation patterns in the cerebral cortex; in CLARK and MEDAWAR Essays on growth and form, pp. 1–22 (Clarendon Press, Oxford 1945).

CLARK, W.E. LE GROS: Fitting man to his environment (King's College Press, Newcastle upon Tyne 1949).

CLAUSEWITZ, *General* KARL VON: Vom Kriege; 10. Aufl. (Behr, Berlin 1915).

CONNOLLY, C.J.: External morphology of the primate brain (Thomas, Springfield 1950).

COON, C.S.: The origin of races (Knopf, New York 1962).

CRAIGIE, E.H.: Studies on the brain of the Kiwi (Apteryx australis). J. comp. Neurol. *49:* 223–357 (1930).

CUNNINGHAM, D.J.: The intraparietal sulcus of the brain. J. Anat. Physiol. *24:* 135–155 (1890).

CUNNINGHAM, D.J.: Contribution to the surface anatomy of the cerebral hemispheres. Mem roy. irish Acad. Sci., 7 (Dublin 1892).

DARESTE, M.C.: Sur les rapports de la masse encéphalique avec le développement de l'intelligence. Bull. Soc. Anthrop. Paris *3:* 26–55 (1862).

DARESTE, M.C.: Sur les circonvolutions du cerveau. C. R. Acad. Sci., Paris *70:* 193 (1870).

DARLINGTON, C.D.: The evolution of man and society (Allen & Unwin, London 1969).

DEJERINE, J.: Anatomie des centres nerveux. 2 vols. (Rueff, Paris 1895, 1901).

DELITZSCH, F.D.: Babylon (Hinrichs, Leipzig 1901).

DEXLER, H.: Zur Anatomie des Gehirns von Elephas indicus. Arb. neurol. Inst. Univers. Wien *15:* 157–281 (1907).

DILLON, L.S.: Comparative studies of the brain in the Macropodidae. J. comp. Neurol. *120:* 43–51 (1963).

DODGSON, M.C.H.: The growing brain. An essay in developmental neurology (Wright, Bristol 1962).

DONALDSON, H.H. and CANARAN, M.M.: A study of the brain of three scholars. J. comp. Neurol. *46:* 1–95 (1929).

DRAKE, S.C. and CAYTON, H.: Black metropolis. A study of Negro life in a Northern city (Harcourt, Brace, New York 1945).

DUGDALE, R.L.: The Jukes (Putnam, New York 1877).

DUNN, L.C. and DOBZHANSKY, T.: Heredity, race and society. Revised edition (Mentor Book, New York 1957).

ECCLES, J.C.: Facing reality. Philosophical adventures by a brain scientist (Springer, New York 1970).

ECONOMO, C. v.: Wie sollen wir Elitegehirne verarbeiten? Z. ges. Neurol. Psychiat. *130:* 419–531 (1929).

ECONOMO, C. v.: Zur Frage des Vorkommens der Affenspalte beim Menschen im Lichte der Cytoarchitektonik. Z. ges. Neurol. Psychiat. *130:* 419–531 (1930).

EDINGER, L.: Vorlesungen über den Bau der nervösen Zentralorgane des Menschen und der Tiere, vol. I: Das Zentralnervensystem der Menschen und der Säugetiere; 8. Aufl. (Vogel, Leipzig 1911).

EDINGER, L.: Einführung in die Lehre vom Bau und den Verrichtungen der Nervensystems; 2. Aufl. (Vogel, Leipzig 1912).

EINSTEIN, A.: Out of my later years (Philosophical Library, New York 1950).

EINSTEIN, A.: Ideas and opinions (Crown, New York 1954).

ELLENBERGER, W. und BAUM, H.: Handbuch der vergleichenden Anatomie der Haustiere (Springer, Berlin 1926).

ESTABROOKS, A.H.: The Jukes in 1915 (Carnegie Institution, Washington 1916).

EUCKEN, R.: Die Einheit des Geisteslebens in Bewusstsein und Tat der Menschheit (Veit, Leipzig 1888).

FISCHER, E.: Die Rehobother Bastards und das Bastardierungsproblem beim Menschen (Fischer, Jena 1913).

FISCHER, E.: Über die Variationen der Hirnfurchen des Schimpansen. Verh. anat. Ges., Erg. W. anat. Anz. *25:* 48–54 (1921).

FLATAU, E. und JACOBSOHN, L.: Handbuch der Anatomie und vergleichenden Anatomie des Zentralnervensystems der Säugetiere. I. Makroskopischer Teil (Karger, Berlin 1899).

FORD, J.: Social deviation (Macmillan, New York 1939).

FOREL, A.: Rückblick auf mein Leben (Europa-Verlag, Zürich 1935).

FRIANT, M.: Le cerveau du marsouin (Phocaena communis cuv.) et les caractéristiques fondamentales du cerveau des cétacés. Acta anat. *17:* 61–71 (1953).

FRIANT, M.: Le cerveau du Balenoptère (Balaenoptera spec.). Acta anat. *23:* 242–250 (1954).

FRIANT, M.: L'extension progressive de l'insula cérébrale, des singes inférieurs, aux anthropoides et à l'homme; in KAPPERS Progress in neurobiology, pp. 317–323 (Elsevier, Amsterdam 1956).

FUSTEL DE COULANGES, N.D.: La cité antique; 17th ed. (Hachette, Paris 1905).

GALTON, F.: Hereditary genius: an inquiry into its laws and consequences; 1st ed., 2nd ed. (Macmillan, London 1869, 1892).

GALTON, F.: The history of twins, as a criterion of the relative powers of nature and nurture. Fraser's Magaz. (1875); quoted after LEVITAN and MONTAGU (1971).

GALTON, F.: Inquiries into human faculty and its development (Macmillan, London 1883).

GENNA, G.E.: Sulla morfologia dei solchi cerebrali dell'uomo. Osservazioni su cervelli di indigeni del Camerun. Riv. Antropol. *26:* 19–173 (1924)

GERLACH, E.J. und WEBER, H.: Über ein menschliches Gehirn mit beiderseitiger Verdoppelung der Zentralfurche. Anat. Anz. *67:* 440–452 (1929).

GIACOMINI, C.: Varieta delle circonvoluzioni cerebrale dell'uomo (Celanza, Torino 1882).

GIBBON, E.: The decline and fall of the Roman empire. 3 vols. (Modern Library, New York n.d.).

GILULA, M.F. and DANIELS, D.N.: Violence and man's struggle to adapt. Science *164:* 396–405 (1969).

GOBINEAU, J.A. *comte de:* Essai sur l'inégalité des races humaines. 4 vols. (Firmin-Didot, Paris 1853–1855, 1884).

GODDARD, H.H.: The Kallikak family (Macmillan, New York 1912).

GOLDSCHMIDT, R.: Die Lehre von der Vererbung (Springer, Berlin 1927).

GOLDSCHMIDT, R.: In and out of the Ivory Tower. The autobiography of *Richard B. Goldschmidt* (University of Washington Press, Seattle 1960).

GRATIOLET, P.: Mémoire sur les plis cérébraux de l'homme et des primates (Bertrand, Paris 1854).

GUILLAIN, G. and BERTRAND, I.: Anatomie topographique du système nerveux central (Masson, Paris 1926).

HAECKEL, E.: Generelle Morphologie der Organismen. 2 vols. (Reimer, Berlin 1866).

Haller v. Hallerstein, Graf V.: Über die Morphologie des Hippocampus, des Gyrus dentatus, Gyrus fasciolaris, Gyrus callosus und des Uncus. Anat. Anz., Erg. H. Bd. *66:* 197–211 (1928).

Haller v. Hallerstein, Graf V.: Zerebrospinales Nervensystem. Äussere Gliederung des Zentralnervensystems; in Bolk *et al.* Handbuch der vergleichenden Anatomie der Wirbeltiere, vol. 2, pp. 1–318 (Urban & Schwarzenberg, Berlin 1934).

Hansemann, D. v.: Über die Gehirne von Th. Mommsen, H. W. Bunsen, und Ad. v. Menzel (Schweizerbarth, Stuttgart 1907).

Harper, J. W. and Maser, J. D.: A macroscopic study of the brain of Bison bison bison, the American plains Buffalo. Anat. Rec. *184:* 187–202 (1976).

Hartland, E. S.: Motherright; in Calverton The making of man, chapt. III, pp. 182–202 (Modern Library, New York 1931).

Hayashi, M. und Nakamura, R.: Über den Hinterhauptlappen des Japanergehirns. Mitt. med. Fak. Kaiserl. Univ. Tokyo 1914; quoted after K. (1927).

Haymaker, W. (ed.): The founders of neurology; 1st ed. (Thomas, Springfield 1953; 2nd ed. 1970).

Hegel, G. W. F.: Vorlesungen über die Philosophie der Geschichte (Reclam, Leipzig n.d.).

Henneberg, R.: Messung der Oberflächenausdehnung der Grosshirnrinde. J. Psychol. Neurol. *17:* 144–158 (1910).

Herre, W.: Neues zur Umweltbeeinflussbarkeit des Säugetiergehirns. Naturwiss. Rdsch. *16:* 359–364 (1964).

Herre, W.: Einige Bemerkungen zur Modifikabilität, Vererbung und Evolution von Merkmalen des Vorderhirns bei Säugetieren; in Hassler and Stephan Evolution of the forebrain, pp. 162–174 (Thieme, Stuttgart 1966).

Herrick, C. J.: The evolution of human nature (University of Texas Press, Austin 1956).

Herrnstein, R. J.: IQ. Atlantic monthly *228:* 44–64 (1971).

Herrnstein, R. J.: IQ: social goals and the genetic heresy. Trans. Coll. Phys. Philad. *40:* 207–218 (1973a).

Herrnstein, R. J.: IQ and the meritocracy (Little, Brown, Boston 1973b).

Herskovits, M. J.: The American Negro, a study in racial crossing (Knopf, New York 1928).

Hines, M.: Studies in the growth and differentiation of the telencephalon in man. The fissura hippocampi. J. comp. Neurol. *34:* 73–171 (1922).

Hines, M.: The development of the telencephalon in Sphenodon punctatum. J. comp. Neurol. *35:* 483–537 (1923).

Hochstetter, F.: Eröffnungsvortag 33. Vers. anat. Ges., Erg. Bd. Anat. Anz. *58:* 2–23 (1924).

Holden, C.: *R. J. Herrnstein:* the perils of expounding meritocracy. Science *181:* 36–39 (1973).

Holl, M.: Über die Insel des Carnivorengehirnes. Arch. Anat. Physiol. anat. Abt. *1899:* 217–266 (1899).

Holl, M.: Über die Insel des Ungulatengehirnes. Arch. Anat. Physiol. anat. Abt. *1900:* 295–334 (1900).

Holl, M.: Über die Insel des Menschen- und Anthropoidengehirnes. Arch. Anat. Physiol. anat. Abt. 1902: 1–44 (1902).

Holl, M.: Über die Insel des Delphingehirnes. Arch. Anat. Physiol. anat. Abt. *1903:* 333–344 (1903).

Holloway, R. L.: On the meaning of brain size. Science *184:* 677–679 (1974).

HOOK, E.B.: Behavioral implications of the human XYY genotype. Science *179:* 139–150 (1973).

HUMPHREY, T.: The development of the human hippocampal fissure. J. Anat. *101:* 655–676 (1967).

HURD-MEAD, K.C.: A history of women in medicine (Haddam Press, Haddam 1938).

IZARD, C.E.: The face of emotion (Appleton-Century, New York 1971).

JACOBS, P.A.: XYY genotype. Science *189:* 1044–1055 (1975).

JAFFÉ, G.: Recollections of three great laboratories. J. chem. Educat. *20:* 230–238 (1952).

JANSEN, J. and JANSEN, J.K.S.: The nervous system of Cetacea; in ANDERSEN, H. The biology of marine mammals, chapt. 7, pp. 175–252 (Academic Press, New York 1968).

JANSSEN, P. et STEPHAN, H.: Recherches sur le cerveau de l'éléphant d'Afrique (Loxodonta africana Blum). Acta neurol. psychiat. belg. *11:* 731–757 (1956).

JENSEN, A.R.: Genetics and education (Harper & Row, New York 1972).

JENSEN, A.R.: Educability and group differences (Harper & Row, New York 1973).

JERISON, H.J.: Evolution of brain size and intelligence (Academic Press, New York 1973).

JOHNSON, R.N.: Aggression in man and animals (Saunders, Philadelphia 1972).

KAMIN, L.J.: The science and politics of IQ (Wiley, New York 1974).

KANT, I.: Zum ewigen Frieden. Ein philosophischer Entwurf (1795; ed. Reclam, Stuttgart 1965).

KAPPERS, C.U.A.: Die vergleichende Anatomie des Nervensystems der Wirbeltiere und des Menschen, vol. 2 (Bohn, Haarlem 1921).

KAPPERS, C.U.A.: The evolution of the nervous system in invertebrates, vertebrates and man (Bohn, Haarlem 1929).

KAPPERS, C.U.A.: Anatomie comparée du système nerveux, particulièrement de celui des mammifères et de l'homme. Avec la collaboration de *E. H. Strasburger* (Bohn, Haarlem & Masson, Paris 1947).

KAPPERS, C.U.A.; HUBER, G.C., and CROSBY, E.C.: The comparative anatomy of the nervous system of vertebrates, including man. 2 vols. (Macmillan, New York 1936).

KARPLUS, J.P.: Über Familienähnlichkeiten an den Grosshirnfurchen des Menschen. Arb. neurol. Inst. Univers. Wien *12:* 1–58 (1905).

KATZ, D.: Psychological atlas (Philosophical Library, New York 1948).

KATZ, P.A.: Towards the elimination of racism (Pergamon Press, New York 1976).

KEEGAN, J.J.: A study of a Plains Indian brain. J. comp. Neurol. *26:* 403–420 (1916).

KEITH, A.: Essays on Human evolution (Putnam, New York 1947).

KEITH, A.: A new theory of Human evolution (Philosophical Library, New York 1950).

KLATT, B.: Über die Veränderung der Schädelkapazität in der Domestikation. S. H. Ges. Naturf. Berlin *1912:* 153–179 (1912).

KLATT, B.: Studien zum Domestikationsproblem. I. Untersuchungen am Hirn. Bibl. genet. *2:* 1–181 (1921).

KLATT, B.: Vergleichende Untersuchungen an Caniden und Procyoniden. Zool. Jb. allg. Zool. *45:* 217–292 (1928).

KLATT, B.: Gefangenschaftsveränderungen bei Füchsen. Jena. Z. Naturwiss. *67:* 452–468 (1932).

KNAPP, P.D. (ed.): Expression of the emotion in man (International Universities Press, New York 1963).

KOHLBRUGGE, J.H.F.: Die Variationen an den Grosshirnfurchen der Affen mit besonderer Berücksichtigung der Affenspalte. Z. Morph. Anthrop. *6:* 191–250 (1903).

Kohlbrugge, J.H.F.: Die Gehirnfurchen der Javanen. Verh. kon. Akad. Wetensch. Amsterdam, Sect. 2, Part 2, No. 4 (1906).

Kooy, F.A.: Over den sulcus lunatus bij Indonesiers. Psychiat. neurol. Bladen *1921:* 145–201 (1921).

Kramer, S.N.: The Sumerians, their history, culture, and character (University of Chicago Press, Chicago 1963).

Kraus, W.M.; Davison, C., and Weil, A.: Measurement of cerebral and cerebellar surfaces: problems encountered in measuring cerebral cortical surface in man. Arch. Neurol. Psychiat. *19:* 454–477 (1928).

Krause, R.: Mikroskopische Anatomie der Wirbeltiere. I. Säugetiere (De Gruyter, Berlin 1921).

Kretschmer, E.: Geniale Menschen (Springer, Berlin 1931).

Kroeber, A.L. (ed.): Anthropology today. An encyclopedic inventory (University of Chicago Press, Chicago 1953).

Krueg, J.: Über die Furchung der Grosshirnrinde der Ungulaten. Z. wiss. Zool. *31:* 297–345 (1878).

Kruska, D.: Vergleichende cytoarchitektonische Untersuchungen an Gehirnen von Wild- und Hausschweinen. Z. Anat. EntwGesch. *131:* 291–324 (1970).

Kruska, D.: Volumvergleich optischer Hirnzentren bei Wild- und Hausschweinen. Z. Anat. EntwGesch. *138:* 265–282 (1972).

Kruska, D.: Domestikationsbedingte Grössenveränderungen verschiedener Hirnstrukturen bei Schweinen; in Matolcsi Domestikationsforschung und Geschichte der Haustiere, pp. 135–140 (Akadémiai Kiado, Budapest 1973a).

Kruska, D.: Cerebralisation, Hirnevolution und domestikationsbedingte Hirngrössenänderungen innerhalb der Ordnung Perissodactyla Owen 1848 und ein Vergleich mit der Ordnung Artiodactyla Owen 1848. Z. zool. Systematik Evolutionsforsch. *11:* 81–103 (1973b).

Kruska, D.: Vergleichend-quantitative Untersuchungen an den Gehirnen von Wander- und Laborratten. J. Hirnforsch. *16:* 469–496 (1975).

Kruska, D. and Röhrs, M.: Comparative-quantitative investigations on brains of pigs from the Galapagos Islands and of European domestic pigs. Z. Anat. EntwGesch. *144:* 61–73 (1974).

Kruska, D. and Stephan, H.: Volumenvergleich allokortikaler Hirnzentren bei Wild- und Hausschweinen. Acta anat. *84:* 387–415 (1973).

Kuhlenbeck, H.: Vorlesungen über das Zentralnervensystem der Wirbeltiere (Fischer, Jena 1927).

Kuhlenbeck, H.: Über die sogenannte Affenspalte des Occipitalhirns. Med. Klinik *24:* 937–938 (1928a).

Kuhlenbeck, H.: Bemerkungen zur Morphologie des Occipitallappens des menschlichen Grosshirns. Anat. Anz. *65:* 273–294 (1928b).

Kuhlenbeck, H.: Brain and consciousness. Some prolegomena to an approach of the problem (Karger, Basel 1957).

Kuhlenbeck, H.: Mind and matter. An appraisal of their significance for neurologic theory (Karger, Basel 1961).

Kuhlenbeck, H.: Gehirn und Intelligenz. Confinia neurol. *25:* 36–62 (1965a).

Kuhlenbeck, H.: The concept of consciousness in neurological epistemology; in Smythies Brain and mind – modern concepts of the nature of mind, pp. 137–161 (Routledge & Kegan Paul, London 1965b).

KUHLENBECK, H.: Weitere Bemerkungen zur Maschinentheorie des Gehirns. Confin. neurol. *27:* 295–328 (1966).

KUHLENBECK, H.: Some comments on words, language, thought and definition; in BUEHNE Helen Adolf Festschrift, pp. 9–29 (Ungar, New York 1968).

KUHLENBECK, H.: *Schopenhauers* Satz 'Die Welt ist meine Vorstellung' und das Traumerlebnis. Schopenhauer Jahrb. *53* (Festschrift Hübscher): 376–392 (1972).

KUHLENBECK, H.: Gehirn und Bewusstsein. Translated by *J. Gerlach* and *U. Protzer*. Erfahrung und Denken: Schriften zur Förderung der Beziehungen zwischen Philosophie und Einzelwissenschaften, vol. 39 (Duncker & Humblot, Berlin 1973).

KUHLENBECK, L.: Im Hochland der Gedankenwelt. Grundzüge einer heroisch-ästhetischen Weltanschauung (Individualismus) (Diederichs, Leipzig 1903).

KUHLENBECK, L.: Giordano Bruno, Gesammelte Werke. Herausgegeben. verdeutscht und erläutert. 6 vols. (Diederichs, Leipzig u. Jena 1904–1909).

KUHLENBECK, L.: Giordano Bruno. Seine Lehre von Gott, von der Unsterblichkeit der Seele und von der Willensfreiheit. Die Religion der Klassiker, vol. 1 (Protestantischer Schriftenvertrieb, Berlin-Schöneberg 1913).

KÜKENTHAL, W. und ZIEHEN, T.: Untersuchungen über die Grosshirnfurchen der Primaten. Jena Z. Naturwiss. *29:* 1–122 (1895).

KURZ, E.: Das Chinesengehirn. Z. Anat. EntwGesch. *72:* 199–382 (1924).

LANDAU, E.: Anatomie des Grosshirns. Formanalytische Untersuchungen (Bircher, Bern 1923).

LANGWORTHY, O. R.: A description of the central nervous system of the porpoise (Tursiops truncatus). J. comp. Neurol. *54:* 437–499 (1932).

LASSEK, A. M.: The human brain from primitive to modern (Thomas, Springfield 1957).

LAYZER, D.: Heritability analyses of IQ scores: science or numerology? Science *183:* 1259–1266 (1974).

LAYZER, D.: Heritability of IQ: methodological questions. Science *188:* 1128–1130 (1975).

LEVITAN, M. and MONTAGU, A.: Textbook of human genetics (Oxford University Press, New York 1971).

LEWIS, J. H.: The biology of the Negro (University of Chicago Press, Chicago 1942).

LIVINI, F.: Il proencefalo di un Marsupiale. Arch. ital. Anat. Embriol. *6:* 549–584 (1907).

LOEHLIN, J. C.; LINDZEY, G., and SPUHLER, J. N.: Race differences in intelligence (Freeman, San Francisco 1975).

LOMBROSO, C.: Genio e follia; 4th ed. (Roma 1882).

LORENZ, K.: On aggression (Methuen, London 1966).

LUBOSCH, W.: Ein merkwürdiger Fall von Affenspaltenrest. Anat. Anz. *67:* 493–497 (1929).

MACCOBY, E. E. and JACKLIN, C. N.: The psychology of sex differences (Stanford University Press, Stanford 1974).

MAIR, R.: Zur Topographie des Tractus und Bulb. olf. beim Menschen. Anat. Anz. *67:* 501–506 (1928).

MARTIN, C. P.: Psychology, evolution and sex (Thomas, Springfield 1956).

MARTIN, M. K. and VOORHIES, B.: Female of the Species (Columbia University Press, New York 1975).

MAURER, F.: Das Gehirn *Ernst Haeckels* (Fischer, Jena 1924).

MAURER, F.: Der Mensch und seine Ahnen (Ullstein, Berlin 1928).

MCHENRY, H. M.: Fossils and the mosaic nature of Human evolution. Science *190:* 425–431 (1975).

Mettler, F.A.: The brain of Pithecus rhesus (M. rhesus). Amer. J. phys. Anthrop. *17:* 309–331 (1933).

Michaels, J.J. and Davison, C.: Measurement of cerebral and cerebellar surfaces. VIII. Measurement of the motor area in vertebrates and man. Arch. Neurol. Psychiat. *23:* 1212–1226 (1930).

Mobilio, C.: Il mantello cerebrale degli equidi. Arch. ital. Anat. Embriol. *13:* 114–271 (1914–1915).

Möbius, P.J.: Ausgewählte Werke. 8 vols. (Barth, Leipzig 1911).

Montagu, A.: Man's most dangerous myth. The fallacy of race; 3rd ed. (Harper, New York 1952).

Montagu, A.: The cultured man (Permabooks, New York 1959).

Montesquieu, C. de Secondat, *Baron* de: Considérations sur les causes de la grandeur des Romains et de leur décadence. Œuvres, nouvelle édition, vol. 5 (Bastien, Paris 1788).

Murakami, S.: Embryological studies on the cerebral sulci in the Japanese fetus (Japanese, with English summary). Acta anat. nippon. *39:* 161–276 (1955).

Myrdal, G.: An American dilemma. The Negro problem and modern democracy. 2 vols., 1st and 2nd ed. (Harper & Row, New York 1944, 1962).

Nauta, W.J.H. and Haymaker, W.: Hypothalamic nuclei and fiber connections; in Haymaker *et al.* The hypothalamus, chapter 4, pp. 136–209 (Thomas, Springfield 1969).

Nei, M.: Molecular population genetics and evolution (Elsevier, New York 1975).

Ngowyang, G.: Über Rassengehirne. Z. Rassenk. *1936:* 26–20 (1936).

Noback, C.R. and Goss, L.: Brain of a gorilla. I. Surface anatomy and brain stem nuclei. J. comp. Neurol. *111:* 321–343 (1959).

Orton, S.T.: Note on an anomaly of the postcentral sulcus, simulating the double Rolandic of Giacomini. Anat. Rec. *5:* 179–181 (1911).

Oxnard, C.E.: Uniqueness and diversity in human evolution. Morphometric studies of Australopithecines (University of Chicago Press, Chicago 1975).

Papez, J.W.: Comparative neurology (Crowell, New York 1929).

Patten, B.M.: Human embryology; 2nd ed. (Blakiston, New York 1953).

Pendell, E. (ed.): Society under analysis. An introduction to sociology (Catell Press, Lancaster 1942).

Pilleri, G.: Die zentralnervöse Rangordnung der Cetacea (Mammalia). Acta anat. *51:* 241–258 (1962).

Pilleri, G.: Morphologie des Gehirnes des 'Southern Right Whale', Eubalaena australis Desmoulins 1822 (Cetacea, Mysticeti, Balaenidae). Acta zool. *45:* 245–272 (1964).

Pilleri, G.: Morphologie des Gehirnes des Seiwals, Balaenoptera borealis Lesson (Cetacea, Mysticeti, Balaenopteridae) J. Hirnforsch. *8:* 221–267 (1965/66a).

Pilleri, G.: Morphologie des Gehirnes des Buckelwals, Megaptera novaeangliae Boronski (Cetacea, Mysticeli, Balaenopteridae). J. Hirnforsch. *8:* 437–491 (1965/66b).

Popper, K.R.: The logic of scientific discovery (Hutchinson, London 1959).

Ranke, J.: Der Mensch. 2 vols. (Bibliograph. Institut, Leipzig 1923).

Ranke, O.: Beiträge zur Kenntnis der normalen und pathologischen Hirnrindenbildung. Beitr. path. Anat. *47:* 51–125 (1910).

Rashevsky, N.: Looking at history through mathematics (M.I.T. Press, Cambridge 1968).

Reade, W.: The martyrdom of man; 1872, new ed. (Watts, London 1925).

RETZIUS, G.: Das Menschenhirn. Studien in der makroskopischen Anatomie. 2 vols. (Norstedt, Stockholm, and Fischer, Jena 1896).

RETZIUS, G.: Biologische Untersuchungen, vol. 8, 9, 10, 11, 12 (Fischer, Jena 1898, 1900, 1902, 1904, 1905).

RICHMAN, D.P.; STEWART, R.M.; HUTCHINSON, J.W., and CAVINESS, V.S., JR.: Mechanical model of brain convolutional development. Science *189:* 18–21 (1975).

RIESE, W.: The cerebral cortex of two prominent scientists in advanced age. J. Neuropath. exp. Neurol. *12:* 92–93 (1953).

RIESE, W.: The brain of Dr. *Trigant Burrow*, physician, scientist and author. Followed by considerations on the scope and limitations of anatomical investigations in relation to mental ability. J. comp. Neurol. *100:* 525–567 (1954).

RIESE, W.: Brains of prominent people; history, facts, and significance. Med. Coll. Virginia Quart. *2:* 106–110 (1966).

RIESE, W. and GOLDSTEIN, K.: The brain of *Ludwig Edinger*. An inquiry into the cerebral morphology of mental ability and left-handedness. J. comp. Neurol. *92:* 133–168 (1950).

RODENWALDT, C.: Die Mestizen von Kisar (Kolff, Batavia 1927).

RÖHRS, M.: Biologische Anschauungen über Begriff und Wesen der Domestikation. Z Tierzücht. ZüchtBiol. *76:* 7–23 (1961).

RÖHRS, M. und KRUSKA, D.: Der Einfluss der Domestikation auf das Zentralnervensystem und Verhalten von Schweinen. Dtsch. tierärztl. Wschr. *75:* 514–518 (1969).

ROMER, A.S.: The vertebrate body (Saunders, Philadelphia 1950).

SALLER, K.: Leitfaden der Anthropologie (Springer, Berlin 1930; 2nd ed. Fischer, Stuttgart 1964).

SANIDES, F.: Representation in the cerebral cortex and its areal lamination patterns; in BOURNE The structure and function of nervous tissue, vol. V, pp. 329–453 (Academic Press, New York 1972).

SCARR-SALAPATEK, S.: Race, social class and IQ. Science *174:* 1285–1295 (1971).

SCHAFFER, K.: Über normale und pathologische Hirnfurchung. Zum Mechanismus der Hirnfurchung. Z. ges. Neurol. Psychiat. *38:* 79–84 (1918).

SCHAFFER, K.: Zum Problem der Hirnfurchung. Arch. Psychiat. Nervenkr. *70:* 452–465 (1923a).

SCHAFFER, K.: Histogenese der Hirnfurchung. Z. Anat. EntwGesch. *95:* 467–482 (1923b).

SCHOBER, W. and BRAUER, K.: Makromorphologie des Gehirns der Säugetiere. Handbuch der Zoologie, vol. VIII (De Gruyter, Berlin 1975).

SCHUMACHER, U.: Quantitative Untersuchungen an Gehirnen mitteleuropäischer Musteliden. Z. Hirnforsch. *6:* 137–163 (1963).

SCHWALBE, G.: Lehrbuch der Neurologie (Besold, Erlangen 1881).

SCHWARTZ, G.E.; DAVIDSON, R.J., and MAER, F.: Right hemisphere lateralization of emotion in the human brain: interaction with cognition. Science *190:* 286–288 (1975).

SEECK, O.: Geschichte des Untergangs der antiken Welt. 6 vols. and supplements (Siememoth, Berlin 1895–1921).

SELIGMANN, K.: The history of magic (Pantheon Books, New York 1948).

SEXTUS EMPIRICUS: Outlines of pyrrhonism. Greek text with translation by the Rev. *R.C. Bury* (The Loeb Classical Library, Heinemann, London 1961).

SHANKLIN, W.M.: The central nervous system of Chameleon vulgaris. Acta zool. *11:* 425–490 (1930).

SHELLSHEAR, J. L.: The brain of the aboriginal Australian. A study in cerebral morphology. Philos. Trans. roy. Soc., Lond., Ser. B. *123:* 469–487 (1937).

SHERRINGTON, C.: Man on his nature; 2nd ed. (Cambridge University Press, London 1951).

SISSON, S. and GROSSMAN, J. D.: The anatomy of the domestic animals (Saunders, Philadelphia 1959).

SMITH, D. H.: History of mathematics. 2 vols. (Dover, New York 1958).

SMITH, G. ELLIOT: On the homologies of the cerebral sulci. J. Anat. Physiol. *36:* 309–319 (1902).

SMITH, G. ELLIOT: On the morphology of the brain in the mammalia with special reference to that of the lemurs, recent and extinct. Trans. Linnean Soc. Lond. *8:* 319–452 (1903).

SMITH, G. ELLIOT: The so-called 'Affenspalte' in the human (Egyptian) brain. Anat. Anz. *24:* 74–83 (1903–1904a).

SMITH, G. ELLIOT: The morphology of the occipital region of the cerebral hemisphere in man and the apes. Anat. Anz. *24:* 436–451 (1903–1904b).

SMITH, G. ELLIOT: New studies on the folding of the visual cortex and the significance of the occipital sulci in the human brain. J. Anat. Physiol. *41:* 198 (1907a).

SMITH, G. ELLIOT: A new topographical survey of the human cerebral cortex, being an account of the distribution of the anatomically distinct cortical areas and their relationship to the cerebral sulci. J. Anat. 237–254 (1907b).

SPÄTH, J.: Studien über Zwillinge. Z. wien. Ges. Ärzte *16:* 225–241 (1860); quoted after LEVITAN and MONTAGU (1971).

SPATZ, H.: Die Evolution des Menschenhirns und ihre Bedeutung für die Sonderstellung des Menschen. Nachr. Giessener Hochschulges. *24:* 52–74 (1955).

SPENGLER, O.: Der Untergang des Abendlandes. 2 vols. (Beck, München 1919, 1927).

SPERINO, G.: L'encefalo dell'Anatomico *Carlo Giacomini.* Riv. Frenetr. Med. legale *27:* 146–171 (1901).

SPITZKA, E. A.: Contributions to the encephalic anatomy of the races (Three Eskimo brains). Amer. J. Anat. *2:* 25–71 (1902).

SPITZKA, E. A.: A study of the brains of 6 eminent scientists. Trans. amer. philos. Soc. *31:* pt. 4, 175–308 (1907).

STEELE, J. H.: Population trends: A historical review and projection as to its effect on veterinary medicine. Milit. Med. *140:* 473–478 (1975).

STEPHAN, H.: Vergleichend-anatomische Untersuchungen an Hirnen von Wild- und Haustieren. III. Die Oberfläche des Allocortex bei Wild- und Gefangenschaftsfüchsen. Biol. Zbl. *73:* 96–115 (1954).

STEPHAN, H.: Allocortex; in v. MÖLLENDORFF Handbuch der mikroskopischen Anatomie des Menschen (fortgeführt von W. BARGMANN), vol. 4, 9. Teil) Springer, Berlin 1975).

STRASSER, H.: Alte und neue Probleme auf dem Gebiet des Nervensystems. Ergebn. Anat. EntwGesch. *2:* 565 (1893).

SUCHIER, W.: *A. W. Amo.* Ein Mohr als Student und Privatdozent der Philosophie in Halle, Wittenberg und Jena 1727/40. Akad. Rdsch. *4:* 441–448 (1915/16).

SUMMER, MONTAGUE *(the Rev. Father):* The history of witchcraft; 1st ed., 2nd ed. (Routledge, London 1926; University Books, New York 1956).

TEILHARD DE CHARDIN, P.: The phenomenon of man (Harper, New York 1959).

TILNEY, F.: The brain from ape to man (Hoeber, New York 1928).

TINBERGEN, N.: On war and peace in animals and man. An ethological approach to the biology of aggression. Science *160:* 1411–1418 (1968).

TOYNBEE, A.J.: A study of history. Abridged by D.C. SOMERVELL. 2 vols. (Oxford University Press, New York 1947, 1957).

VATTER, E.: Die Rassen und Völker der Erde (Quelle & Meyer, Leipzig 1927).

VILLIGER, E.: Gehirn und Rückenmark; 7th ed. (Engelmann, Leipzig 1920).

WEININGER, O.: Geschlecht and Charakter; 22nd ed. (Braumüller, Wien 1921).

WELKER, W.I. and CAMPOS, G.B.: Physiological significance of sulci in somatic sensory cerebral cortex in mammals of the family Procyonidae. J. comp. Neurol. *120:* 19–26 (1963).

WELLS, H.G.: The outline of history; revised ed. (Garden City Publishing, Garden City 1949).

WHITEHEAD, A.N.: Science and the modern world (Macmillan, New York 1925).

WHITNEY, D.D.: Family treasures. A study of the inheritance of normal characteristics in man (Cattell Press, Lancaster 1942).

WILSON, E.O.: Sociobiology. The new synthesis (Harvard University Press, Cambridge 1975).

WOOLLARD, H.H.: The Australian aboriginal brain. J. Anat. *63:* 207–223 (1929).

WOOLLARD, H.H.: The growth of the brain of the Australian aboriginal. J. Anat. *65:* 221–224 (1931).

WOOLSEY, C.N.: Some observations on brain fissuration in relation to cortical localization of function; in TOWER and SCHADÉ Structure and function of the cerebral cortex, pp. 64–68 (Elsevier, Amsterdam 1960).

YENI-KOMSHIAN, G.H. and BENSON, D.A.: Anatomical study of cerebral asymmetry in the temporal lobe of humans, chimpanzees, and rhesus monkeys. Science *192:* 387–389 (1976).

YOUNG, J.Z.: The life of vertebrates (Oxford University Press, Oxford 1955).

ZIEHEN, T.: Über die Grosshirnfurchung der Halbaffen und die Deutung einiger Furchen des menschlichen Gehirns. Arch. Psychiat. Nervenkr. *28:* 898–930 (1896).

ZIEHEN, T.: Das Zentralnervensystem der Monotremen und Marsupialier. Teil 1, 2(I), 2(II), 3 (Fischer Jena, 1897, 1901, 1908, 1905).

ZIRKLE, C.: The death of a science in Russia (University of Pennsylvania Press, Philadelphia 1949).

ZUCKERKANDL, E.: Zur Morphologie des Affengehirnes. Z. Morph. Anthrop. *4:* 463–499 (1902); *6:* 285–321 (1903); *7:* 223–260 (1904a); *8:* 100–122 (1905).

ZUCKERKANDL, E.: Über die Affenspalte und das Operculum occipitale des menschlichen Gehirnes. Arb. neurol. Inst. Univers. Wien *12:* 207 (1904b).

ZUCKERKANDL, E.: Zur Oberflächenmodellierung des Atelesgehirns. Arb. neurol. Inst. Univers. Wien *18:* 60–100 (1910).

XV. The Mammalian Cerebral Cortex

1. Ontogeny and Presumptive Phylogeny

Both the ontogenetic and the phylogenetic origin as well as the further development of the Mammalian cortex cerebri may be considered from two different viewpoints, namely with respect to *histogenesis* and to *morphogenesis*.

The *histogenetic* aspect concerns the differentiation of the superficial cortical griseal plate from a periventricular matrix, the differentiation of the relevant cortical cellular elements, and their arrangements in layers, respectively columns, resulting in so-called architectural characteristics such as e.g. *cyto-* and *myeloarchitecture.*[1]

In contradistinction to these topics, essentially but not exclusively related to '*structure*', the *morphogenetic aspect* concerns the expansion, location, and subdivision of *distinctive cortical topological neighborhoods* as definable *areas* or *regions*, and their distribution upon hemispheric 'lobes'. Morphogenesis thus evidently involves the problem of homologies.

It will here be recalled that *cortex cerebri* was defined as a suprasegmental neural correlation tissue developed as a sheet of superficial gray matter within the wall of the telencephalon, and located peripherally to some or most of its fiber pathways representing a subcortical medulla or *subcortikales Mark* (cf. e.g. chapter VI, section 6, p. 476, vol. 3/II).

Mutatis mutandis, this definition[1a] also applies to cortex tecti mesen-

---

[1] Other such characteristics are '*fibrilloarchitecture*' (based on rather uncertain criteria) and *angioarchitecture*. Studies on '*chemoarchitecture*' as undertaken in recent years generally concern not so much 'architecture' than the distribution of particular substances, e.g. enzymes, over various brain regions (cf. e.g. the publication by WÄCHTLER, 1973). BRAAK (1972) has recently recorded regional differences in the distribution of neurolipofuscins *(pigment architecture)* for allocortical regions of the Human brain.

[1a] Said definition evidently implies at least three layers, namely (1) the zonal or plexiform layer, in which some fiber pathways remain located, intermingled with relatively few nerve cells, (2) a cortical cell plate (stratified or essentially non-stratified), and (3) a subcortical medulla with input and output fibers. A 'radial organization' or a 'columnar arrangement' are not an obligatory criterion, although such sort of organization may be displayed by highly differentiated cortices.

cephali and cortex cerebelli, dealt with in chapter XI and X of volume 4. The Vertebrate brain thus displays, at three different levels of the neuraxis, three different types of cortex, all of which may, in some lower forms, remain at the ependymal precortex stage. In the everted telencephalon of Ganoids and Teleosts, moreover, the homologa of the cortical grisea in other Vertebrates are not differentiated as cortex or precortex, but rather as 'nuclei' or diffuse grisea. This is also the case for one pallial neighborhood of the inverted Avian telencephalon (cf. below, footnote 4).

The early ontogenetic stages of Mammalian corticogenesis, with reference to comparable stages of presumptive phylogeny, were dealt with in chapter VI, section 6, pp. 608–619 of volume 3/II on the basis of our own studies (K. and v. DOMARUS, 1920; K., 1922, 1924b) and those of JAKOB and ONELLI (1911), KAPPERS (1928), LOO (1929) and KAHLE (1951).

In *statu nascendi*, the pallial cortical plate, derived from the migrated periependymal matrix cells, may display differences in its thickness, such that this latter decreases with a fairly steep laterobasal gradient from neocortex to anterior piriform lobe cortex, and with a rather low latero-dorso-medial gradient within the neocortex, moreover from neocortex to parahippocampal cortex, with a further decremental gradient to hippocampal cortex, as can be seen in Figures 319 A, B and 320 A, B of volume 3/II (pp. 584–586). The overall latero-dorso-medial gradient of the early cortical plate is correlated with a similar gradient in the thickness of the pallial hemispheric wall.

HOCHSTETTER (1919) did not concern himself with problems of regional cortex subdivisions or details of corticogenesis, but these gradients are well illustrated, although not pointed out, in the excellent photomicrographs of his plates XIX, XX, XXIII, XXIV and XXV. LAISSUE (1963), who uses a terminology somewhat different from that adopted by myself, has described and attempted to interpret these early stages of the cortical plate in Man and another Primate (Microcebus murinus).

Some histogenetic aspects, pertaining to cellular displacements of matrix elements in the developing neural tube at early stages, and particularly pointed out by FUJITA (1966 and other publications), as well as by other authors, were discussed in chapter V, section 1 of volume 3/I. No further elaborations on these topics are therefore required in the present context. Additional studies on histogenesis at fetal stages are those by CAJAL (1911), POLIAKOV (1961), BERRY and ROGERS

(1966), and HOLMES and BERRY (1966). Postnatal processes of maturation were particularly investigated by CONEL (1939–1959), SARKISOV *et al.* (1966), SUGITA (1917–1918), as well as by contributors to the publication on brain maturation edited by PURPURA (1974). Comments on these questions, about which much uncertainty still remains, are also included, with further bibliographic references, in STEPHAN's recent treatise (1975).

Of particular importance concerning the Human cerebral cortex are the pioneering studies on myelogeny undertaken by FLECHSIG (1920, 1927), and summarized in that author's last publications. A recent study on myelogenesis in the Rat was published by JACOBSON (1963).

DE CRINIS (1932) investigated the postnatal development of the dendrites in the Human cerebral cortex *(cytodendrogenesis)* and obtained results supplementing some of the data recorded by FLECHSIG and essentially corroborating this latter author's basic concepts.

Data on the ontogenetic differentiation into layers were recorded by HALLER (1910), by BRODMANN (1909), by LOO (1929), and more recently by KAHLE (1969). The two last named authors investigated fetal stages of the Human brain,[2] while BRODMANN also referred to various other Mammals.

As regards both ontogeny and presumptive phylogeny of the Mammalian cortex cerebri, and for an appropriate classification of its diverse fundamental types, to be dealt with further below in section 2, it appears of fundamental relevance to distinguish two main cortical subdivisions, namely (1) *cortex pallii* and (2) *cortex basalis.*

It will be recalled that, in Anamnia, the pallium is represented by the dorsal D-zones, and the basis by the B-zones. In Amniota, however, because of the internation of the D_1-zone, the secondary pallium consists only of the D_2 and D_3 zones, while the secondary basis includes, in addition to the B-zones, the now hypopallial zone D_1. This particular morphogenetic change, pointed out by the author more than fifty years ago (K., 1925–26), is a special case of a more general type of formative event designated as *internation* by REMANE (1966) or as introversion by SPATZ (1966), and is a most characteristic feature distinguishing the Amniote telencephalon from that of Gnathostome Anamnia with comparably inverted lobus hemisphaericus. It represents a highly significant step in the evolution of the Vertebrate telencephalon.

[2] References to additional reports by other authors can be found in these publications as well as in STEPHAN's (1975) treatise, and in two papers by KIRSCHE (1972, 1974).

In Gnathostome Anamnia with inverted lobus hemisphaericus (Selachians, Dipnoans, Amphibians) a *cortex pallii* is clearly displayed by Selachians and Dipnoans. This cortical plate, however, does not manifest a distinct stratification[3] and does not exhaust the periventricular matrix, which is retained as a conspicuous griseum. This cortex pallii is thus *merogenic* or, in ROSE's (1926a) terminology, a *cortex semiparietinus*. In addition, Selachians, Dipnoans and at least some Amphibians (Gymnophiona) also display a likewise merogenic or semiparietine *cortex basalis* derived from the B-zones and without cortical stratification. In Amniota (Reptiles, Birds, Mammals), the merogenic cortex basalis, without, or, at most, with minimal, nondescript stratification, and derived from the B-zones, is also present.

The Amniote cortex pallii, derived from the D_2 and D_3 zones, tends, on the other hand, to exhaust the matrix and to become *hologenic (totoparietine* or *holoparietine* in ROSE's terminology). This is the case with most of the three (lateral, dorsal, medial) pallial cortical plates in Reptiles, and with the parahippocampal and hippocampal cortex of Birds.[4] In Mammals, the entire pallial cortex also becomes essentially hologenic (totoparietine), with the following two qualifications.

Because of the complications resulting from the internation of D_1 and the formation of a secondary augmented basis, lateral regions of cortex pallii (anterior piriform lobe cortex and parts of adjacent neocortical insular cortex) arise partly by hologenic contribution from D_2, and partly by merogenic contribution from D_1. This latter component increases with a basolaterally directed gradient, and said cortices can thus be designated as *hekaterogenic* or, in ROSE's (1928) terminology, as cortex *pallio-striatalis sive bigenitus*.

The morphologically fundamental distinction between the essentially hologenic dorsal cortex pallii and the merogenic cortex basalis, involving a relevant difference *qua 'morphologische Wertigkeit'*, whereby most of the basal matrix remains paraventricular respectively non-cortical as basal and paraterminal grisea, was pointed out and stressed in a paper on the origin of the basal ganglia (K., 1924a). In Mammals, these latter arise from the augmented basis D_1 representing the transi-

---

[3] There obtains, of course, a stratification into zonal layer, cortical plate, and subcortical layer, which latter, poor in cells, separates the cortical plate from the periventricular griseum.

[4] In Birds, the homologa of neocortex and of part of anterior piriform lobe cortex do not display a cortical structure but are represented by the grisea of nucleus diffusus dorsalis and n. diffusus dorsolateralis.

tory *lateral ganglionic hill*, and from the primordial basal B_{1+2} zones displaying the transitory *medial ganglionic hill* (cf. Figs. 312, 313, p. 576, vol. 3/II). The paraterminal grisea arise from the basimedial B_{3+4} zones.

The second qualification, bearing upon the fully justifiable classification of cortex pallii as hologenic, concerns a very variable minor remnant of pallial non-exhausted periventricular matrix represented by the *subependymal cell plate* extending along parts or even most of the dorsal ventricular wall. In some instances it may be negligible or missing. The subependymal cell plate, moreover, is likewise generally present or suggested not only internally to the differentiated telencephalic basal grisea, but along the ependymal lining of the brain's entire ventricular system.[5]

As regards the presumptive *phylogenetic development* of the Mammalian cerebral cortex it should be emphasized that nothing whatsoever is known about said cortex in extinct primordial Mammalian forms evolved from Submammalian ones. In addition, nothing certain, despite various speculations, is known about the origin of Mammals, which is merely inferred on the basis of a comparison of recent forms, combined with ambiguous interpretations of fossil remnants.

Tetrapods possibly may have arisen in the Silurian period (roughly 360 to 325 million years ago), and were present in the Devonian period (about 325 to 280 million years ago). In the Carboniferous period (roughly 280 to 230 million years ago) three distinct orders of Tetrapods seem to have been present, namely *Labyrinthodonts (Stegocephalia)*, *Lepospondyls*, and *Phyllospondyls*, considered to be Amphibian forms. The *Embolomeri*, which can be subsumed under the Labyrinthodonts, may have evolved on one hand into primitive Reptiles, and on the other hand into 'higher' Amphibians (NOBLE, 1931, 1954). In the Triassic Period of the Mesozoic era (roughly about 205 million years ago), the *Theriodonts*, perhaps evolved in the late Permian Palaeozoic era, may have been primitive Mammals, but it is difficult to decide whether they should be classified as Mammals or Reptiles. Again, Branchiosaur Amphibia, presumably derived from Labyrinthodonts, are not clearly distinguishable from some *Cotylosaur Reptiles* of late Palaeozoic and early Mesozoic. It is, moreover, not impossible that the Mammalian 'phylo-

[5] The subependymal cell plate, whose significance was particularly pointed out by GLOBUS and myself, is dealt with in section 1, chapter V of volume 3/I, which also includes references to the observations by ALLEN (1912), HIKIJI (1933), and OPALSKI (1934).

genetic radiation' originated directly from Branchiosaur Amphibians independently of the specifically Reptilian 'radiation'. Thus, as pointed out in volume 1, p. 144, the connection of Mammals to recent Amphibians could, in some respects, perhaps be less remote than that to recent Reptiles.

Be that as it may, all recent Mammals are separated from all recent Reptiles, *qua* cytoarchitecture of isocortex and parahippocampal cortex, by an apparently abrupt step or unbridgeable gap *('unüberbrückbare Kluft'*; K., 1927). Birds, on the other hand, which are doubtless phylogenetically unrelated to Mammals with respect to their separate origin from Reptilian forms, and altogether lack a cortical differentiation of the neocortical (D_{2al} and D_{2+1b}) homologon, display, in some forms, a well stratified parahippocampal cortex, which can even be described as hexalaminar (cf. vol. 5/I, chapter XIII, section 9, p. 657) and seems thereby perhaps better differentiated than parahippocampal cortical regions in some 'lower' Mammals.

As regards the hippocampal cortex, the anterior piriform lobe cortex, and the basal cortex, however, there is much less of a difference between their Reptilian and their Mammalian type of cytoarchitectural differentiations.[6]

The overall features of Mammalian pallial corticogenesis, beginning with a periventricular matrix, from which, by migration, followed by matrix exhaustion, a cortical plate develops, and subsequently becomes stratified, can be interpreted as a manifestation of HAEKKEL's biogenetic law *(biogenetisches Grundgesetz)*,[7] regardless of various deviations from the presumptive phylogenetic sequence. The limitations qualifying the interpretations in accordance with said law were clearly recognized by HAECKEL, who pointed out the *cenogenesis* modifying the *palingenetic* aspects. The overall validity of the biogenetic 'law' or perhaps better 'principle' with regard to corticogenesis was also recently upheld by STEPHAN (1975).

Although a pallial cortex is present in recent Selachians and Dipnoans, it seems not improbable that the early inverted Gnathostome Anamniote telencephalon displayed merely a periventricular arrange-

---

[6] With respect to the hippocampal cortex, however, this statement applies only to the rudimentary *precommissural hippocampus* of all Mammals, and to the *reduced supracommissural hippocampus* in 'higher' Mammals with expanded corpus callosum (cf. Fig. 223C, vol. 5/I).

[7] K. and v. DOMARUS (1920) and K. (1922). Cf. also the comments on the *'biogenetic law'* in chapter III, section 6 of volume 1. As regards the interpretation which I suggested in terms of KAPPER's neurobiotaxis, cf. chapter VI, section 6, pp. 610–618 of volume 3/II.

ment of its neuronal elements, comparable to that obtaining in Urodele Amphibians. Whether said arrangement in these latter forms does or does not represent a 'regressive feature' remains an open question, but this dubious point needs not contradict the above-mentioned phylogenetic interpretation.

Concerning the presumptive phylogenetic evolution of the Human cortex cerebri within the Mammalian series, STEPHAN (1975) justly states that this '*muss fast ausschliesslich auf dem indirekten Wege des Vergleichs rezenter Arten erschlossen werden*'. The cited author points out that, in this respect, the series Anthropoid Apes, Monkeys, Prosimii, and Insectivores provides suitable material for comparison. In his treatise on the allocortex, STEPHAN particularly considers the telencephalon of the Insectivore Erinaceus, of the Prosimian Galago, and of the Monkey Cercopithecus in comparison with that of Man.

2. The Basic Zonal Pattern and the Fundamental Cortical Types: Hippocampal Cortex, Parahippocampal Cortex, Neocortex, Anterior Piriform Lobe Cortex, and Basal Cortex

The Gnathostome Vertebrate lobus hemisphaericus telencephali displays, in all forms, a fundamental pattern or *bauplan*, characterized by seven longitudinal primordial griseal zones of the telencephalic alar plate, of which three (D_3, D_2, D_1) are *pallial*, and four (B_1, B_2, B_3, B_4) are *basal*. This pattern remains permanent in Anamnia, and is clearly recognizable at certain ontogenetic stages in all Amniota. In these latter, moreover, D_1 by *internation*, as stated above in section 1, joins the basilateral components, thereby becoming included in a *secondary, augmented basis lobi hemisphaerici* characteristic for all Amniota. In addition, during further ontogenetic formative processes, the primordial zonal arrangement becomes blurred by fusions of the D_1 and basilateral components, and by further differentiations within the D_2 and the D_1 components.

On the basis of ontogenetic stages succeeding the key stages at which the primordial longitudinal zones are most conspicuous, it is, nevertheless, possible to ascertain the derivation of the five main cortical types from distinctive topologic neighborhoods of the zonal pattern, and to establish, by topologic mapping, rigorously definable morphologic homologies.

The *five main cortical types* resulting as the outcome of these morpho-

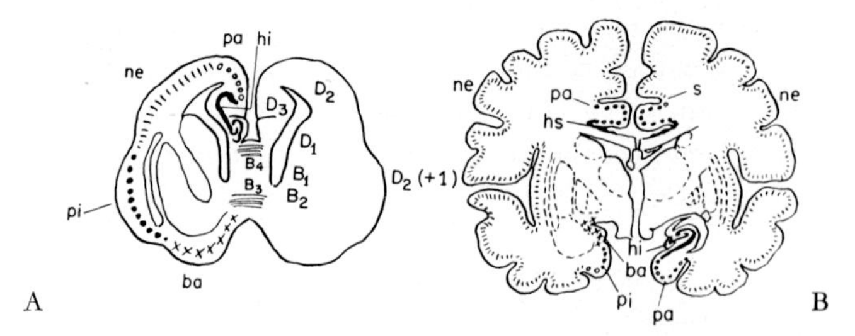

Figure 74. Diagrammatic cross-sections through the telencephalon of (A) a Metatherian Marsupial (Didelphys) at the level of commissura anterior and commissura hippocampi, and (B) through the Human forebrain at level of amygdaloid complex (left) and (right) level of postcommissural hippocampus (from K., 1957). ba: basal cortex; hi: hippocampal cortex; hs: supracommissural hippocampus in Man; ne: neocortex; pa: parahippocampal cortex; pi: cortex anterior lobi piriformis; s: sulcus cinguli. The topologic D and B notations are indicated in A.

genetic events, and clearly distinguishable by rigorous formanalytic procedure, are, in dorsomedial-dorsolateral-basilateral and basimedial sequence, the following (cf. Fig. 74).

1. The *hippocampal cortex*, derived from the zone D_3, is homologous to the medial cortical plate (or lamina) of Reptiles, and to the '*primordium hippocampi*' of Anamnia. It consists of a *precommissural*, a *supracommissural*, and a *postcommissural* subdivision. This latter displays two distinctive components, *cornu Ammonis* and *gyrus dentatus (sive fascia dentata)*, to be dealt with in section 3. These components are also differentiated in the supracommissural subdivision of some lower Mammals, but are merely represented by a gradient in most other Mammals (cf. Fig. 223C, vol. 5/I, and Fig. 100 of the present volume). Such gradient, comparable to that displayed by the medial cortical plate of Reptiles, also obtains in the rudimentary precommissural hippocampus[8] (cf. Fig. 222B, vol. 5/I).

2. The *parahippocampal cortex*, derived from a medial neighborhood of D_2, namely D_{2am}, is homologous to most of the dorsal cortical plate of Reptiles. It adjoins, as its name indicates, the hippocampal cortex, following the caudal limb of the hemispheric bend. The caudobasal

[8] In other words, rudimentary differentiation of Mammalian hippocampal cortex, as manifested by precommissural, and predominantly also by supracommissural hippocampus, is characterized by continuity between cornu Ammonis and gyrus dentatus, merely involving a gradual transition (gradient) from larger cellular elements (cornu Ammonis) to smaller ones (gyrus dentatus).

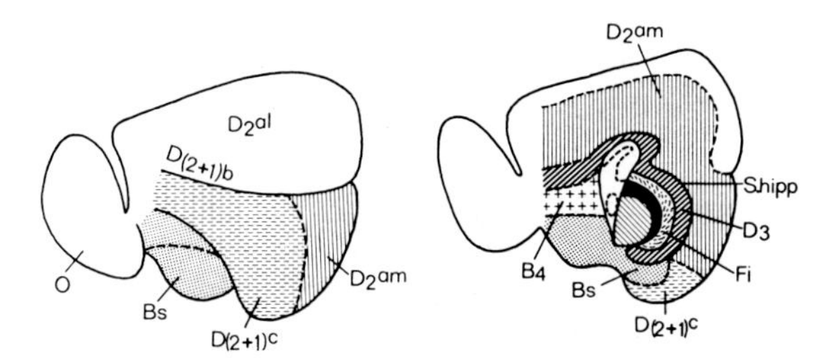

Figure 75. Mappings of the main cortical types in the Insectivore Mammalian Erinaceus (modified after K., 1927 by labelling in accordance with the topologic notation, from STEPHAN, 1975). O: olfactory bulb; Fi: fimbria fornicis; Shipp: sulcus hippocampi.

portion of the parahippocampal cortex, which also expands lateralward basally to pars posterior of sulcus rhinalis lateralis, thereby occupies the posterior part of the piriform lobe (cf. Fig. 75). This region corresponds approximately to BRODMANN's field 28 (area entorhinalis) and might be designated as cortex posterior lobi piriformis.

3. The *neocortex*, derived from lateral neighborhoods of D_2, namely D_{2al} and D_{2b}, which latter receives minor contributions from D_1, and thus becomes $D_{2(+1)b}$, is homologous to a small lateral neighborhood of the dorsal cortical plate in Reptiles. This Reptilian homologon of Mammalian neocortex was first pointed out by ELLIOT SMITH (1910a) as here shown in Figure 77 and subsequently was also recognized in the Alligator by CROSBY (1917). The term *neopallium*, from which the designation *neocortex* was subsequently derived, was introduced by ELLIOT SMITH (1901). This author defined the neopallium as a great unlimited area (far removed from the disturbing influences of the purely administrative parts of the nervous system), where impulses of diverse nature coming from all regions of the body and from all the sense organs[9] may meet and play upon each other, and which is in a greater degree the organ of the mind. It can be seen (Fig. 76) that ELLIOT SMITH at first included the parahippocampal cortex into the neopallium. In his subsequent *Arris and Gale lectures*, ELLIOT SMITH (1910a) excluded the subiculum hippocampi,[10] namely the cortex adjacent to the hippocam-

[9] Excluding here the sense of smell.

[10] The term *subiculum*, as here used by ELLIOT SMITH in accordance with the terminology of the older authors (cf. also vol. 5/I, chapter XIII, section 10, p. 696) refers to the cortical region adjacent to the hippocampal cortex and includes thus the *gyrus cinguli*. This is clearly indicated e.g. by Figures 24 and 26 of ELLIOT SMITH's (1910a) *Arris and Gale lectures*. Most contemporary cytoarchitecturalists, on the other hand, use the therm subi-

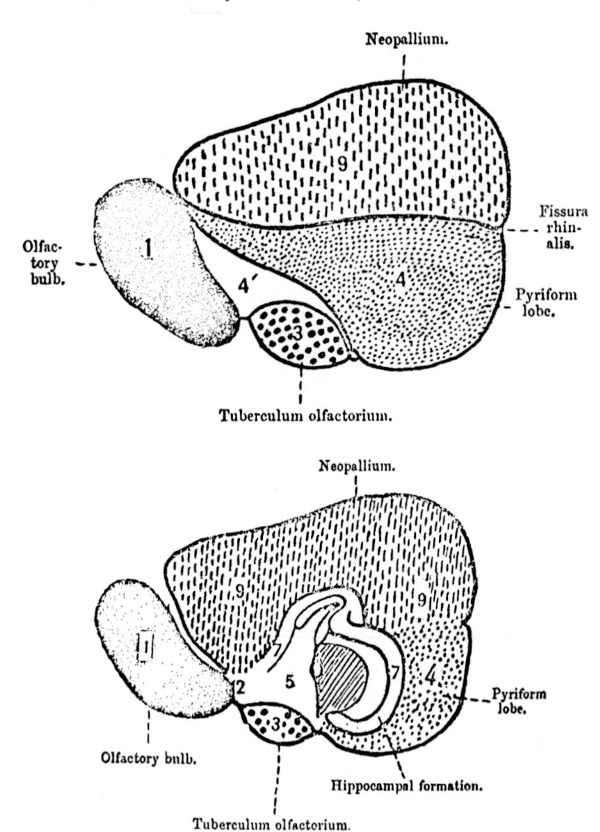

Figure 76. Subdivisions of hemisphere in 'typical' lower Mammals (Insectivore) in lateral and medial view, as originally proposed by ELLIOT SMITH (from ELLIOT SMITH, 1901). 1: olfactory bulb; 2: olfactory peduncle; 3: tuberculum olfactorium; 4: piriform lobe; 4′: lateral olfactory tract; 5: paraterminal body; 7: hippocampal formation; 9: neopallium.

pal cortex, from the neopallium (cf. Fig. 77). In a still later paper, ELLIOT SMITH (1919) introduced the designation *parahippocampal cortex* for that of the subiculum. Said designation, based on sound morphological relationships, was adopted by the present author (K., 1924c). The term *isocortex*[11] was introduced by O. VOGT about 1910 and fur-

---

culum for that portion of cornu Ammonis which adjoins the parahippocampal cortex (cf. Fig. 86 D).

[11] As in the case of ELLIOT SMITH's neocortex, the VOGTS' concept of isocortex underwent some changes concerning the classification of parahippocampal components, some of which, particularly in higher Mammalian forms, tend to display a stratification

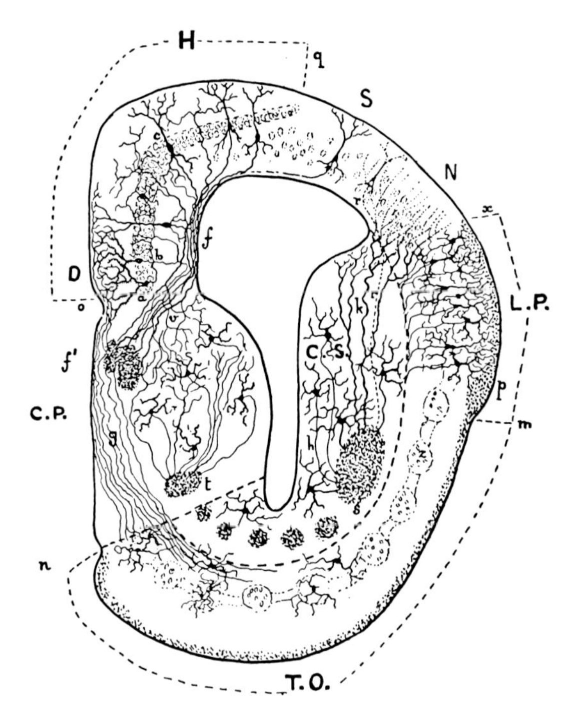

Figure 77. Diagrammatic sketch representing cross-section through the hemisphere of an hypothetical 'promammalian' form, intermediate between Reptiles and true Mammals, as suggested by ELLIOT SMITH (from ELLIOT SMITH, 1910a). C.S.: corpus striatum; C.P.: corpus paraterminale; D: fascia dentata; H: hippocampal cortex; L.P.: lobus piriformis (pars anterior; N: earliest rudiment of neopallium; S: subiculum (parahippocampal cortex); T.O.: tuberculum olfactorium; a: cells of gyrus dentatus; b: 'intermediate type of cell'; c: cell of cornu Ammonis; f, f': fornix; g: fibers connecting tuberculum olfactorium with medial pallium; h, k: fibers joining lateral forebrain bundle; m: sulcus endorhinalis; n–x: extent of superficial fibers from olfactory tracts; o–q: extent of hippocampal cortex; p: part of lateral olfactory tract joining cortex of anterior piriform lobe; r: cell and fiber of neocortex; s: lateral forebrain bundle; t: medial forebrain bundle; v: hippocampal fibers joining paraterminal body; z: *islet of Calleja.*

ther qualified in the paper by C. and O. VOGT (1919) summarizing the overall concepts of these investigators. Roughly speaking, isocortex can be regarded as synonymous with neocortex, while the term *allocor-*

---

analogous to that of neocortex, especially in *gyrus cinguli.* This tendency toward '*neocorticalization*' introduces certain difficulties for a suitable classification. These difficulties may be resolved, as already intimated by ELLIOT SMITH (1910a), if an additional distinction is made between *neocortex sensu strictiori* and *sensu latiori*, such that this latter might include either part or the whole of *parahippocampal cortex.*

tex, likewise introduced by the VOGTS, subsumes all non-neocortical telencephalic cortices.

4. The *anterior piriform lobe cortex*, derived from the most lateral neighborhood of D_2, namely D_{2c}, which receives some contributions from D_1, thereby becoming $D_{(2+1)c}$, is homologous to the lateral cortical plate of Reptiles. It extends basally to sulcus rhinalis lateralis and dorsally to sulcus endorhinalis, and represents the cortex of a large part of piriform lobe, except for a variable portion of that lobe's posterior part, which contains parahippocampal cortex. The anterior piriform lobe cortex is doubtless related to the olfactory system and represents a component of the rhinencephalon, except in the few Mammalian forms which are entirely anosmatic, but in which said cortex is, nevertheless, still recognizable, and may perhaps subserve non-olfactory functions based on connections presumably already obtaining in the rhinencephalic anterior piriform lobe. The cortex of that lobe is also commonly designated as '*prepiriform cortex*', but this term seems less appropriate since that cortex is a conspicuous component of a large portion of piriform lobe.[12]

5. The *basal cortex* is a merogenic (or semiparietine) derivative of the zones B_1 to B_4 becoming thereby B_s, and is homologous to the basal cortex of Reptiles. The two main subdivisions of the basal cortex are the *tuberculum olfactorium* and its surroundings, located basally to the sulcus endorhinalis, and the *cortex of nucleus amygdalae (nucleus amygdalae corticalis)*, likewise separated from the piriform lobe by a caudal extension of sulcus endorhinalis, designated in Primates as sulcus semiannularis (cf. Fig. 175D, vol. 5/I, and Fig. 56A of the present volume). These two main subdivisions of the basal cortex, also displayed by various Anamnia with inverted lobus hemisphaericus, were also designated as *area ventralis anterior* and area *ventrolateralis posterior* (K., 1927). The former includes, in addition to the cortex of tuberculum olfactorium with its lateral, rostral, and caudal gradients (such e.g. as the gri-

[12] There is, moreover, no clear-cut rostral delimination of the piriform lobe, which can be considered to reach rostrad as far as olfactory bulb, stalk, or nucleus olfactorius anterior, as the case may be. The 'prepiriform cortex' lies thus entirely within the piriform lobe. One might, however, distinguish, within the anterior piriform lobe cortex *sensu latiori*, an anterior *(sensu strictiori)* and an intermediate subdivision. Both essentially correspond to subdivisions of BRODMANN's area 51, while the posterior piriform lobe cortex is parahippocampal, corresponding to the so-called entorhinal cortex or to BRODMANN's area 28. *In toto*, the cortex anterior lobi piriformis seems to be represented by BRODMANN's field 51 with additional subdivisions (a, b, etc.).

seum of substantia perforata anterior) variable superficial, i.e. 'cortical' components of the paraterminal complex B_3 and B_4, that is, 'septal' grisea or 'nuclei' as well as interstitial cell groups of *Broca's diagonal band.* There is little doubt that the basal cortex functionally represents a component of the 'rhinencephalon' although with connections respectively relationships to other systems.[13] With respect to entirely anosmatic Mammals, the qualifications concerning the anterior piriform lobe cortex, pointed out above, likewise apply.

The foregoing subdivision of the Mammalian cerebral cortex, based on the application of topological concepts to the formulation of morphologic homologies, appears as rigorously established as the inherent weaknesses of abstractions, the limitations of semantics, and the intrinsic difficulties in drawing boundaries for open topologic neighborhoods permit. The most relevant clue for a solution of the pertinent problem was given by the fact that the permanent (adult) telencephalic zonal configuration of the Amniote lobus hemisphaericus can be detected at key stages of Amniote ontogenesis, thus allowing for a recording of the subsequent transformations producing the Amniote telencephalic patterns displayed by Reptiles, Birds, and Mammals. The internation of the zone D_1 is here a most significant formative event resulting in the considerable gap between the configuration of lobus hemisphaericus in Anamnia and of that of Amniota. Without a proper understanding of these morphogenetic events, an accurate establishment of homologies, and an appropriate subdivision of Mammalian respectively Amniote cortex cerebri cannot be achieved.

The solution of the pertinent problems was indicated in the author's papers of 1924 and 1925–1926, and further developed in those of 1929 and 1938. The relevant implications of topology were then elaborated in chapter III of volume 1, and in chapter VI, section 1A, pp. 56–67 of volume 3/II. Some concluding remarks on the controversial homology concept will be given in chapter XVI of the present volume.

With regard to the various different concepts and classifications concerning the subdivision of the Mammalian cerebral cortex elaborated by other authors, the interpretations proposed by Edinger and by Kappers may be mentioned first.

Edinger (1908, 1911, 1912 *et passim)* introduced the term *archipallium*, from which the expression *archicortex* became derived, to desig-

[13] Possibly including the mechanism of the *oral sense* postulated by Edinger (cf. Fig. 237E, vol. 5/I).

nate the *hippocampal cortex*. ELLIOT SMITH, to whom this term was attributed, since he had coined the designation neopallium, strongly objected to EDINGER's term, particularly because it applied to only one of the non-neopallial components.[14]

As regards Reptiles, EDINGER considered the dorsal cortical plate homologous to the Mammalian cornu Ammonis, and the medial plate to the fascia dentata of Mammals. EDINGER, moreover, interpreted the lateral cortical plate of Reptiles as being homologous to the Mammalian neopallium respectively neocortex. *Prima facie*, if the *topographic* location of said Reptilian cortical plate is assumed to represent its *topologic locus*, this interpretation appears quite plausible, regardless of differences in fiber connections. Impressed by EDINGER's fundamental work, upon whose results my first investigations were based, I accepted this interpretation (K., 1922) before an intensive study of *key ontogenetic stages* made me realize the significance of the displacements resulting from the 'infolding of the epibasal nucleus' (i.e. from the *internation* of D_1), which is manifested by the prominent ridge of the *lateral ganglionic hill* of Mammalia (K., 1924c, 1925–26).

KAPPERS (1921) and KAPPERS *et al.* (1936) followed EDINGER's interpretation of the Reptilian dorsal and medial cortex and retained the term archicortex. The lateral cortex of Reptiles, however, he considered to be homologous to the Mammalian prepiriform cortex and designated it as *palaeocortex*, pertaining to his *palaeopallium*. KAPPERS did not explicitly elaborate upon the extent of his palaeopallium or palaeocortex, but included all or most of the basal cortex into it.[15] His delimitation of the piriform lobe is not unequivocably defined. The Reptilian neocortical primordium is interpreted by this author in agreement with ELLIOT SMITH (1910a) and CROSBY (1917) as a rostral lateral differentiation of the dorsal cortical plate in the region of *superpositio lateralis*. The results of my own formanalytic studies (K., 1924c, 1927, 1929) confirmed the validity of this neocortical homology.

My esteemed old friend HUGO SPATZ retained EDINGER's original view concerning neocortex and archicortex combined with KAPPER's

---

[14] Cf. ELLIOT SMITH (1910b): The term 'archipallium' – a disclaimer.

[15] Thus, KAPPERS *et al.* (1936, p. 1517) clearly state that the sulcus endorhinalis 'lies within the palaeocortex-piriform lobe cortex – and therefore is an axial fissure'. It will be recalled that I consider it a limiting sulcus between the *pallial piriform lobe cortex* and the *basal cortex*. BECCARI (1943) retains KAPPERS subdivision into archicortex, palaeocortex and neocortex, and, like KAPPERS, does not specifically refer to the cortex basalis and its status in said scheme of cortex classification.

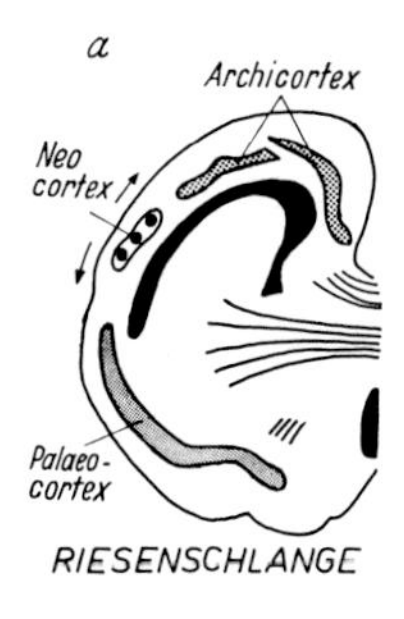

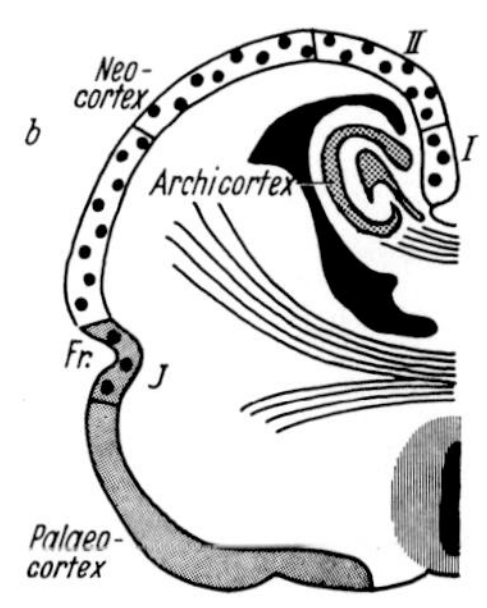

Figure 78. Main types or regions of cortex cerebri (a) in a Reptile, and (b) in a lower Mammal, as interpreted by H. SPATZ (adapted and modified after EDINGER, 1912, from SPATZ, 1966). Fr.: sulcus rhinalis lateralis; J: insular cortex; I: 'splenial segment' of neocortex; II: 'ectomarginal segment' of neocortex. I and II represent parahippocampal cortex in the terminology which I have adopted. It can also be seen that in this interpretation, the 'palaeocortex' would be entierly basal. The unlabelled surface groove at the bottom presumably represents the sulcus endorhinalis.

concept of palaeocortex, as here shown in Figure 78. He conceived, moreover, the insular neocortex as intermediate between neocortex and palaeocortex. In addition, he regarded two segments of neocortex, displaced *('supprimiert')* into the fissura interhemisphaerica (fissura longitudinalis cerebri) as splenial (I) and ectomarginal (II) segments, of which he considered the former, together with the archicortex, as pertaining to the limbic system.

As regards the terms palaeopallium and archipallium, respectively palaeocortex and archicortex, I share ELLIOT SMITH's misgivings about these designations and indicated on pp. 663–668, section 6 of chapter VI in volume 3/II the several reasons why said terms preferably should be dropped, while the concept of 'neocortex' can be upheld on more solid grounds. The concept of palaeocortex is particularly objectionable since it lumps together two morphologically quite different cortices, namely the pallial cortex of anterior piriform lobe, and the basal cortices.

BRODMANN (1909) whose important studies on cortical cytoarchitectonics in Mammals concerned ontogenetic as well as adult aspects, distinguished two main subdivisions, namely *homogenetic* and *heterogenetic* cortex. The former is said to display, at certain ontogenetic stages, a basic hexalaminar 'genetic type' which may subsequently undergo some regional changes; it corresponds to the isocortex or neocortex.

The heterogenetic cortex, corresponding to the allocortex, which, beginning at relevant ontogenetic stages, manifests an architecture differing from the hexalaminar type, consists of three main subdivisions, namely *cortex heterogeneticus primitivus*, *rudimentarius*, and *striatus*.

The *cortex primitivus* displays no distinctive lamination of its cell plate and comprises bulbus olfactorius, substantia perforata anterior and nucleus amygdalae.

The heterogenetic *cortex rudimentarius* includes hippocampus with fascia dentata, subiculum,[16] indusium griseum, septum pellucidum, and area praeterminalis.[17]

The heterogenetic *cortex striatus* develops a variety of laminar stratifications which may either be less or in some instances even more in number than those of homogenetic cortical areas, but which do not derive from the hexalaminar pattern. The cortex striatus is said to include area praepiriformis, area entorhinalis, area praesubicularis,[18] area retrosubicularis and perhaps area ectosplenialis. In addition to his main, homogenetic and heterogenetic cortex types, and to his parcellation of the cortex into numbered areae, BRODMANN also distinguished eleven regions containing the diverse areae, namely (1) regio postcentralis, (2) praecentralis, (3) frontalis, (4) insularis, (5) parietalis, (6) occipitalis, (7) temporalis, (8) cingularis, (9) retrosplenialis, (10) hippocampica, (11) olfactoria (cf. also Fig. 103).

BRODMANN's concepts, which greatly impressed me and influenced my early investigations, do not sufficiently take into consideration the morphologic data provided by comparative neuroanatomy of submammalian forms (Anamnia, Sauropsida) and may be evaluated as '*neopalliocentric*'. Although I attempted to bring my own findings in agreement with BRODMANN's views until about 1929, I finally became convinced that many of that author's interpretations cannot be upheld.[19]

---

[16] This is, in BRODMANN's terminology, the region of cornu Ammonis adjacent to the parahippocampal cortex, i.e. to the subiculum of ELLIOT SMITH and other older authors.

[17] I was unable to identify with certainty what BRODMANN meant by his area praeterminalis or field 25. It may perhaps even include the preterminal paraterminal grisea (B_{3+4}).

[18] This is the parahippocampal subiculum of older authors as mentioned in footnote 16.

[19] Already in 1927, I reached the conclusion '*dass* BRODMANN *in seiner Auffassung, die Urarchitektonik des Cortex cerebri der Mammalier auf den sechsschichtigen tektogenetischen Grundtypus zurückzuführen, manchmal zu weit geht*' (K., 1927, p. 308). The delimitation of cortical laminae (layers) is, moreover, highly arbitrary and subjective. Even KAHLE (1966), who follows BRODMANN, is willing to admit that the hexalaminar stratification of the fetal

As regards the cited author's main subdivisions, the homogenetic cortex evidently corresponds to the neocortex. Cortex primitivus corresponds to cortex basalis, namely area ventralis anterior and ventrolateralis posterior (cortical nucleus amygdalae). The olfactory bulb, however, although it may display, in Mammals, a cortex-like arrangement with at least two (mitral and granular) cell layers, does not, in my opinion, represent a suprasegmental correlation tissue corresponding to the definition of cortex given above in section 1. The olfactory bulb and its stalk, moreover, can be evaluated as a telencephalic *grundbestandteil sui generis*, distinctive from the lobus hemisphaericus with its longitudinal zonal pattern. Thus, neither olfactory bulb nor nucleus olfactorius anterior would represent 'true' cortical regions of the hemispheric lobe. It will also be recalled that the olfactory bulb and tract may entirely be missing in some anosmatic forms (cf. Fig. 34), while the 'rhinencephalic' cortex basalis *sive* 'cortex primitivus' nevertheless persists.

Again, insofar as paraterminal grisea are superficial and cortex-like (e.g. *Broca's diagonal band)*, they should be included in the basal cortex, rather than with BRODMANN's 'septum pellucidum' in the 'cortex rudimentarius'. Much the same could be said about part of the indusium, namely the indusium verum (cf. vol. 5/I, chapter XIII, section 10 and Fig. 223C).

BRODMANN's cortex heterogeneticus rudimentarius, discounting his 'septum pellucidum', clearly represents the hippocampal cortex, to which also the indusium spurium pertains.

The cortex heterogeneticus striatus of the cited author subsumes two entirely different cortices, namely the cortex anterior lobi piriformis ('area praepiriformis') and the parahippocampal cortex (e.g. areae entorhinalis, praesubicularis, retrosubicularis).

With regard to the main regions distinguished by BRODMANN and enumerated above, regions 1 to 7 represent neocortex, regions 8 to 10 parahippocampal cortex, while regio olfactoria (11) lumps together pallial cortex anterior lobi piriformis and basal cortex, as well as, apparently, hippocampal cortex.[20]

---

neocortex '*ist in vielen Bezirken nur ein flüchtiges Durchgangsstadium, das keineswegs im ganzen Isocortex das gleiche stereotype Bild bietet, sondern von Anfang an lokale Variationen aufweist*'.

[20] BRODMANN (1909) is not quite explicit as to the inclusion of hippocampal cortex in his scheme of regions. On p. 208 however, he states that said region '*wird durch rudimentäre Rindenformationen gebildet*'. Misled by some ambiguities in BRODMANN's Human brain

Contemporary schemes of main cortical subdivisions in Mammals tend toward a combination of the concepts neocortex (isocortex) and allocortex, respectively archicortex and palaeocortex, with concepts derived from cytoarchitectural and myeloarchitectural parcellations, as well as from ontogenetic stages. Studies and interpretations concerning this topic are those by FILIMONOFF (1947, 1964), KIRSCHE (1972, 1974), LAISSUE (1963), SANIDES (1962, 1972), and STEPHAN (1956, 1960, 1975, and others). In the *Nomina Histologica* of 1972, an attempt was made to formulate a standardized nomenclature.

Although, in my opinion, the morphological distinction of the five main cortical types, the overall arrangement of their surface distribution, and their homologies with submammalian telencephalic regions seem rigorously established on the basis of topologic one-many and many-one mappings, difficulties *qua* terminology result if these cortical regions are not properly understood as representing topologically *open* neighborhoods,[21] which, being *open* sets, interlock, and do not allow for the drawing of 'precise' linear boundaries.

These difficulties have led to the formulation of diverse terminologies by various authors who, as also indicated by the above-mentioned *Nomina Histologica* of 1972, have introduced specific designations for certain gradients obtaining between or within some of the main cortical regions.[22] If said gradients are appropriately included into the relevant main regions, special terminologic subdivisions such as *mesoarchi-*

charts, I mistakenly included on p. 317 of my '*Vorlesungen*' (1927) his field 51 ('prepiriform area') into his regio hippocampica (10). Said field is actually included by BRODMANN into regio olfactoria (11).

[21] Cf. the elementary introduction to topologic concepts in section 2, chapter III of volume 1, and their application to a 'relaxed' morphologic topology, where a *bauplan* is conceived as a topologic space. Additional comments, with some emendations kindly pointed out to me by a noted mathematical topologist (Dr., now Professor BETTINGER), can be found on pp. 60–67, section 1 A of chapter VI of volume 3/II.

[22] Very steep gradients, approaching abrupt transiting, and providing hair-sharp *(haarscharfe)* boundaries, are rare and the extension *(Umfang)* of this sort of boundary is based on subjective, arbitrary viewpoints. As regards the five main cortical types, only the boundary between parahippocampal and postcommissural hippocampus (as well as in supracommissural hippocampus of some 'lower' Mammalian forms) is almost linear. Within the neocortex of some 'higher' Mammals, the transition between area striata (field 17 of BRODMANN) and area parastriata (field 18) is indeed practically 'hair-sharp'. Within the fully differentiated hippocampal cortex, the transition of cornu Ammonis to gyrus dentatus is likewise quite 'sharp', and fairly steep gradients can be recognized within the fully developed cornu Ammonis as e.g. characteristic for postcommissural hippocampus.

cortex, *mesopalaeocortex* etc. may be considered unnecessary. The designation '*mesocortex*' was introduced by ROSE (1927a, b).

Detailed discussions on the different contemporary subdivisions and terminologies concerning the Mammalian cerebral cortex and its homologies, together with the relevant bibliography, can be found in the paper by KIRSCHE (1974) and in the treatise by STEPHAN (1975). This latter author justly remarks: '*Eine einheitliche, allgemein anerkannte Umgrenzung des Allocortex gibt es nicht. Unterschiedliche Auffassungen bei der Abgrenzung gegenüber den nichtcorticalen Strukturen beruhen auf dem Fehlen einer einheitlichen Cortexdefinition. Schwierigkeiten bei der Abgrenzung gegenüber dem Isocortex entstehen durch Übergangsgebiete, die Merkmale beider corticaler Grundtypen haben.*'

STEPHAN, on the basis of his fundamental studies concerning the allocortex,[23] formulates the following overall subdivisions:

Allocortex	Allocortex primitivus	
	Periallocortex	Mesocortex
Isocortex	Proisocortex	
(sensu lato)	Isocortex maturus	

The allocortex primitivus comprises (A) a palaeocortex I *sive* semicortex, and a palaeocortex II *sive* eupalaeocortex, and (B) the hippocampal cortex with its precommissural, supracommissural and retrocommissural (postcommissural) subdivisions.

3. Problems of Structure and Architectonics

For a brief review of overall structural and architectural aspects of the five fundamental types of the Mammalian cerebral cortex, it appears convenient, in the aspect here under consideration, to deal with these cortices in the following order: (a) basal cortex, (b) cortex of anterior piriform lobe, (c) hippocampal cortex, (d) neocortex, and (e) parahippocampal cortex. This latter will be taken up last, following the neocortex, because the peculiar stratification of the parahippocampal cortex may appropriately be discussed with reference to conventional lamination concepts applied to the neocortex.

[23] STEPHAN subsumes the olfactory bulb (including the accessory one) under his concept of allocortex, but as 'allocortex bulbi olfactorii', distinct from his palaeocortex I *sive* semicortex. In this latter, however, he includes the 'regio retrobulbaris', apparently corresponding to the nondescript grisea of 'nucleus olfactorius anterior'.

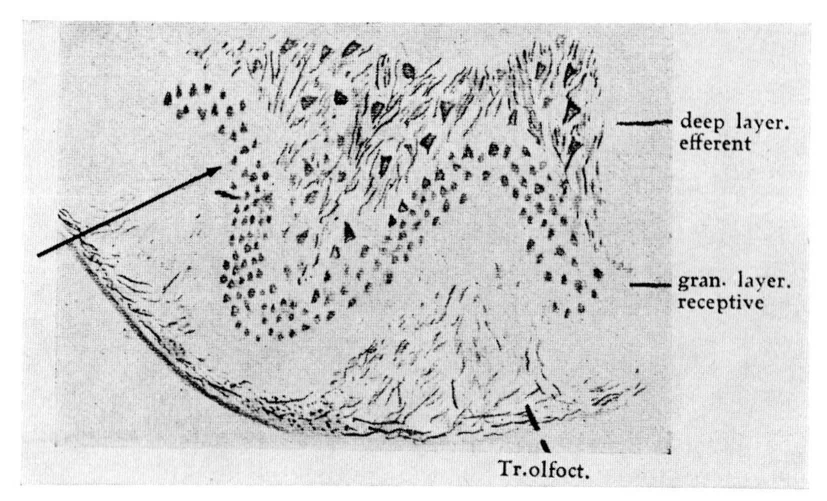

Figure 79. Basal cortex of tuberculum olfactorium in the Marsupial Hypsiprymnus rufescens (from KAPPERS, 1928). The added arrow at left indicates transition to cortex anterior lobi piriformis. KAPPERS includes the basal cortex into his palaeocortex.

The architecture of the merogenic *basal cortex* is relatively simple. Generally speaking it displays a molecular (or zonal) layer with a few scattered neuronal elements, a dense cortical plate of small or medium-sized cells, and a deeper layer of scattered larger cells, which is vaguely delimited from the non-cortical basal grisea. The molecular layer includes fibers of the lateral olfactory tract, and deeper fibers of the intermediate olfactory tract may run internally to, or through the deeper layer. Medialward, in the paraterminal region, the two cellular layers of the basal cortex extend as a less dense single nondescript cell band along the surface of the basimedial grisea.

Although a parcellation into various additional regional subdivisions related to rostro-caudal gradienst of dubious significance could be made, it seems at present sufficient to distinguish two main regions, namely *area ventralis anterior* and *area ventralis posterior*.[24]

In macrosmatic Mammals, the area ventralis anterior represents the tuberculum olfactorium and its immediate caudal and medial surroundings, the medial one being the paraterminal basal cortex, including the griseum of *Broca's diagonal band*. The tuberculum olfactorium in

[24] In *Anamnia*, the area ventralis posterior assumes a lateral position and is thus designated as *area ventrolateralis posterior*. In Mammals, because of the pallium's expansion, it becomes displaced into a medial position. This area is the cortical 'nucleus amygdalae'.

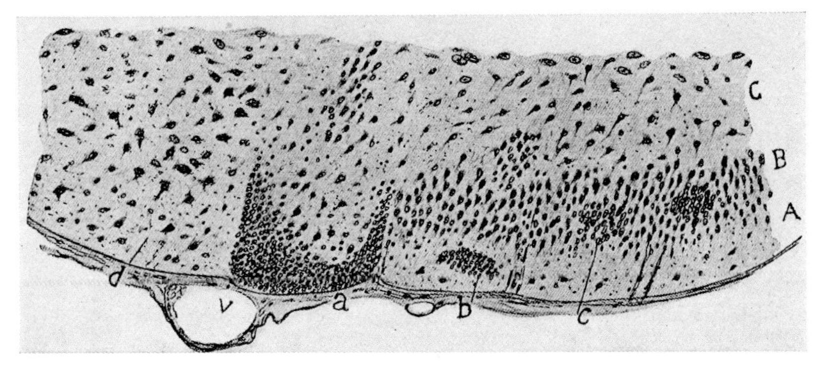

Figure 80. Parasagittal section *(Nissl stain)* through the basal cortex in a 2 months old Cat (from CAJAL, 1911). A: molecular (plexiform) layer; B: 'medium-sized pyramids'; C: layer of 'polymorphous cells'; a: area ventralis posterior (cortical amygdaloid nucleus; b: *island of Calleja* in molecular layer; c: *island of Calleja* in layer B; d: probably part of hypothalamic lateral preoptic area; v: blood vessel (its position roughly corresponds to telodiencephalic boundary; CAJAL interprets d as part of tuberculum olfactorium rostral to optic chiasma).

macrosmatic forms, which frequently displays scalloped foldings of its dense cortical plate which do not affect the tubercle's surface, is, moreover, characterized by clusters of small or medium-sized cells either within the said cortical plate or, more peripherally, within the molecular layer, the so-called *islets of* CALLEJA (1893).[25]

The area ventralis posterior is the cortical nucleus amygdalae[26] dealt with in section 10 of chapter XIII in volume 5/I. Several nondescript subdivisions of this area can be arbitrarily delimited. Of these latter, however, an anterior medial portion, the nucleus amygdalae delta, (cf. chapter XIII, vol. 5/I) is a rather distinctive subarea.[27] Figures 79–82 illustrate relevant configurational and architectural aspects of cortex basalis.[28]

Structurally, the basal cortex consists of rather irregular multipolar cells of small and somewhat larger size, as mentioned above. Some

---

[25] The superficial position of such islets suggests a 'neurobiotactic effect' of the olfactory tract fibers.

[26] This is ROSE's (1929b *et passim)* area periamygdalaris 2. The cited author designates as area periamygdalaris 1 a ventral portion of cortex anterior lobi piriformis.

[27] This is apparently ROSE's area periamygdalaris 3.

[28] Functional aspects were dealt with in section 10 of chapter XIII, volume 5/I. The relationship of tuberculum olfactorium to the oral sense postulated by EDINGER for at least some Mammals should again be pointed out.

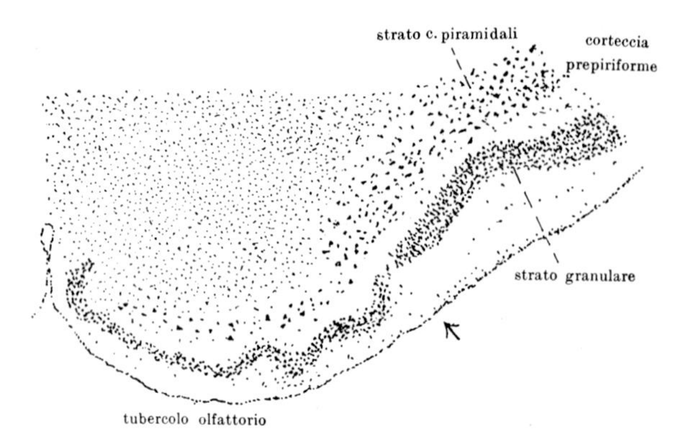

Figure 81. Basal cortex (tuberculum olfactorium) and cortex anterior lobi piriformis in the Mouse, as drawn from a *Nissl stain* preparation (from BECCARI, 1943). The arrow indicates boundary between basal and pallial cortex.

of these elements, however, tend to assume the shape of 'pyramidal cells'.

The architecture of the *cortex anterior lobi piriformis* is likewise relatively simple and does not significantly differ from that of well-differentiated regions of basal cortex. A zonal (molecular, plexiform) layer, mainly, but perhaps not exclusively containing fibers of the lateral olfactory tract, a dense outer cell layer, and a less dense inner cell layer are present. The outer cell layer is generally thicker than that of area ventralis anterior. The inner cell layer is, as a rule, more clearly delimited by a subcortical layer, relatively poor in cells, from the internal grisea. The ventral edge of the inner cell layer, however, is generally continuous with the ventral portion of the claustrum, particularly in ma-

---

Figure 82 A. Basal cortex with adjacent pallial cortices (*Nissl stain*) at preterminal level in the Mouse (from ROSE, 1929b). The outer arrows and numbers have been added in accordance with my interpretation. 1: insular neocortex; 2: cortex anterior and intermedius lobi piriformis; 3: cortex basalis with *islets of Calleja;* 4: cortex basalis of paraterminal region; 5: precommissural hippocampus; 6: precommissural parahippocampal cortex; Ce: capsula externa; Cl: claustrum (the capsula extrema is not developed, but the claustrum can easily be identified.

Figure 82 B. Caudal portion of cortex anterior lobi piriformis and area ventralis posterior of basal cortex *(Nissl stain)* at post-terminal level in the Mouse (from ROSE, 1929b). The outer arrows and numbers have been added as in A. 7: rostral end of parahippocampal cortex of lobus piriformis (cf. Fig. 75); 8: area ventralis posterior of basal cortex (three nondescript subdivisions of this 'cortical amygdaloid nucleus' can be seen; 9: telodiencephalic sulcus and boundary; 10: lateral preoptic grisea and (the dense cell group at surface) supraoptic nucleus.

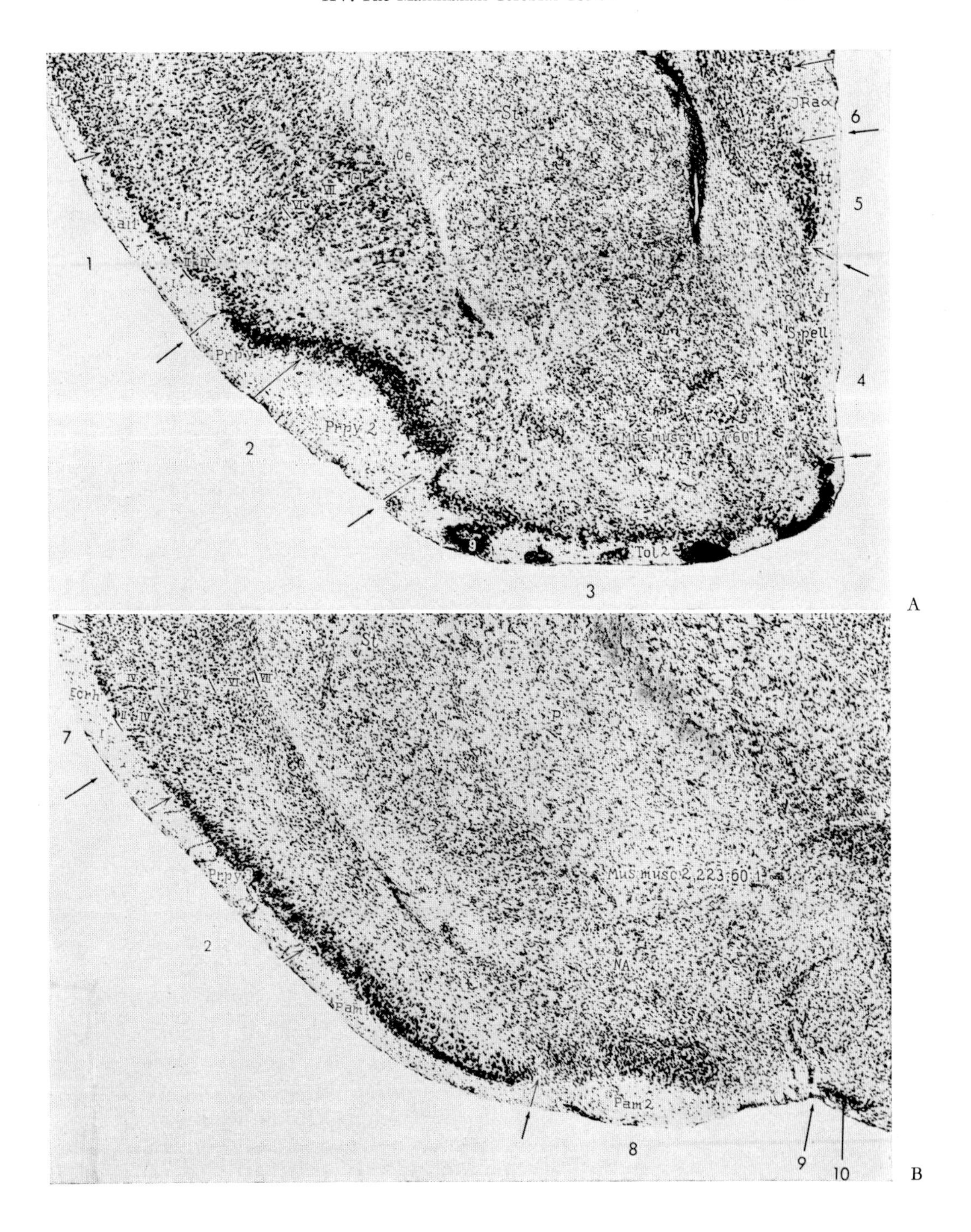

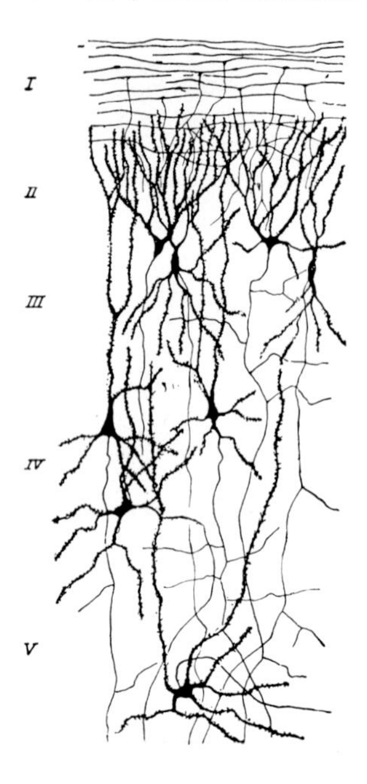

Figure 82 C. Structure of cortex anterior lobi piriformis *(Golgi impregnation)* in a young Rabbit (after CAJAL, 1911, from BECCARI, 1943). I: layer of fibers from lateral olfactory tract; II: plexiform layer; III: layer of superficial polymorph cells': IV: layer of pyramidal cells; V: layer of 'deep polymorph cells'.

crosmatic 'lower' Mammals, the just mentioned subcortical layer being the capsula extrema. In microsmatic higher Mammals, such as Primates, the cortex anterior lobi piriformis is relatively reduced and rather indistinctly differentiated. Figures 81 and 82 illustrate cytoarchitectural and configurational aspects of cortex anterior lobi piriformis. Slight differences may obtain between its rostral subregion (subregio anterior *sensu stricto)* and its more posterior subregio (regio intermedia), which borders, through a not very steep gradient, on the parahippocampal cortex of posterior piriform lobe (BRODMANN's area 28), the anterior and intermediate piriform lobe cortex corresponding to that author's area 51.

As regards structural aspects, the cells in the dense outer cell layer tend to assume a more distinctly pyramidal shape than those of the basal cortex, but fusiform and stellate cells are also present. The inner cell layer contains mostly irregular multipolar cells, but some pyramidal cells are also present. As can be seen in Figure 82C, CAJAL distinguished five layers of cortex anterior lobi piriformis, layer I being represented by fibers of the lateral olfactory tract, layer II by dendrites of cortical cells, intermingled with said tract, layer III by 'polymorphous superficial cells', layer IV by pyramidal cells, and layer V by larger

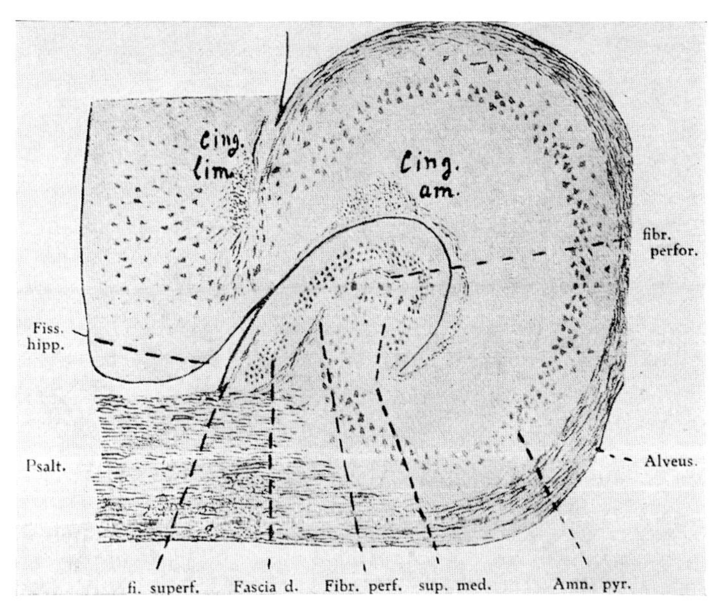

Figure 83. Fully developed supracommissural hippocampus in the Marsupial Hypsiprymnus rufescens (from KAPPERS, 1928). cing.am.: molecular layer of cornu Ammonis; cing.lim.: cingulum fibers of parahippocampal cortex; Psalt.: commissura hippocampi (psalterium); sup.med.: hilus of gyrus dentatus (interpreted by KAPPERS as corresponding to Reptilian superpositio medialis); other abbreviations self-explanatory. The lineal continuation of 'fissura hippocampi' is presumably not an extension of the more shallow hippocampal sulcus, but the boundary between molecular layers of gyrus dentatus and cornu Ammonis, through which zone branches of hippocampal artery run (cf. Fig. 89C).

deep polymorphous cells. In the simplified lamination concept which I favor, I and II represent the molecular layer, III and IV the dense outer cell lamina, and V the more loosely arranged inner lamina.

As regards the architecture of the *hippocampal cortex*, three distinct regional subdivisions must be distinguished, namely, *precommissural*, *supracommissural*, and *postcommissural hippocampus. In toto*, the Mammalian hippocampal cortex can be evaluated as homologous to the D_3 zone of all Gnathostome Vertebrates, and, in particular, to the medial cortical plate of Reptiles, respectively to the hippocampal cortex of Birds.

The *precommissural hippocampus* is rather rudimentary and displays a simple architectural arrangement, similar to that obtaining in Sauropsida. It represents the topologically basimedial edge of the pallial cortex, and is not sharply demarcated from the adjacent parahippocampal cortex, being continuous with this latter through a vague gradient. At

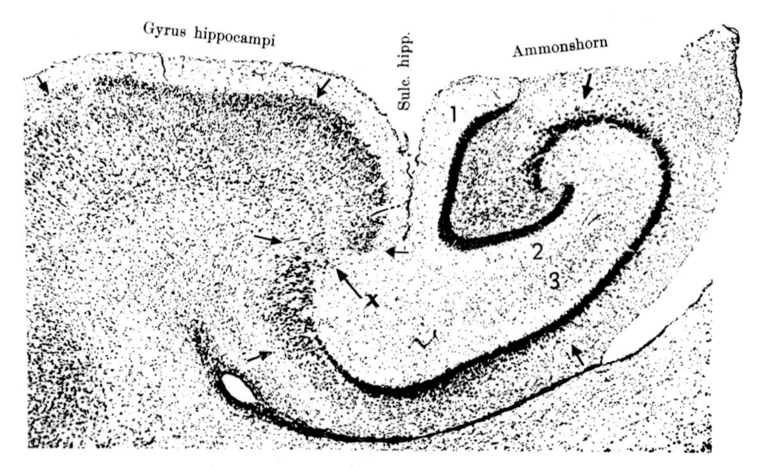

Figure 84 A. Cross-section (*Nissl stain*) through postcommissural hippocampus and adjacent parahippocampal cortex in a young Kangaroo of the species Onychogale frenata (after BRODMANN, 1909, from K., 1927). 1, 2: outer and inverted molecular layer of gyrus dentatus; 3: molecular layer of cornu Ammonis; arrow x indicates boundary between parasubicular portion of cornu Ammonis and parahippocampal cortex of subiculum *sive* gyrus hippocampi. Other arrows indicate subdivisions made by BRODMANN. 1, 2, 3, and x have been added in accordance with my interpretation.

its edge, however, the precommissural hippocampus is demarcated by a rather steep and well defined gradient from the adjacent basal paraterminal grisea. In appropriate sections, the precommissural hippocampus appears as a tapering wedge, whose more or less pyramidal cells are larger near the parahippocampal cortex, and become smaller toward the beveled edge of the cortical plate. The portion with the larger cells corresponds to the cornu Ammonis, and the small-celled portion to the gyrus dentatus. Figures 222A, B of volume 5/I, and 82A of the present volume illustrate this arrangement, which can be compared with Figures 210A, C, and 212A, as well as Figures 217C and 218A–D of volume 5/I, illustrating the hippocampal cortex in Sauropsida.

The *supracommissural hippocampus* in various lower Mammals (cf. section 10 of chapter XIII, vol. 5/I) is fully developed, with differentiated gyrus dentatus and cornu Ammonis, and with steep gradient at the transition to parahippocampal cortex (Fig. 222C, vol. 5/I, and Fig. 83 of the present volume). In higher Mammals, however, it is greatly reduced by the expansion of corpus callosum, and becomes the indusium spurium, which is comparable to the rudimentary precom-

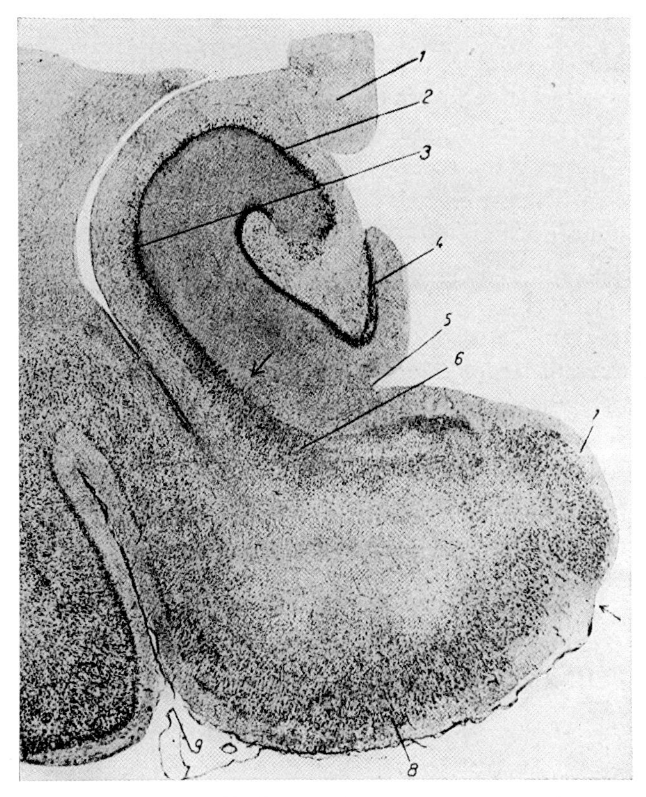

Figure 84 B. Cross-section *(Nissl stain)* through postcommissural hippocampus and adjacent parahippocampal cortex in the Dog (from BECCARI, 1943). 1: fimbria; 2: paradentate (CA 2–3) portion of cornu Ammonis; 3: pars prominens (CA 1) of cornu Ammonis; 4: gyrus dentatus; 5: sulcus hippocampi; 6: pars parasubicularis cornu Ammonis; 7: parahippocampal subiculum cornu Ammonis; 8: parahippocampal cortex of lobus piriformis, pars posterior; 9: sulcus thinalis lateralis, pars posterior. BECCARI's designations are here partly modified in accordance with my interpretation.

missural hippocampus, being frequently still more stunted than this latter (cf. Fig. 223C, vol. 5/I, and Fig. 100 of the present volume).

The *postcommissural hippocampus*, which follows the curve of the hemispheric bend, whereby the basal extremity of the hippocampus becomes directed basorostralward and adjoins the amygdaloid complex, is the main, fully developed portion of the hippocampal formation.[29] The *cornu Ammonis* consists of an essentially single lamina of

[29] The term *hippocampus* was apparently introduced by ARANTIUS in 1587, who compared the gross aspect of that configuration in the Human brain to the aspect of the *Sea-*

fairly large pyramidal cells which becomes wider and more diffuse at the transition to parahippocampal cortex. Its opposite end dips into a hilus formed by the cell plate of *gyrus dentatus*, where pyramidal and irregular cells of cornu Ammonis are scattered within the concavity of said hilus forming the so-called '*Endblatt des Ammonshorns*'. Even in the regions where the cornu's cell band is narrow, some scattered cells may form an indistinct diffuse inner sublayer. In the Human hippocampus, the cell band of cornu Ammonis tends to be wider than in lower forms (cf. Figs. 84A–C).

According to Brodmann and to Kappers (1928), the cell plate of cornu Ammonis represents a continuation of the deep layers of the adjacent parahippocampal cortex. These deep layers Kappers designates as subgranular (or infragranular), corresponding to Brodmann's neocortical layers V and VI. Although the architectural arrangement in the adult condition (cf. Figs. 84A, 85A) might give that impression, ontogenetic stages indicate that a continuity of comparable layers cannot be assumed. It seems evident that the embryonic cell plate of cornu Ammonis is, *in toto*, continuous with the entire cell plate of adjacent parahippocampal cortex, including that latter's dense outer cell layer (cf. Fig. 85B). Subsequently, in the cornu Ammonis, the differentia-

---

Horse Hippocampus (the Lophobranch Teleost Hippocampus antiquorum L.), or to the form of a white Silk-Worm). Arantius, in a chapter entitled '*De cerebri particulis Hippocampum referentibus*', stated: '*nascitur substantia, quae*'... '*flexura figura praedita est, quae Hippocampi, hoc est marini equuli effigiem refert, vel potius, bombycini vermis candidi.* Winslow, about 1732, introduced the term *Ammon's horn (Ram's horn)* for the hippocampus (cf. Lewis, 1923). At present, the designation cornu Ammonis has become restricted to the plate of large pyramidal cells. Shepherd (1974) refers to 'an S-shaped structure which reminded the early histologists of a sea horse (Hippocampus) or a ram's horn (Ammon's horn)'. This author evidently refers to the aspect as seen in histologic cross-sections, but fails to realize that the early descriptions by Arantius, Winslow, and others were exclusively based on anatomical gross dissections. Strangely enough, Burr (1960), late Professor of anatomy at *Yale University*, states: 'This structure exhibits one of the most complicated differentiations of cortical gray. Grossly, it presents in the floor of the descending horn of the lateral ventricle, an elongated eminence which to the ancients, was likened to an elephant's foot. The descending ridge terminates in a broadened area with three or four grooves marking off what looked to them like toes of a clumsy foot. Why this structure is called hippocampus is unknown. The term can be roughly translated as a race track for horses' (p. 105). 'The neurons which end in the cortical areas of the brain eventually run into an exceedingly complicated structure in the descending horn of the lateral ventricle, known as the hippocampus. This is a curious ridge of gray matter in the floor of the ventricle which ends in a club-shaped foot. No one knows why this is called the hippocampus; it looks most of all like an elephant's foot' (Burr, 1960, p. 180).

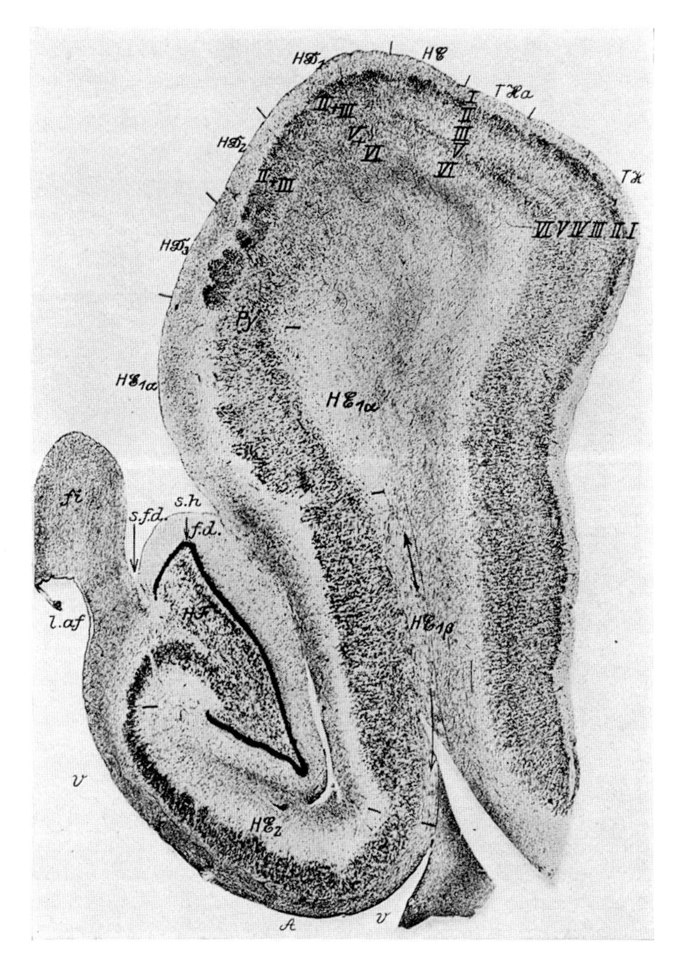

Figure 84 C. Cross-section *(Nissl stain)* through middle portion of Human gyrus hippocampi (from v. ECONOMO, 1927). A: ventricular prominence of cornu Ammonis; HC, HD_{1-3}, $HE_1\alpha$: inner portion of parahippocampal cortex; $HE_1\beta$: parasubicular portion of cornu Ammonis; HE_2: pars prominens and pars paradentata of cornu Ammonis; HF: intradentate portion of cornu Ammonis; TH, THa: outer portion of parahippocampal cortex; Py: pyramidal cells; f.d.: gyrus dentatus; fi.: fimbria; l.af.: taenia fimbriae; s.f.d.: sulcus fimbrio-dentatus; s.h.: sulcus hippocampi. Area notations of v. ECONOMO are here explained in accordance with my interpretation. HE_Z is HE_2.

tion in depth remains at the stage of an essentially single cellular lamina, while in the adjacent parahippocampal cortex a stratification in depth results from further abventricular migration and lamina differentiation. This is also still suggested at the adult stage by scattered cells

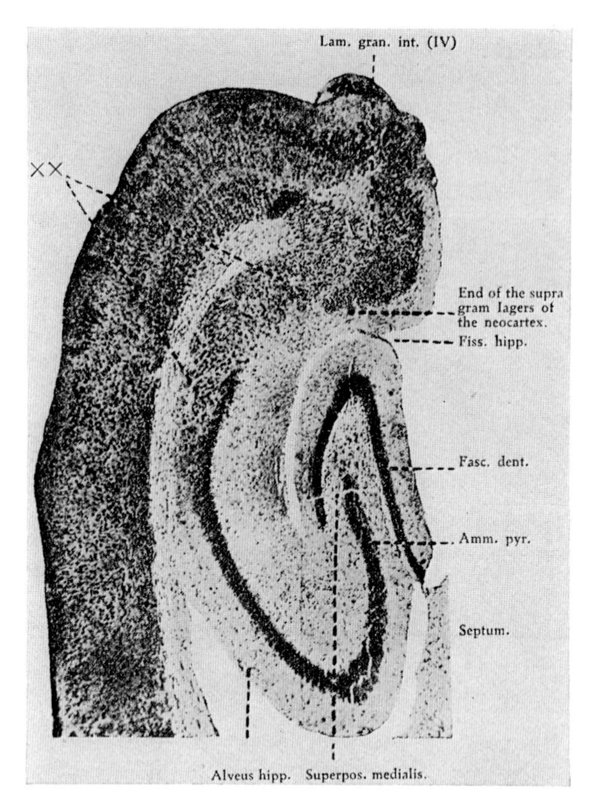

Figure 85 A. Horizontal section *(Nissl stain)* through postcommissural hippocampus and adjacent cortical regions in the adult Rat (from KAPPERS, 1928). xx: indicates the parasubicular portion of cornu Ammonis, interpreted by KAPPERS as transition of the archicortex 'into the deep layers of the neocortex'. The cited author includes the parahippocampal cortex into his concept of neocortex *sensu latiori*.

from the superficial parahippocampal cell layers forming a nondescript transition to the cell plate of cornu Ammonis.

The *gyrus dentatus* is a curved and frequently scalloped cell plate of densely crowded smaller 'granular' nerve cells, which can be regarded as representing modified pyramidal type. The *hilus* of said plate surrounds the peripheral edge *(Endblatt)* of cornu Ammonis. The molecular (plexiform) layer of gyrus dentatus facing the brain surface represents the serrated *fascia dentata* of gross anatomy. Within the sulcus hippocampi, this molecular layer bends inward along the convexity of its cell plate as far as the internal edge of the hilus.

KAPPERS (1928) regards the granular cell plate of the gyrus dentatus

B

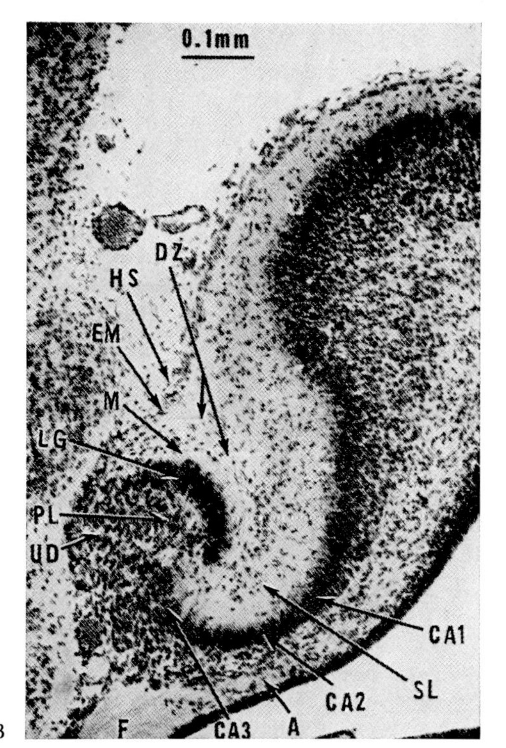

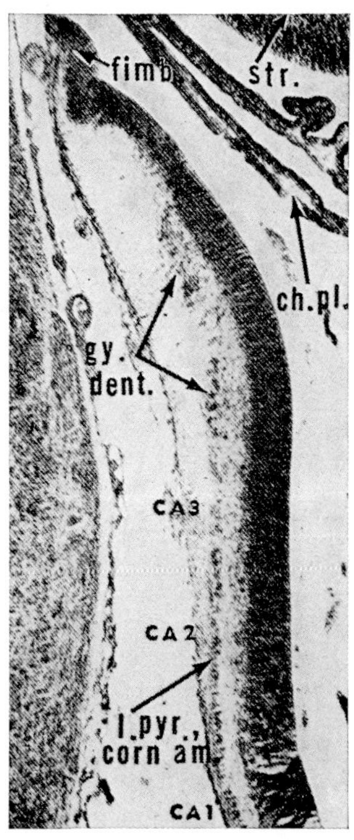

C

Figure 85 B. Cross-section through postcommissural hippocampus at a developmental stage in a 12.5 mm embryo of the Chiropteran Tadarida mexicana showing separation of gyrus dentatus from cornu Ammonis and the overall continuity if hippocampal cell plate with layers of parahippocampal cortex (from Brown, 1966). A: alveus; CA1–3: subdivisions of cornu Ammonis; DZ: 'diffuse zone'; EM: 'external limiting membrane'; F: fimbria; HS: sulcus hippocampi; LG: differentiated (granular) portion of gyrus dentatus; M: molecular layer; PL: 'polymorphic layer' (intradentate end of cornu Ammonis); SL: 'stratum lacunosum' (molecular layer of cornu Ammonis); UD: undifferentiated portion of gyrus dentatus (and of intradentate cornu Ammonis).

Figure 85 C. Early stage of differentiation of postcommissural hippocampal formation in a 37 mm CR Human embryo as seen in an horizontal section, showing periventricular matrix and migrated cortical cell plate of gyrus dentatus and cornu Ammonis (from Humphrey, 1966). ch.pl.: choroid plexus of lateral ventricle; fimb.: fimbria; l.pyr.corn. am.: early lamina pyramidalis of cornu Ammonis; str.: anlage of corpus striatum. Other abbreviations self-explanatory.

as being homologous to the granular layer IV of the neocortex (including parahippocampal cortex), and to the superficial layer of his palaeocortex. As in the case of the pyramidal cell plate of cornu Ammonis, a homologization of the gyrus dentatus cell lamina with neocortical or

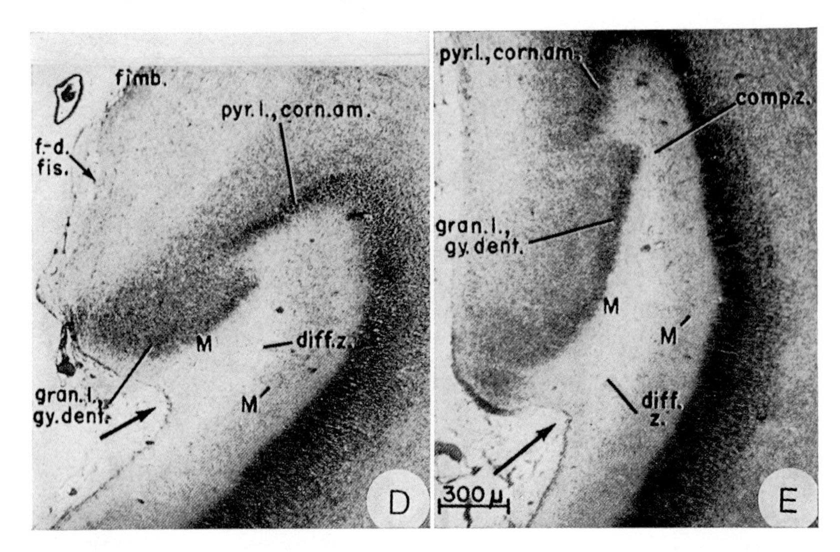

Figure 85 D, E. Cross-sections (toluidin blue-erythrosin stain) through the post-commissural hippocampus in a $15^{1}/_{2}$ weeks old Human fetus of 112 mm CR length, showing the beginning separation of gyrus dentatus and cornu Ammonis by a shearing displacement (from HUMPHREY, 1967). comp.z.: 'compression zone'; diff.z.: 'diffuse zone'; f.-d.fis.: sulcus fimbrio-dentatus; fimb.; fimbria; gran.l.: granular layer of gyrus dentatus; M, M': molecular layer of gyrus dentatus, respectively of cornu Ammonis; pyr.l., corn.am.: pyramidal layer of cornu Ammonis. The unlabelled arrow points to the shallow sulcus hippocampi.

parahippocampal cell layers does not seem justified. Like the cell lamina of cornu Ammonis, that of gyrus dentatus develops ontogenetically from a differentiation of the entire embryonic cell plate's edge, which is flush with the primordium of cornu Ammonis, but subsequently becomes separated from that latter by a shearing displacement (cf. Fig. 85 B, D, E), aspects of which are suggested by adult stages of precommissural hippocampus in some lower Mammals (cf. e.g. Figs. 222 A, B, vol. 5/I). Relevant ontogenetic stages of the Mammalian hippocampal cortex have been described with diverse interpretations of the recorded details by HUMPHREY (1966, 1967), KAHLE (1969), and others, and are reviewed in STEPHAN's (1975) treatise on the allocortex.

The molecular layer of cornu Ammonis dips around the bottom of sulcus hippocampi and faces, as it were, respectively adjoins, the inverted molecular layer of gyrus dentatus. Into this latter extend, at the boundary between the two adjacent molecular layers as well as into the portion of molecular layer pertaining to the superficial fascia dentata,

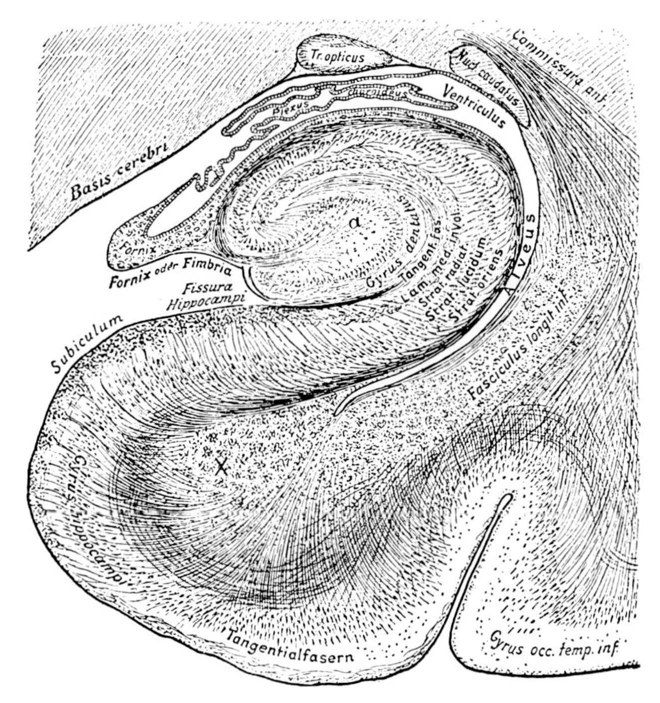

Figure 86 A. Cross-section through the Human postcommissural hippocampus and gyrus hippocampi as interpreted by EDINGER (from EDINGER, 1912). a.: hilus of gyrus dentatus. The system here designated as 'Commissura ant.' presumably contains also some components of fasciculus uncinatus. The system labelled 'Fasciculus longit.inf.' includes the fasciculus occipitalis inferior at top, and the optic radiation below. The added x indicates components of the cingulum system. 'Gyrus occ.temp.inf.' is the gyrus fusiformis of standard terminology.

tangential fibers from the zonal layer of homolateral parahippocampal cortex and, by way of the hippocampal commissure, from contralateral parahippocampal and presumably also hippocampal cortex (cf. Fig. 87 A). This conspicuous fiber layer forms the so-called *lamina medullaris circumvoluta*[30] which, in part, is an extension of the medullated

[30] Fibers reaching lamina circumvoluta and outer face of dentate molecular layer have been designated as fibrae perforantes (respectively tractus perforans) because they seem to run through an obliterated portion of sulcus hippocampi. This, however, is not the case. The sulcus hippocampi is, *ab initio*, always a very shallow groove, and not a 'fissure'. The deceiving appearance of an obliterated or partly obliterated deep 'fissure' is given by lumina of blood vessels running along the lamina circumvoluta. This vascular layer corresponds presumably to the *stratum lacunosum* of older authors.

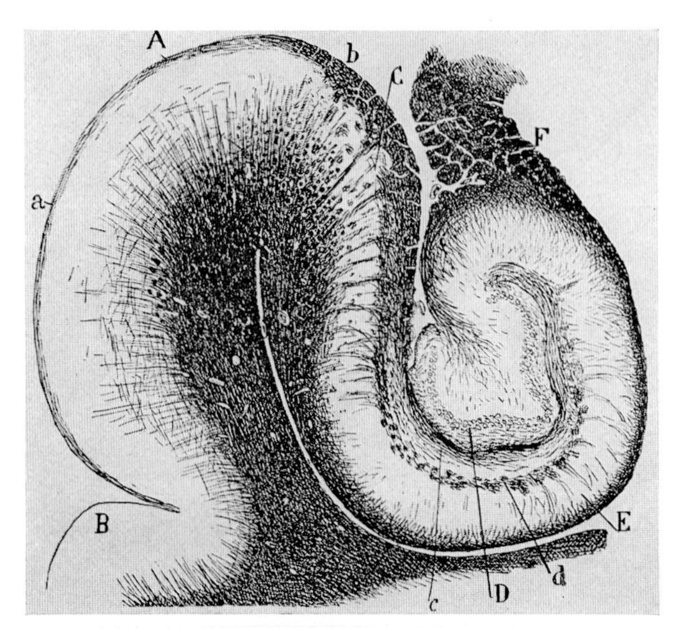

Figure 86 B. Cross-section *(Weigert carmine stain)* through the Human postcommissural hippocampus as interpreted by CAJAL (from CAJAL, 1903, 1911). A: gyrus hippocampi; B: gyrus fusiformis; C: 'subiculum' (probably end of parasubicular portion of cornu Ammonis, i.e. of subregion CA 1); D: granule cells of gyrus dentatus; E: cortex of cornu Ammonis; F: fimbria; a: 'plexiform layer' (medullated fiber system of parahippocampal cortex); b: superficial fiber system of 'subiculum'; c: 'fiber system of cornu Ammonis, continued from b' (lamina medullaris circumvoluta); d: 'deep fiber tract, continuous with collaterals of large cornu pyramids' *(stratum medullare medium of Obersteiner, with collaterals of Schaffer)*.

stratum zonale subiculi. Said lamina circumvoluta is shared by the adjacent zonal (molecular) layers of both gyrus dentatus and cornu Ammonis.

A complex *stratification* of cornu Ammonis was described by the older authors (e.g. MEYNERT, 1872; GANSER, 1882; CAJAL, 1911; EDINGER, 1911; OBERSTEINER, 1912), whose terminologies, interpretations, and concepts do not entirely coincide, and these terminologies have been adopted, again with some differences, by contemporary authors. A detailed discussion of this topic can be found in STEPHAN'S (1975) treatise.

In the aspect here under consideration, it seems convenient, beginning with the ependymal lining of the cornu Ammonis, to distinguish next the *alveus*, which is the cornu's medulla. This is followed by the

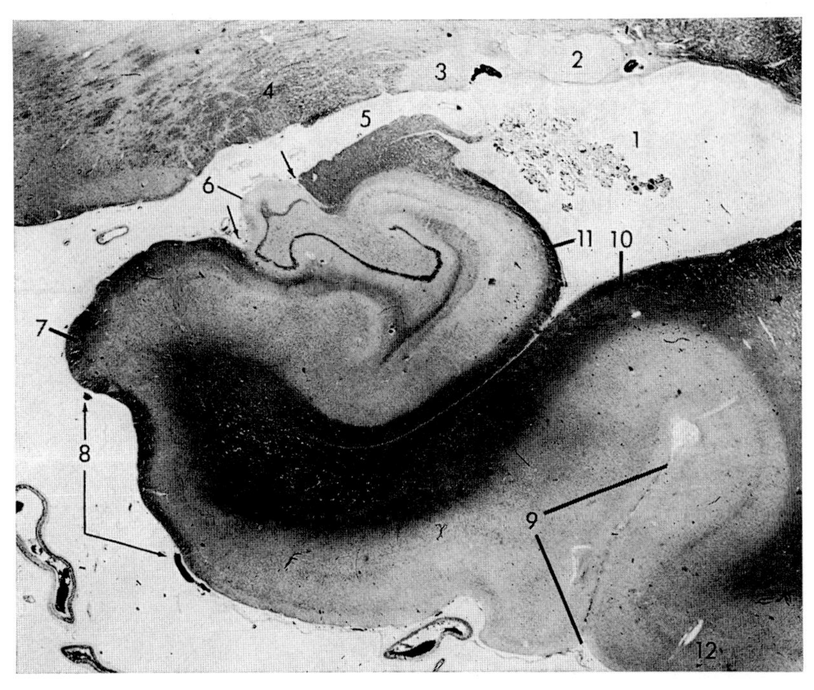

Figure 86 C. Cross-section (myelin stain) through hippocampus and gyrus hippocampi of adult Man, to be compared with Figures 84 C and 86 A, B. 1: inferior horn of lateral ventricle with choroid plexus; 2: tail of caudate nucleus; 3: stria terminalis with its interstitial griseum; 4: basis pedunculi at transition to retro- and sublenticular portion of capsula interna; 5: fimbria fornicis; 6: gyrus dentatus with sulcus fimbrio-dentatus; 7: medullated fiber system in molecular layer of parahippocampal cortex; 8: gyrus hippocampi (parahippocampal cortex, subiculum of older authors); 9: sulcus (fissura) collateralis; 10: eminentia collateralis sive trigonum collaterale; 11: alveus; 12: gyrus fusiformis. Lower arrow: sulcus hippocampi; upper: s. fimbrio-dentatus.

stratum oriens, which contains basal dendrites of the cornu's pyramidal cells and some scattered 'multiform' nerve cells. The next layer ist the *stratum pyramidale sive lucidum*,[31] formed by the pericarya of the pyramidal cells. Externally to this lies the *stratum radiatum*, characterized by said cells' apical dendrites. Next follows the molecular layer of the cornu Ammonis with several and in part variable sublayers. The

[31] STEPHAN (1975), referring to an early paper by GANSER (1882) objects to the identification of stratum pyramidale with stratum lucidum in the terminologies used by LORENTE DE NÓ (1934) and others. I am, nevertheless, inclined to prefer that identification, since, in myelin stains (cf. e.g. Figs. 86B, C) the pyramidal cell layer of cornu Ammonis assumes a translucent, gelatinous aspect.

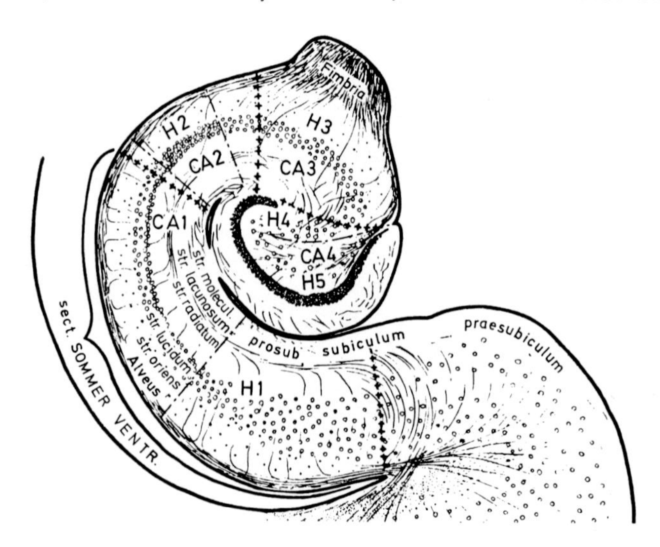

Figure 86 D. Comparison of hippocampal parcellation by Rose (H, ++++) and (CA, —·—·—·—) by Lorente de Nó (modified after Gastaut and Lammers, 1961, from Stephan, 1975).

innermost one is a tangential fiber layer (Obersteiner's *stratum medullare medium)* which contains the so-called *collaterals of Schaffer*, to be dealt with further below, but also some fibers of lamina medullaris circumvoluta with which it may or may not be fused.[32] A *stratum lacunosum*[33] may separate said lamina from stratum medullare medium, or be present internally to that latter. The lamina circumvoluta pertains to the molecular layers of both cornu Ammonis and of involuted leaf of gyrus dentatus. Into a stratum of this molecular layer, between the granular cells of gyrus dentatus and the lamina circumvoluta, extend dendrites of the granular cells (cf. Fig. 87 A). Figure 86 A shows Edinger's (1911) concept of hippocampal layers and may be compared with the myelin stain pictures of Figures 86 B, C.

The cell plate of cornu Ammonis has been parcellated into diverse subregions, with different notations and arbitrary boundary lines, by Rose (1927a, b) and by Lorente de Nó (1934), as shown in

[32] Externally to the 'tangential' stratum medullare medium, there runs also a thin stratum of longitudinal fibers.

[33] The 'lacunar' aspect, which may have led to the designation 'stratum lacunosum' by the older authors, is presumably due to the dense vascular and capillary network between lamina circumvoluta and stratum medium or within this latter (cf. Fig. 89 A).

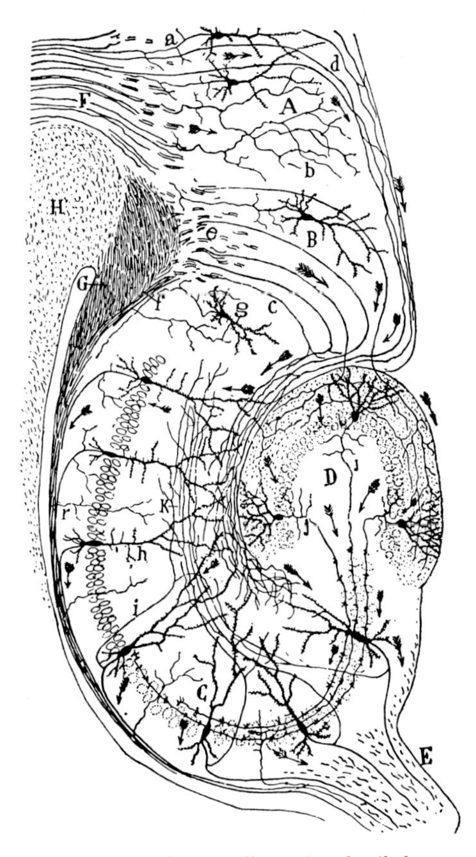

Figure 87 A. Semidiagrammatic drawing illustrating detailed structure and connections of Mammalian postcommissural hippocampus formation (from CAJAL, 1911). A: '*ganglion de la pointe occipitale*' (parahippocampal cortex); B: 'subiculum' (perhaps still parahippocampal cortex); C: cornu Ammonis; D: hilus of gyrus dentatus; E: fimbria; F: cingulum; G: '*faisceau angulaire ou temporo-ammonique croisé*' (crossed bundle, from contralateral parahippocampal cortex, having passed through commissura hippocampi); H: corpus callosum; K: *collaterals of Schaffer;* a: neurite joining cingulum; b: end arborizations of cingulum fibers; c: perforating fibers from G; d: 'perforating fibers from cingulum'; e: layer of 'superior fibers' from G; f: fibers related to alveus; g: '*cellule du subiculum*' (pertaining to parasubicular portion of cornu Ammonis); h: apical dendrite of pyramidal cell; i: *collateral of Schaffer;* j: origin of mossy fibers (these latter run only through CA 2–4); k: *stratum medullare medium of Obersteiner;* r: collateral (or ending) of alveus fiber. Cf. also Figure 201 J, p. 298 of vol. 3/I. k same as K.

Figure 86D. It will be seen that these authors, like other recent ones, designate the region of hippocampal cortex adjacent to parahippocampal cortex as 'subiculum'. CAJAL (1903, 1911) also seems to follow this terminology, although he apparently fails to recognize a clear-cut boun-

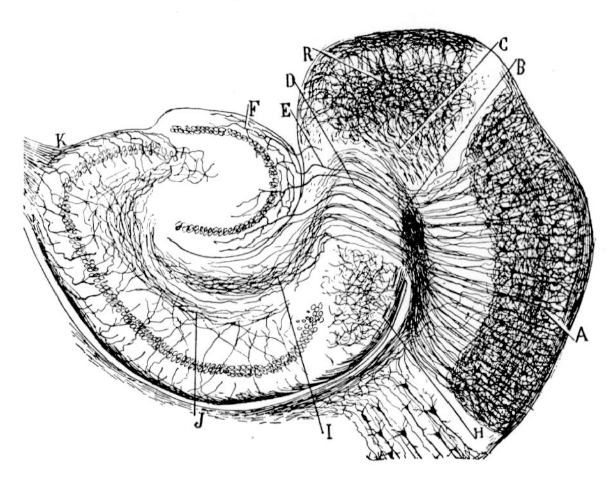

Figure 87 B. Horizontal section *(Golgi impregnation)* through hippocampal formation and parahippocampal cortex in a young Mouse of 15 days (from CAJAL, 1911). A: *'ganglion spheno-occipitale sive temporale superius'* (parahippocampal posterior piriform lobe, cf. Fig. 221D, vol. 5/I); B: junction of perforating (homolateral) parahippocampal-hippocampal fibers *('faisceaux temporo-ammoniques inférieurs')*; C: afferent fibers to inner region of parahippocampal subiculum; D: fibrae perforantes; E: perforating fibers to gyrus dentatus; F: terminals in gyrus dentatus; H: plexus of 'temporo-ammonic' fibers which have reached alveus; I: plexus of perforating fibers in lamina medullaris circumvoluta; J: plexus of *Obersteiner's stratum medullare medium (collaterals of Schaffer)*; K: fimbria; R: deep plexus of parahippocampal subiculum.

dary between cornu Ammonis and parahippocampal cortex, whereby his 'subiculum' includes adjacent regions of both (cf. Figs. 86B and 87A). There is, however, no doubt that the early authors (MEYNERT, 1872; SCHWALBE, 1881; EDINGER, 1911; OBERSTEINER, 1912; VILLIGER, 1920) designated the entire (parahippocampal) gyrus hippocampi as *subiculum cornu Ammonis*,[34] and I have preferred to retain this terminology introduced by the pioneer investigators.

Despite the *prima facie* relatively uniform cytoarchitectural arrangement of the cell plate formed by the cornu Ammonis, there is no doubt that, even *qua* cytoarchitectural characteristics, the following four main subregions can be distinguished: (1) The scattered cells of the '*Endblatt des Ammonshorns*', representing the *intradentate portion* within the hilus of gyrus dentatus.[35] (2) The adjacent portion of cornu Ammonis,

[34] Cf. e.g. SCHWALBE (1881, p. 734) and OBERSTEINER (1912, p. 127).

[35] This is area CA 4 of LORENTE DE NÓ, while ROSE again subdivides it into H 5 and H 4.

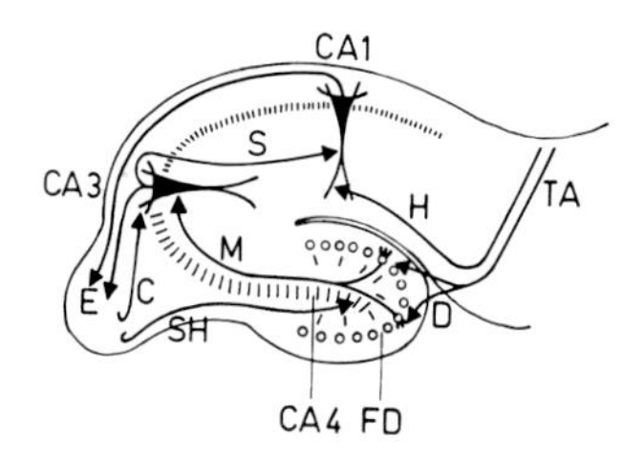

Figure 87 C. Simplified diagram of hippocampal fiber systems (from C. v. EULER, 1961). C: 'commissural afferents'; CA 1–4: subregions of cornu Ammonis; D: components of TA to gyrus dentatus; E: efferent fibers of cornu Ammonis; FD: gyrus dentatus; H: components of TA to cornu Ammonis (lamina medullaris circumvoluta); M: mossy fibers; S: *Schaffer's collateral;* SH: 'septal-hippocampal afferents'; TA: 'direct or perforant temporo-ammonic tract' (parahippocampal-hippocampal channel).

characterized by the mossy fibers to be dealt with further below, represents the *paradentate portion.*[36] (3) The next portion of cornu Ammonis, corresponding to the bulge of the hippocampal convexity protruding into the lateral ventricle, represents the *pars prominens.*[37] (4) The adjacent portion of the cornu's cell plate, generally becoming wider and less densely packed, is the pars *parasubicularis,*[38] demarcated by a commonly fairly steep gradient (cf. Fig. 84 A) from the parahippocampal subiculum.

At the distal (rostrobasal) end of the hippocampal formation, this latter forms a bulging configuration adjacent to cortical amygdaloid griseum and lobus piriformis. In 'lower' Mammals, this is the *tuberculum hippocampi* of ELLIOT SMITH (1898), which in 'higher' Mammals assumes a hook-like appearance, the *uncus hippocampi*, in which the gyrus dentatus, becoming the *benderella* (or *band of Giacomini*), bends dorsad across the terminal bulbous expansion of the cornu Ammonis (cf. Fig. 57 C), separating this latter into *gyrus uncinatus* and *gyrus intralimbicus* (cf. Fig. 175 D, vol. 5/I, and Fig. 57 C, I of the present volume). The rather complex and somewhat variable configuration of the uncus in Man and other Primates is not easily visualized in microscopic section. It has been described by v. ECONOMO and KOSKINAS (1925),

---

[36] Essentially CA 3 and CA 2 of LORENTE DE NÓ, and H 3 and H 2 of ROSE. Aspects of chemoarchitecture shall be pointed out further below in discussing problems of structure.

[37] This is CA 1 of LORENTE DE NÓ, and the distal part of ROSE's H 1.

[38] This is the 'prosubiculum' and partly the 'subiculum' of LORENTE DE NÓ, and the proximal part of ROSE's H 1.

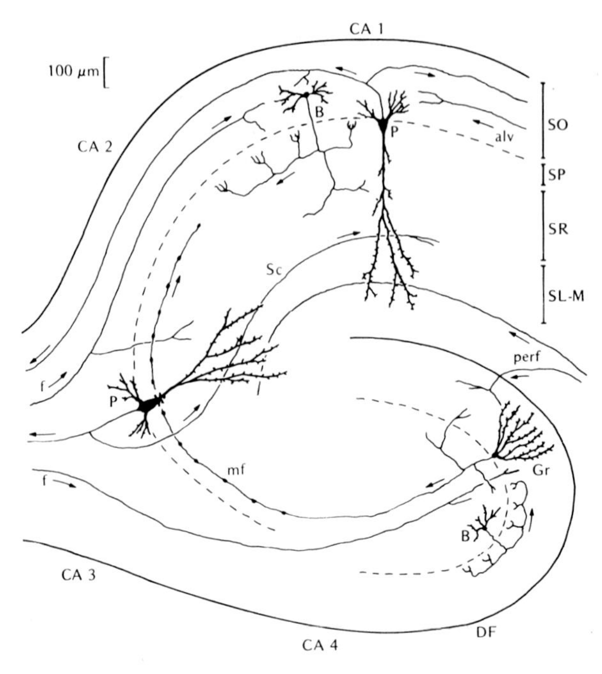

Figure 87 D. Diagram illustrating neuronal elements and fiber connections of post-commissural hippocampus according to contemporary views (from SHEPHERD, 1974). B: 'intrinsic neurons' ('basket cells'), CA 1–4: subregions of cornu Ammonis; DF: gyrus dentatus; Gr: 'granule cells' of gyrus dentatus; P: pyramids of cornu Ammonis; Sc: *Schaffer's collaterals;* SO: alveus and stratum oriens; SP: stratum pyramidale; SR: stratum radiatum; SL-M: 'stratum lacunare' and moleculare; alv: alvear pathway; f: fornix pathway; mf: mossy fiber; perf: perforating pathway.

v. ECONOMO (1927) and others, and also is dealt with in STEPHAN'S (1975) treatise.

Fundamental studies on the structure of the hippocampal cortex were initiated and carried out by CAJAL (1903, 1911 *et passim*). By means of the *Golgi method*, this author demonstrated the dendritic and axonal expansions of the 'granular' cells in gyrus dentatus as well as of the pyramids in cornu Ammonis. He showed, moreover, the arrangement of the main fiber connections (cf. Fig. 87 A, B, C). It can be seen that the fibers of the tractus perforans, coming from the molecular layer of the (parahippocampal) subiculum and, through the commissura hippocampi, from the contralateral parahippocampal cortex, provide the bulk of the *lamina medullaris circumvoluta*, effecting synaptic connections with dendrites of gyrus dentatus and cornu Ammonis cells.

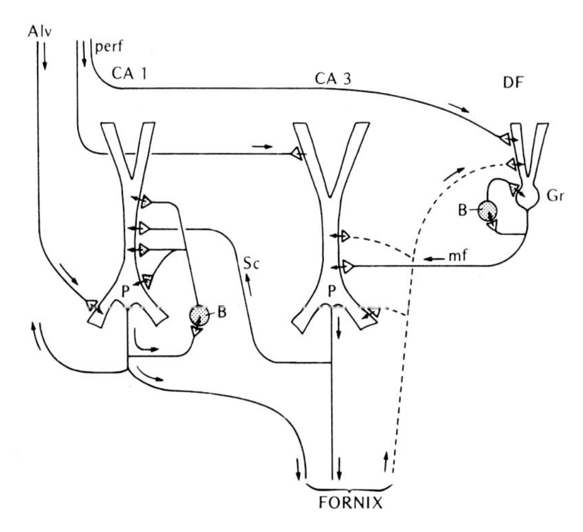

Figure 87 E. Diagram purporting to illustrate the 'basic circuit' arrangement of the hippocampal formation according to contemporary views (from SHEPHERD, 1974). Abbreviations as in Figure 87 D.

The neurites of the '*granular*' elements in *gyrus dentatus* gather to form a bundle of '*mossy fibers*' connecting with the cell bodies or the origin of apical dendrites of the cornu's pyramidal cells. These mossy fibers[39] extend only through the intradentate and paradentate sectors of cornu Ammonis, their 'mossy' bulbous synaptic endings being particularly conspicuous in the paradentate sector.

The *pyramids of cornu Ammonis* send their neurites through the alveus into the fimbria fornicis. *Recurrent collaterals*, described by SCHAFFER (1892) run back through the cell plate into the inner stratum of the cornu's molecular layer, and seem to form the bulk of the *stratum medullare medium*, which latter may or may not merge with lamina medullaris circumvoluta. Some other collaterals, particularly in the parasubicular region, seem to run through the alveus into the parahippocampal cortex.

---

[39] The synaptic varicosities of these fibers were described by CAJAL (1911) as '*appendices divergents, courts et épais*', or as '*filaments tenus, assez longs et terminés par une nodosité. Cet aspect reproduit, en définitive, en traits peut-être moins marqués, celui que nous avons découvert sur certaines fibres du cervelet, auxquelles nous avons donné le nom de fibres moussues.*' He then adds: '*nous appellerons aussi fibres moussues les cylindres-axes des grains de la fascia dentata*'. McLARDY (1960) has described a '*syncytial skeining arrangement*' formed by the periaxoplasm of the mossy fiber bundles.

CAJAL's investigations also disclosed two sorts *of Golgi II type* nerve cells, now generally called '*basket cells*', represented by multiform elements in stratum oriens of cornu Ammonis, and by comparable cells in gyrus dentatus. The short neurite effects connections with several pyramidal respectively granular cell bodies (cf. Figs. 87D, E). Additional irregular cell forms, presumably of *Golgi II type*, were also shown in the molecular layers, as well as internally to the main cell plates of cornu Ammonis and gyrus dentatus.

There is, moreover, evidence that the *fornix* also contains some input for hippocampal cortex, reaching both cornu Ammonis and gyrus dentatus (cf. Fig. 87D). This input presumably comes from the paraterminal grisea ('septum') and the hypothalamus, perhaps also from anterior thalamic grisea.

Numerous additional details of intrinsic and extrinsic hippocampal fiber connections have been recorded in the investigations of HJORT-SIMONSEN (1971, 1973 and others), of RAISMAN *et al.* (1966), and other authors as summarized by SHEPHERD (1974) and STEPHAN (1975). Ultrastructural details, obtained by electron microscopy, have been investigated by NIKLOWITZ (1964, 1966), and NIKLOWITZ and BAK (1965).

The hippocampal formation, consisting of cornu Ammonis and gyrus dentatus,[40] mainly seems to represent the efferent cortical griseum of the parahippocampal cortex which pertains to the limbic lobe. Generally speaking, it could be said that the hippocampal formation despite some complexities and regional subdivisions, is characterized by a remarkable 'stereotyped' structure. In this latter respect it is comparable to the likewise 'stereotyped' cerebellar cortex.

As emphasized by KAPPERS (1928), the 'granular' cells of gyrus den tatus represent an essentially receptive 'correlative' griseum, discharging into the cornu Ammonis. The large pyramidal cells of this latter are the efferent elements discharging into the fornix system. To a lesser extent, some of the cornu's output is directed back to the homo- and controlateral parahippocampal cortex. In addition, the pyramids of cornu Ammonis receive input, not mediated through gyrus dentatus, from the parahippocampal cortex by way of alveus and lamina circum-

---

[40] Discounting here the rudimentary precommissural hippocampus and those likewise rudimentary types of supracommissural hippocampus reduced to an indusium. No relevant data as regards detailed structure and synaptology of these rudimentary hippocampal cortices seem to have been recorded.

voluta, and from paraterminal, diencephalic, and mesencephalic grisea by way of ascending fornix fibers.[41] Some of these latter also reach the gyrus dentatus. Figures 87C–E illustrate present-day concepts concerning the circuitry of the hippocampal formation.[42]

FLECHSIG (1920), moreover, recorded in the Human brain a fiber system characterized by early myelinization about the time of birth, and extending through the 'hippocampic zone' from the region of amygdaloid complex adjacent to uncus along alveus and subiculum of parahippocampal gyrus hippocampi. FLECHSIG designated it as the system β of the uncus, and stated that it pertains, together with fornix longus and cingulum, to the hippocampal formation.[43] This system (cf. Figs. 252D, E, vol. 5/I) may be an important channel of the limbic system and seems to suggest still poorly defined connections between amygdaloid complex, hippocampal formation, and parahippocampal cortex.[44]

As regards secondary architectural details of the hippocampal formation *in toto*, it should be pointed out that, within its rather constant overall pattern in the Mammalian series, numerous taxonomical and also individual variations are displayed. Additional topics of interest about the hippocampal formation concern histochemical aspects, angioarchitecture, and selective vulnerability.

With respect to *histochemistry*, MASKE (1955) and FLEISCHHAUER and HORSTMANN (1957) reported that gyrus dentatus as well as intradentate and paradentate regions of cornu Ammonis were selectively stained red by diphenylthiocarbozone *(dithizone)*, probably by the formation of *zinc-dithizonate*. A sulfide-silver reaction suggesting the presence of *zinc* was introduced by TIMM (1958). MCLARDY (1960, 1962) assumed that the presence of said metal was related to the system of mossy

[41] It will be recalled that the bulk of the fornix system seems to be descending, and presumably represents the main output channel of cornu Ammonis.

[42] Details on the ultrastructural aspects of the various sorts of synapses found in the hippocampal cortex are reviewed, with bibliographic references, in the treatise by STEPHAN (1975).

[43] It is said (FLECHSIG, 1920, p. 55) to be the first medullating long association system in the hemisphere: *'es verläuft zwischen Uncus und Ammonshorn'*.

[44] The so-called 'entorhinal area' represents a particularly conspicuous region of parahippocampal cortex extending into, and covering the posterior part of lobus piriformis (cortex posterior lobi piriformis, field 28 of BRODMANN). Its arbitrary delimitation from more dorsal and rostral parahippocampal regions remains uncertain and controversial (cf. below in the discussion of parahippocampal cortex).

fibers.[45] VON EULER (1961) substantiated by diverse methods that the metal displayed by the not entirely specific reactions was indeed zinc, and concluded that a zinc component is present in the synaptic apparatus of the mossy fiber system. Said metal was interpreted as probably the constituent of an enzyme. HAUG (1967, 1973) and others recorded the zinc within the synaptic boutons, apparently externally to the synaptic versicles. Figure 88A illustrates the distribution of zinc as shown by the sulfide-silver reaction.

The distribution pattern of *cholinesterase* is illustrated by Figure 88B. As regards *serotonin*, recent studies (e.g. MOORE and HALARIS, 1975) seem to indicate that 'serotonin neurons' of the midbrain raphe nuclei[46] in tegmentum and central gray provide input to the hippocampal formation. The serotonin is here particularly but not exclusively distributed to part of the molecular layer of CA 1. The pathway seems to run both through fornix and cingulum.

With respect to *angioarchitectonics*, PFEIFER (1930, 1940) recorded in his investigations[47] some aspects of the vascularization pattern in the hippocampal formation as shown in Figures 89A and B. It will be seen that a rather dense capillary network is found in the pyramidal layer of CA 2–3, corresponding to the mossy fiber system, and is also noticeable in the adjoining molecular layers of gyrus dentatus and CA 1.

The *blood supply* of the Human postcommissural hippocampus is provided by the posterior cerebral artery, from which, at right angle, hippocampal arteries enter the hippocampal formation along the sulcus hippocampi.[48] Variable anastomoses with posterior and anterior choroidal arteries obtain, as illustrated in Figure 386, p. 714 of volume 3/II. The arrangement of the hippocampal blood supply has been described by ALTSCHUL (1938), HEIMAN (1938), LINDENBERG (1955), and UCHIMURA (1928); its overall pattern, discounting diverse variations, is here shown in Figure 89C.

---

[45] References to further studies on the hippocampal formation by MCLARDY (1963, 1964, 1974) are included in the detailed review of the relevant topics by STEPHAN (1975), which contains an extensive bibliography.

[46] The 'nucleus centralis superior' indicated by the cited authors may be CASTALDI's median 'nucleo lineare', and their central gray nucleus CASTALDI's nucleus ventralis and ventrolateralis grisei centralis (cf. vol. 4, pp. 954–955).

[47] Cf. the comments on angioarchitecture and PFEIFER's investigations in chapter V, section 5, pp. 320–326, with Figures 212–214 of volume 3/I.

[48] Longitudinal clefts, which actually represent the lumina of these vessels, may give the erroneous impression of a deep and partly obliterated 'fissura hippocampi' (cf. above. footnote 30, p. 197).

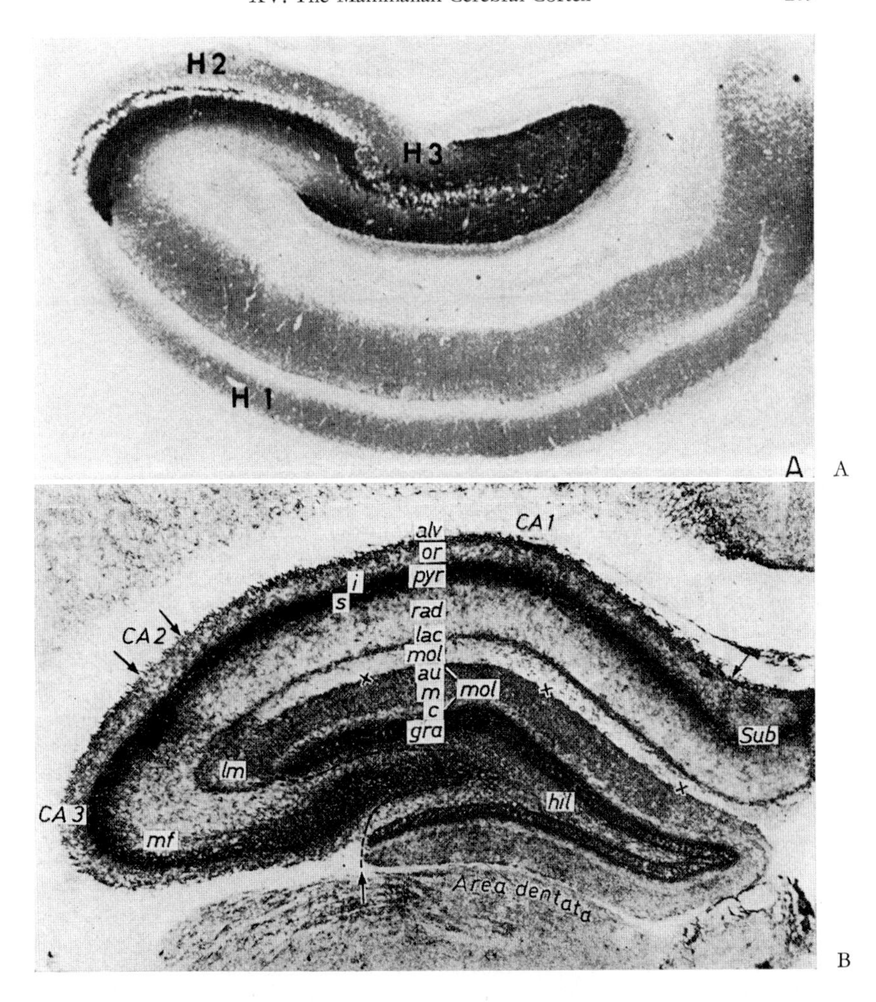

A

B

Figure 88 A. Histochemical silver sulfide reaction demonstrating distribution of zinc in the hippocampal formation of the Rabbit (after Friede, 1966a, from Stephan, 1975). H 1–3: subregions of cornu Ammonis in Rose's notation (cf. Fig. 86 D).

Figure 88 B. Acetylcholinesterase reaction in the hippocampal formation of the Rat (after Ritter *et al.*, 1972, from Stephan, 1975). CA 1–3: subregions of cornu Ammonis; Sub: parasubicular portion of cornu; alv: alveus; au: external zone of dentate molecular layer; c: inner zone of dentate molecular layer; gra, hil: granule cells and hilus of gyrus dentatus; i: infrapyramidal zone; lac: 'stratum lacunare'; lm: 'substratum eumoleculare-lacunosum'; mf: substratum lucidum, zone of mossy fibers; m: middle zone of dentate molecular layer; mol: molecular layer (of cornu Ammonis respectively of gyrus dentatus); or: stratum oriens; pyr: stratum pyramidale (with subzones i, s); rad: stratum radiatum; s: suprapyramidal zone; x: additional boundaries in dentate molecular layer.

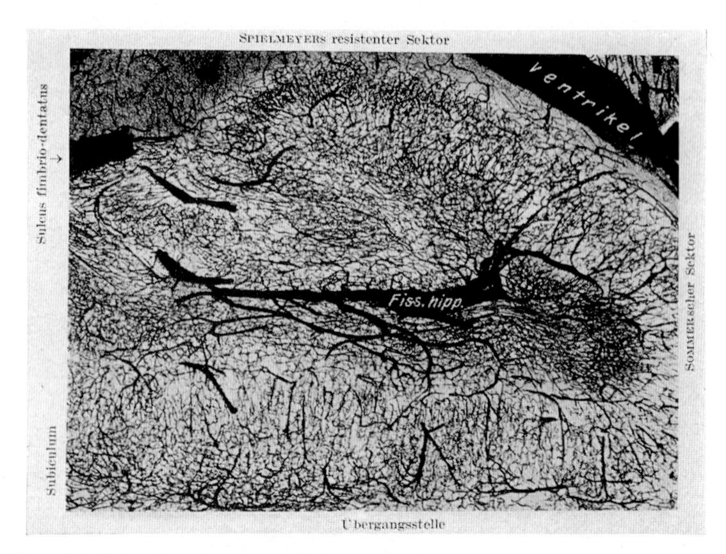

Figure 89 A. Angioarchitecture of the Human hippocampal formation (from PFEIFER, 1930; ×15.8, red.3/4). The dense capillary network in *Spielmeyer's resistant sector* is noteworthy. 'Fiss.hipp.' is not the non-existent deep hippocampal 'fissure', but the lumen of an hippocampal vessel.

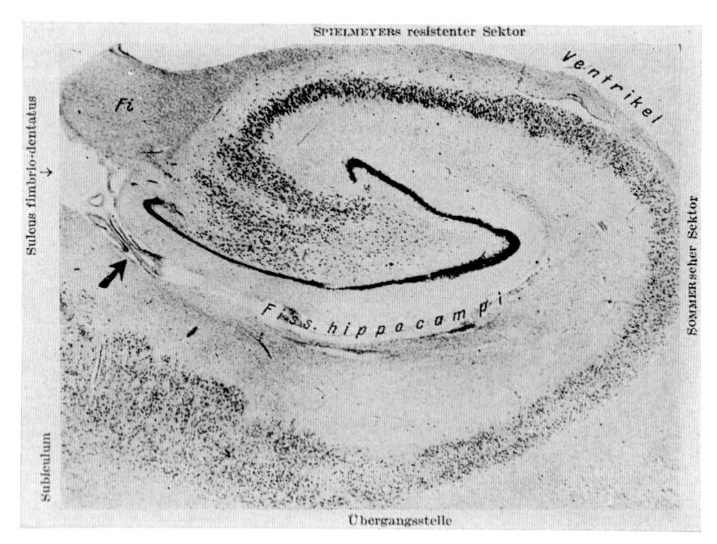

Figure 89 B. Cytoarchitecture of Human hippocampal formation for comparison with preceding angioarchitectural picture (from PFEIFER, 1930; ×18.7, red. 3/4). As regards 'Fiss. hippocampi', cf. legend of Figure 89 A. The added arrow approximately points to bottom of shallow true sulcus hippocampi.

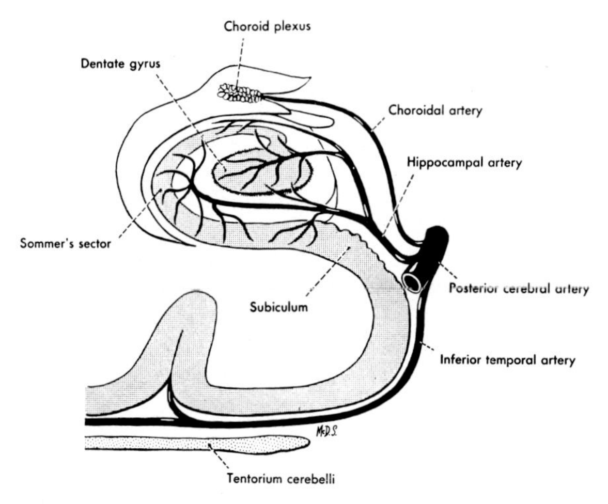

Figure 89 C. Arterial supply of Human hippocampal formation (slightly modified after LINDENBERG, 1955, from HAYMAKER, 1969). Compare this Figure with Figure 386, p. 714 of volume 3/II.

The hippocampal formation displays a *selective vulnerability (pathoklisis)* in diverse pathologic conditions,[49] such that one sector of cornu Ammonis, approximately corresponding to CA 1, and known as *Sommer's sector*[50] is particularly sensitive and easily degenerates, while regions CA 2–3, representing *Spielmeyer's resistant sector*,[51] and gyrus dentatus commonly remain unaffected (cf. Fig. 89D).

Various theories based on the type of vascularization have been propounded, including a 'hydrodynamic explanation suggested by SCHARRER (1940), who investigated carbon monoxide poisoning in the Opossum. C. and O. VOGT (1929), on the other hand, assumed that physicochemical peculiarities represented the relevant factors. In view of the particular distribution of zinc, mentioned above, and roughly corresponding to Spielmeyer's resistant sector, the VOGTS' interpretation seems indeed justified. It is also noteworthy that in rabies, the *Negri bodies*, intraprotoplasmatic inclusion bodies containing aggregates of the virus, are generally found in the pyramidal cells of cornu Ammonis, but hardly ever in other nerve cells with the exception of the

---

[49] E.g. epilepsy, dementia paralytica, pertussis encephalitis (cf. Fig. 89D), carbon monoxide poisoning, diverse circulatory disturbances, and rabies.

[50] SOMMER (1880).

[51] SPIELMEYER (1925). The subregion CA 4 seems to be somewhat less resistant.

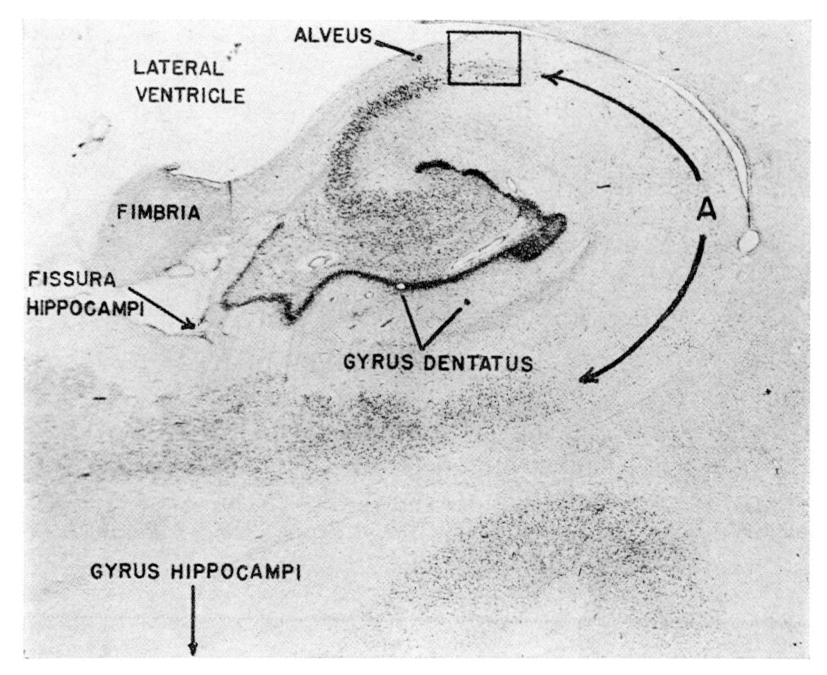

Figure 89 D. Complete degeneration and disappearance of pyramidal cells in *Sommer's sector* of the Human hippocampal formation caused by fatal pertussis encephalitis associated with severe generalized convulsive seizures of so-called pertussis eclampsia in a 16 months old boy (courtesy Army Institute of Pathology, from material prepared by HAYMAKER, NEUBUERGER, and HURTEAU). A: *Sommer's sector.* The rectangular inset marks a zone of transition between degenerating and relatively normal pyramidal cell at the boundary region betwen *Spielmeyer's and Sommer's sectors.*

Purkinje cells.[52] Nevertheless, since multifactorial effects obtain in normal as well as pathologic biological events, the vascular and the physicochemical hypotheses do not appear mutually exclusive (cf. also the discussion of this topic by STEPHAN, 1975).

Generally speaking, the hippocampal formation, which can be evaluated perhaps as the *main output griseum of the 'limbic system'*, is notable for its generation of rhythmic discharges. This rhythmic activity has been investigated in detail by numerous authors, but the multitudinous obtained data, recently reviewed by SHEPHERD (1974) do not lead

[52] The analogy, qua certain structural aspects, of hippocampal formation and cerebellar cortex was mentioned above on p. 206. This analogy might also hold, *qua* generalized aspects of '*modulating*' circuitry.

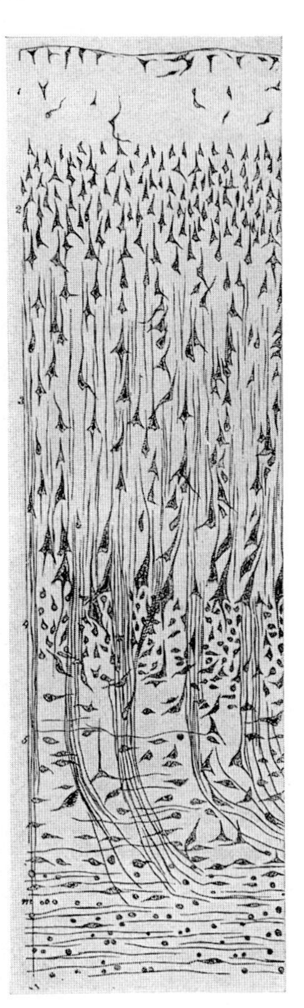

Figure 90. Section through cortex of Human third frontal convolution (apparently cortex frontalis granularis) depicted by MEYNERT from a carmine stain preparation at about ×100 (from MEYNERT, 1872). 1: 'layer of scattered small cortical corpuscles' (molecular layer, I); 2: 'layer of dense small pyramidal cortical corpuscles' (external granular layer, II); 3: 'layer of large pyramidal corpuscles like those of cornu Ammonis' (pyramidal layer, III); 4: 'layer of small dense irregular corpuscles', '*körnerartige Formation*' (internal granular layer, IV); 5: 'layer of spindle-shaped corpuscles', '*Vormauerformation*' (apparently ganglionic layer V, and multiform layer VI, taken together; m: medulla.

to very definite conclusions, except to an overall interpretation of the hippocampal formation as an important link in a '*modulating circuit*' with a diversity of functional implications, some of which shall be considered in section 6. In addition to SHEPHERD's review, and to STEPHAN's (1975) treatise, diverse aspects of the hippocampal formation have been dealt with by FRIEDE (1966a, b) and by HASSLER (1967 and other publications). A very detailed treatise on the hippocampus, in two volumes, was edited by ISAACSON and PRIBRAM (1975). It contains contributions by numerous authors and includes, besides the synopsis of the available recorded data concerning the different relevant structural and functional aspects, a wide variety of interpretations and speculations. For commissural systems cf. p. 744, vol. 5/I.

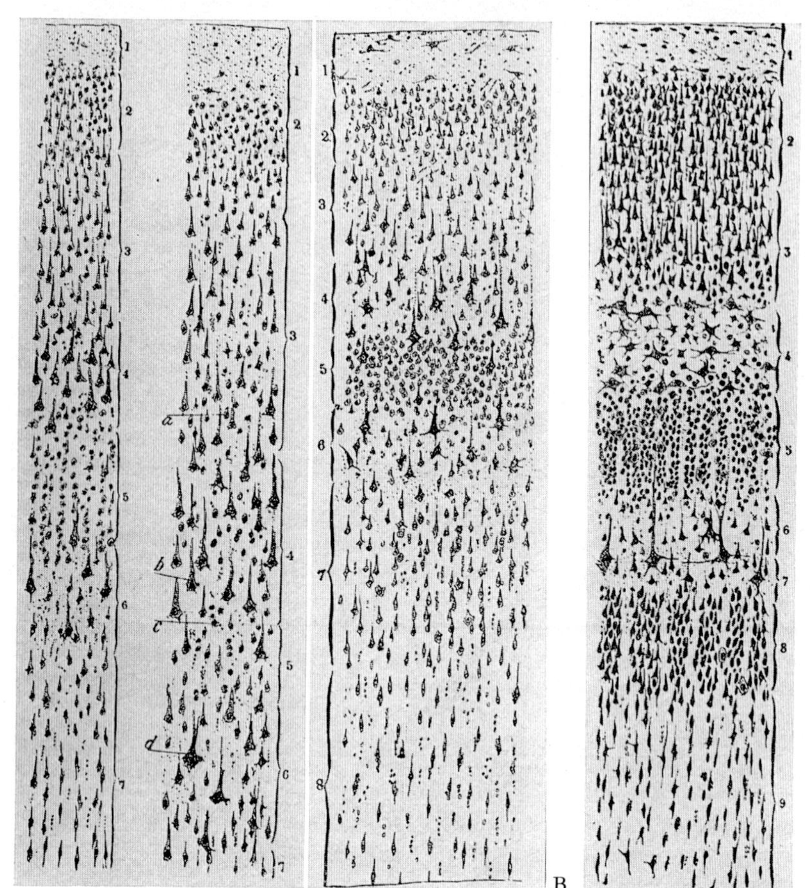

Figure 91 A. Section through precentral motor cortex (right) and postcentral sensory cortex (left) of Man, as drawn by CAJAL from *Nissl preparations* (from CAJAL, 1900b, 1911). 1: plexiform layer (I); 2: 'small pyramids' (II); 3: medium-sized pyramids (upper part of III); 4: 'large superficial pyramids' (lower part of III); 5: 'small star cells' (IV); 6: large deep pyramids (V); 7: 'layer of spindle-shaped and triangular cells' (VI). CAJAL's drawing does not take into consideration the substantial difference in width between the narrower postcentral and the very wide precentral motor cortex (cf. Fig. 114 A).

Figure 91 B. Section through convex portion of Human postcentral convolution, as drawn by CAJAL from a *Nissl preparation* (from CAJAL, 1900b, 1911). 1–4 as in A; 5: 'layer of small pyramids and star cells' (IV); 6, 7 as in A; 8: 'deep layer' of 7, 'particularly conspicuous in convex portion of the convolution'. This latter is called '*hintere Centralwindung*' in 1900b, and '*circonvolution pariétale ascendante*' in 1911.

Figure 91 C. Section through Human calcarine (visual) cortex, as drawn by CAJAL from a *Nissl preparation* (from CAJAL, 1911). 1–2 as in A; 3: 'layer of medium-sized pyramids' (this apparently includes lower part of III together with IVa); 4: 'layer of large star cells' (IVb); 5: 'layer of small star cells' (IVc); 6: 'deep plexiform layer of small pyramidal

With respect to the *neocortex*, BAILLARGER (1840) undertook early microscopic studies of its architecture before the import of the cell theory, propounded in a first approximation about that same time,[53] was properly understood. Amplifying previous observations by VICQ D'AZYR, GENNARI, and others, BAILLARGER examined rather thick, unstained sections of Human neocortex[54] and described six layers alternately gray and white, proceeding from inside out. He found that layers 2, 4, 6 were white and 1, 3, 5 gray. The relationships of his white layers to neocortical lamination shall be pointed out further below in dealing with myeloarchitecture.

Relevant *cytoarchitectural* studies were begun by MEYNERT (1868, 1872) and by BETZ (1874, 1881). The latter author recorded the giant pyramidal cells in the precentral motor cortex. In his description of the cortical lamination, MEYNERT distinguished a general type characterized by five layers as shown in Figure 90. In the occipital (visual) cortex, he distinguished eight layers. Another pioneer investigator was BEVAN LEWIS (1878, 1882) who suggested a six-layered lamination pattern. Subsequent fundamental investigations relevant to cytoarchitecture were undertaken by CAJAL (1900–1906, 1911), CAMPBELL (1905), BELA HALLER (1908, 1910), HAMMARBERG (1895), v. KOELLIKER (1896), MOTT (1907) and others listed and considered in the publications by CAJAL, BRODMANN, and v. ECONOMO. Of particular import, however, are the detailed and systematic studies by BRODMANN (1909 and others) concerning numerous Mammals including Man, and those by v. ECONOMO and KOSKINAS (1925) and v. ECONOMO (1927) on the Human cerebral cortex.

CAJAL distinguished seven layers in most neocortical regions, but nine layers in the occipital cortex. Yet, although he depicts seven layers in precentral and postcentral cortex, he described in his text (1911) only six layers in the former, and depicts, in another chapter, eight layers in the postcentral one, by separating an additional deeper sublayer from his seventh (cf. Figs. 91 A, B, C).

[53] As regards the early formulations of the cell theory by SCHLEIDEN and SCHWANN in 1838 and 1839, cf. pp. 500–503 of section 7, chapter V, volume 3/I.

[54] At that time, of course, the concept of neocortex had not yet been formulated.

---

cells with ascending axon' (probably deep part of IVc with neighborhood of V); 7: 'layer of giant pyramidal cells' (V); 8: 'layer of pyramids with arciform and ascending axon' (apparently outer sublayer of VI); 9: fusiform cells (inner VI).

BRODMANN (1909) adopted, with slight modifications in terminology,[55] the concept of six layers proposed by BEVAN LEWIS (1878). BRODMANN, whose notation became widely accepted, designated the entire neocortex (isocortex) as *homogenetic*,[56] since he believed that, despite local structural changes in the course of the 7th and 8th months of Human ontogenesis, the neocortical lamination derived *in toto* from a six-layered fundamental pattern.

Although various doubts obtain concerning this mode of development, it seems indeed possible to interpret the stratification of the neocortex in terms of the postulated six layers. BRODMANN's viewpoint, nevertheless, can, with some justification, be qualified not only as '*neopalliocentric*' but also as '*anthropocentric*', since, as will be pointed out further below, the six-layered pattern is much less clearly displayed in the neocortex of 'lower' Mammals. Even in the neocortex of Man and other Primates, it is generally not possible to draw sharp boundaries between the layers, which represent '*open neighborhoods*' of topology. The uncertainties inherent in a delimitation of cortical laminae are discussed by SHOLL (1967), who concludes that at the best, laminae have only a limited topographical interest insofar as they may give a rough indication of the cortical depth to which one may wish to refer.[57] Yet, although the laminae contain only the pericarya of neuronal elements whose dendrites or neurites extend into other layers, it appears convenient, for practical purposes, to retain the conventional cytoarchitectural *hexalaminar* concept.

In BRODMANN's notation the neocortical layers (cf. Fig. 92) are designated as follows from the surface inward:

I. Lamina zonalis (molecular layer).
II. Lamina granularis externa (outer granular layer).
III. Lamina pyramidalis (pyramidal cell layer).
IV. Lamina granularis interna (inner granular layer).
V. Lamina ganglionaris (ganglion cell layer).
VI. Lamina multiformis (layer of spindle, fusiform, or polymorph cells).

---

[55] A tabulation comparing his lamination concept with that of previous authors can be found on p. 15 of BRODMANN's (1909) monograph.

[56] Cf. p. 179, section 2 of the present chapter.

[57] As regards the arbitrariness of laminar deliminations, reference may also be made to the comments on so-called laminae of the spinal cord grisea, dealt with in chapter VIII of volume 4 (pp. 8–9; 210–216).

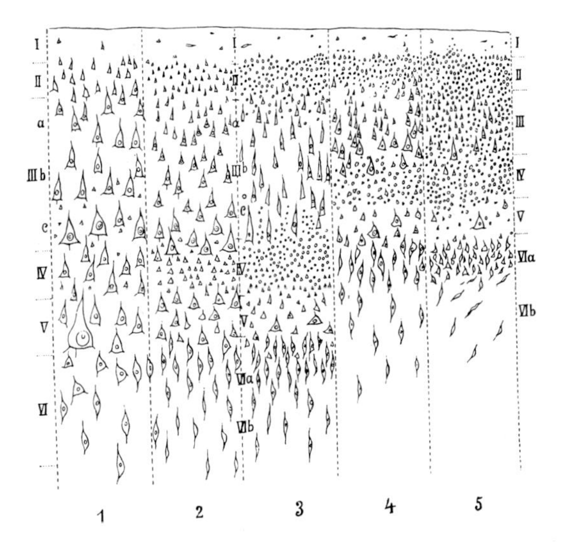

Figure 92 A. The five cytoarchitectural types of the neocortex according to v. Economo's classification (from v. Economo, 1927). 1: agranular (gigantopyramidal); 2: frontal granular; 3: parietal granular; 4: polar granular; 5: koniocortex; a, b, c: further subdivisions of main layers.

The investigations of v. Economo (1927) and v. Economo and Koskinas (1925), concerning the Human cerebral cortex, are of particular importance because that author, who retained Brodmann's notation for the laminae, clearly distinguished several distinctive cytoarchitectural types of the neocortex, characterized by easily recognizable differences of some of its layers (Fig. 92A). Arbitrarily assuming *five architectural types* which he again subdivided into *heterotypic* and *homotypic neocortices*, the former displaying conspicuous modifications of the hexalaminar arrangement which, on the other hand, is clearly manifested by the homotypic cortex. The three sorts of *homotypic* cortex are designated as frontal (2), parietal (3), and polar (4).

The *heterotypic cortices* are of the *agranular type* (1), in which the two granular layers, particularly the inner one, are indistinct, respectively more or less reduced, and of the *granular type* (5) in which the layers II to IV are especially developed, with especial emphasis on layer IV. Von Economo uses the term pyramidalization *(Verpyramidisierung)* as characterizing type 1, and granularization *(Verkörnelung)* as characterizing type 5, which, because of its multitudinous small granular cells, appear dust-like in microscopic preparations seen at moderate microscopic magnification. The cited author coined the appropriate designation *koniocortex (Staubrinde)* for such cortices, characteristic for the sensory projection areas of the neocortex, namely the postcentral (somesthetic), transverse temporal (acoustic), and calcarine occipital (visual) cortices. In this latter, moreover, lamina IV is triplicated, consisting of two koniocortical sublayers IVa and IVc, separated by a less dense cell layer with a few larger elements, and corresponding to the *stria of Gennari*, which is the layer 4 of BAILLARGER, now also called *outer stria of Baillarger*.

Two subdivisions of heterotypic *cortex agranularis* are recognized, namely *cortex agranularis gigantopyramidalis*, displaying the *giant pyramids of Betz* in layer V, characteristic for the motor area praecentralis, and the *cortex frontalis agranularis*, lacking *Betz cells*, and representing the so-called premotor cortex, extending to a variable degree of expansion rostrad from the motor area, gradually merging with the homotypic frontal cortex.

One could, as an appropriate and perhaps more useful simplification, merely distinguish three types of neocortex, namely *cortex agranularis* (1), *cortex granularis* (2, 3, 4), and *koniocortex* (5). Figure 92B illustrates, as reproduced from v. ECONOMO's photomicrographs, actual samples of cortical lamination types corresponding to the cited author's classification. It will be seen that, although a hexalaminar stratification according to BRODMANN's notation is indicated, the layers III to VI are again variously subdivided into sublayers a, b, and occasionally c.

While, in other Primates, the differentation of the laminar pattern is closely similar to that obtaining in Man (cf. Fig. 93A), it becomes, generally speaking, less distinctive in the Subprimate group. This tendency may be illustrated by comparing the lamination of the visual projection area of a Primate with that of an 'intermediate' Mammal (Carnivore Cat), and that of a lower Mammal (Lagomorph Rabbit and

---

Figure 92 B. Cytoarchitectural features of various Human neocortical regions as seen in photomicrographs of *Nissl preparations* (from v. ECONOMO and KOSKINAS, 1925, as reproduced by RANSON, 1943). 1: premotor cortex from superior frontal gyrus; 1a: precentral motor cortex; 2: prefrontal (frontal granular) cortex from middle frontal gyrus; 3: parietal cortex from supramarginal gyrus; 4: occipital (peristriate) cortex; 5: transverse temporal (auditory) koniocortex; 5a: calcarine (visual) koniocortex (OC) with steep gradient at transition to peristriate cortex (OB). All sections reproduced at identical scale, thereby showing the obtaining differences in thickness.

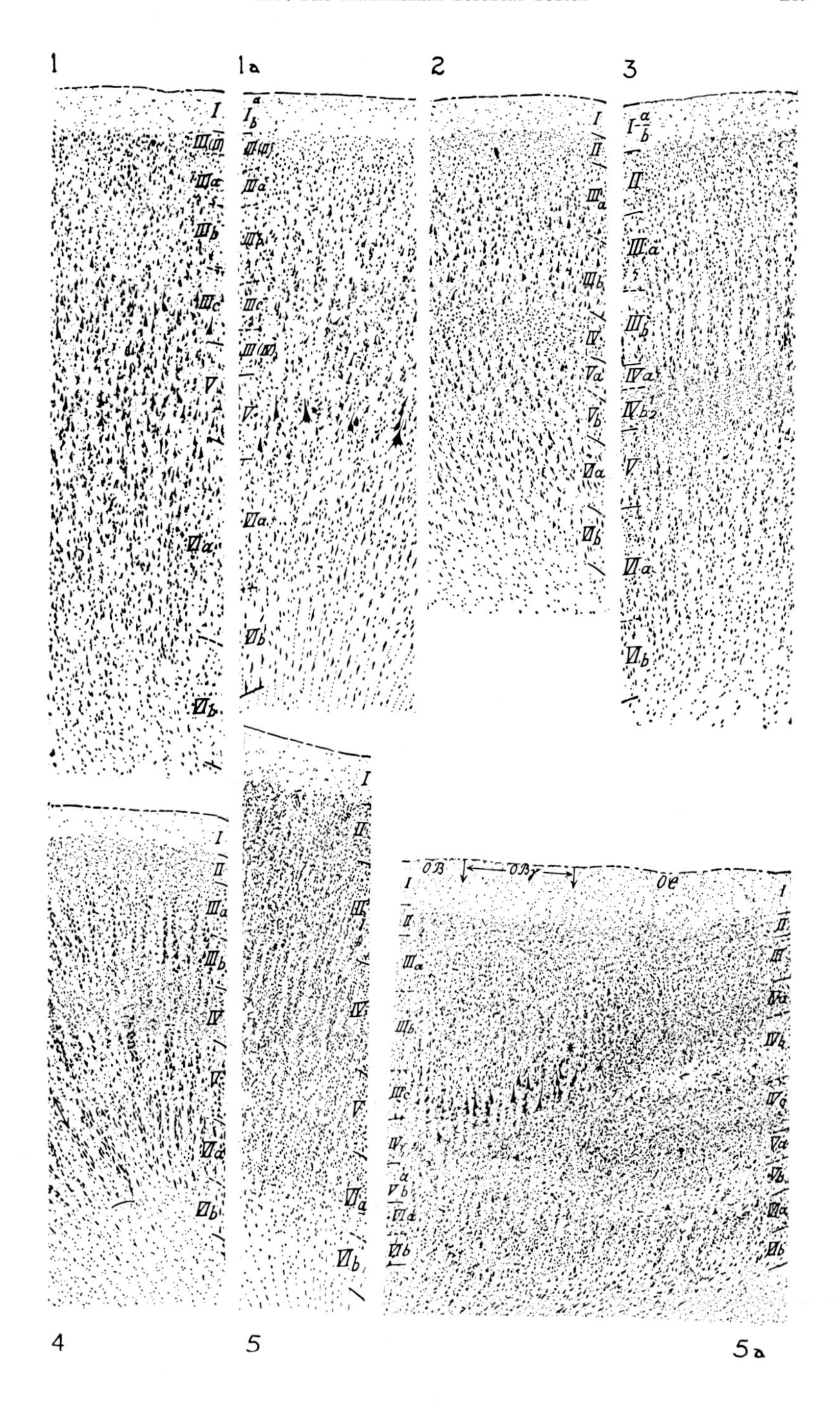
1
1a
2
3
4
5
5a

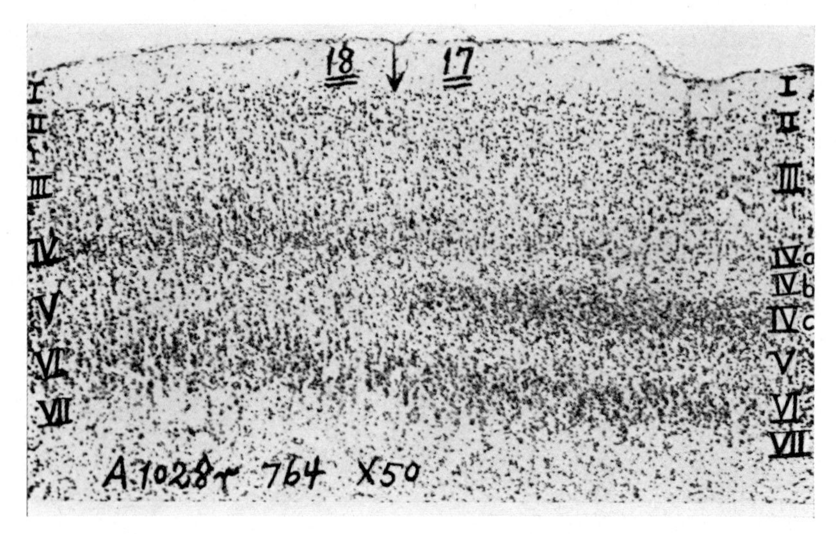

Figure 93 A. Photomicrographs of cross section at boundary of area striata (17) and parastriata (18) in a *Nissl preparation* of occipital lobe cortex in the Macacus monkey (from NGOWYANG, 1937).

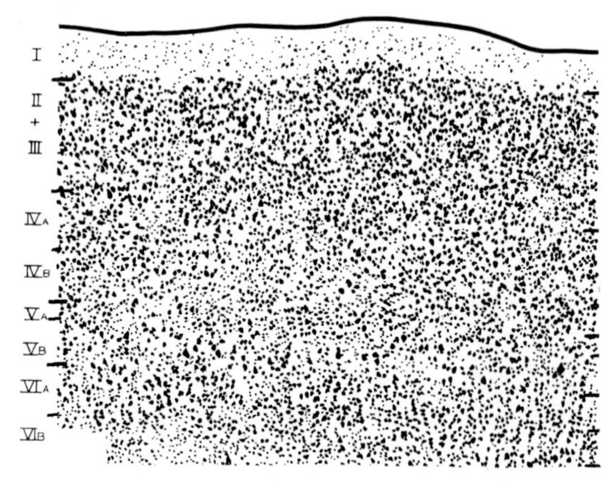

Figure 93 B. Portion of the Cat's visual cortex showing local variations in a given area (compare left with right halves) as seen in a *Nissl preparation.* The 'area striata' is here at most bistriate with IV A and B (from O'LEARY, 1941).

Rodent Rat) as shown in Figures 93A–E. Thus, v. VOLKMANN (1928a, b, c) distinguishes three types of visual cortex, namely tristriate (IVa, b, c) in Primates, bistriate (IVa, b) in Carnivores such as the Cat, and unistriate (IV) as e.g. in small Rodents. As an exception, he pointed

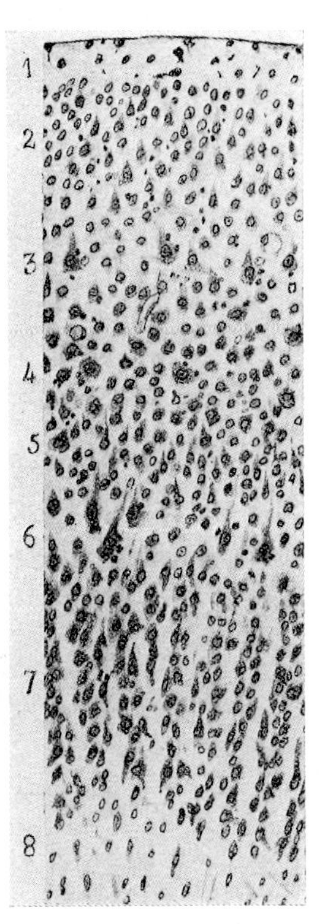

Figure 93 C. Portion of the Cat's visual cortex as seen in a *Nissl preparation* and drawn by CAJAL (from CAJAL, 1921). 1: plexiform layer; 2: 'layer of small pyramids' (II); 3: 'layer of large superficial pyramids' (III); 4, 5: layer of granular cells with two sublayers (IV A, IV B); 6: 'layer of large solitary pyramids of Meynert' (V); 7: 'layer of polymorphous cells' (VI A); 8: 'layer of fusiform cells and white substance' (VI B).

out that part of the visual cortex in the Rodent Squirrel is, however, tristriate.[58]

Examples of lamination in other neocortical areas of Subprimates are shown in Figures 94 A–E. It is here of interest to note the relatively simple, somewhat diffuse and in part less dense neocortical arrangement in Cetaceans, as e.g. pointed out by RIESE (1925). LANGWORTHY (1932), who does not refer to RIESE's investigation, likewise repeatedly stresses the 'very primitive architecture' of the cerebral cortex, 'with few cells and poor differentiation of the cell layers'. Additional observations on the Cetacean telencephalon including its cortex were published by Jansen and JANSEN (1968) and by PILLERI (1965/66a, b and

[58] Tristriate, bistriate, and unistriate refer here to sublaminae of IV, but the term 'area striata' refers to the conspicuous medullated *stria of Gennari*, which is not clearly shown in Subprimates. Thus, in this sense, a true 'area striata' is only present in Man and some Primates. The hypothesis of KLEIST, accepted by v. VOLKMANN, relating the tristriate aspect respectively laminae IVa and IVc to binocular vision, has not been corroborated.

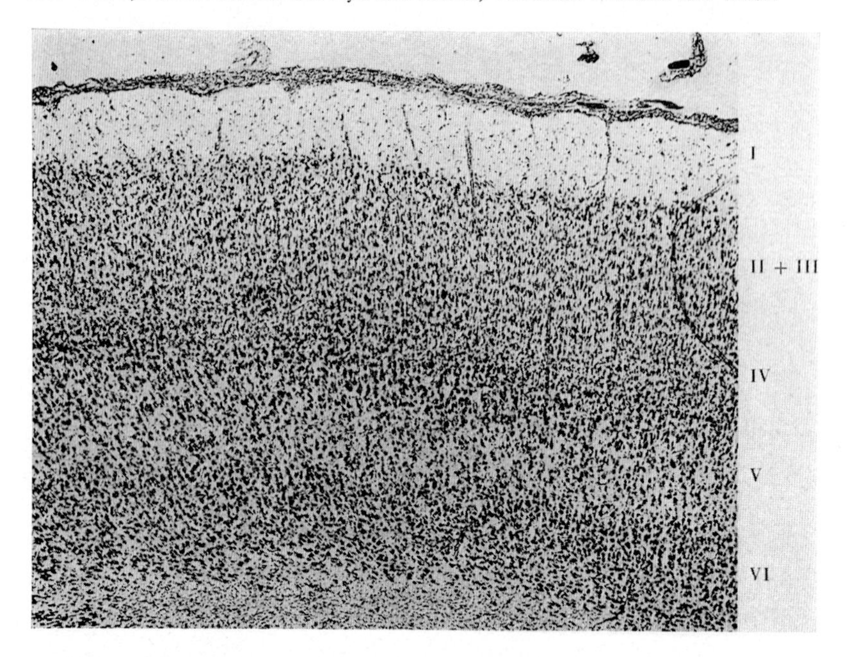

Figure 93 D. Occipital (visual) cortex in the Rabbit as seen in a *Nissl preparation* (from K. *et al.*, 1960; ×42, red. $^4/_5$).

other reports). It is, moreover, of interest that the cytoarchitecture of the Elephant's neocortex is strikingly similar to that of Cetaceans, being, in particular, also characterized by the lack of a clearly recognizable layer IV.

On the other hand, both these Mammals possess a very extensive neopallium, surpassing that of Man. The poorly clarified problems concerning the relationship of Cetacean neopallium to a high degree of 'intelligence'[59] were briefly pointed out on pp. 741–742, chapter VI of volume 3/II and require here no additional comments.

For further details on the cytoarchitecture of the Mammalian cortex, the reader may be referred to the publications by ABBIE (1940a, b, 1942), v. BONIN (1942, 1945), BRODMANN (1909), CAJAL (1911, 1921, 1922), CRAIGIE (1925), GRAY (1924), GUREWITSCH and BYCHOWSKY

[59] LILLY (1963, 1975) and LILLY and MILLER (1961) suspect that the Dolphin intelligence is comparable to the Human one, and that these Cetaceans communicate with each other by means of a complex speech pattern. LILLY's views on this topic remain controversial but deserve attention, and the results of further studies on these questions must be awaited.

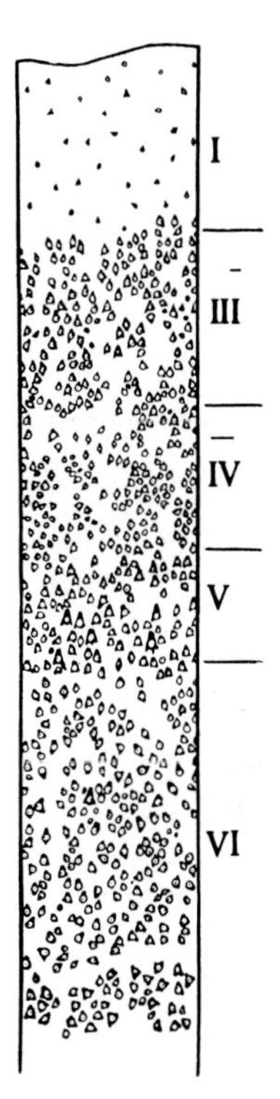

Figure 93 E. Cytoarchitectural lamination of occipital (visual) neocortex in the adult albino Rat as drawn by CRAIGIE (from CRAIGIE, 1925).

(1928), GUREWITSCH and CHATSCHATURIAN (1928), HASSLER (1962), HASSLER and MUHS-CLEMENT (1964), HERRICK (1926), VAN'T HOOG (1920), JANSEN and JANSEN (1968), KOIKEGAMI *et al.* (1943), KRIEG (1946), KUHLENBECK *et al.* (1960), LORENTE DE NÓ (1922), NGOWYANG (1937), O'LEARY (1941), PILLERI (1962, 1963, 1964, 1965/66a, b), RIESE (1925, 1927, 1943), ROSE (1912, 1926b, 1929a, b, 1931), ROYCE *et al.* (1975), RYZEN and CAMPBELL (1955), TOWER (1954), v. VOLKMANN (1926, 1928a, b, c), and ZEMAN and INNES (1963). Useful summaries with references to the older literature can also be found in the treatises by BECCARI (1943), by KAPPERS (1947), and by KAPPERS *et al.* (1936).

It should here be added that, generally speaking, the lamination of neocortex in 'lower' Mammals, such e.g. as in the Rabbit or in Rat and Mouse, is not strictly identical with that of Primates and Man. Second and third layer are not clearly separated. The fourth layer, although definitely recognizable in various regions, is likewise poorly separated from the inner part of layer III and from the outer part of layer V. This latter, as a whole, and particularly its inner part, stands out in relatively

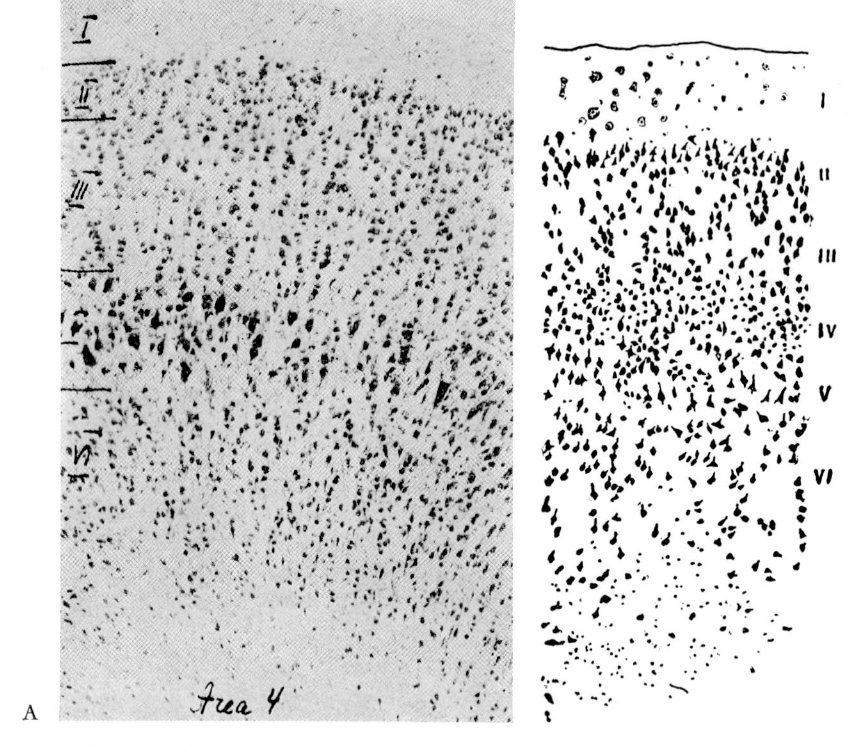

Figure 94 A, B. Cytoarchitectural lamination (*Nissl stain*). *A* Motor neocortex in the Carnivore Dog (from Gurewitsch and Bychowsky, 1928). *B* Somatosensory cortex in the Marsupial Opossum (after Gray, from Herrick, 1926).

thick sections as a clear band, characterizing the extent of the neocortex. In some regions of the Mouse, Rose (1931) subdivided it into two or even three sublayers. The sixth layer is again dense and, rather randomly, displays a somewhat less dense innermost sublayer, called 'ental' sublayer in contradistinction to an 'ectal' one by Sugita (1917). Rose (1931), in conformity with a notation formulated by the Vogts, designates the ental sublayer as layer VII or lamina infima. Yet, on the whole it is perhaps possible to agree, as regards the Rabbit and at least various other 'lower' Mammals, on the conventional notation of six layers generally conceived as characteristic for Human and the 'higher' Mammalian neocortex (K. *et al.*, 1960).

As regards *myeloarchitecture*, O. Vogt (1903, 1927) and C. and O. Vogt (1919) have undertaken very detailed and systematic investi-

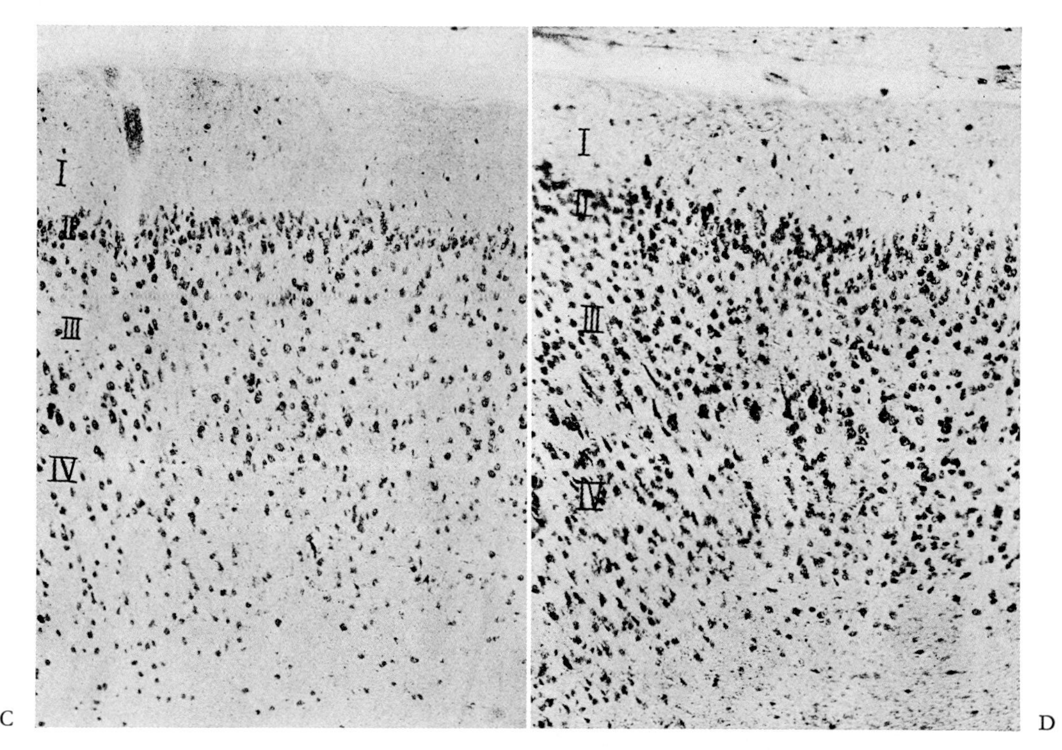

Figure 94 C, D. Cytoarchitectural lamination *(Nissl stain)*. *C* Rostral neocortex (*'vordere Frontalrinde'*) in the Cetacean Dolphin (from RIESE, 1927). *D* 'Occipital' neocortex of the Dolphin (from RIESE, 1927).

gations establishing a fundamental pattern for the isocortex. The main medullated laminae were described as corresponding to the cellular layers of Brodmann's notation in the following manner (cf. Fig. 95 A).

1. Lamina tangentialis (I. zonalis)
   1 o. Pars fibrosa laminae tangentialis
   1 a. Pars superficialis l.t.
   1 b. Pars intermedia l.t.
   1 c. Pars profunda l.t.
2. Lamina dysfibrosa (II. granularis externa)
3. Lamina suprastriata (III. pyramidalis)
   3a 1. *Stria Kaes-Bechterewi*
   3a 2. Pars typica
   3b. Pars profunda s. interna

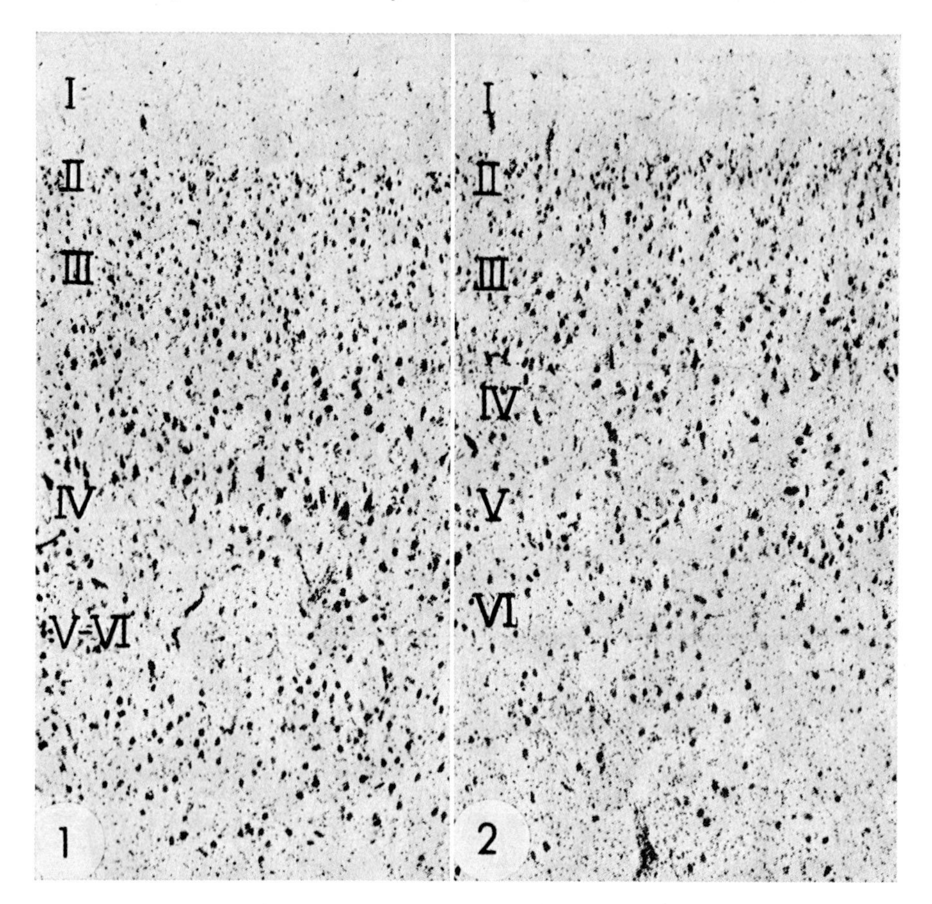

Figure 94 E. Cytoarchitectural lamination (modified *Nissl stain*) of the neocortex in the Cetacean Fin Whale, Balaenoptera physalus (from JANSEN and JANSEN, 1968). 1: frontal convexity near sulcus ectolateralis; 2: parietal region between s. suprasylvius and ectolateralis. The evident difficulty in delimiting the layers is disclosed by comparing the JANSENS' notation with that of RIESE in Figures 94 C and D. It will be noted that, although I-III agree in both notations, RIESE combines the layer internally to the large III pyramids into one lamina IV, while the JANSENS distinguished IV and V–VI in 1, and IV, V, and VI in 2.

4. *Stria Baillargeri externa* (IV. granularis interna)
5a. Lamina interstriata } (V. ganglionaris)
5b. *Stria Baillargeri interna* } (V. ganglionaris)
6. Lamina infrastriata (VI. multiformis)
 6a α. Lamina substriata
 6a β. Lamina limitans externa

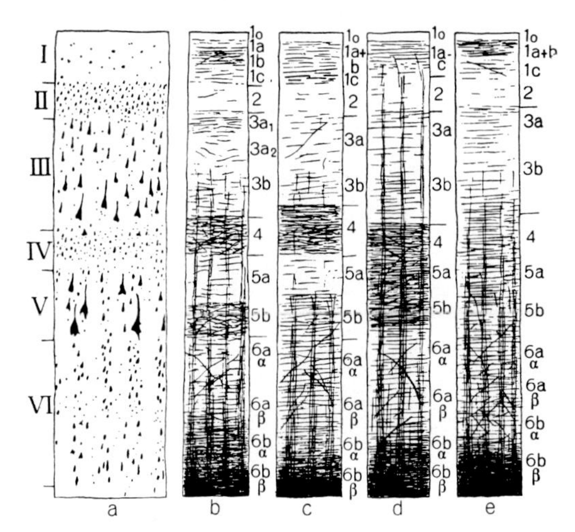

Figure 95 A. Diverse types of Human neocortical myeloarchitecture (b–e) compared with the corresponding cytoarchitectural lamination (based on Figures of BRODMANN and the VOGTS, from K., 1927). a: homogenetic fundamental cytoarchitectural plan according to BRODMANN; b: typus bistriatus quadrizonalis euradiatus; c: typus unistriatus trizonalis infraradiatus; d: typus unitostriatus bizonalis infraradiatus; e: typus astriatus trizonalis euradiatus. Sublayers are explained in the text.

6b α. Lamina limitans interna

6b β. Zona corticalis albi gyrorum

Of these laminae 1–3 are combined as external main layer, and 4–6 as the internal main one *(äussere und innere Hauptschicht)*. It can easily be seen that VOGTS' layers 5b, 4, and 1a–c correspond to BAILLARGER's original layers 2, 4, 6 respectively, while layer 4, described by KAES (1907) and also seen by BECHTEREW,[60] was not seen by BAILLARGER, who included it in his gray layer 5. It should also be added that, in their later publications, the VOGTS have also designated the internal portion of layer 6 as a separate layer 7.

The main criteria for a distinction of local differences in myeloarchitecture are provided by the behavior of radii, by the *striae of Baillarger*, and by the tangential fibers in the molecular layer. The *radii* are formed

[60] This is quite ununderstandable, since the *stria of Kaes-Bechterew* is the most diffuse and least conspicuous of the medullated laminae, requiring meticulously processed myelin stains in order to be visualized. On the other hand, the reader of BAILLARGER's paper will be surprised to find how much of the myeloarchitectural details described by the VOGTS were already accurately recorded by BAILLARGER. A translation of said author's report is included in the collection of papers on the cerebral cortex selected by v. BONIN (1960).

B

C

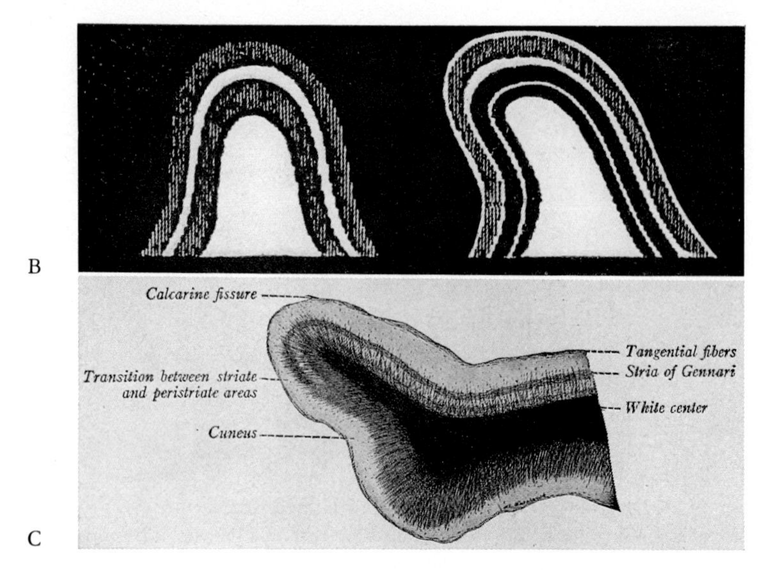

Figure 95 B. Drawings showing the naked-eye aspect of sections of the fresh Human cerebral cortex, displaying (left) *external stria of Baillager* and (right) *inner and outer striae of Baillarger*, and tangential white stria (adapted after BAILLARGER and QUAIN's *anatomy*, from RANSON, 1943).

Figure 95 C. Myeloarchitecture *(Weigert-Pal stain)* of Human area striata in rostral part of cuneus (from RANSON, 1943). It will be noted that, in addition to *stria of Gennari (outer stria of Baillarger)*, the *inner stria of Baillarger* (not labelled) is recognizable in most of the here shown portion of area striata.

by bundles of cortical input and output fibers. These bundles, to a greater or lesser extent, tend to separate the cellular elements of the deeper laminae into more or less distinct columns whose significance for cortical structure and function will be pointed out further below.

In accordance with the relative length of the medullated *radii*, three architectural types can be distinguished: (1) *typus mediradiatus sive euradiatus*, in which the radii reach to levels of lamina suprastriata; (2) *typus supraradiatus* with radii extending to lamina tangentialis; (3) *typus infraradiatus*, in which the radii do not extend beyond the *outer stria of Baillarger*.

With respect to the *striae of Baillarger*, four types can be distinguished: (1) *typus bistriatus* with two distinct striae separated by lamina interstriata, and considered to be the 'fundamental tectogenetic type'; (2) *typus unitostriatus* with fusion of the two striae into one; (3) *typus unistriatus*, displaying only the external stria, the internal one merging with the infrastriate medulla; (4) *typus astriatus* in which 4th, 5th and 6th layer display a diffuse arrangement of the striae, merging with the medulla, and without distinct individual striae.

Again, different types can be distinguished in accordance with the relative density and width of the two striae *(typus externidensior, internilatior, aequilatus, aequidensus)*.

Finally, the lamina zonalis, in addition to its pars afibrosa, may include one, two, or three medullated sublaminae, displaying thus a typus *bizonalis, trizonalis* or *quadrizonalis*.

It should, however, be pointed out that identical cortical regions in different Human brains can manifest considerable variations in the distinctiveness and differentiation of the medullary laminae, either *qua* macroscopic appearance in the fresh, unfixed state, or in myelin stain preparations carefully processed under standardized conditions. Figure 95B roughly shows aspects of neocortex as seen in the fresh aspect. It is here of interest that, depending on circumstances, the area striata may be seen as depicted either by the left or by the right section in that Figure.[61] In myelin stain preparations, the area striata may appear as shown in Figure 95C, displaying a conspicuous *external stria of Baillarger (stria of Gennari)* and a much less conspicuous inner stria. This latter can be seen missing in the whole or in part of the Human area striata despite careful staining technique, and seems to be individually variable. It is likely that these differences are related to the individual variability in the degree of myelinization of said stria's fibers rather than to their absence. There may also be some variability with respect to that stria's compactness.

As a rule, the Mammalian cerebral cortex at birth displays very few medullated fibers restricted to particular locations,[62] or no medullated fibers at all. In contradistinction to cytoarchitectural differentiation, which is commonly already quite distinctively established at birth, the development of myeloarchitecture essentially seems to progress in the

[61] BAILLARGER (1840) states that the conspicuous stria of VICQ-D'AZYR (or of GENNARI) may indeed seem unique, but that, with greater attention, above and below this main white line, another hardly visible one can be recognized (i.e. his 6th and 2nd). Yet, despite BAILLARGER's statement, and even with much attention, I have been occasionally quite unable to see in some specimens of fresh area striata any other line than that shown on the left side of Figure 95B. Cf. the further comments in the text with respect to myelin stained sections.

[62] E.g. in Man, as shown by the investigations of FLECHSIG, demonstrating the fiber systems related to the projection areas (cf. the comments in section 10 of chapter XIII, volume 5/I, with Figures 252A–E.

postnatal period. In this respect, the observations made by KAES (1907) indicate that, in Man, the amount of medullated fibers appears substantially to increase during the first two thirds of postnatal life, until approximately the 40th or 50th year, and then to decrease in the course of senescence, perhaps not necessarily through actual loss of fibers, but possibly, at least in part, merely by loss of or reduction of their medullary sheath. Such inferences concerning progressive respectively regressive age changes, including also senile cell loss, are evidently based on extrapolations by comparing findings at different ages in different individuals. A *farceur*, questioning 'side by side illustrations demonstrating age changes', stated that they suffer from what he calls the '*Kara Mustapha fallacy*'.[63] Yet, one could maintain that, cautiously interpreted, such extrapolations may be considered reasonably valid and not essentially differing from 'side by side illustrations' of embryonic stages in different individuals demonstrating the well substantiated ontogenetic sequences.

As regards structure in contradistinction to architecture or 'general tectonics', the pictures obtained by means of *Golgi impregnations* have provided some data concerning the diverse types of neurons, their dendritic and neuritic processes, and general aspects of cortical interconnections. Detailed and fundamental studies on this topic were undertaken by CAJAL (1900, 1902, 1903, 1906, 1911, 1922). Among additional reports are those by LORENTE DE NÓ (1922, 1949), and by more recent authors using electron microscopy as well as various experimental methods such as extracellular and intracellular single unit recordings by means of microelectrodes. The resulting contemporary concepts of neocortical synaptic organization have been summarized, with bibliographic references, in a publication by SHEPHERD (1974).

The *molecular (zonal or plexiform) layer* contains relatively few cells. In addition to the glial elements, some of which contribute with their pedicles to the formation of the membrana limitans externa, there are various sorts of neuronal elements, either stellate or pertaining to the

---

[63] Cf. Nature *233:* 60 (1971). The skull of the grand vizier *Kara Mustapha*, unsuccessful at the siege of Vienna (1683) and ordered by the sultan to commit suicide, is said to be on display both at Vienna and Budapest, the specimen at Budapest allegedly being of an earlier period than the Vienna specimen. Long ago, I heard in Japan a similar joke concerning a collector proudly displaying a rather small skull as that of the famed Shôgun *Minamoto Yoritomo (floruit* ca. 1180 A.D.). Upon being asked how such a small skull could be that of a warrior reputed for his mighty body size, the collector replied: this is the skull of his childhood time *(kodomo no toki no hone desu)*.

type of *horizontal cells of Cajal.*[64] Some of these cells seem to have unspecific processes *(apotiles, neurotendrils)*, others display several dendrites and two or three neurites, and still others various dendrites and one neurite.[65] The axons of these neuronal elements seem to be mostly rather short, but even the long ones appear to remain intracortical and restricted to the molecular layer Some such fibers may become medullated and contribute to the tangential fibers of myeloarchitecture. Tangential fibers are also formed by ascending neurites respectively collaterals from the lower cortical layers and perhaps by a few extensions of thinly medullated input fibers entering the cortex from the medulla and reaching the molecular layer by way of the radii. COLONNIER (1966) denies that the first layer receives specific afferent and extrinsic intercortical association fibers, but NAUTA (1954) recorded some (although 'only few') such fibers, including callosal ones, as reaching layers I and II. The molecular layer is also reached by the terminal, usually branching expansions of the apical dendrite of pyramidal cells located in the deeper layers.

The neuronal elements in the *second (or external granular) layer* are either small pyramidal cells with long neurite, or small stellate cells of *Golgi II type* with short neurite.

The *third (or pyramidal) layer* is characterized by pyramidal cells whose size tends to increase with the depth of their location below the surface, although displaying considerable variations at any one depth.

Pyramidal cells, as described and depicted in section 2, chapter V of volume 3/I, display a long apical dendrite with some side branches, and basilateral dendritic expansions. The neurite arises from the basis, roughly opposite to the apical dendrite, and is directed towards the medulla, joining one of the radii. Numerous intracortical horizontal and also some ascending collaterals commonly branch off from the neurite (Figs. 96 A, B).

The third layer also contains bipolar (or double bush) cells with dendrites extending upward and downward, and, a laterally arising neurite. In addition, diverse types of stellate cells are present, the neu-

---

[64] Cf. Figures 58–60, p. 81 in section 2, chapter V of volume 3/I, and the comments on p. 79, section 2, and pp. 547–548, section 7 of the cited volume.

[65] Although CAJAL had originally described horizontal cells with two or three neurites, he subsequently retracted this interpretation (1911, p. 526), stating that, among the processes of these cells only '*un seul mérite d'être considéré comme un cylindre-axe*'. On the basis of my own observations, however, I am inclined to believe that CAJAL's original interpretation could perhaps be upheld.

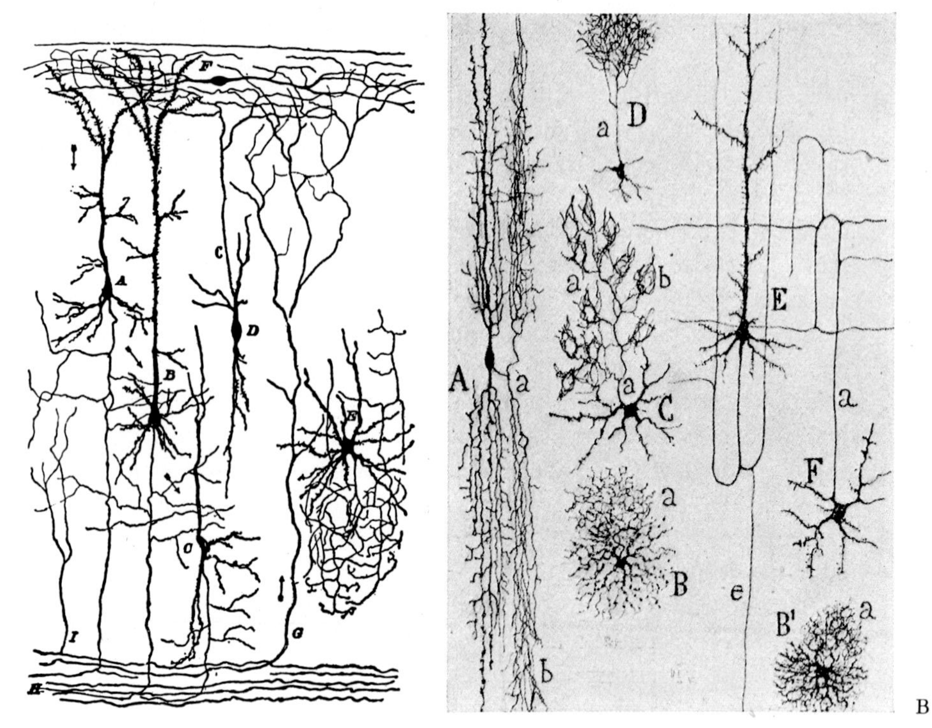

Figure 96 A. Structural aspects of the Mammalian neocortex as seen in *Golgi preparations* (from CAJAL, 1933). A: medium sized pyramidal cell; B: large pyramidal cell; c: cell of the polymorphous layer; D: cell of *Martinotti;* E: *Golgi II type* star cell; F: *Cajal cell* of the plexiform (or molecular) layer; G: afferent ascending to layer I; H: white substance; I: collateral from white substance. The upper C is the neurite of D. Cf. also, in vol. 3/I, Figs. 188 A, B, p. 275–276, Figs. 198–200, p. 292–293, and Figs. 201 A–I, p. 295–298.

Figure 96 B. Various neuronal types, mostly with short axon *(Golgi II cells)* of the cerebral cortex in the Human neocortex of an a few months old infant (from CAJAL, 1923). A: cell with double dendritic bushes; B, B′: dwarf cell *(elemento enano)* with short axon; C: basket cell *(célula de cestas);* D: dwarf element with axonic bushes; E: pyramidal cell with arciform ascending collaterals; a: axon; b: axonic baskets; e: descending axon collateral.

rites of some such cells *('basket cells')* forming basket-like terminal arborizations (cf. Fig. 97 A). In the third as well as in other layers (fifth and sixth) there are fairly large multipolar elements designated as *Martinotti cells*[66] whose neurites ascend as far as the molecular (or plexiform) layer.

[66] First described by MARTINOTTI (1890).

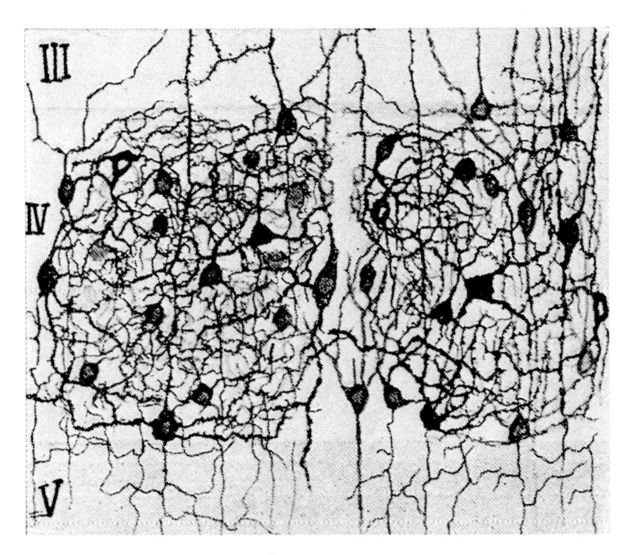

Figure 96 C. Two 'glomeruli' *(Golgi impregnation)* of the internal granular layer in the temporal (acoustic) neocortex of the Mouse (from LORENTE DE NÓ, 1922).

The *fourth (or internal granular) layer* whose density respectively degree of development varies in accordance with diverse taxonomic forms, and with different regions, as pointed out above with reference to v. ECONOMO's classification, contains stellate cells (star cells, spider cells) of diverse sorts. In the koniocortex, these are mostly small *Golgi II type cells*, but larger ones, so-called star pyramids with a long ascending apical dendrite and a fairly long axon are present in the cortex of the visual projection area and possibly also in a few other regions. It is doubtful whether their axon leaves the cortex.[67] The *outer stria of Baillarger*, respectively the *stria of Gennari or Vicq d'Azyr* is located within the internal granular layer. It seems to be formed by neurites of stellate cells, collaterals of pyramidal cells of the third layer, and presumably also by 'horizontal' ramifications of cortical input fibers. Many such afferents to the cortex seem to end in the fourth layer, where groups of granule cells with their dendritic plexuses, together with end-arborizations of cortical afferents, may form so-called *glomeruli* (cf. Fig. 96 C) as described by LORENTE DE NÓ (1922). In sections cut at a plane parallel to that of the cortical surface, these glomeruli within

[67] Such cells may represent *Golgi II elements* with relatively long axon, or, as it were, an intermediate type between *Golgi I (Deiters type)* and *Golgi II cells*.

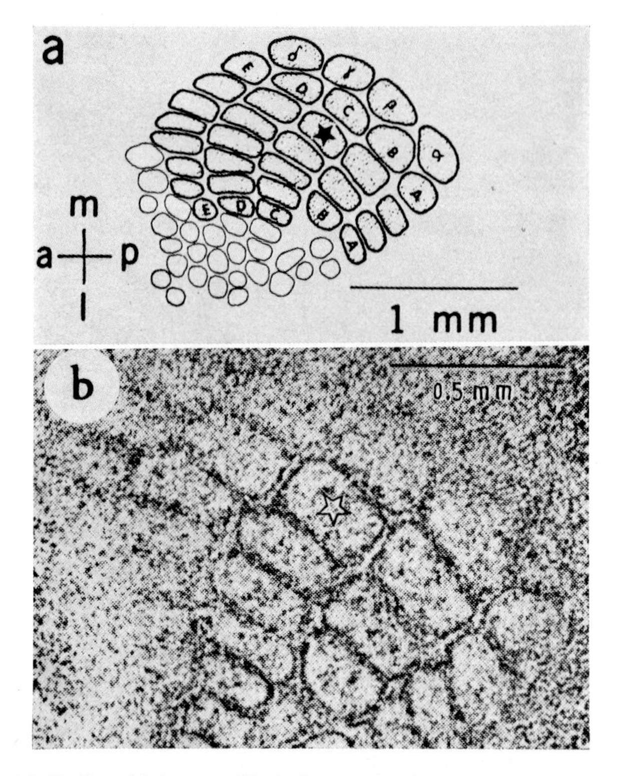

Figure 96 D. 'Barrels' (presumably in layer IV) of the somatosensory neocortex of the Mouse as seen in a cell-stained section parallel to cortical surface (from VAN DER LOOS and WOOLSEY, 1973). The designations of the 'barrels' refer to corresponding designations for the vibrissae, and are not relevant in the present context. The orientation is indicated by a: anterior, 1: lateral, m: medial, and p: posterior.

the fourth layer of the somatosensory cortex of the Mouse can be identified as cytoarchitectural units which have been called *'barrels'* (cf. Fig. 96D) and apparently correspond to particular input loci for neural signals originating in sensory vibrissae of the muzzle (VAN DER LOOS and WOOLSEY, 1973; PASTERNAK and WOOLSEY, 1975). *'Barrels'* are likewise described in the somatosensory cortex of the Rat, where some of the smaller barrels could be noted outside of the head area in regions receiving input from forepaw and hindfoot (WELKER, 1976) as shown in Figure 96E.

The *fifth (or ganglion cell) layer* is characterized by the presence of scattered rather large pyramidal cells, including, in the motor cortex, the *giant pyramids of Betz*. This layer which among other elements may

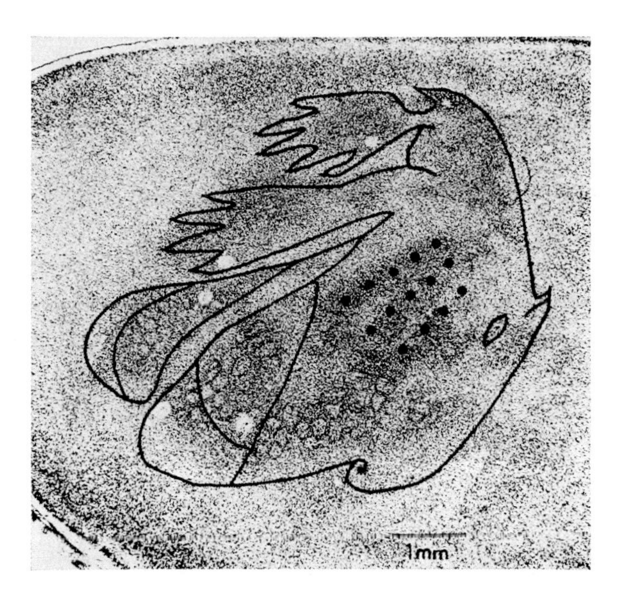

Figure 96 E. 'Barrels' and related cytoarchitectural organization of the Rat's somatosensory cortex as seen in a 'flat' section, partly passing through the plane of layer IV. The architectural subareas conform to the distorted body figure (*'animalculus'*) of the animal, as shown by the outline drawn upon the photomicrograph. The black dots correspond to the projection of five rows of 'mystacial vibrissae' (from WELKER, 1976).

also contain a few *Martinotti cells*, is generally recognizable by its lesser density. It includes the *inner stria of Baillarger.*

The *sixth layer*, in which the radii generally become most conspicuous and thereby separate the neuronal elements into rather distinctive columns, contains spindle-shaped, fusiform, and irregular polymorphous elements of diverse sorts, including *Martinotti cells.* Most neurites, however, seem to leave the cortex.

The *radii* represent bundles of cortical input and output fibers. As regards input channels, at least two different sorts can be distinguished, namely (1) input fibers from the thalamic sensory relay nuclei to the cortical sensory projection centers. Such fibers are generally called *specific afferents*, and seem to end mostly in the fourth layer, although apparently also reaching layer III. In addition, LORENTE DE NÓ (1949) and others describe (2) *unspecific (or nonspecific) afferents*, which are said to ascend as far as the upper cortical layers (cf. Figs. 97 A, B). Such nonspecific afferents might originate in the thalamic grisea conceived as *cortical modulators* (K., 1954) and could perhaps also be in part

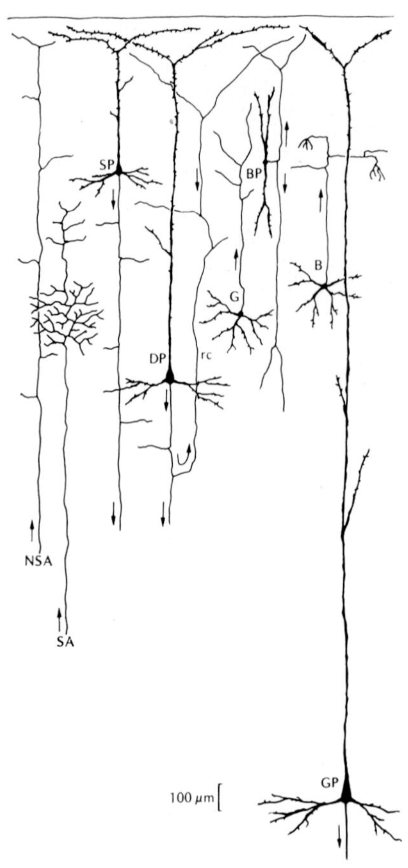

Figure 97 A. Neuronal elements and afferents of Human neocortex in a contemporary interpretation (from SHEPHERD, 1974), B: basket cell; BP: bipolar cell; DP: deep pyramidal cell; G: granule cell; GP: giant pyramid *(Betz cell)*; NSA: non-specific 'sensory' afferents; SA: specific sensory afferents; SP: superficial pyramidal cell; rc: recurrent collateral. It is not quite clear why an arrow pointing downward has been affixed to the upper portion of fiber rc.

related to the postulated so-called centrencephalic and activating reticular systems, being presumably also relayed by diencephalic grisea.

The arrangement of cortical elements either between or surrounding the radii seems to indicate a *columnar (vertical) organization* of the neocortex, comprising neuronal elements of all subzonal laminae as suggested, on the basis of structural features, by LORENTE DE NÓ (1949), K. (1957, 1973),[68] v. BONIN and MEHLER (1971), and other authors. These columns, of which the above-mentioned barrels might be components, could perhaps include afferent or efferent fibers, i.e. radii, as an axis. Subsequent investigations using microelectrode techniques (e.g. HUBEL and WIESEL, 1969) have substantiated the concept of col-

[68] K. (1957, pp. 253–254; 1973, pp. 278–279, 318–319).

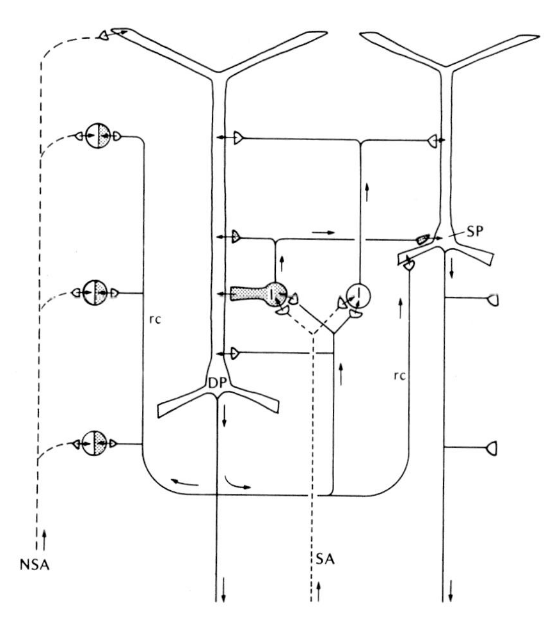

Figure 97 B. Diagram purporting to illustrate 'basic circuit' connections in the neocortex (from SHEPHERD, 1974). I: intrinsic neurons; assumed inhibitory elements respectively synapses are stippled. Other designations as in A.

umns. Further aspects of this structural and functional organization into small vertical units will be considered in section 6 of the present chapter.

With respect to the laminar differentiation, VAN'T HOOG (1918, 1920), KAPPERS (1928) and others have suggested a functional localization in depth *(Tiefenlokalisation)*, such that the granular layer (IV) is receptive, the infragranular layers (V, VI) are efferent, and the supragranular layers (III, II) subserve correlative associations of higher order. The infragranular elements are supposed to provide projection fibers to the subcortical grisea, and callosal commissural fibers. The supragranular pyramids, on the other hand, are believed to provide association fibers between cortical regions.

LORENTE DE NÓ (1949) assumes that, save for layer I and II, all cortical layers contain cells having synaptic contacts with specific afferents, so that it would be improper to call any one layer 'receptor'. Again, every layer except I seems to originate axons reaching the medullary center, and therefore, according to the cited author, no particular layers may be called the 'effector'.

Be that as it may, the inference that at least most of the projection fibers to subcortical grisea of the neuraxis originate in the infragranular layers seems reasonably well supported, but it still remains a moot question to which extent axons from the other layers, particularly from layer III, may contribute to these projection channels.

Summarizing, in a simplified manner, the main and well-recognized neocortical cell types, several classifications can be used. The most commonly accepted one distinguishes (1) pyramidal cells, (2) granule or stellate cells, (3) horizontal cells of lamina I, (4) *Martinotti cells*, and (5) fusiform and polymorphous cells, particularly in lamina VI.

Another overall classification, used by LORENTE DE NÓ (1949), recognizes four main cell types, namely (1) cells with descending axon reaching the medullary center as fiber of projection or of association, (2) cells of *Golgi II type*, often ending 'within a homogeneous zone of the dendritic plexus', (3) cells with ascending axon ramified in one or several cortical layers, and (4) cells with horizontal axons.

Still another classification is proposed by SHOLL (1967), who distinguishes four types of pyramidal cells (P_{1-4}) and three types of stellate cells (S_{1-3}). As regards the former, P_1 gives off an unbranched axon to white matter, P_2 a branched axon to white matter, and P_3 likewise, but with recurrent collaterals, while P_4 gives off an axon forming recurrent collaterals and branches only. The stellate cells S_1 distribute their axon within their own dendritic field, S_2 cells give off an axon to white matter, and S_3 elements send their axon to the 'outmost cortical zone'.

Various authors, in addition, distinguished so-called '*special cells*', such, e.g. as the *giant pyramids of Betz*, the large pyramids *(Solitärzellen)* in the fifth layer of the calcarine (visual cortex), the large (giant) pyramids in the third layer of the parastriate cortex in a strip adjacent to the area striata, the giant star cells (also *Solitärzellen)* of MEYNERT in layer IVb of the area striata. NGOWYANG (1932, 1934, 1936) demonstrated 'fork-cells' *(Gabelzellen)* in the fifth layer of insular cortex and in the cornu Ammonis of Man and anthropoid Apes. Each fork-cell has a straight basal dendrite and two large apical ones, directed toward the surface of the cortex. Rather large, but otherwise typical spindle cells in the fifth layer of the Human insula were also considered to represent 'special cells' by v. ECONOMO (1927 *et passim)*.

Of all the grisea in the Mammalian neuraxis, the neocortex presents perhaps the greatest difficulties for an intelligible analysis of synaptic organization. SHEPHERD (1974) mentions, in this respect, two obvious reasons. First, cortical input and output fibers run through the depth

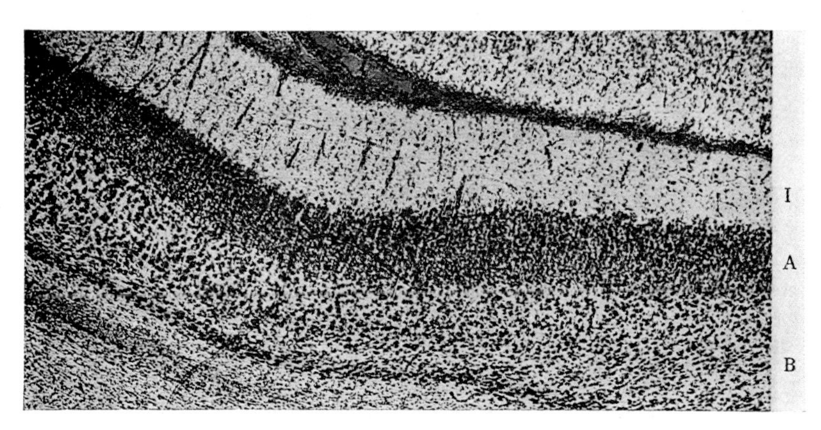

Figure 98 A. Lamination of parahippocampal cortex in the postcommissural region *(Nissl stain)* of the Rabbit (from K. *et al.*, 1960). Gradual transition between outer part (at left) and inner part begins in the right third of the Figure ($\times$42, red. $^{2}/_{3}$). Further explanations in text. Some apparent areal differences, as e.g. in Figs. 109a, b are related to curvature of brain surface vs. plane of section.

of the cortex, where they become inextricably intermingled with each other and with the labyrinthic intracortical networks. Since input and output channels do not form separate and discrete bundles, electrophysiological analysis is very difficult and gives ambiguous results. Second, with the exception of cortical regions receiving input from well-known specific sensory channels, the types of input information reaching the cortex are difficult to characterize. Much the same applies to cortical output. Although the corticobulbar and corticospinal channel arising in the motor and premotor cortex is, to some degree, understood, the understanding of this best known output channel still remains very limited *qua* specific details. The association areas are evidently the most difficult to analyze in terms of input, output, and intrinsic organization.

It might be added that, as regards other cortices, and with the necessary restrictive qualifications *qua* satisfactory understanding, the cerebellar cortex, because of its stereotype organization, is perhaps the one best understood. *Mutatis mutandis*, this also applies, with further qualifications, to cortex tecti mesencephali, and to the telencephalic hippocampal cortex.

Turning now to the architecture of the *parahippocampal cortex* it sems appropriate to distinguish, as in the case of the hippocampal cortex, at least three comparable main regional subdivisions, namely pre-

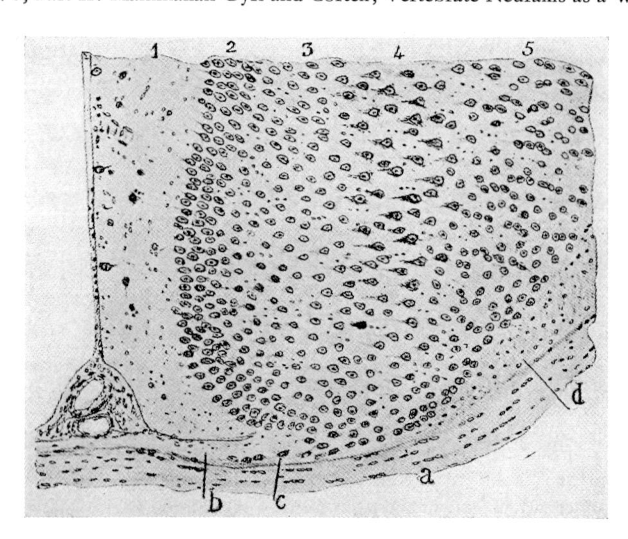

Figure 98 B. Supracommissural parahippocampal and hippocampal cortex *(Nissl stain)* in the Guinea Pig (from CAJAL, 1911). 1: plexiform (molecular) layer; 2: 'layer of spindle cells'; 3: 'deep plexiform layer'; 4: 'large pyramids'; 5: polymorph cells; a: corpus callosum; b: perhaps indusium verum; c: reduced supracommissural hippocampus; d: cingulum system.

commissural,[69] supracommissural, and postcommissural. This latter subdivision, moreover, at least in many forms, could be further subdivided into a retrosplenial and a posterior piriform ('entorhinal') subregion.

In addition to this rostrocaudal subdivision, the parahippocampal cortex displays an outer zone, adjacent to neocortex, and an inner one, adjacent to hippocampal cortex (K. *et al.*, 1960). All these subdivisions and zones merge through rather indistinct gradients, but, by their combination, e.g. precommissural outer and inner, supracommissural outer and inner, etc., a number of ill-defined areae could perhaps be described.

Particularly in lower Mammals, it is rather difficult to classify the layers of this cortex which, *qua* stratification, appears 'simpler' than that of neocortex (cf. Fig. 98 A as compared with Fig. 93 D). The molecular layer tends to be wider, and a rather dense and relatively small-celled layer A is continuous with layers II, III, and IV of the neocor-

[69] The precommissural parahippocampal cortex is shown in Figures 222 A, B of volume 5/I, Figure 82 A of the present volume.

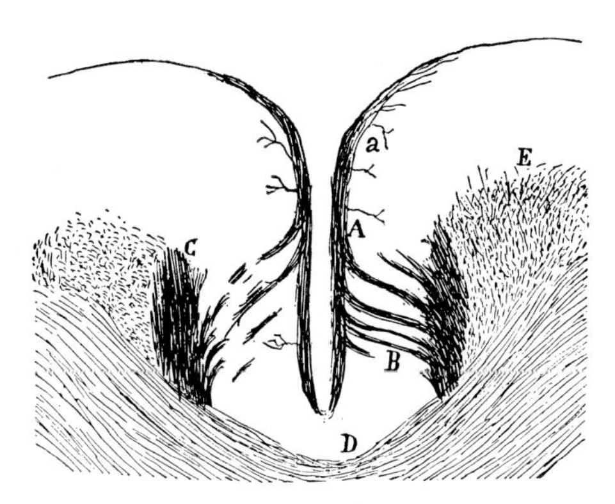

Figure 98 C. Fiber systems *(Golgi impregnation)* related to supracommissural parahippocampal cortex of the Mouse (from CAJAL, 1911). A: fiber system of plexiform layer; B: perforating fibers from plexiform layer to cingulum system or vice-versa; C: cingulum; D: corpus callosum; E: medulla of neocortex, probably fasciculus arcuatus; a: collaterals from plexiform fiber system.

tex. An inner layer B, less dense, and with larger cells, is continuous with layer V and VI of the neocortex.[70] Layer B shows an outer sublayer, rather poor in cells, which separates layers A and B, and could also be considered as separate layer (substratum dissecans, STEPHAN, 1975), more or less corresponding to the neocortical layer V. Regional, individual, as well as taxonomically related variations in the differentiation of these layers seem to obtain.[71] In the supracommissural parahippocampal cortex of the Guinea pig, CAJAL (1911) subdivided our layer A into layers 2 and 3, and our layer B into layers 4 and 5 (cf. Fig. 98 B). The molecular layer of the parahippocampal cortex (CAJAL's layer 1) tends to contain conspicuous, generally medullated fiber bundles, which join the cingulum system (cf. Fig. 98 C) or, in the postcommissural parahippocampal cortex, directly join the hippocampal formation (cf. Figs. 86 B, C, 87 A).

The postcommissural parahippocampal cortex corresponds essentially to the retrosplenial region of BRODMANN, and to that author's entorhinal cortex (area 28), which latter is the cortex of the posterior piriform lobe (K., 1927). This was also implied but not sufficiently indicated or stressed in our report of the cortex cerebri in the Rabbit (K. *et al.*,

[70] This subdivision was tentatively adopted in a report on the zonal pattern of cortex cerebri in the Rabbit (K. *et al.*, 1960).

[71] Cf. e.g. Figure 10, p. 356 of SANIDES (1972), in which the conditions obtaining in Rat and in Hedgehog are compared.

Figure 99 A. Lamination of (parahippocampal) posterior piriform lobe cortex *(Nissl stain)* as interpreted by CAJAL (after CAJAL, from BECCARI, 1943). I: stratum plexiforme; II: stratum of stellate cells; III: stratum of medium sized pyramidal cells; IV: internal plexiform stratum; V: stratum of horizontal fusiform elements, and granular cells; VI: stratum of polymorph cells. II–III are layer A, IV–VI are layer B of our notation in Figure 98 A.

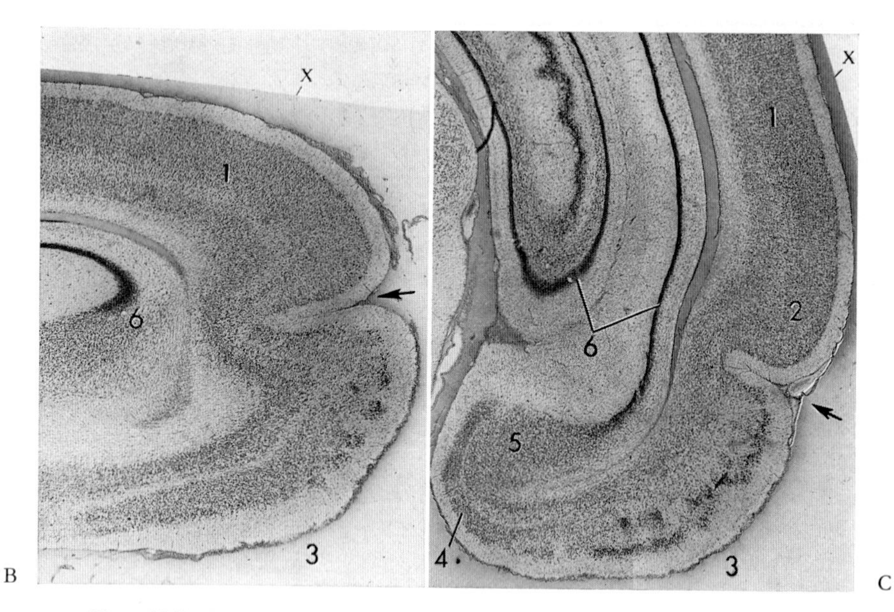

Figure 99 B, C. Posterior piriform lobe cortex ('entorhinal area') in the Rabbit *(Nissl stain,* ×20, red. 1/2). B: sagittal section; C: transverse section. 1: occipital (visual) neocortex; 2: temporal (auditory) neocortex; 3: posterior piriform lobe; 4: subicular area of posterior piriform lobe cortex; 5: parasubicular portion of cornu Ammonis (obliquely cut in B); x: edge of celloidin imbedding. The arrow indicates sulcus rhinalis, pars posterior. 6 is cornu Ammonis.

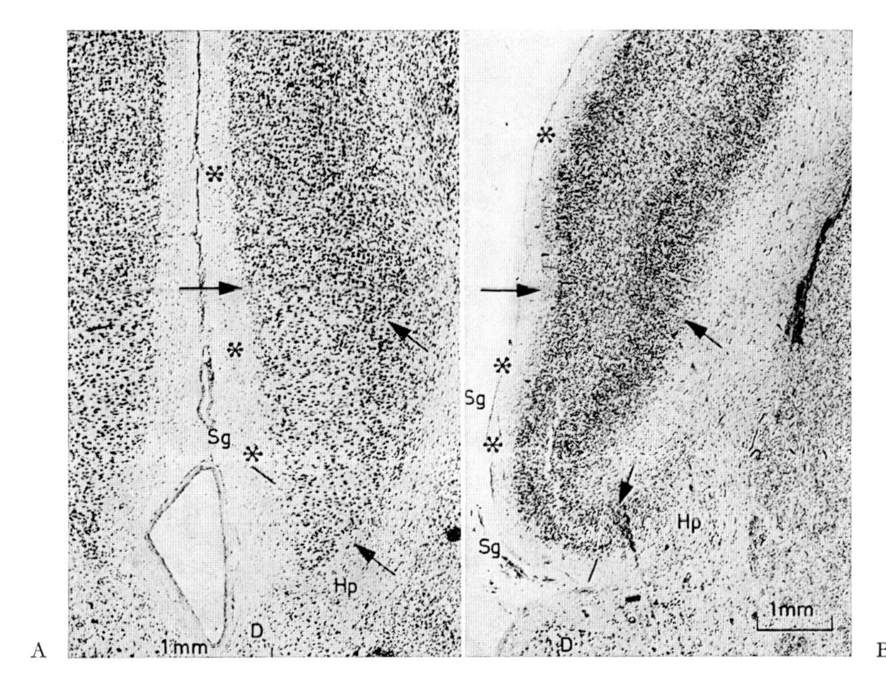

Figure 100, A, B. Horizontal sections *(Nissl stain)* through precommissural parahippocampal (and hippocampal) cortex in Primates (from STEPHAN, 1975). A: Cercopithecus ascarinus; B: Homo; D: paraterminal grisea *(Broca's diagonal band)*; Hp: precommissural hippocampus; Sg: parahippocampal cortex. The lower arrow approximately indicates transition of parahippocampal to hippocampal cortex. Asterisks and other arrows refer to tentative subdivisions of parahippocampal cortex. Low gradient to neocortex at top. The deformed (unlabelled) sulcus on left side of A is sulcus parolfactorius posterior, which is likewise identifiable in B.

1960). Again, substantial taxonomically related differences in the development of the above-mentioned layers obtain in retrosplenial[72] and in entorhinal (posterior piriform lobe) parahippocampal cortex. In the entorhinal region, the cells of the outer layer A tend to increase in size (cf. Fig. 99A). In addition, there is frequently, in this layer, an arrangement of the cells in clusters, distinctly reminiscent of the *islets of Calleja* in the basal cortex (cf. Figs. 99B, C). These clusters may be separated by mostly medullated fibers from the molecular layer. In the Human parahippocampal cortex of the subiculum cornus Ammonis such clusters are often conspicuous and provide together with the medullated

[72] Cf. e.g. Figures 38–41, pp. 63–64 of BRODMANN's (1909) treatise.

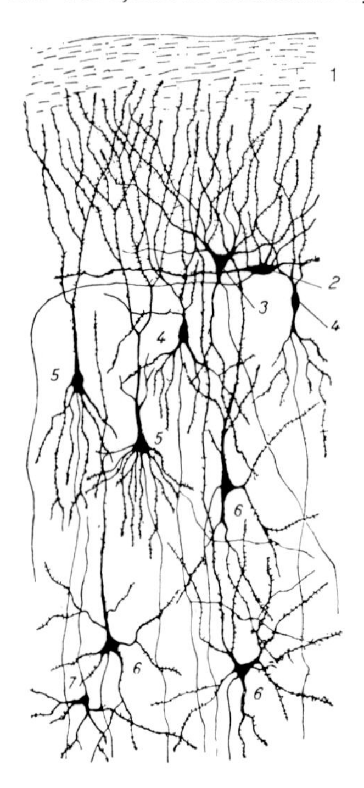

Figure 101. Structural details *(Golgi impregnation)* of parahippocampal posterior piriform lobe cortex (after CAJAL, 1911, from BECCARI, 1943). 1: 'olfactory fibers of lateral olfactory tract' (fibers of plexiform layer, their olfactory origin is doubtful); 2: 'semilunar nerve cell'; 3: triangular nerve cell; 4: fusiform cells; 5: pyramidal cells; 6: polymorph cells; 7: deep 'granulimorph' cell with short neurite.

fibers from layer I, the so-called *substantia reticularis alba Arnoldi* (ARNOLD, 1851, cf. Fig. 86 B.)

Generally speaking, the myeloarchitecture of the parahippocampal cortex displays substantial tangential fibers, and, to a variable degree, the *outer and inner stria of Baillarger*, as well as likewise variable radii. Except for the tangential fibers, the parahippocampal myeloarchitecture seems to appear, on the whole, more nondescript or blurred than the neocortical one.

Concerning Primates and particularly Man, whose precommissural parahippocampal (and hippocampal) cortex is shown in Figure 100, it should be added that the cytoarchitecture of the outer parahippocampal zone, particularly in the supracommissural (cingular) region, tends to approach that of neocortex, becoming, as it were, 'neocorticalized'. This tendency accounts for the various contemporary attempts at cortical classification distinguishing intermediate cortical types, such e.g. as 'proisocortex'. I believe, however, that the simpler basic classification into five main cortical types (cf. section 2 of this chapter) is preferable. A proper understanding of steep and low gradients eliminates the necessity for the more detailed classifications with their unduly more complex nomenclature.

With regard to structure, the parahippocampal cortex, with still poorly understood regional differences, essentially contains the same sort of neuronal elements as found in the neocortex, but perhaps with a somewhat more random distribution of the diverse types and possibly with a lesser complexity of the neural network. The large triangular cells and the horizontal elements in the outer part of the A layer are conspicuous in Figure 101 which exemplifies the structure of the posterior piriform ('entorhinal') parahippocampal cortex. For additional details, and the numerous still ambiguous recorded data on said cortex, the interested reader is referred to STEPHAN's (1975) comprehensive treatise.

4. Problems of Localization and Parcellation, Particularly Concerning the Neocortex

The early study of unstained, fresh Human cerebral cortex by BAILLARGER (1840), referred to in the preceding section of the present chapter, already disclosed regional differences in the cortical structure. These differences were subsequently emphasized by MEYNERT (1867, 1868) who initiated the first systematic cytoarchitectural investigations and suggested that the structural regional differences were related to differences in function. Specific areas of functional localization, as already inferred for articulated speech by BROCA (1861), were first clearly demonstrated in the Dog by the electrical stimulation experiments of FRITSCH and HITZIG (1870), who established the presence of an excitable cortical motor area.

FERRIER (1876), MUNK (1881) and others recognized additional sensory cortical areas, and the pioneering investigations of FLECHSIG (1876, 1896, 1920, 1927) established the relevant difference between projection and association areas of the Human cerebral cortex as well as their location.

Because experimental studies disclosed that even widespread destruction of cortical regions may cause comparatively minor functional disturbances or may be followed by functional restitution while the structures remain destroyed beyond repair, GOLTZ (1881) and others strongly attacked the concept of distinctive cortical localization and elaborated what is now known as an '*holistic*' viewpoint such as e.g. later

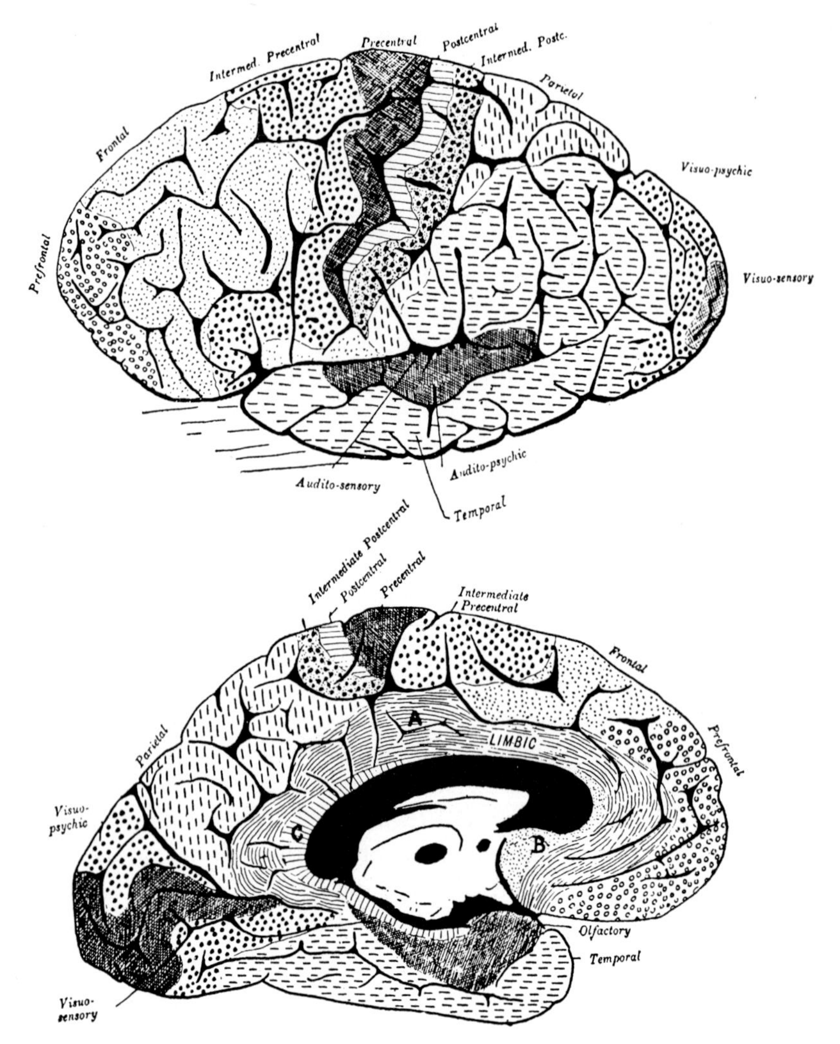

Figure 102. CAMPBELL's cytoarchitectural map of the Human cerebral hemisphere in lateral and medial view (from CAMPBELL, 1905).

stressed by LASHLEY. Nevertheless, the numerous subsequent experimental and clinical studies corroborated the validity of the concepts propounded by FRITSCH and HITZIG, MUNK, FLECHSIG and others, and demonstrated, beyond reasonable doubt, the existence of localized sensory and motor representation[73] in the cortex, as well as the pres-

[73] The term 'representation' was apparently introduced by HUGHLINGS JACKSON (1835–1911). Cf. volume 1 of JACKSON's selected writings (1932). The '*holistic*' concept is

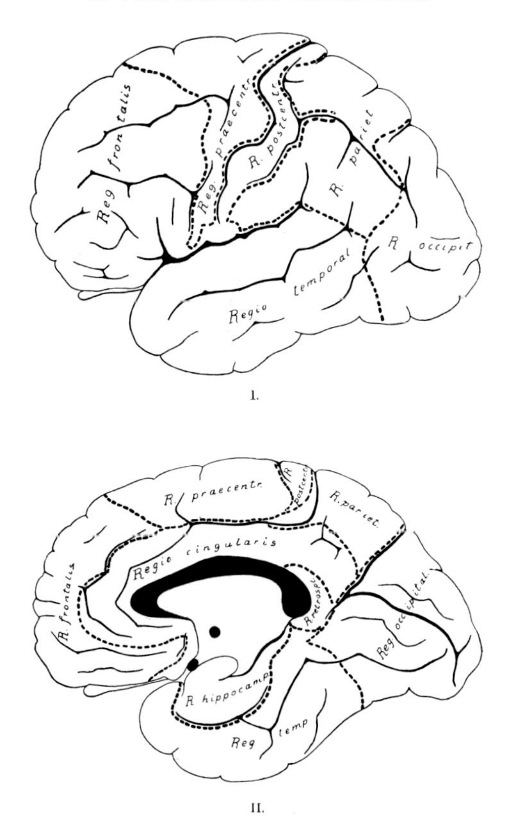

Figure 103. Main cytoarchitectural regions of the Human cerebral hemisphere according to BRODMANN (from BRODMANN, 1909).

ence of some specific regions related to particular aspects of 'higher' integrating neural activities.[74] Relevant details of this functional localization respectively representation in the cortex were dealt with in section 10 of chapter XIII in volume 5/I.

Concerning the Human brain, a detailed cytoarchitectural map dis-

---

evidently to some extent justified, in view of the redundancy and of the complex network properties displayed by the cerebral cortex as well as the correlated subcortical grisea, but should not be overemphasized as contradicting the concept of localization. Even GOLTZ, despite his 'holistic' viewpoint, felt constrained to admit diverse overall aspects of localization. In other words, '*holistic*' and '*localizatory*' concepts are not necessarily mutually exclusive.

[74] A compilation and translation of a few (12) selected historically relevant papers on the cerebral cortex, from FLOURENS (1824) to LEYTON and SHERRINGTON (1917) was published, with an introductory comment, by v. BONIN (1960).

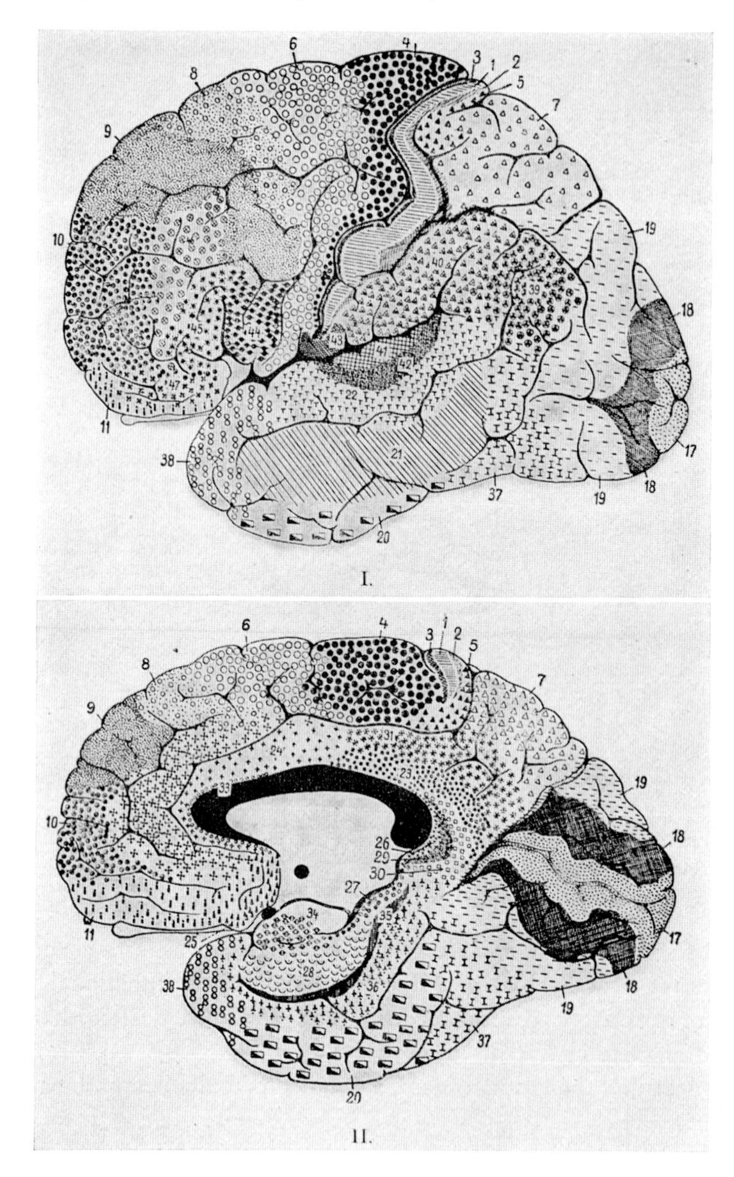

Figure 104. BRODMANN's detailed map of Human cortical areae (from BRODMANN, 1909).

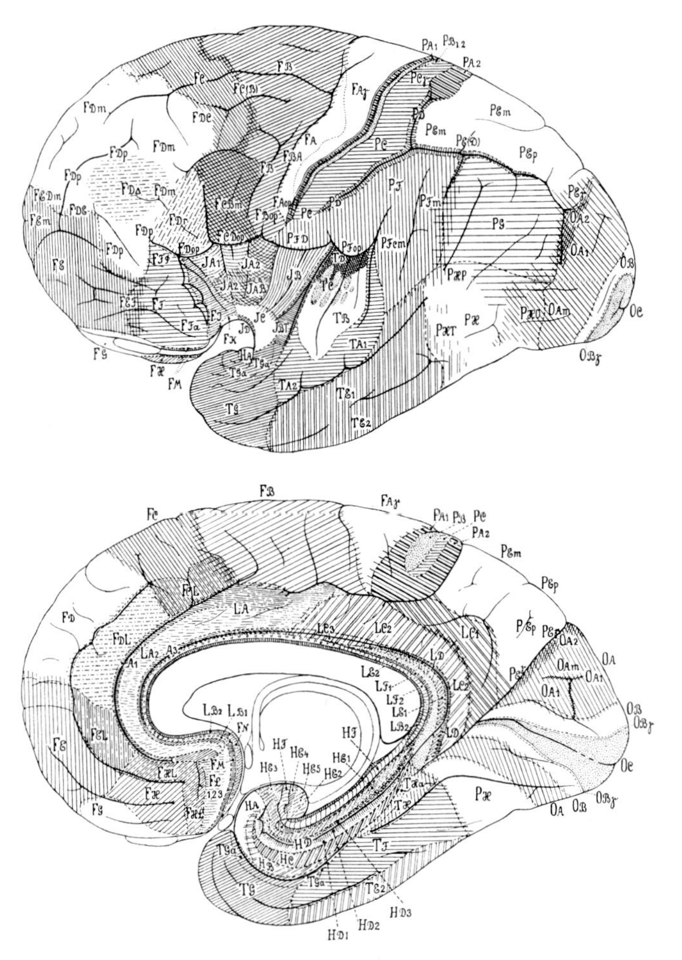

Figure 105. Cytoarchitectural cortical map of the Human brain according to VON ECONOMO, in lateral and medial view (from V. ECONOMO, 1927).

tinguishing about 16 areae was elaborated by CAMPBELL (1905) as here shown in Figure 102. About the same time, BRODMANN had initiated his comprehensive cytoarchitectural studies on the Human and Mammalian cerebral cortex, summarized in his treatise of 1909. In accordance with his classification concept discussed in section 2 of the present chapter, he distinguished 11 main cortical regions and about 50 areae, for which he introduced a widely accepted numerial notation. Figures 103 and 104 show his cortical mappings for the Human brain.

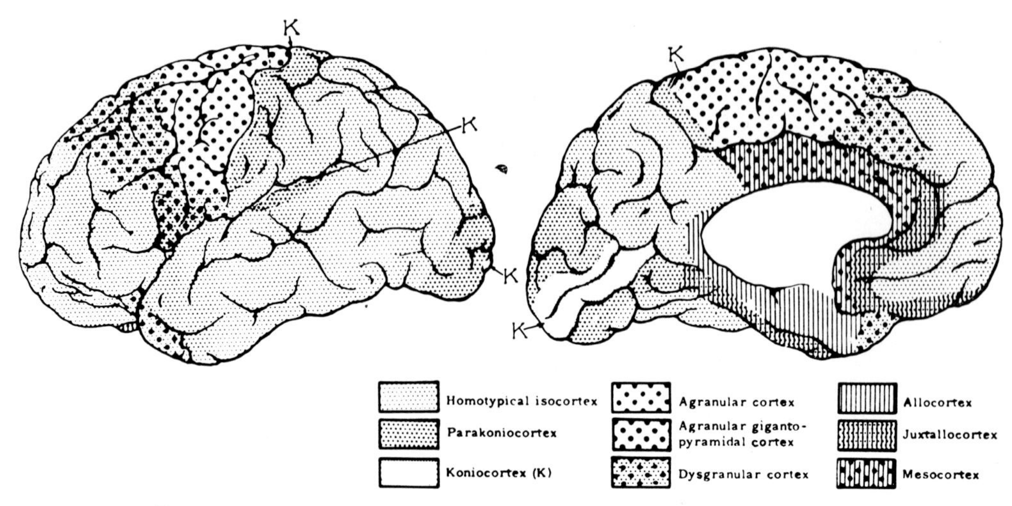

Figure 106. BAILEY's and v. BONIN's cytoarchitectural map of the Human brain, as illustrated in black-and-white (adapted after BAILEY and v. BONIN, from CHUSID, 1964).

As regards this latter, v. ECONOMO and KOSKINAS (1925) published a detailed work on cytoarchitecture, including an atlas with large photomicrographs. These findings were then summarized in a smaller treatise by v. ECONOMO (1927). This author likewise distinguishes roughly about 50 main areae,[75] using a notation by letters referring to the lobes, as shown in Figure 105. It can be seen that, as regards the parahippocampal cortex, he distinguished a limbic portion L, and a hippocampal one H, which both include parahippocampal and hippocampal cortex. Because of this and some other lack of distinction between isocortical and allocortical components, STEPHAN (1975) justly points out that, as v. ECONOMO and KOSKINAS themselves concede, their notation blurs a relevant appreciation of allocortical subdivisions.[76]

Still more recently, BAILEY and v. BONIN (1951), who are critical of earlier cytoarchitectural studies, have mapped the isocortex of Man,

[75] Cf. v. ECONOMO (1927), Tabulations of '*wichtigste Areae*' on p. 24. Further subdivisions of these main areae are, however, described by this author, being indicated by affixes to his FA, OB, etc. notations, and result in a substantially increased parcellation exceeding 100 areas.

[76] '*Ein zusammenhängender Überblick über den Allocortex wird dadurch sehr erschwert*' (STEPHAN, 1975, p. 15).

using blending colors to emphasize low gradients. Because of the difficulty of accurate color reproduction, CHUSID (1964) developed a black-and-white supplement of their map, shown here as Figure 106. SHOLL (1967) points out that this map is not very different from that made by CAMPBELL 50 years ago (cf. Fig. 102).

Although BAILEY and v. BONIN are mainly concerned with the isocortex *sensu strictiori*, these authors also distinguish an allocortex, a juxtallocortex, and a mesocortex, but these three subdivisions as conceived by the cited authors seem rather questionable and have been criticized by STEPHAN (1975).[77]

A detailed documentary work on the differentiation of cortical architectonics in Man was undertaken by CONEL (1939–1963). This author studied seven stages of development (newborn, 1 month, 3 months, 6 months, 15 months, 2 years, 4 years) and brings, as regards selected frontal, parietal, occipital, temporal, and insular areae, large comparable photomicrographs of *Nissl*, *Cajal*, and *Golgi preparations*, supplemented by a number of myeloarchitectural pictures. As regards the allocortex, he also includes some illustrations of parahippocampal and hippocampal areae. CONEL considers width of cortex and of cortical laminae, chromophil substance, neurofibrillar development (from the *Cajal preparations*), cell shape and cellular processes (from the *Golgi preparations*), and myelogenetic aspects. The cited author, who has adopted v. ECONOMO's areal notation, does not depict a cortical map based on his own studies, but brings, in volume I, the maps elaborated by CAMPBELL (1905), BRODMANN (1909), v. ECONOMO and KOSKINAS (1925), and v. ECONOMO (1927). CONEL's work may be considered a relevant documentary supplement to that of ECONOMO and KOSKINAS.[78]

With respect to the width of the Human neocortex, particularly investigated by *v. Economo*, it will here suffice to mention the following data. Generally speaking, this width varies between 4.5 and about 1.8 mm. The widest cortex is that of the precentral motor area (about 4.5 mm), and the thinnest ones are parts of the koniocortices of post-

---

[77] '*Diese Veröffentlichung, die sich vorwiegend mit dem Isocortex befasst, enthält nun, zumindest was den Allocortex betrifft, auf wenigen Seiten so viele Unklarheiten und Fehler, dass eine Klarstellung zweckmässig erscheint...*' (STEPHAN, 1975, p. 23).

[78] Although CONEL brings a chapter of 'general summary' (vol. I), 'general comments' (vol. II), respectively 'comments' (vol. III–VII) at the end of his detailed descriptions in each volume, he does not present an overall succinct formulation of his findings and conclusions (cf. also STEPHAN, 1975, p. 151: '*Eine zusammenfassende Darstellung seiner Befunde gibt* CONEL *nicht*').

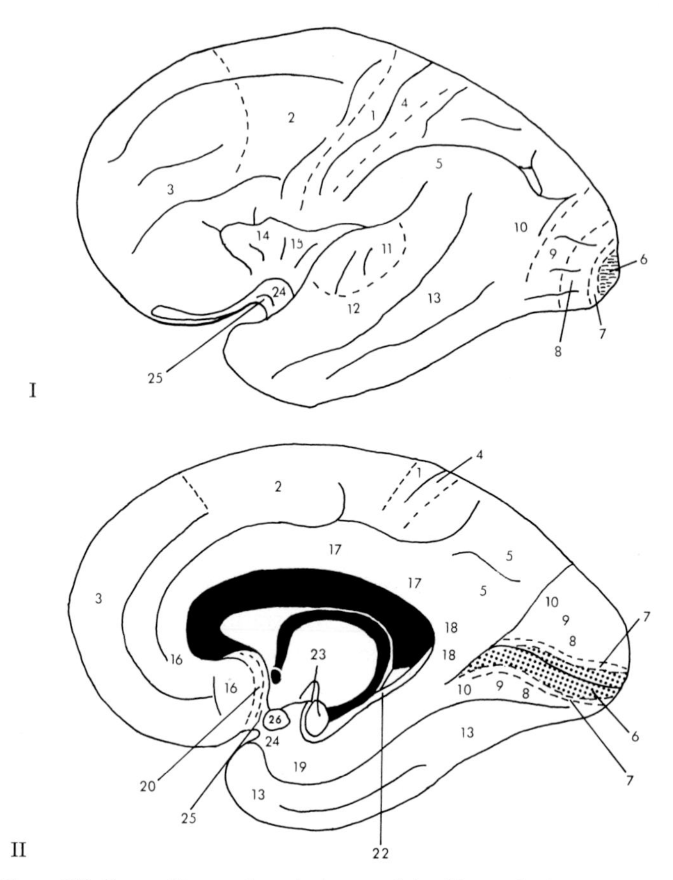

Figure 107. Cytoarchitectural cortical map of the Human brain as conceived by the present author. The neocortices 1–15, the parahippocampal cortices 16–19, the basal cortices 25 and 26 are named and explained in the text. As regards the hippocampal cortices, only 20 (precommissural), 22 (postcommissural), and 23 (uncus areae) are here indicated, omitting the narrow supracommissural one. The three uncus areae 23 are, in rostrocaudal sequence, (a) area uncinata, (b) *Giacomini's band* or *benderella* and (c) area intralimbica. Further relevant areal subdivisions of hippocampal cortex are dealt with in the text. The numbers are not intended as areal notations replacing those introduced by BRODMANN, but merely to indicate the cortices as discussed in the text. It will also be seen that, in many instances, I have avoided to draw areal boundaries.

central somatosensory area (about 2.0 mm) and optic sensory area striata (about 1.8 mm).

Figure 107 shows the results of my own attempts at mapping the Human cortical areas, mainly on the basis of cytoarchitecture, supplemented by myeloarchitectural data. These unpublished attempts were initiated in 1920 and intermittently continued for more than 50 years on a large number of brains. Originally strongly influenced by the work of BRODMANN, and often perplexed by my inability to recognize many of that author's areal boundaries or to link his descriptions with my own findings, I have also carefully considered the reports by other investigators of cortical architecture, especially the documentary data published by v. ECONOMO.

It will be seen that I was able to distinguish with reasonably sufficient certainty only about 15 neocortical 'areae' or 'regions', and that, despite some differences, the resulting map corresponds more closely to CAMPBELL's than to BRODMANN's interpretation.[79] The 'areae' considered more or less constantly recognizable are the following. In the frontal lobe (1) area praecentralis agranularis gigantopyramidalis or motor cortex, (2) area frontalis agranularis or premotor cortex, and (3) area frontalis granularis or 'prefrontal' cortex. In the parietal lobe (4) the postcentral koniocortex or somatosensory area, and (5) the parietal cortex granularis. In the occipital lobe (6) the koniocortex of the area striata or visuosensory cortex, (7) the granular internal parastriate area, characterized by giant pyramids in the deep part of layer III, (8) the granular external parastriate area, (9) the granular peristriate area, and (10) the granular occipital area. In the temporal lobe (11) the koniocortex of the transverse temporal gyri or 'auditosensory' area, (12) the parakoniocortical superior temporal area, and (13) the granular general temporal area. Within the insula, (14) anterior (dysgranular or agranular), and (15) posterior (granular or eugranular) insular areae.

As regards the boundaries between neocortical areae respectively regions, BRODMANN claimed that they were quite definite, although conceding the existence of transitional zones. VON ECONOMO, who admits some sharp boundaries, points out that numerous areal transitions are blurred (*'verwaschene Grenzen'*). The VOGTS and their school, however, emphasize 'hairsharp', linear boundaries between all cortical areas.[80] In my own studies, I was unable to recognize sharp areal

---

[79] In various respects, the cortical map based on my own observations also rather closely resembles that drawn by BAILEY and v. BONIN for the *isocortex* and, more particularly, that prepared by LE GROS CLARK on p. 986 and 993 (Figs. 866 and 873) in his contribution to the 9th edition of CUNNINGHAM's textbook of anatomy (Oxford University Press, London 1951).

[80] Cf. the controversy between the *Vogt school* and v. ECONOMO (1930) dealt with by the latter author in the reply to a severe criticism of his work.

boundaries in the Human neocortex with one single exception, namely the indeed 'hairsharp' boundary between area striata and parastriata (cf. Figs. 92B, 5a, and 95C). I was, moreover, impressed by the considerable individual variations and found that no two brains displayed exactly identical minute details of laminar differentiation and of more or less definable areal boundaries. I am, moreover, uncertain whether the above mentioned areae (5), (9), (10) and (13) are unambiguously separable and do not represent a more or less architecturally continuous parieto-occipito-temporal large 'association area'.

Concerning the *parahippocampal cortex* as dealt with in section 3, I am inclined to distinguish four main areae, each of which could again be subdivided in an outer and an inner subarea, the former being adjacent to neocortex, the latter to hippocampal cortex. The main areae are (16) precommissural, (17) supracommissural, (18) retrosplenial, and (19) posterior piriform (entorhinal). With respect to the boundaries, the gradient to hippocampal cortex is steep, that between neocortex and entorhinal cortex fairly steep, all others being rather low, particularly those between the supracommissural (cingular) and precommissural parahippocampal cortex on one hand and the neocortex in the other hand.[81]

The *hippocampal cortex* can be subdivided into (20) precommissural, (21) supracommissural, (22) postcommissural, and (23) uncus areae. The two first ones are rudimentary, the two last ones clearly display the separation between fascia dentata and cornu Ammonis. This latter, at least in (22), again can be further subdivided into subarea parasubicularis, prominens, paradentata, and intradentata, and, at the uncus, into gyrus uncinatus, and gyrus intralimbicus, separated by the dentate *benderella* (cf. Fig. 175D, vol. 5/I, and Fig. 57C of the present volume).

The Human *anterior piriform lobe cortex* is rather poorly developed, and shows (24a) a rostral portion, including the limen insulae, and (24b) an intermediate portion[82] represented by the gyrus ambiens (Fig. 175D, vol. 5/I, and 56A of present vol.).

---

[81] This might explain as mentioned in section 3, the inclusion of these cortices, e.g. the outer cingular cortex, into the neocortex *sensu latiori* respectively as 'mesocortex' or 'juxtallocortex' etc.

[82] Since the piriform lobe includes rostrally anterior piriform lobe cortex $D_{(2+1)c}$, and caudally parahippocampal (entorhinal) cortex $D_{2\,am}$, some confusion has arisen by occasionally designating the caudal portions of anterior piriform lobe cortex as 'posterior piriform lobe cortex'. This confusion can be avoided by calling this latter 'intermediate piriform', or still better 'intermediate anterior piriform lobe cortex', and restricting the

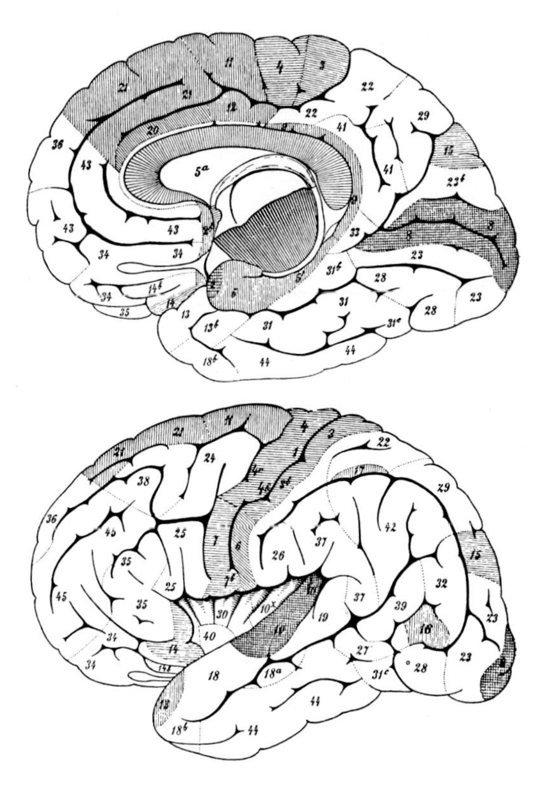

Figure 108 A. The myelogenetic areae of the Human hemisphere (from FLECHSIG, 1920). The primordial areae of the cited author (1–21) are indicated by hatching. Area 10 (posterior insula) should have been likewise so indicated.

The Human *basal cortex* is likewise rather poorly developed, and displays (25) a *rostral portion* (area ventralis anterior), representing the cortex of substantia perforata anterior with its immediate neighborhoods, including, medially, the griseum of *Broca's diagonal band.*[83] The *caudal area* of the basal cortex is (26) the cortical amygdaloid griseum *sive* gyrus (semi)lunaris (Fig. 175D, vol. 5/I), corresponding to the area ventrolateralis posterior of Amphibians.

Thus, roughly speaking, there are, in addition to about 15 neocorti-

---

term posterior piriform lobe cortex to the parahippocampal one. I must confess that, in view of these complications, I have not always been sufficiently consistent. Using BRODMANN's notation, $D_{(2+1)c}$ is essentially field 51, and $D_{2\,am}$ is essentially field 28.

[83] Thus, the field (25) could again be further subdivided, but I doubt the usefulness of such additional parcellation.

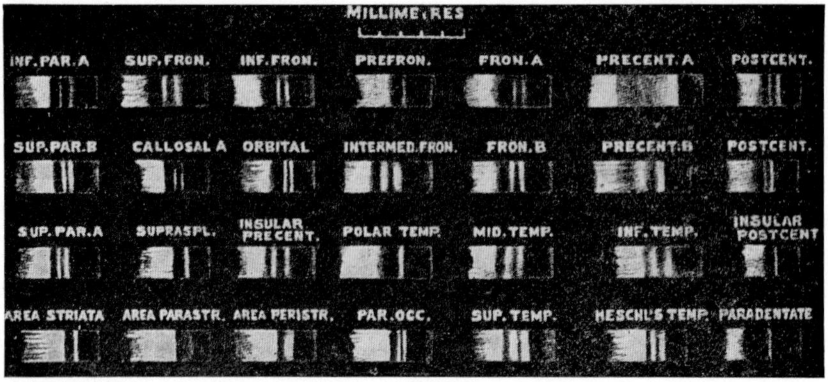

Figure 108 B. Drawings showing the naked-eye appearance of sections of the Human cerebral cortex in different regions, as seen by ELLIOT SMITH and corresponding to the cortical areas of that author's brain chart (from ELLIOT SMITH, 1907).

cal 'areae', at least 11 allocortical ones. If further distinctions are made as regards these latter, there are 8 parahippocampal subareae, 10 hippocampal ones,[84] besides 2 anterior piriform (rostral and intermediate), and at least 2 basal cortical areae, thereby bringing the total of allocortical subdivisions to 22, and that of the overall cortical subdivisions to 37.

FLECHSIG (1920), on the basis of his myelogenetic studies, distinguished about 45 successively differentiated cortical fields,[85] as here indicated in Figure 108A. However, while his early map of 1896, as shown in Figure 252A of chapter XIII, volume 5/I, and particularly emphasizing the important projection centers, may be considered of fundamental importance, I believe that his parcellated mapping of the diverse association areas characterized by additional progressive myelinization could result from the recording of a variety of poorly understood complex events not necessarily implying an unambiguous and functionally significant delimitation of specific areas.

As regards *myeloarchitecture*, ELLIOT SMITH (1907) attempted to draw a map based on the macroscopic aspect of cortical sections from fresh or freshly fixed brains studied with a hand lens. He distinguished about 28 areas, as shown in Figure 108 B and his cortical map is, on the

[84] Namely, inner and outer each *qua* (16)–(19), and, in addition to (20) and (21), the above-mentioned subdivisions of (22) and (23).

[85] In 1905, he mentioned 35 fields.

whole, somewhat similar to that of CAMPBELL. In my own attempts at a cortical mapping of the Human brain, I also made use of ELLIOT SMITH's procedure, but found it highly unreliable,[86] both because of individual variations and of differences presumably related to freshness as well as to fixation effects on the material, and I was unable to confirm most of the cited author's findings. The area striata, however, could always easily be identified. Nevertheless, even here, I sometimes saw it as depicted by ELLIOT SMITH in Figure 108B (left bottom), and sometimes as depicted by BAILLARGER (Figure 95B, right), namely with external and at least fairly distinct internal stria.

The VOGTS and their school (cf. e.g. BECK, 1925), using a modification of the *Weigert method*, and in accordance with the myeloarchitectural criteria described in section 3, subdivided the Human cerebral cortex in a very large number, exceeding 200, of distinctive areas.

SHELLSHEAR (1927, 1933) attempted to show that the distribution fields of Human cerebral surface arteries correspond to the outlines of cortical architectural areas, but in view of the considerable variations in the details of the vascular distribution, the descriptions of this author can be evaluated as unconvincing. A critique of the detailed cortical parcellation concepts will be given further below, following the comments on cortical maps in diverse Mammals.

Summarizing the architectural cortical localization and its presumable functional significance in the Human brain, it seems obvious that, as regards the neocortex, three sensory projection areas in FLECHSIG's terminology are clearly indicated by architecture, fiber connections, and functional respectively clinical aspects, namely the somatosensory, the visual (striate) and the acoustic koniocortices. The architecturally likewise conspicuous area agranularis gigantopyramidalis represents the relevant motor projection area from which the important pyramidal tract seems mainly to arise.

There is much less certainty about the architectural and functional localization of the assumed gustatory and general visceral afferent projection areas, which can be suspected to be present in the posterior (granular) insular cortex, the gustatory area being perhaps located dorsally, adjacent to the transition of somatosensory to insular cortex.

---

[86] Others, apparently, shared the same experience: SHOLL (1956, 1967) comments twice (p. 1 and 24) by remarking that the differences described by ELLIOT SMITH 'are difficult to confirm', and that 'those who have tried this method have found great difficulties in making a similar map'.

With regard to the functionally established second somatomotor and somatosensory areae, it can be surmised that they are represented in the lower part of the precentral and postcentral areae.

Concerning FLECHSIG's association areas, this author pointed out the significance of particular border zones *(Randzonen)* adjacent to his projection areas. Such zones are the parastriate and peristriate,[87] the superior temporal parakoniocortical[88] zone, perhaps also a parakoniocortical zone of parietal cortex granularis,[89] and the premotor cortex.

The peristriate and parastriate cortices may receive input from both (right and left) striate areae rather than direct input from the geniculo-calcarine tract, and presumably represent the second and third optic areae for hierarchically higher optic signal integration, to be dealt with in section 6.

The parakoniocortical temporal zone might perhaps correspond to the second auditory area identified in various Mammals.[90] The significance of the parakoniocortical postcentral zone remains still less well understood. The premotor cortex, presumably also includes PENFIELD's supplementary motor area on the medial surface, dorsally to the limbic lobe (cf. Figs. 254B, C, vol. 5/I), moreover the frontal 'eye field' from which conjugated eye movements can be elicited by electrical stimulation, and *Broca's area* in the caudal portion of the inferior frontal convolution of one (usually the left) hemisphere.

The prefrontal and the temporo-parietal-occipital cortices pertaining to the terminal regions *(Terminalgebiete)* of FLECHSIG's association areas, which have also been called 'interpretative' cortices, are apparently concerned with functions of 'higher order', to be dealt with in section 6, in which the functional significance of the limbic lobe shall also be considered.

As regards the *olfactory cortices*, pertaining to the 'rhinencephalon' *sensu stricto*, it will here be sufficient to state that olfactory input doubtless reaches the basal cortex (area ventralis anterior) of substantia perforata anterior and its surroundings, and of cortical nucleus amygdalae (area ventrolateralis posterior), moreover the cortex of anterior and in-

---

[87] These areae are not clearly distinguished by CAMPBELL (cf. Fig. 102), but roughly correspond to that author's 'visuopsychic area'.

[88] CAMPBELL's 'auditopsychic' area, v. ECONOMO's TB.

[89] CAMPBELL's 'intermediate postcentral', which I was unable to delimit with sufficient certainty.

[90] Unless first and second auditory areae are represented by v. ECONOMO's fields TD and TC.

termediate piriform lobe. It is, however, not improbable that rostral parts of the parahippocampal cortex of posterior piriform lobe (gyrus hippocampi adjacent to gyrus ambiens, cf. Fig. 175D, vol. 5/I), that is, part of BRODMANN's entorhinal field 28, receive olfactory input. Whether some of this latter is also transmitted to the uncus hippocampi remains questionable, but not impossible.

Since the weak and nondescript medial olfactory tract seems to reach the precommissural and perhaps even supracommissural hippocampus, these regions, possibly with an adjacent parahippocampal fringe, are perhaps also part of the 'rhinencephalon' in a functional sense. Again, it is understood that the foregoing discussion of cortical regions and areae essentially refers to the Human brain.

The investigations of CAMPBELL, BRODMANN, and others concerned with a mapping of the Human cerebral cortex, preceded, as it were, extensive similar studies in other Mammals and thereby resulted in what could be called an 'anthropocentric' or 'neopalliocentric' tendency *qua* areal terminology and interpretation.

Turning now to the cortical localization in 'lower' Mammals, e.g. Methatherian Marsupials, and Eutherian Insectivores, Rodents, and Lagomorphs, I found it impossible to parcellate their *neocortex* into definitely circumscribable cyto- or myeloarchitectural areas separated by recognizable boundaries. Although differences in the laminar arrangement doubtless obtain, these differences are far less distinctive than in Man and other Primates. Nevertheless, very low rostro-caudal and medio-lateral or dorso-basal gradients can be detected (cf. e.g. Figs. 222B–D, 229B, D, vol. 5/I; 109 present volume), and ambiguous cytoarchitectural differences between rostral and caudal neocortex indeed exist. On the basis of such differences, a few regions, such as frontal (rostral), parietal, insular, occipital, and temporal may be distinguished, but definite boundaries cannot be assigned to these regions which the author designated as '*Primärgebiete*' (K., 1927). It was suggested that, in the course of phylogenetic evolution, these primary neocortical regions might have undergone a further differentiation, resulting in the various territories classified by FLECHSIG as 'projection' and as 'association' centers.

In the lower Mammals, whose neocortex doubtless performs functions such as pattern recognition (abstraction of invariants, symbolizing activities), memory and various aspects of 'learning', the primary regions are not exclusively 'projection areas' but also represent, presumably jointly, as a continuum, primordial 'association areas'. This

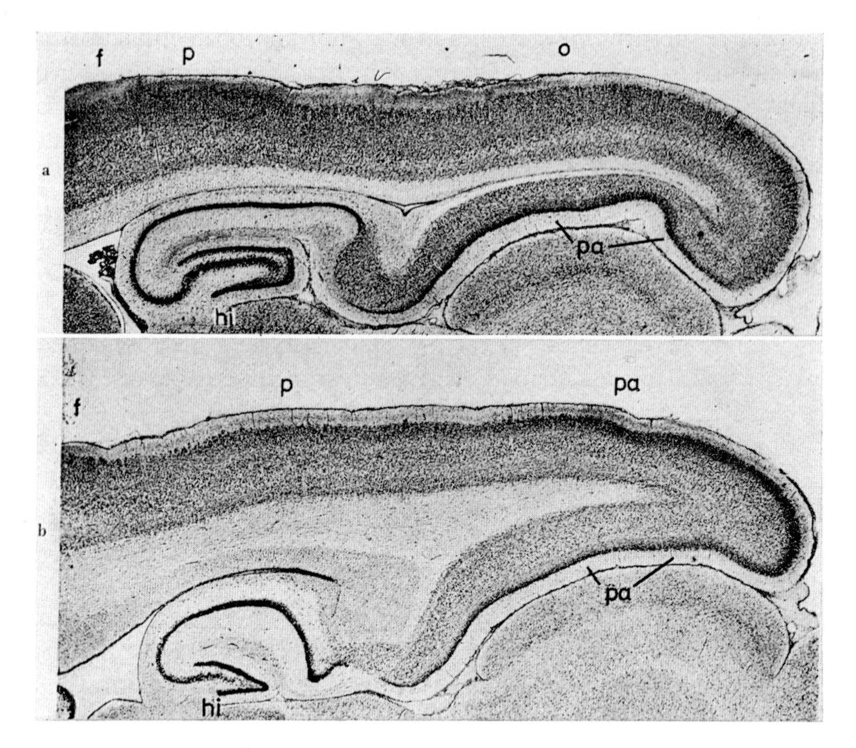

Figure 109. Sagittal sections *(Nissl stain)* through the region of caudal pole of the Rabbit's telencephalon (about ×11, red. $^{2}/_{3}$; from K. *et al.*, 1960). a: appr. 3.8 mm from midline; b: appr. 2.2 from midline; f: frontal (rostral) neocortex; hi: hippocampal cortex; o: occipital neocortex; p: parietal neocortex; pa: parahippocampal cortex. No attempt has been made to indicate exact boundaries. The sections were 30 μ thick.

would agree with the findings of LASHLEY, discussed in section 10 of chapter XIII in volume 5/I).

Again, the occipital area, although receiving the optic input, is neither functionally nor architecturally equivalent to the Primate area striata, or BRODMANN's area 17, although commonly so designated. Much the same could be said about temporal, insular, and parietal areas. As regards the rostral or frontal area, from which pyramidal and corticopontine tract fibers arise,[91] pyramidal cells of layer V are indeed fairly large, but hardly stand out as 'giant pyramids'.

---

[91] Some pyramidal tract fibers seem likewise to originate from the rostral part of parietal region. From parieto-occipito-temporal neighborhoods there may also originate some fibers of that tract, together with the corticopontine fibers.

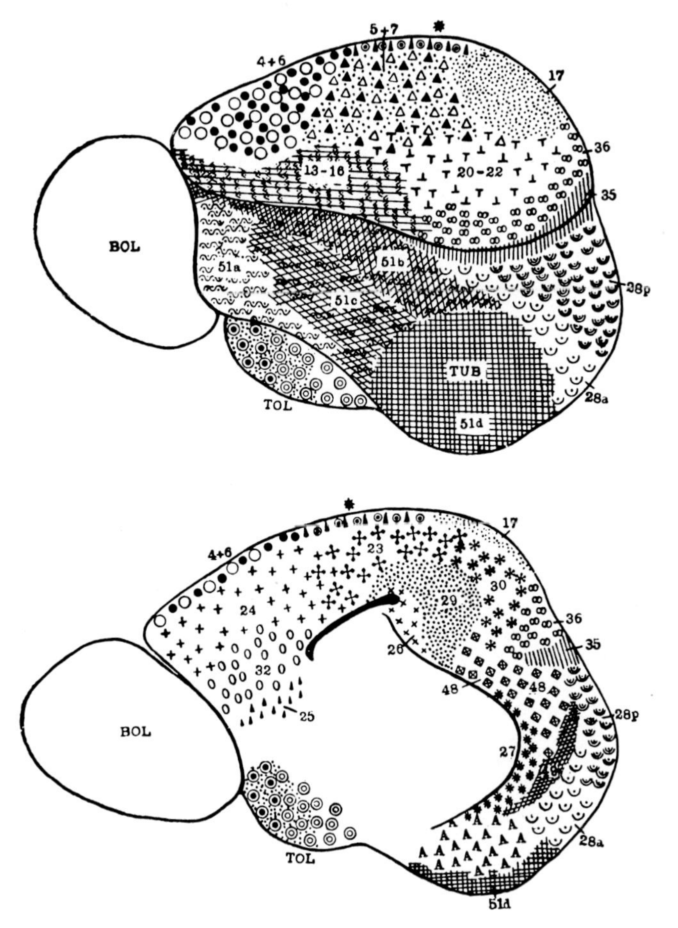

Figure 110 A. Brodmann's cytoarchitectural areae in the hemisphere of the Insectivore Erinaceus europaeus (from Brodmann, 1909) This map should be compared with my own mapping as shown in Figure 221 C of chapter XIII, volume 5/I.

Thus, the separation of typical 'projection' and 'association' areas, stressed by Flechsig, and doubtless obtaining in 'higher' Mammals, could be the result of a 'division of labor' or further 'specialisation' that occurred in the course of phylogenesis.

Concerning the *allocortex*, there is, in these lower forms, an extensive *parahippocampal region*, delimited by low gradients, with outer and inner anterior, posterior and entorhinal (posterior piriform) subregions. The *hippocampal cortex* is well differentiated except for its precommissural subregion. Depending on taxonomic forms, the supra-

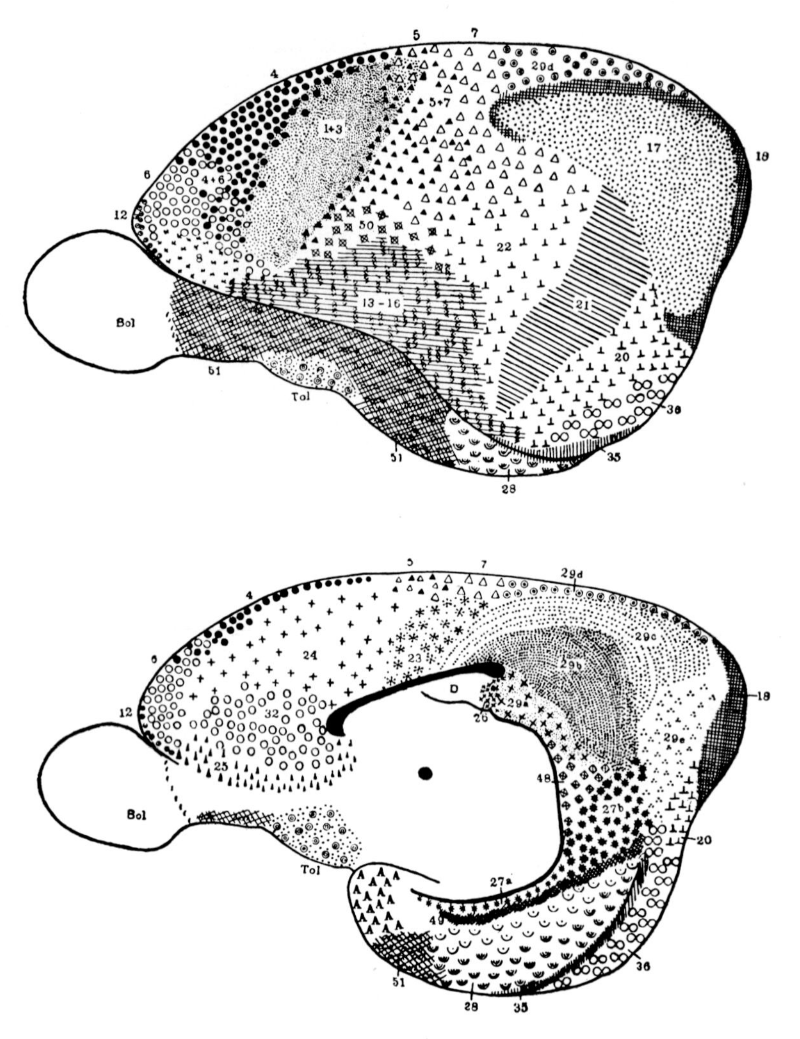

Figure 110 B. Brodmann's areal map of the hemisphere in the Lagomorph Rabbit (from Brodmann, 1909).

commissural hippocampus may or may not be well differentiated into its typical subareae (cf. e.g. Figs. 222 C, 223 A, C, vol. 5/I). The *anterior piriform lobe cortex* and the *basal cortex* with its two subareae (area ventralis anterior and cortical amygdaloid nucleus) are likewise well differentiated.

Brodmann (1909) himself felt constrained to abandon his detailed parcellation in 'lower' Mammals such e.g. as the Insectivore Erina-

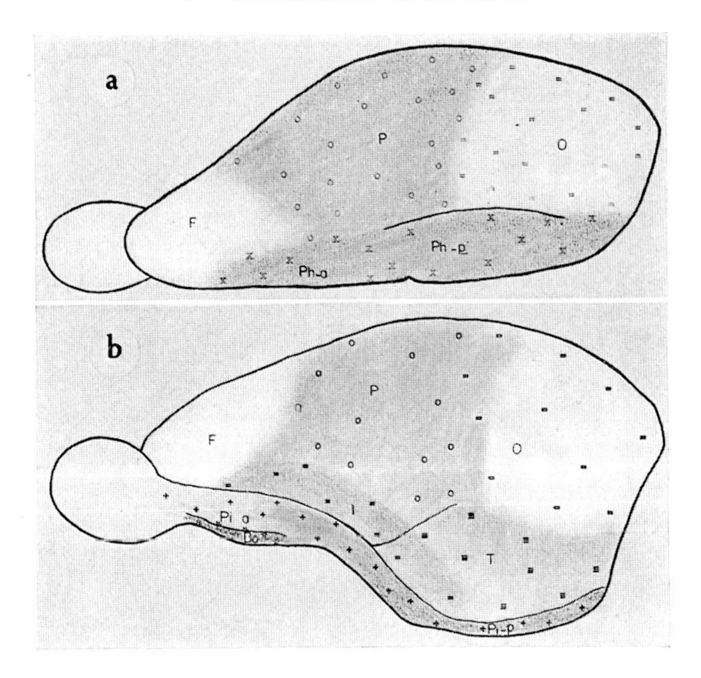

Figure 110 C. Our own cytoarchitectural cortical map of the Rabbit's hemisphere (from K. *et al.*, 1960). Ba: basal cortex; F: frontal (or rostral) neocortex; I: insular neocortex; P: parietal neocortex; Ph-a, Ph-p: anterior and posterior parahippocampal cortex; Pi-a: anterior piriform lobe cortex; Pi-p: parahippocampal cortex of posterior piriform lobe; O: occipital neocortex; T: temporal neocortex.

ceus, and designated the frontal area as 4+6, the parietal as 5+7, the temporal as 20+22, the insular as 13+16, while retaining 17 ('area striata') for the occipital area. With the qualifications and restrictions mentioned above, and discounting distinctive boundaries, my own cortical mapping of Erinaceus (Fig. 221 C, vol. 5/I) fully corresponds to that by Brodmann (Fig. 110A). Figure 110B illustrates that author's mapping of the Lagomorph Rabbit's cortex, to be compared with our own mapping shown in Figure 110 C. This latter, again, is rather similar to, but, as regards the neocortex, somewhat less parcellated than, the Rabbit's cortical map drawn by *Elliot Smith* and published as Figure 793, p. 999 in his contribution to the 7th edition (1937) of Cunningham's text-book of anatomy.

It is also of interest to compare these cytoarchitectural maps with the functional maps in Figures 240 A, B, 241 B, C, D, E, and 243 D in section 10 of chapter XIII of volume 5/I. Although the cytoarchitectural map of a Marsupial obtained by our own studies is quite similar to that of the Insectivore Erinaceus and of the Lagomorph Rabbit, it will be seen that LENDE denies a rostral somatomotor area in Marsupials (Figs. 241 B, 243D, vol. 5/I) and claims that it is included in a single parietal somatosensory-motor area, while GRAY and TURNER (Fig. 240 A, vol. 5/I) as well as ABBIE (1940b) recorded a rostral excitable motor area. I am inclined to accept the findings by the latter authors. It is also possible that, in these 'lower' Mammals, but to a lesser extent than in the Monotremes, which are briefly dealt with in the next paragraphs, the parahippocampal cortex, whose relatively low gradient toward neocortex results here in an indistinct boundary, may, with individual and taxonomical variations, occasionally occupy a region extending toward and including the rostral pole, thereby displacing the frontal neocortical ('motor') area slightly caudalward.

As regards the Prototherian *Monotremes* Ornithorhynchus and Echidna, their cytoarchitecture was studied by ABBIE (1940a), who also prepared cortical maps. Although conforming to the overall Mammalian pattern, the Prototherian hemisphere is characterized by a peculiar predominance of the parahippocampal cortex,[92] which expands far lateralward and also occupies not only the rostral pole but also a large rostral region of the hemisphere, whereby the neocortical fields are, as it were, pushed caudalward and basad (cf. Fig. 241 A, vol. 5/I). The anterior piriform lobe cortex is not very extensive in both Monotreme forms (cf. e.g. Fig. 226 C, vol. 5/I) and the basal cortex is likewise of quite moderate extent. Regardless of their insufficiently understood phylogenetic relationship to presumed common Mammalian ancestors, the extant recent Prototherians can be considered highly aberrant Mammalian forms.

Turning now from the 'lower' to some of the 'intermediate' Mammals, cytoarchitectural maps, by BRODMANN (1909), of a Carnivore and of an Ungulate brain are illustrated in Figures 111 A and B, which should be compared with the functional maps of Figures 241 D, 242 A to D, and 243 A, B shown in section 10 of chapter XIII, volume 5/I. As regards *Primates*, Figures 112 A and B illustrate BRODMANN's map-

---

[92] ABBIE (1940a), although using the term parahippocampal cortex, evaluates it as one of the two major subdivisions of the neocortex, the other one (i.e. the neocortex *sensu stricto* of the present treatise) being the parapiriform cortex. Some of the details of ABBIE's cytoarchitectural interpretations and mappings appear unconvincing, but the overall aspects of his findings can be evaluated as providing an adequate description of the peculiar conditions displayed by the Monotreme cortex cerebri.

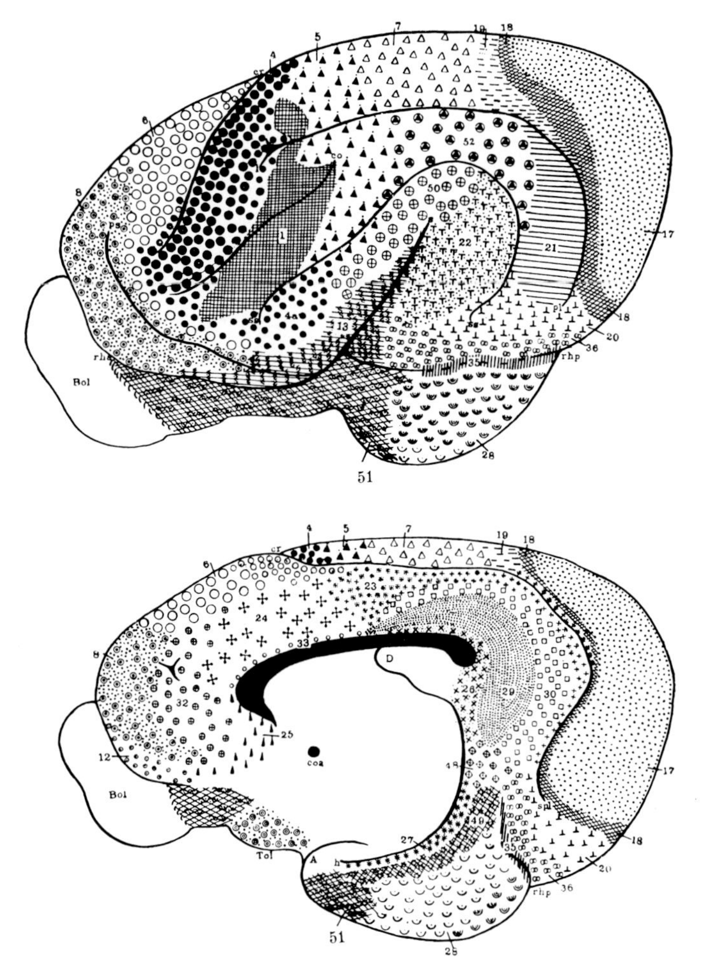

Figure 111 A. Brodmann's cortical areal map of the Carnivore Ursid Cercoleptes (from Brodmann, 1909).

pings in a 'lower' (Lemuroid), and in an 'intermediate' (Cercopithecoid) Primate form.

Under the influence of Brodmann's work, which strongly impressed me at the beginning of my cytoarchitectural and to some extent also myeloarchitectural studies more than fifty years ago, I made a determined effort to bring my own findings in accordance with that author's concepts. Initially, I assumed that the discrepancies which I encountered were due to insufficient training and to lack of familiarity

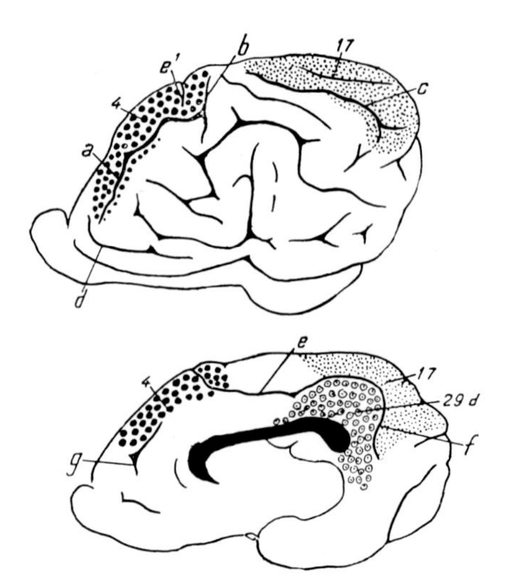

Figure 111 B. BRODMANN's partial cortical areal map of the Ungulate Capra (after BRODMANN, 1909, from BECCARI, 1943). a: sulcus coronalis; b: sulcus ansatus; c: sulcus lateralis; d: sulcus praesylvius; e: sulcus splenialis and its ascending branch (e′); f: sulcus retrosplenialis; g: sulcus genualis.

with the technical methods and the proper interpretation of the thereby obtained pictures. With gradually progressing experience, however, I finally became convinced that the obtained cyto- and myeloarchitectural parcellations resulting in the well-known and here illustrated checkered or '*crazy quilt*' maps with linear boundaries were highly artificial and devoid of intrinsic significance.

Psychological factors akin to those evidenced by the *Rorschach test*, the multitudinous individual variations, and last but not least, in the gyrencephalic brains, the distortions caused by the curvature of the brain surface along the gyri and sulci, as pointed out by BOK (1929)[92a] are presumably decisive factors for an elaborate parcellation. KAWATA (1927) who also anticipated some of the observations of BOK (1929) as

[92a] BOK, although pointing out the fallacy of parcellation into subareas as particularly elaborated by M. ROSE and by BECK, nevertheless upholds the parcellation into areas as carried out by BRODMANN. Yet, as regards many of these, and especially their alleged linear boundaries, the distortions recognized by BOK likewise preclude the type of areal mapping outlined by BRODMANN. Moreover, in lissencephalic brains, whose surface curvature involves a less significant degree of distortion, except in the region along the sulcus rhinalis lateralis, many aspects of BRODMANN's cytoarchitectural parcellation can be considered highly dubious.

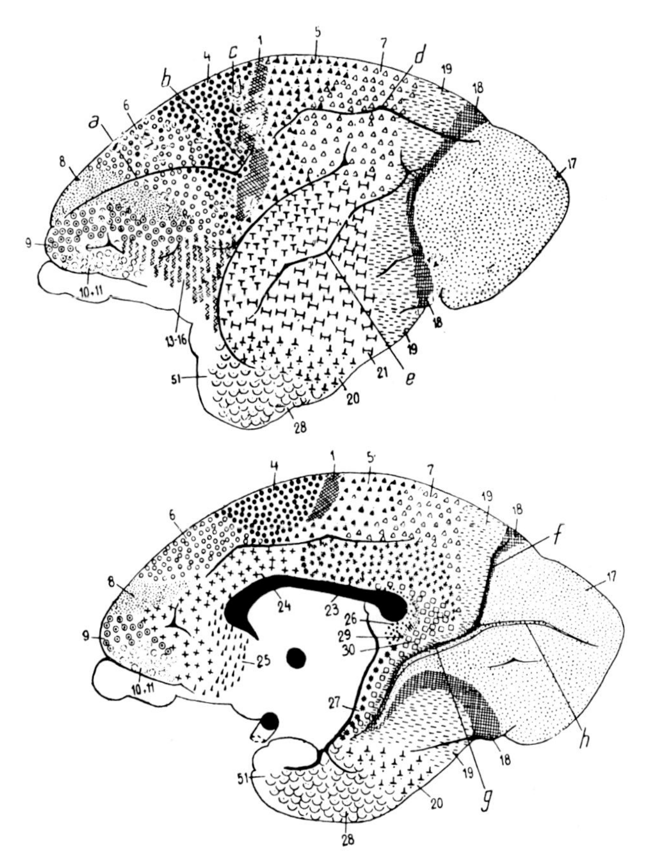

Figure 112 A. BRODMANN's cortical areal map of the Primate Lemur (after BRODMANN, 1909, from BECCARI, 1943). a: sulcus rectus; b: sulcus coronalis (in KAPPERS' interpretation); c: sulcus ansatus; d: sulcus lateralis; e: sulcus temporalis superior; f: sulcus perpendicularis (parieto-occipitalis); g: sulcus calcarinus anterior; h: sulcus calcarinus posterior.

well as the viewpoints of LASHLEY and CLARK (1946) furthermore showed that in myeloarchitectural studies the varying degree of staining or of subsequent differentiation of myelin-stained sections, regardless of exact technical standardization in laboratory procedures, may simulate considerable differences in architecture. The so-called 'objective registration' of cortical myeloarchitecture described by HOPF (1966) is as much subject to the vagaries of myelin staining techniques as the descriptive evaluation of microscopic sections upon inspection. All these circumstances explain the manifest contradictions in the de-

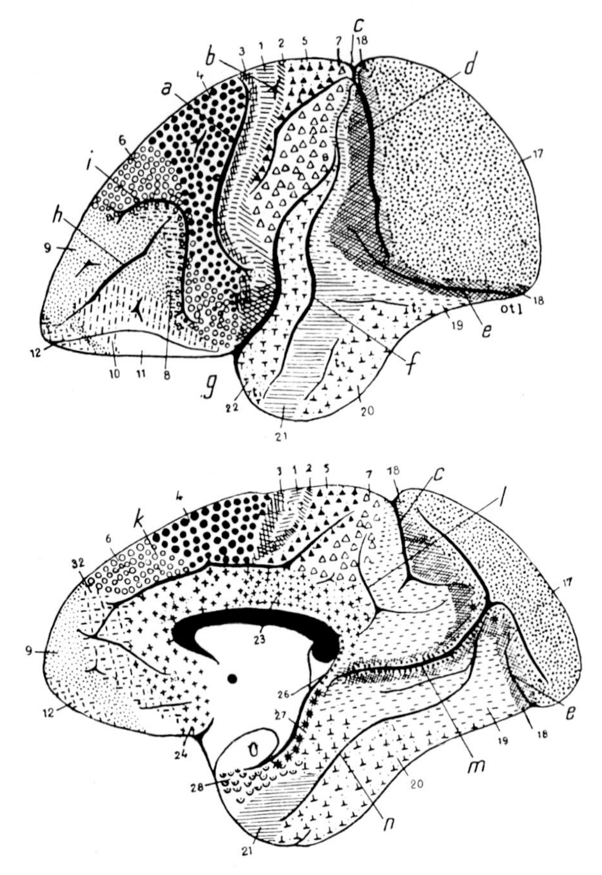

Figure 112 B. BRODMANN's cortical areal map of the Primate Cercopithecus (after BRODMANN, 1909, from BECCARI, 1943). a: sulcus centralis; b: sulcus postcentralis; c: sulcus perpendicularis (parieto-occipitalis); d: sulcus lunatus; e: sulcus occipitotemporalis lateralis; f: sulcus temporalis superior; g: fissura lateralis Sylvii; h: sulcus principalis; i: sulcus arcuatus; k: sulcus marginalis; l: sulcus subparietalis; m: sulcus calcarinus; n: sulcus occipitotemporalis medialis.

tailed cortical mappings of brains, belonging to the same species, by different investigators.

While I frequently discussed this matter with colleagues at various meetings, and emphasized this viewpoint in my lectures, my heavy commitments prevented me from completing a much needed well documented presentation of this topic. I hoped, nevertheless, that somebody would finally initiate a revision of the parcellation concept. Fortunately, the important critical study by LASHLEY and CLARK (1946) on

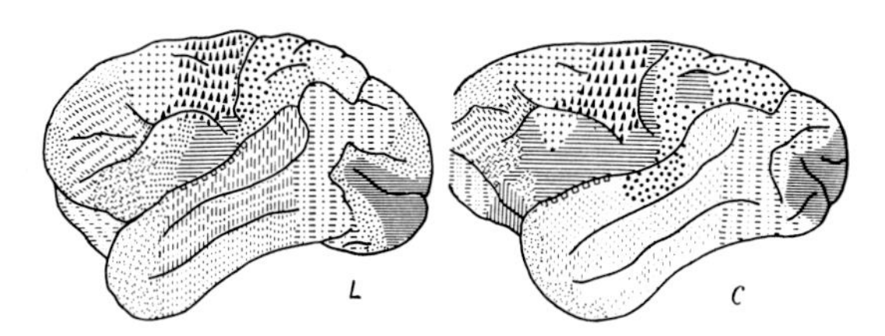

Figure 113. Cytoarchitectural parcellations of the neocortex in the Primate Platyrrhine Spider Monkey Ateles, made independently by LASHLEY (L) and by CLARK (C) (from LASHLEY and CLARK, 1946).

the cytoarchitecture of the Primate Ateles fulfilled this expectation and opened the path to a better evaluation of cortical architectonics.

Attempts of parcellation of the Ateles cortex made independently by LASHLEY and by CLARK (Fig. 113) disclosed inconsistencies leading to the critical examination, by the cited authors, of the relevant architectural criteria. There is no doubt that within the neocortex of Primates frontal granular, precentral agranular, precentral gigantopyramidal cortices, postcentral koniocortex, calcarine, parieto-occipital, temporal and insular regions have characteristic architectural structures. In the case of calcarine area striata, there is even a hair-sharp boundary. Further subdivisions of these regions, however, appear based 'upon variable characters which have no probable functional significance. On the basis of concomitant variations in functionally unrelated regions it is suggested that many structural differences are a product of developmental variations, unrelated to the ultimate functions of the areas. Marked local variations in cell size and density among individuals of the same species may constitute a basis for individual difference in behavior' (LASHLEY and CLARK, 1946). These authors also re-affirm a previous statement by LASHLEY: 'The empirical evidence does not confirm the assumption that dissimilar psychological functions necessarily imply architectonic differences.' For further details of their critical comments, the interested reader is referred to LASHLEY and CLARK's original paper (1946) and to the comments on their findings made by SHOLL (1956, 1967).[93]

---

[93] SHOLL, e.g., evaluates the myeloarchitectural parcellations be the *Vogt school* as carried out 'to an extreme that can only be described as fantastic'. Again, when the myth of the 'suppressor zones' or 'bands' based on a misinterpretation of LEÃO's spreading depression became popular, some cartographers of detailed parcellation promptly looked

Other authors followed the lead of LASHLEY and CLARK; some investigators of detailed cytoarchitectonics seem even to have been converted. CAMPBELL (1951) stressed the necessity for a 're-interpretation' of the structure of the cerebral cortex, with less emphasis on laminar pattern and minutiae of its variation. He conceived the cortex as a continuum exhibiting a rather simple but meaningful pattern of gradients of architectural structure. This concept leads to a mapping differing from the highly parcellated ones.[94] A revised map should show, in Man and other Primates, the sulcus centralis, the edge of the striate area, and the rhinal sulcus as regions of steep gradients. Large areas of the frontal, parietal and temporal regions would exhibit low, but patterned gradients.

About the same time, BAILEY and v. BONIN (1951) published their monograph on the Human isocortex, likewise criticising[95] and discounting the parcellation concept. While STEPHAN's (1975) critique of BAILEY's and v. BONIN's views concerning the allocortex (cf. above, footnote 77) seems justified, I would agree with the essential aspect of the neocortical mapping carried out by the cited authors.[96]

The concept of the cortex as a *continuum* seems most appropriate in order to counteract the tendency to which LE GROS CLARK (1952) alludes in a perspicacious note on cytoarchitectonics, namely the questionable tendency 'to focus attention on differences in cortical structure (which in some cases may be very slight) rather than on the all-pervading similarity of the cortex as a whole'. The 'griseal continuum' concept does not in any way preclude some fairly well localized distribution of input and output, including even a 'point to point' (better: 'neighborhood to neighborhood') projection in certain instances.

---

for, found, and outlined a 'suppressor zone' 4s (cf. the comments by LASHLEY and CLARK, 1946, pp. 247–248).

[94] I recall that, in presenting his paper at the 1951 meeting of the American Association of Anatomists, B. CAMPBELL jokingly compared the highly parcellated cortical maps with '*the Emperor's new clothes*' in ANDERSEN's well-known Fairy Tale.

[95] BAILEY and v. BONIN (1951, p. 189) state that 'the drawing of sharp areal boundaries, on the basis of varying distinctiveness and significance, is the fundamental defect of most maps and has been carried to absurd lengths...'.

[96] BAILEY's and v. BONIN's failure appropriately to deal with the allocortex, as pointed out by STEPHAN (1975), resulted, in my opinion, not from any unfamiliarity with Primate cortical architecture, but from a lack of appreciation of the fundamental morphologic *bauplan* features based on data resulting from comprehensive formanalytic studies of all Vertebrate groups from Cyclostome to Man, that is, including the Submammalian Amniotes and Anamnia.

However, the methodology proposed by B. CAMPBELL (1954) for an analysis of that continuum's organization, namely (1) cell frequency profiles, (2) use of scanning densitometer, and (3) electroarchitectonics, does not seem very promising as far as (1) and (2) are concerned. Electroarchitecture (3) as conceived by KORNMÜLLER (1937) and dealt with further below, cannot be upheld, while diverse methods of electric recording and stimulation have indeed yielded relevant results with respect to cortical mapping. Further aspects of electrophysiologic investigations concerning cortical functions are discussed in section 6 of the present chapter.

A determined defense of the detailed architectural parcellation concept was presented by SANIDES (1972). This author is particularly critical of LASHLEY and CLARK, stressing that, as psychologists, they were preconceived 'for basically psychologic reasons'. SANIDES also implies that they did not have the specific 'long training and experience' required for specialized architecturalists. I believe, however, that CLARK and particularly LASHLEY, who had also undertaken painstaking work concerning the architecture and connections of thalamic grisea, were fully qualified, regardless of various differences between my own views and those of the late Professor LASHLEY.

SANIDES also conveniently ignores the gist of the criticisms of parcellation expressed by KAWATA (1927), BAILEY and v. BONIN (1951), B. CAMPBELL (1951), LE GROS CLARK (1952), SHOLL (1967), and others. I would consider all these investigators to be fully competent and definitely familiar with the relevant aspects of cortical architectonics.

A frequently quoted paper supporting the architectural parcellation concept is that by KORNMÜLLER (1937) who reported that, in the Rabbit, different, distinct and identifiable electrocorticograms can be recorded from the diverse cytoarchitectural areas of ROSE's (1931) highly parcellated map. According to KORNMÜLLER, specific patterns of potential variations, so-called '*Feldeigenströme*' are characteristic, in the resting condition, for ROSE's different areas. Since, despite our rejection of highly parcellated maps, we did not deny regional cytoarchitectural differences, nor boundary zones represented by gradients, we carefully re-examined the electrical activity of the Rabbit's cortex with respect to different territories (K. *et al.*, 1960). Our results[97] unequivo-

[97] It is amusing that, although we were, in 1959, officially invited to contribute a paper to the '*Spiegel Festschrift*' in the *Journal of Nervous and Mental Diseases*, our manuscript was returned many months later, in June 1960, just about as the *Festschrift* was to appear in

cally showed that, in contradiction to the findings reported by KORNMÜLLER, different cortical regions did not display, either within their respective territories, or with respect to each other, typical or definable differences in the patterns of the electrocorticographic recordings. This statement applies to recordings obtained in (1) the 'awake resting state', (2) in periods of apparent 'sleep', (3) during ether narcosis, and (4) in nembutal narcosis.

Reverting to BRODMANN's widely accepted and perused numerical areal notation, which I originally also adopted in my diverse unpublished attempts at cortical mappings of the Human brain, and in the '*Vorlesungen*' of 1927, the following comments seem perhaps appropriate.

As regards the neocortex, fields 17, 4 and 6 represent indeed distinctive areae, with the following qualifications. Field 17 of Primates, although kathomologous to field 17 of 'lower' Mammals, and representing the visual projection area, is neither structurally nor in all respects functionally analogous within the entire Mammalian taxonomic series. Field 6, moreover, should, in my opinion, include field 8, which I was unable clearly to distinguish from the former.

Fields 41 and 42 apparently represent areas of the temporal koniocortex, but there is little doubt that homologous respectively kathomologous temporal koniocortices are present in all Mammals, approximately corresponding to the fields which BRODMANN here variously designates as 20, 21, 22, and even 36. These fields do not seem comparable with the identically numbered areas in Man (cf. e.g. Figs. 104, 110 A, B, 111 A, 112 A, B), nor, in all cases with each other in diverse Mammals. Again, I was unable to distinguish distinctively different fields 1, 2, 3 within the postcentral somatosensory koniocortex in Primates (Man, Macacus), and within the corresponding somatosensory cortex of 'lower' forms.

As regards field 18, BRODMANN fails clearly to distinguish the nar-

---

print, with the following comment: 'this investigation, while certainly technically competent, does not represent a significant contribution to the literature' and would 'not be suitable for publication in the Journal of Nervous and Mental Diseases'. It is evident that our paper, which was then forthwith published in *Confinia Neurologica*, was rejected and then unnecessarily held back for many months, by an anonymous reviewing board (whose identity I have reasons to suspect) prejudiced in favor of the parcellation concept, and attempting to suppress contradictory evidence. This trivial incident illustrates the dishonesty and prejudices frequently vitiating the system of so-called '*peer-review*'.

row but characteristic gigantopyramidal (in layer III) strip in Man from the remainder of his large field 18, which, again, is difficult to distinguish from his field 19. Within the insular region (BRODMANN'S fields 13–16) I could, at most, vaguely distinguish an anterior and a posterior insular cortex.

All other neocortical fields of BRODMANN could, as far as the Human brain is concerned, at best be evaluated as vague topographic designations for nondescript surface regions related to particular gyri. Their exact identification in other Mammals seems to me most unconvincing.

Turning now to the *parahippocampal cortex*, only area 28 of BRODMANN, namely the posterior piriform lobe cortex (area entorhinalis) represents a delimitation with which I could more or less agree, while all the other fields of BRODMANN (23, 24, 26, 27, 29, 32 etc.) represent regions that, at most, can only in a very unsatisfactory manner be compared with my own subdivisions of the parahippocampal cortex as dealt with above. BRODMANN'S field 25 was to me particularly puzzling.[98] In Man (Fig. 104) it seems to include adjacent regions of precommissural parahippocampus and hippocampus as well as paraterminal body. In the Hedgehog (Fig. 110A), in the Rabbit (Fig. 110B), and in the Carnivore Cercoleptes (Fig. 111A) it could be restricted to the precommissural hippocampus and the paraterminal body, in the Lemur (Fig. 112) it seems again to include both parahippocampal and hippocampal components, as well as paraterminal body, while in Cercopithecus all these regions are included in field 24. BRODMANN deals rather cursorily with the hippocampal and with the basal allocortices. As regards the anterior piriform lobe cortex, however, BRODMANN'S field 51 accurately corresponds to that cortical region clearly recognizable in the Mammalian brains which I could examine.

In concluding the present section, and in order to illustrate some ambiguous aspects of the parcellation problem, three arbitrarily selected examples shall briefly be pointed out, namely the 'boundary' of precentral and postcentral areae (1) in Man, (2) in the Dog, and (3) a comparison of cortical maps of the Mouse as drawn (a) by ROSE (1929b), (b) by a recent author from the Harvard group (CAVINESS, 1975), and (c) by myself.

(1) Figure 114A shows an outline sketch of Human sulcus centralis with adjacent precentral agranular gigantopyramidal cortex and post-

[98] Cf. K. (1927, pp. 311–312, footnote 3).

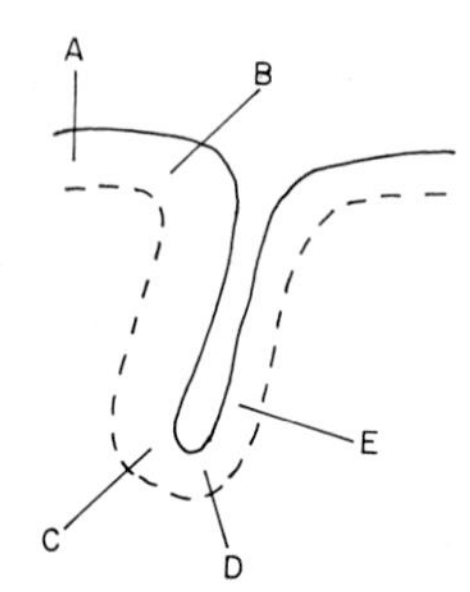

Figure 114 A. Sketch of section through sulcus centralis and adjacent gyri in Man, showing the variable extension of the typical gigantopyramidal precentral cortex. A, B: variable approximate limits of rostral extension; C, D, E: variable limits of posterior extension.

central somatosensory koniocortex. The lines A and B approximately indicate the variable rostral extension of typical *Betz cell* rows in layer V, while the lines C, D, E show their variable caudal extensions. It is of interest that these variations obtain (a) in different brains at comparable levels of the central sulcus, and (b) in one and the same brain at different levels of that sulcus. As regards this latter finding, an accurate drawing of the 'boundary' between precentral and postcentral cortex would not represent a fairly smooth line, but rather a quite irregular zigzag. Quite frequently, moreover, the *Betz cell* rows do not end sufficiently abruptly, but with scattered large elements of various sizes, thereby resulting in highly blurred 'boundary zones' of 'area 4'.

(2) Figure 114 B shows a photomicrograph by GUREWITSCH and BYCHOWSKY (1928) illustrating the boundary between gigantopyramidal motor cortex (area 4) and somatosensory cortex (area 3) in the Dog. The cited authors, competent specialists in cytoarchitectonics, have indicated said boundary at the bottom of sulcus coronarius *(sive* coronalis). The two lines X and Y, which I have added, indicate two different delimitations any of which could likewise be conceived as a reasonably justified 'boundary'. The interested reader may be left to draw his own conclusions in accord with his personal viewpoints.[99]

It should also be added that GUREWITSCH and BYCHOWSKY quite competently discuss the problem of boundaries with reference to the opposite views of the *Vogt school* and of ECONOMO and KOSKINAS. GUREWITSCH and BYCHOWSKY remark that even the VOGTS admitted transitions as so-called '*limotrophe Adaptation*', while ECONOMO and KOSKINAS find '*einen allmählichen Übergang der Felder ineinander*', '*wobei*

[99] Cf. also, with regard to 'laminar boundaries' in the spinal cord, Figures 124 -128 of chapter VIII, volume 4 with the comments on pp. 210–216 of that volume.

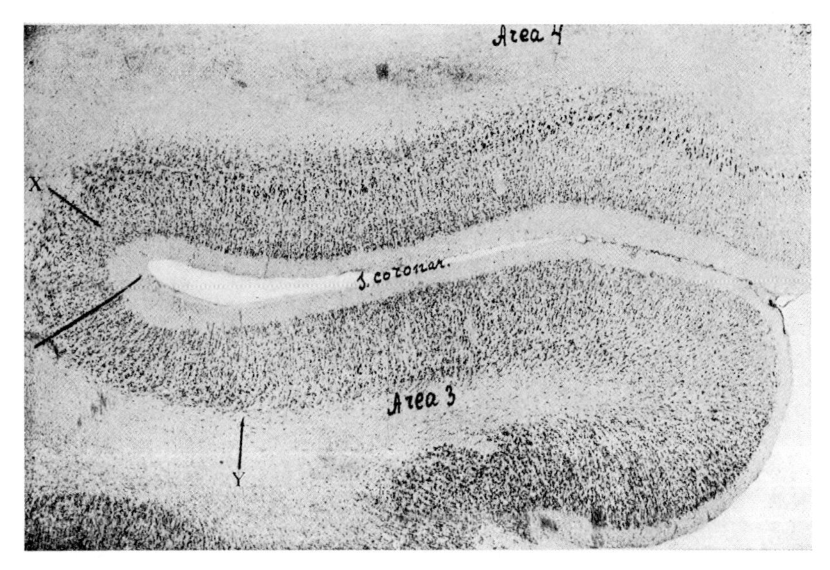

Figure 114 B. Photomicrograph *(Nissl stain)* of section through somatic motor and somatosensory cortex in the Carnivore Dog (from GUREWITSCH and BYCHOWSKY, 1928). The straight line indicates the boundary between areae 4 and 3 as drawn by the cited authors. The added arrows X and Y indicate two alternate versions of this boundary which I would likewise consider justifiable. s. coronar.: sulcus coronarius *sive* coronalis.

das Feststellen der Grenzen schwierig und bis zu einem gewissen Grade willkürlich erscheint'.[100]

(3) Figures 115 A, B, C show three different cortical mappings of the Mouse, as (A) drawn by ROSE (1929b), (B) by CAVINESS (1975), and (C) by myself. Despite numerous differences in opinions, I would consider the late M. ROSE to be a competent, and for that matter highly specialized cytoarchitecturalist with long and wide experience. His cited atlas of the cerebral cortex in the Mouse, from which the maps of Figure 115 A are taken, contains numerous and excellent large photomicrographs (cf. e.g. Figs. 82 A, B). Nevertheless I can bring my own map, based on original preparations and photomicrographs, in complete agreement with the photomicrographs published by ROSE (1929b) as well as by CAVINESS (1975), merely differing from both cited authors as regards their drawing of boundaries. Here, again, further comments

[100] GUREWITSCH and BYCHOWSKY (1928) added the further comment: '*Auf Grund unserer Untersuchungen schliessen wir uns im allgemeinen eher der Ansicht* VON ECONOMOS *und* KOSKINAS *an; dabei muss aber im Auge behalten werden, dass bei verschiedenen Tieren und auch für verschiedene Felder eines und desselben Tieres die Sachlage in dieser Hinsicht eine verschiedene ist.*'

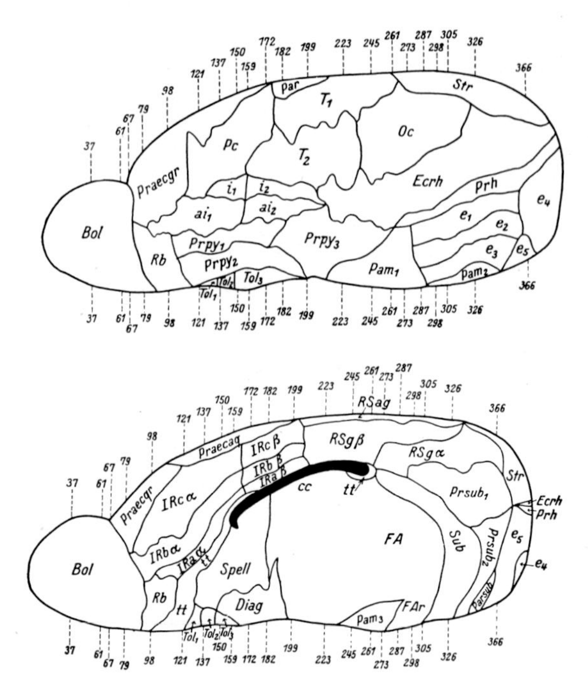

Figure 115 A. Highly parcellated cytoarchitectural cortical map of the Mouse as conceived by M. ROSE seen in lateral and medial view (from ROSE, 1929b). The areal designations are irrelevant in the aspect here under consideration.

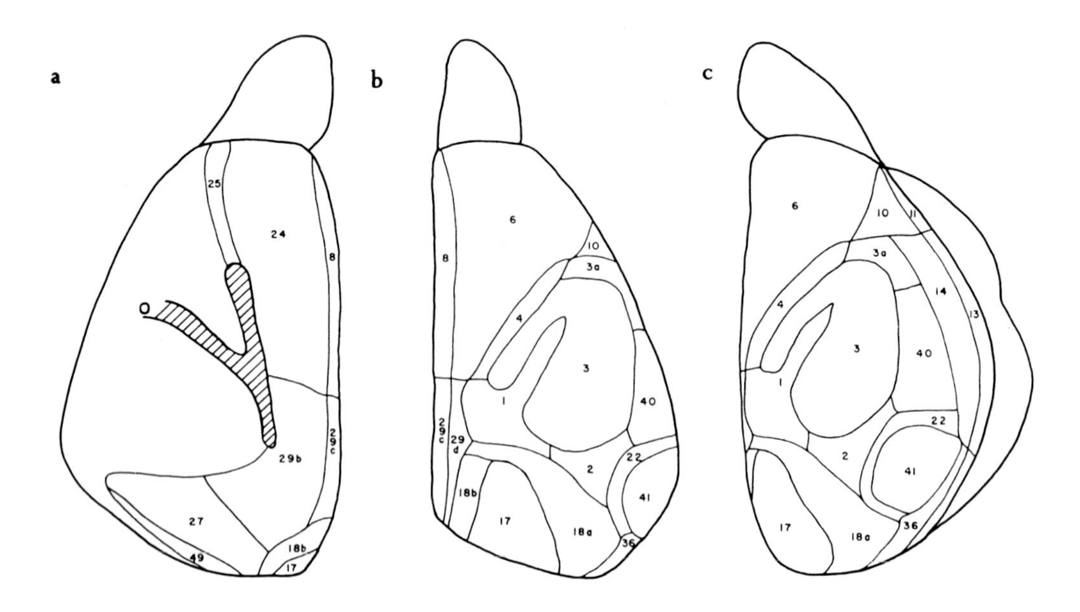

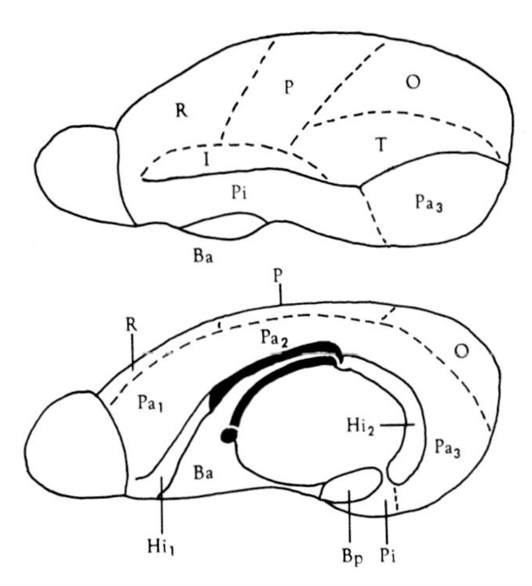

Figure 115C. Architectural cortical map of the Mouse, as conceived by the present author, and seen in lateral and medial view. Ba: anterior basal cortex; Bp: posterior basal cortex (cortical nucleus amygdalae); Hi_1: precommissural hippocampal cortex; Hi_2: postcommissural hippocampal cortex (the narrow strip of supracommissural hippocampus, adjacent to parahippocampal cortex, is not shown); I: insular neocortex; O: occipital neocortex; P: parietal neocortex: $Pa_{1,\ 2}$: precommissural and supracommissural parahippocampal cortex; Pa_3: parahippocampal cortex of posterior piriform lobe; Pi: anterior and intermediate piriform lobe cortex; R: rostral (or frontal) neocortex; T: temporal neocortex. Most boundary lines are merely approximations.

are superfluous and the reader may be left to the conclusion of his choice.

With respect to areal differences, it could also be added that, in two recent papers, JONES and BURTON (1976) and BURTON and JONES (1976) stress the coincidences of neocortical areas in Rhesus and Squirrel monkeys with the boundaries of cortical projection fields of individual thalamic nuclei. Such coincidence, as regards the Human brain, was already long ago demonstrated by the myelogenetic studies of FLECHSIG

Figure 115 B. Architectonic map of neocortex in the Mouse according to CAVINESS, as shown in medial, dorsal, and lateral view (from CAVINESS, 1975). The notation of the areae is essentially adapted to that of BRODMANN. It will be seen that this author includes the parahippocampal cortex in his 'neocortex'. The photomicrographs published by said author show that his field 25 is the inner subdivision of the precommissural parahippocampal cortex. The peculiar outline of field 3 was perhaps (consciously or unconsciously) suggested to the author by the reported animalculus outlines (cf. Fig. 96 E of the present volume, and Figs. 241 B, D, E of vol. 5/I) which, however, display a caudal instead of a rostral convexity.

(1920, 1927), which showed the visual, somesthetic and acoustic projection fields to correspond with definite cytoarchitectural areas, such as area striata (field 17), koniocortex of region postcentralis, and koniocortex of *Heschl's transverse temporal gyri* (cf. also K., 1927, p. 327). These projection fields, in turn, were known to receive input, as also demonstrated by FLECHSIG, from lateral geniculate, posterolateral and posteromedial thalamic, and medial geniculate grisea of the diencephalon. In various Primates and in lower Mammals, such coincidences are likewise well known since many years and were demonstrated, *inter alia*, by the experimental studies of WALKER (1938, Primates), and LASHLEY (1941, Rat).

5. Fiber Systems

The present section summarizes the relevant connections of the cerebral cortex dealt with in section 10 of chapter XIII in volume 5/I in which the main communication channels of the Mammalian telencephalon were pointed out with respect to the diverse griseal configurations. In addition, various cortical fiber systems not dealt with in said chapter shall here briefly be considered.

Generally speaking, the fiber connections of the cerebral cortex could be classified as comprising two main categories, namely (A) *extrinsic*, that is to say, by fibers passing through the medulla, and (B) *intrinsic*, namely by intracortical fibers remaining within the cortical griseum.

The *extrinsic* fiber systems can again be classified as (1) *projection fibers*, which are either corticofugal or corticopetal, connecting the cortex with subcortical grisea, (2) *association fibers* interconnecting different homolateral cortical regions, and (3) *commissural fibers* effecting connections between cortical regions of the antimeric hemispheres.

Publications particularly concerning, evaluating, and summarizing Mammalian extrinsic cortical fiber systems are those by POLIAK (1932), LUDWIG and KLINGLER (1956), KNOOK (1965), and KRIEG (1947, 1954, 1963, 1973).[101] The treatise by CROSBY *et al.* (1962) likewise contains

[101] POLIAK (1932) deals with afferent fiber systems in the Monkey, LUDWIG and KLINGLER (1956) bring an atlas of the Human brain based on a method of gross dissection which can rather well display a number of telencephalic fiber systems, KNOOK (1965) has presented a very detailed and critical report on all fiber connections of the forebrain, based on his own experimental studies in the Rat, but also evaluating those described in other Mammals

detailed statements on cortical and other telencephalic fiber connections in Man, compared with data obtained in other Mammals, and includes numerous bibliographic references.

Among the essentially *corticofugal channels the pyramidal (corticospinal and corticobulbar) tract* originates mainly from the precentral giganto-pyramidal cortex, as particularly well shown in anthropoid Apes by LEYTON and SHERRINGTON (1917). Although some doubts remain, its fibers mainly seem to be neurites of cells in layers V and VI, of which the former layer includes the large *Betz cells*.[102] In addition to the precentral motor cortex, pyramidal tract fibers doubtless also originate from the premotor area. No evidence for the origin of pyramidal tract fibers from the postcentral sensory cortex was found by LEYTON and SHERRINGTON (1917), while other authors (e.g. KRIEG, 1963) assume that 'a few' may arise in that region.

The *corticopontile tracts* seem to arise from the premotor cortex *(frontopontine tract)* and from parieto-occipito-temporal cortices *(parieto-occipito-temporopontine tract)*.

Corticohypothalamic channels arising from the prefrontal cortex are assumed by various authors (cf. e.g. CROSBY *et al.*, 1962) and denied by others.

Among *corticothalamic fibers*, it will here suffice to mention those of the reciprocal connections between prefrontal cortex and nucleus medialis thalami, as well as those of motor and premotor cortex with nucleus ventralis lateralis thalami, those of parietal cortex with the nucleus lateralis posterior thalami, and those of occipitoparietal cortex with pulvinar (cf. Fig. 141, chapter XII, vol. 5/I).

Corticotectal fibers are generally presumed to exist, of which an 'auditory' component from temporal cortex, with or without possible connections to medial geniculate grisea, reaches the posterior colliculus. An optic component, from occipital cortex, again with or without possible connections to lateral geniculate grisea, reaches the superior colliculus. This component likewise includes *corticopretectal fibers*.

Corticotegmental fibers from parietal cortex, *corticonigral* and *corticorubral fibers* from various regions, including motor or premotor cortex seem likewise to be present.

---

by the authors concerned with this topic. KRIEG deals in great detail with cortical connections in Rat and Macacus (1947, 1954, 1963) as well as with the fiber systems in the Human brain (1973).

[102] As regards further details concerning the pyramidal tract cf. the comments in chapter VIII, section 11, and chapter IX, section 10 of volume 4.

Although *corticostriate and corticopallidal fibers* are assumed by many authors to exist, much uncertainty and controversy obtains as regards the details of these connections. Collaterals of pyramidal tract fibers to grisea of corpus striatum have been described (cf. e.g. Fig. 117B).

The hippocampal cortex, which, as pointed out in section 3 of this chapter, essentially represents an efferent griseum of the parahippocampal limbic lobe, gives off the complex, to a large extent corticofugal fornix system, originating from the cornu Ammonis, and reaching, in addition to paraterminal grisea, a variety of diencephalic ones (in particular the mammillary complex of the posterior hypothalamus). Reblet (1976) recently reported, on the basis of degeneration techniques in the Cat, that hippocampal efferents also reach nucleus entopeduncularis, tegmental raphe grisea, and nucleus of locus coeruleus.

The connections between parahippocampal cortex and hippocampus, as well as the components of the fornix, were dealt with in section 10 of chapter XIII, volume 5/I, and partly also in section 3 of the present chapter. Details concerning a topographic organization of projections from the parahippocampal 'entorhinal' cortex to the hippocampal formation in the Rat, interpreted on the basis of autoradiographic techniques, can be found in a recent report by Steward (1976a).

In addition to the problems concerning reciprocal connections in many if not most efferent channels, it should also be kept in mind that few Mammalian forms have been systematically investigated with regard to detailed hodology. In this respect, the Rat and a number of Primates have been most intensively studied by means of experimental and other methods. The fiber systems of the Human forebrain were likewise abundantly investigated on the basis of clinical and pathologic observations, and of data obtained from normal material by anatomic, myelogenetic and histologic methods.

The cortical organization in the Rat, and even that in 'lower' Primates, as discussed in the previous sections of this chapter, can be assumed to differ in various relevant aspects from that characteristic for Man. The attempt by Crosby *et al.* (1962) to bring a coherent description of hodology reveals the high degree of unresolved uncertainty except for some overall aspects concerning a few major communication channels.

Turning now to the essentially *corticopetal* but also partly reciprocal projection fibers, there are those of the thalamic sensory relay nuclei to

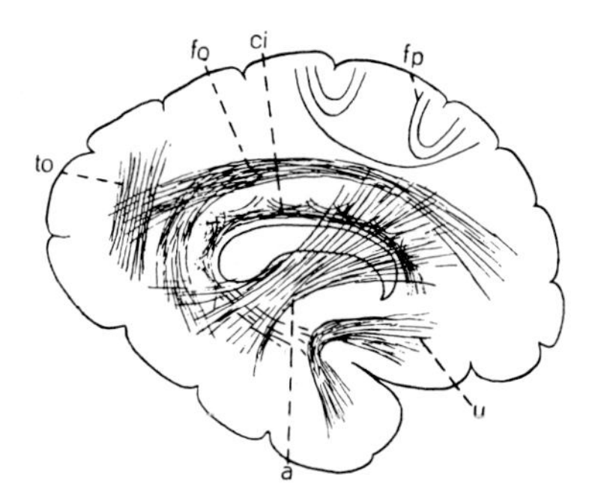

Figure 116 A. Simplified sketch showing the approximate arrangements of diverse association channels in the Human brain (modified after EDINGER, 1912, from K., 1927). a: fasciculus arcuatus; ci: cingulum; fo: fasciculus fronto-occipitalis superior; fp: fibrae propriae (U-fibers); to: fasciculus transversus occipitalis; u: fasciculus uncinatus.

the occipital (visual), temporal (acoustic), postcentral (somatosensory) koniocortices, and the less well identified special visceral (taste) and general visceral projection fibers from the grisea of nucleus ventralis posteromedialis thalami dorsalis presumably reaching basal neighborhoods of postcentral gyrus at transition to insula (taste ?), and posterior insular cortex. There is, moreover, olfactory input from bulbus olfactorius respectively 'nucleus olfactorius anterior' through lateral and medial olfactory tracts as well as 'intermediate olfactory radiation' to basal cortex, piriform lobe cortex[103] and precommissural hippocampus respectively its parahippocampal neighborhood.

Another important essentially corticopetal projection system is that from the anterior thalamic grisea to the parahippocampal cortex of the limbic lobe, particularly of gyrus cinguli. Again, the corticopetal (reciprocal) connections from the above-mentioned presumably modulating dorsal thalamic nuclei (e.g. nucleus lateralis posterior and pulvinar) and the assumed 'activating' or 'unspecific cortical afferents' from intralaminar and ventral thalamic grisea (e.g. nucleus reticularis thalami) should be kept in mind (cf. the comments by CROSBY *et al.*, 1962).

As regards the *extrinsic association fibers* of the cerebral cortex, taking their course through the medulla, it is customary to distinguish relatively short ones, represented by the so-called U-fibers, and long ones, more or less gathered into distinctive association tracts.

The U-fibers, also known as *fibrae propriae sive arcuatae*, of which

[103] Perhaps including a neighborhood of posterior piriform (entorhinal) cortex.

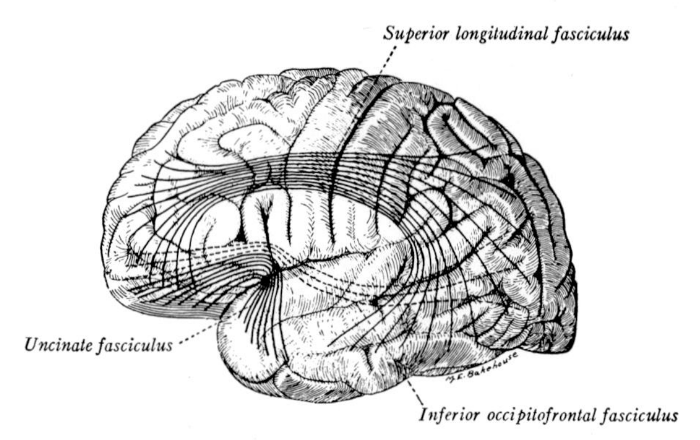

Figure 116 B. Some of the association channels projected upon the lateral aspect of the Human cerebral hemisphere (from RANSON, 1943). It is likely that the bundles shown in this and in Figure 116 C give off collaterals along their course.

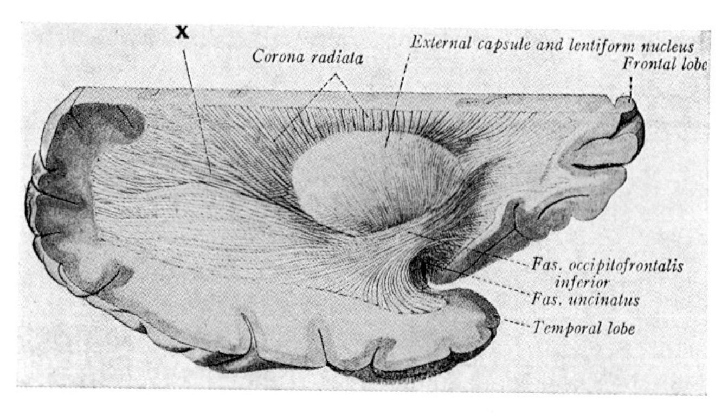

Figure 116 C. Lateral view of a dissection of a Human cerebral hemisphere, showing diverse fiber systems (from RANSON, 1943). Insula, opercula, and adjacent parts have been removed. x: internal sagittal stratum.

there are again short and somewhat longer ones, seem to interconnect either adjacent or not widely separated gyri respectively cortical neighborhoods (cf. Fig. 116A).

With respect to the '*association tracts*', those of the Human brain appear much better developed and definable than those in other Mammals. The following tracts are generally recognized in Man.

The *cingulum*, taking its course dorsally to corpus callosum represents the important association system of the parahippocampal cortex (limbic lobe). It begins rostrobasally to rostrum of corpus callosum, dorsally encircling this latter, and bending basally to splenium, reach-

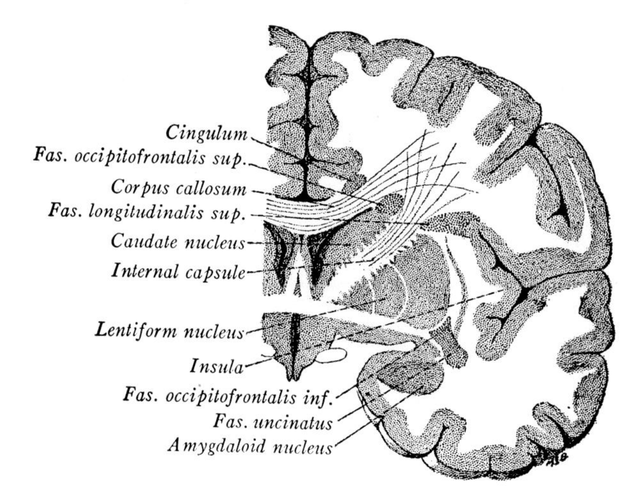

Figure 117 A. Frontal section of a Human cerebral hemisphere at level of commissura anterior, showing the location of long association tracts (from RANSON, 1943).

ing the gyrus hippocampi, when, close to the uncus hippocampi, it is continuous with the fiber bundle described by FLECHSIG (1920) as system β (cf. section 3, p. 207). The cingulum can be assumed to include short and long intralimbic fibers, as well as connections between limbic lobe and adjacent neocortical areas, such as prefrontal and medial portions of parietal and occipitotemporal cortices. Fibers connecting parahippocampal with hippocampal cortex are presumably likewise present. The cingulum is, moreover, an association tract also rather well recognizable in the 'lower' Mammals.

The *fasciculus fronto-occipitalis (or occipitofrontalis) superior* of FOREL-ONUFROWICZ,[104] separated from the cingulum by the corpus callosum (cf. Fig. 117 A) seems to connect occipital and temporal cortical regions with frontal (premotor and prefrontal) cortices. In the occipital lobe, it joins mostly the lateral tapetum of the ventricle's posterior horn but to some extent also the medial tapetum. This explains why, as pointed out by FOREL (1907), a distinct tapetum is preserved in cases of congenital agenesis of corpus callosum.

The *fasciculus longitudinalis superior sive arcuatus*, separated from the preceding one by the bundles of capsula interna, appears to interrelate frontal, parietal, insular, and occipitotemporal regions. Some of its

---

[104] Cf. FOREL (1907, p. 228), who designates it as the 'true' fasciculus longitudinalis superior.

basal fascicles run through dorsal portions of both capsula externa and extrema.

The *inferior fronto-occipital or occipitofrontal fasciculus* runs through basal portions of capsula externa and seems to provide connections between occipitotemporal cortices (including fusiform and lingual gyri) and basilateral regions of the frontal lobe. This bundle can also be fairly well recognized in some Ungulate brains.

The *uncinate fasciculus* lies basally and closely adjacent to inf. occipitofrontal fasciculus, from which it cannot be clearly separated, and runs in part through capsula extrema. It appears to interrelate parahippocampal cortex of gyrus hippocampi, and probably also grisea of the amygdaloid complex, with prefrontal cortex, particularly also with that of the orbital surface (basal neocortex of SPATZ). At cross-sectional levels of the amygdaloid complex, fibers of fasciculus uncinatus, of occipitofrontal bundle, and of caudal limb of anterior commissure become intermingled, and disrupt the griseal continuity between amygdaloid complex and claustrum. The basal portion of this latter griseum becomes broken up into nondescript gray fragments, some of which may either pertain to claustrum (pars diffusa *sive* claustrum parvum) or represent components of the main amygdaloid complex (nucleus amygdalae β).[105]

The so-called *fasciculus longitudinalis inferior*, originally demonstrated by REIL and BURDACH on the basis of gross dissections, was shown by FLECHSIG to represent the *geniculocalcarine tract*, forming the *stratum sagittale externum* of the occipital lobe. There are two additional sagittal strata in the Human occipital lobe, namely the intermediate *stratum sagittale internum* and the innermost *tapetum*, adjacent to the posterior ventricular horn's ependyma (cf. Figs. 246E, 247A, 251D, chapter XIII, vol. 5/I). The stratum sagittale internum includes fibers between occipital cortex and pulvinar, pretectal region, and superior colliculus, while the tapetum contains the fibers of fasciculus fronto-occipitalis superior and of forceps posterior corporis callosi.

Within the frontal and the occipitoparietal lobe, *dorsobasal (transverse) association systems* have been described, e.g. as orbitofrontal fibers respectively fasciculus transversus occipitalis. Figures 116A–117A illustrate the approximate arrangement of the diverse association systems as presumed to be present in the Human hemisphere.

Turning now to the *commissural channels*, interhemispheric cortical

[105] Cf. p. 703 of section 10 in chapter XIII of volume 5/I.

'associative connections' are provided by *commissura anterior*, *corpus callosum*, and *commissura hippocampi*.[106]

Details of the *anterior commissure* were dealt with in section 10 of chapter XIII in volume 5/I. It will here be sufficient to recall that, as far as intercortical associations are concerned, this commissure includes fibers interconnecting antimeric grisea of the anterior piriform cortex, of the parahippocampal posterior piriform lobe, and of neocortex. In Mammals without 'true' corpus callosum (Prototheria and Metatheria), in which the commissura dorsalis *sive* pallii is essentially a commissure of hippocampal and parahippocampal cortex ('commissura hippocampi'), the neocortical component of commissura anterior, interconnecting the antimeric neopallial grisea, is very substantial. In 'higher' forms, particularly in Primates including Man, the neocortical component of that commissure becomes essentially restricted to more or less extensive regions of the temporal lobe.

The Eutherian *corpus callosum*,[107] whose gross anatomical details were taken up in section 10 of chapter XIII in volume 5/I, is the main commissural system of the neocortex, and also that of much of the parahippocampal cortex. The fibers of corpus callosum interconnect antimeric (symmetric) grisea as well as, particularly in 'higher' Mammals, also different (asymmetric) cortical areas. The calcarine cortex of Primates, and part of the occipital (visual) cortex in 'lower' forms, seem to lack callosal commissural connections. Among reports on callosal fiber distributions are those by CURTIS (1940), McCULLOCH and GAROL (1941), v. BONIN *et al.* (1942) concerning Primates, and of YORKE and CAVINESS (1975) concerning the Mouse.

The role played by corpus callosum and to some extent also by anterior commissure in the interhemispheric transfer of information has been extensively studied in recent years following experimental transections of the corpus callosum in animals, and, for therapeutic purposes, in Man. Clinical cases of agenesis of the Human corpus callosum, diagnosed by pneumoencephalography, have likewise provid-

---

[106] To which extent interhemispheric cortical communications may be shunted through commissura posterior, commissura habenulae (and other diencephalic commissures) remains an open question.

[107] Data referring to ontogenesis and presumptive phylogenesis of the corpus callosum were discussed in section 6, chapter VI of volume 3/II. Additional data concerning commissura pallii dorsalis of Prototheria and Metatheria, and its morphologic relationship to the corpus callosum as developed in various Eutheria, were also dealt with in section 10, chapter XIII of volume 5/I.

ed a variety of pertinent observations (cf. the symposium edited by ETTLINGER, 1965, and the summary on 'split-brain studies' by CUÉNOD, 1972).

Generally speaking, a training exercise, learned by Primates with one hand, can be fairly well performed by the other hand on the basis of interhemispheric information transfer. In the 'split-brain animal' however, such task learned by one hand cannot be performed by the other, unless this latter is subsequently separately trained. In other words, the hemispheres become relatively independent of each other in their capacity to receive, process, and store sensory information as well as to control the performance of adequate motor responses. Thus, it seems that, under normal conditions, each hemisphere receives both a direct and an indirect, commissural input. On the other hand, interhemispheric transfer and integration are neither perfect in intact organisms nor completely prevented by commissurotomy. Numerous and diverse parameters appear to obtain, such as animal species, extent of midline sections, type and location of stimuli, age of animal, and delay between surgery and tests (cf. CUÉNOD, 1972). Bimanual skills acquired before the operation are not affected.

As regards Man with a dominant left hemisphere for speech, a patient with transected corpus callosum is unable to describe, with eyes closed, an object held in the left hand or seen in the left visual field, although the 'nature' of the object is apparently understood. The right hemisphere seems here to have no access to language engrams in the left hemisphere. There is, however, evidence that in Man the dominant hemisphere for language, usually but not always the left, is not dominant in all respects, the contralateral hemisphere being superior in other respects, e.g. in the perception of three-dimensional spatial relationships. Observations on these topics have led to a modification or re-evaluation of the concept of hemisphere dominance in Man (cf. e.g. the short summary given by BARR, 1974).

Although the 'split-brain' studies have yielded numerous data on commissural functions, many poorly understood problems remain. CUÉNOD (1972) refers to the diverse discrepancies reported in the literature, which make it difficult to directly compare observations made under different conditions. The cited author justly remarks that these studies 'have been developed around a technique more than around a problem'.

Another insufficiently clarified problem concerns the cortical elements (perhaps mostly pyramidal cells of the third layer ?) from which

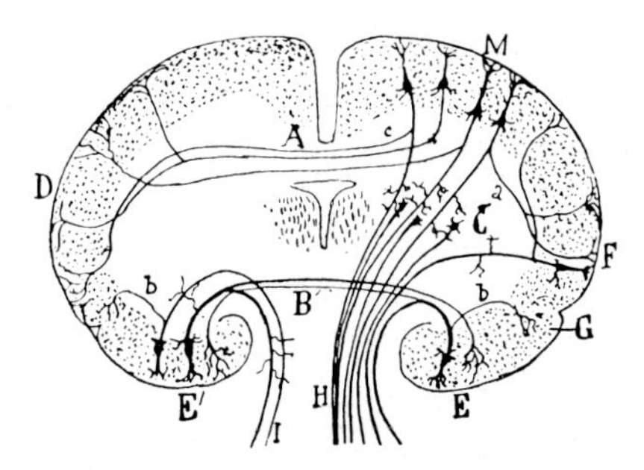

Figure 117 B. Diagram of projection and association fibers in the telencephalon of a Rodent (from CAJAL, 1911). A: corpus callosum; B: anterior commissure; C: corpus striatum; D: 'commemorative optic center'; E: 'olfactory perceptive center'; F: 'commemorative optic center'; G: 'olfactory commemorative center'; H, I: projection fibers; M: 'perceptive optic center'; a, b, c: collateral branches of neurites. Cf. the comments on CAJAL's diagram given in the text.

the callosal fibers arise. Figure 117B shows an older concept elaborated by CAJAL (1911). Discounting its diagrammatic aspect and the details of CAJAL's interpretation, this illustration is of interest since it is based on that author's highly competent observations in *Golgi preparations* of the Rodent brain, showing (1) that collaterals of cortical projection fibers may join the corpus callosum, while (2) other collaterals seem to provide homolateral association fibers, and (3) that still other collaterals of such corticofugal projection fibers may effect connections with the striatal grisea.

The Eutherian *commissura hippocampi*,[108] as particularly investigated by CAJAL (1911), includes efferent channels of the postcommissural parahippocampal cortex reaching the contralateral postcommissural hippocampal cortex (CAJAL's *voie temporo-ammonique croisée* of his psalterium dorsale). In addition, fibers interrelating the antimeric hippocampal cortices seem to be present (CAJAL's *commissure interammonique* in his psalterium ventrale). The commissura hippocampi may, moreover, also contain fibers interrelating the antimeric postcommissural parahippocampal cortices.

Turning now to the *intrinsic*, *intracortical fiber connections*, as mentioned above to represent category (B), and discounting the structural details of synaptology dealt with in section 3 of the present chapter, it will be sufficient to point out the fiber systems directed at a right angle

[108] Cf. the comments of footnote 107.

to the radii, and provided by *inner and outer stria of Baillarger*, by *stria of Kaes-Bechterew*, and by the *tangential fibers* of the molecular layer. Although sufficiently convincing data as to the provenance of these fibers, and as regards the extent of their course parallel to the cortical surface have not been recorded, it can be assumed that the *inner stria of Baillarger* and the *stria of Kaes-Bechterew* contain collaterals of pyramidal cell neurites and of extrinsic cortical afferents. While some such fibers may also be present in the *external stria of Baillarger*, this latter, particularly related to the cytoarchitectural lamina IV (granularis interna) seems to mainly contain neurites of that layer's stellate cells.

The *tangential fibers* of the molecular layer are presumably provided by neurites of *Cajal's horizontal cells* and by ascending intracortical axons, such e.g. as recurrent axon collaterals of pyramidal cells or neurites of *Martinotti cells* reaching the molecular layer and spreading horizontally.[109]

As regards the postcommissural parahippocampal cortex, however, its inner region adjacent to the hippocampal cortex contains, in its wide molecular layer, an important medullated fiber system providing parahippocampal output reaching the hippocampal cortex and dealt with in section 3 of the present chapter. It may also be added that, according to STEWARD (1976b), granule cells of the Rat's gyrus dentatus 'denervated' by unilateral destruction of the parahippocampal ('entorhinal') cortex are 're-innervated' by proliferation of surviving pathways from the contralateral (antimeric) parahippocampal cortex. Retrograde labelling with horseradish peroxidase is said to indicate that the here relevant contralateral neuronal elements are of the same cell type which normally project to the homolateral gyrus dentatus.

6. Functional Problems, including Engraphy and Consciousness

In attempting to elucidate the functions of the cerebral cortex, two opposite and apparently contradictory overall viewpoints have been adopted by some of the authors concerned with this topic. According to the concept stressing *localization*, the cortex consists of a diversity of distinctive areae representing functional '*centers*'. According to the' *holistic*' viewpoint, the cortex is, to a significant degree, *an equipotential continuum*, acting as a '*whole*'. During the early period of

[109] Cf. also the comments on p. 232, section 3.

systematic investigations on cortical functions, the localization concept was particularly upheld by HITZIG and MUNK (e.g. MUNK, 1881), and the 'holistic' viewpoint was championed by GOLTZ (1881). At the present time, it could be said that both viewpoints are by no means mutually exclusive, but merely emphasize two actually obtaining different aspects of cortical mechanisms.

From still another viewpoint, and particularly with regard to the Human cerebral cortex, it seems perhaps appropriate to distinguish, amplifying the important concepts of FLECHSIG, the following types of cortical functional activities.

1. Direct (upper motor neuron) *motor output* to primary somatic and 'special visceral' efferent spinal and cranial nerve nuclei.

2. Registration and primary cortical processing of *sensory input* (somesthetic, optic, auditory, taste, general visceral, olfactory).

3. Further processing activities by additional abstraction of invariants *(symbolization)* including logical mechanisms and storage *(engraphy)*. These activities also involve the '*programming*' necessary for the execution of learned, skillful, patterned motor performances, including writing and articulated speech.

4. Mechanisms of *affectivity* respectively *emotion* and, *qua* introspection, 'will'. These mechanisms, moreover, participate in 'programming' and 'selecting' logical activities, as well as in the 'recall' *(ecphory)* of stored information (engrams). As justly emphasized by the valid aspects of the *James-Lange theory* of emotions, manifestations of affectivity are intimately correlated with visceral activities controlled by the vegetative nervous system.

In a simplified formulation we have thus (a) the motor and sensory cortices *(Flechsig's projection centers)*, (b) the higher interpretive, directing, and storing cortices *(Flechsig's association centers* of various hierarchic orders), and (c) the 'emotional' or 'affective' parahippocampal cortex with the hippocampal cortex as its main output griseum.[110] In addition, the prefrontal neocortex appears likewise related to affectivity, of which there seem to obtain two components, namely a parahippocampal and a neocortical one.

---

[110] The 'emotional cortex' (c) corresponds, of course, to the 'limbic lobe'. The limbic system, moreover, also includes subcortical configurations such as amygdaloid and paraterminal (septal) grisea. Likewise, the striatal grisea seem related to motor activities as well as perhaps to those of limbic lobe. As regards the presumed peculiar phylogenetic relationship between olfactory system and limbic lobe cf. p. 542, section 2, chapter XIII, volume 5/I.

In other words, the cortical mechanisms can be conceived as *sensorimotor*, as *storing*, as *predictive (extrapolating)* and *logical*, and as *emotive*. All these activities appear to be closely interdependent and interwoven (*'miteinander verschränkt'*).

In still more abbreviated and oversimplified formulation there is a *sensorimotor*, a *logical*, and an *emotional* cortex. In lower Mammalian forms, sensorimotor and logical cortices (i.e. projection and association 'centers' seem to be not clearly separated from each other, but represented by the primordial neocortical areas which, however, are well distinguishable from the parahippocampal-hippocampal allocortex, and from the sensory allocortex of the olfactory system.

As regards the *electrically excitable somatomotor cortex*, whose demonstration by FRITSCH and HITZIG in 1870 firmly established the concept of cortical localization, numerous poorly understood problems nevertheless remain. Quite evidently, electrical stimulation of the cortex represents a rather coarse, 'unnatural' or 'unphysiological' interference if compared with normal neural impulse conduction. Although electrical stimulation has disclosed a fairly constant topographic respectively topologic representation as e.g. illustrated by Figures 240 A–242 E, 253 A, and B of chapter XIII in volume 5/I, a variability, instability or even 'reversal' of responses to stimulation of identical spots has been reported by most investigators[111] (e.g. LEYTON (formerly GRÜNBAUM) and SHERRINGTON (1917). Such observations suggest that a relatively predominant, probabilistic, but not a rigid localization obtains. In other words, instead of a detailed mosaic with boundary lines, an arrangement of overlapping open neighborhoods can be assumed. These questions have been discussed by numerous authors, but it will here be sufficient to refer to the comments by WALSHE (1943, 1947), to the investigations by LANDGREN *et al.* (1962), and by ASANUMA and ROSEN (1973), as well as to the brief summary on some aspects of cerebral localization by DALY (1975).

LEYTON and SHERRINGTON (1917) stress that the anterior 'edge' of the typically excitable precentral motor area seems to 'fade away somewhat gradually into inexcitable cortex'. This zone of *'gradatim merging'* was observed to expand forward following repeated stimulation. It is of interest that this functional zone of transition corresponds to the

[111] Thus, arrest of movement or prevention of movement may at times be induced by stimulation. It should also be mentioned that, in profound anesthesia, the motor cortex becomes inexcitable.

poorly delimitable, vague boundary zone between precentral and premotor cytoarchitectural regions.

As regards the Human *corticospinal tract*, dealt with in chapter VIII of volume 4, it should be recalled that its unilateral amount of fibers is estimated at about one million, the number of very large, coarse medullated fibers (about 30 000–40 000) roughly corresponding to the number of *Betz cells*, being thus less than the estimated unilateral number of spinal motoneurons (between 80 000 and 160 000). Thus, discounting the probability of short spinal internuncials, as well as the complications introduced by the gamma cells, and merely considering overall relationships of pyramidal tract fibers to 'typical' motoneurons, a ratio of roughly 10 : 1 obtains, and with respect to *Betz cells*, the ratio is reversed, becoming about 1 : 2 or 1 : 3. Nothing certain is known concerning the details of synaptic connections and the arrangements related to the opposite excitatory as well as inhibitory effects of the corticospinal (respectively also the corticobulbar tract). These tract fibers presumably give off collaterals and end-arborizations with synaptic endings on perhaps several neurons, and more than one synapse, perhaps even very many, on one and the same nerve cell.

Generally speaking, it can be assumed that specific 'colonies' of neuronal elements, controlling flexor and extensor motoneurons, are represented by columns, some elements of which might effect monosynaptic connections with primary motoneurons. Such columns seem to be intermingled with other ones of diverse functional significance.

Since the precentral cortex is concerned with motor output, an 'irritating' cortical lesion in this region may lead to a spread of excitation expressing itself in convulsions of the type designated as '*Jacksonian epilepsy*', which usually begins in one of three foci, namely thumb and index finger, angle of the mouth, or great toe. The attack generally starts with clonic movements, more rarely with tonic spasm, on the opposite side of the body, subsequently spreading to the entire musculature, and finally becoming bilateral. In contradistinction to '*idiopathic*' or '*constitutional*' *epilepsy*,[112] there is often no loss of consciousness, except usually upon generalized, bilateral spread.

---

[112] *Epilepsy*, of which there are numerous forms with a diversity of still poorly understood causal factors, may be regarded as an uncontrolled neural discharge involving the cerebral cortex. It is a paroxysmal and transitory disturbance with sudden loss of consciousness, which may or may not be associated with tonic spasm and clonic contractions of the muscles. A generalized attack is known as '*grand mal*', an attack characterized by loss of consciousness only, without falling and without, as a rule, motor accompani-

In addition to the main somatomotor area, electrically excitable regions, *qua* motor responses, and with a variable degree of distinctive localization, include the second and the supplemental motor areas, and the premotor cortex. There are, moreover, the 'eye-fields' of so-called area 8 (presumably pertaining to premotor cortex) and of lateral occipital lobe, from which conjugate contralateral deviation can be elicited.

Movements evoked from stimulation of the premotor cortex tend to be more complex than the rather isolated movements elicited from the precentral cortex. Conscious patients commonly do not feel that they have 'willed' a movement resulting from electrical stimulation.[113] The premotor region presumably includes *Broca's convolution* of the hemisphere dominant for language and, generally speaking, seems to be involved in the processing of neural signals organizing the output of the motor region required for particular skillful activities.[114]

Typical motor responses upon electrical stimulation of the somatosensory region were not observed by LEYTON and SHERRINGTON (1917), although these authors noted occasional or exceptional reactions which they evaluated as 'facilitation' or as 'echo-responses' obtained when stimulation of the postcentral gyrus quickly and directly follows that of corresponding loci in precentral gyrus. Other authors, e.g. PENFIELD, BOLDREY, and VOGT, nevertheless reported motor responses from postcentral gyrus upon strong stimulation (cf. the comments and bibliographic references by FULTON, 1949). The sensory

---

ments, as '*petit mal*' or as '*absence*' ('*blank spell*'). Cf. BRAIN and WALTON (1967). From the viewpoint of neurochemistry, epileptic seizures are presumed to result (a) from imbalance between excitatory and inhibitory transmitters or (b) from synaptic membrane abnormalities including electrolyte derangements, or from both conditions (a and b) combined. Neuroglia, in particular astrocytes, may be involved in these metabolic processes. Again, disturbances of GABA metabolism seem related to vitamin B_6 deficiencies, since this vitamin is required for GABA synthesis.

[113] This seems readily understandable if the motor regions are conceived as output grisea, through which motor activities are 'funneled', and which receive their 'willed' directive signals from distant cortical regions. ECCLES (1970), however, who has reintroduced 'will' as a little ghost or soul inhabiting the brain and directing its 'higher' activities, quotes said involuntary aspect of cortical stimulation in support of his '*little-ghost-in-the-machine*' theory.

[114] BROCA (1861) aptly stated that in 'aphemia' caused by damage to the cortical region which he had identified, the patient's brain looses the ability '*du procédé qu'il faut suivre*', that is to say, becomes incapable of initiating the procedure that must be followed in order to articulate the word.

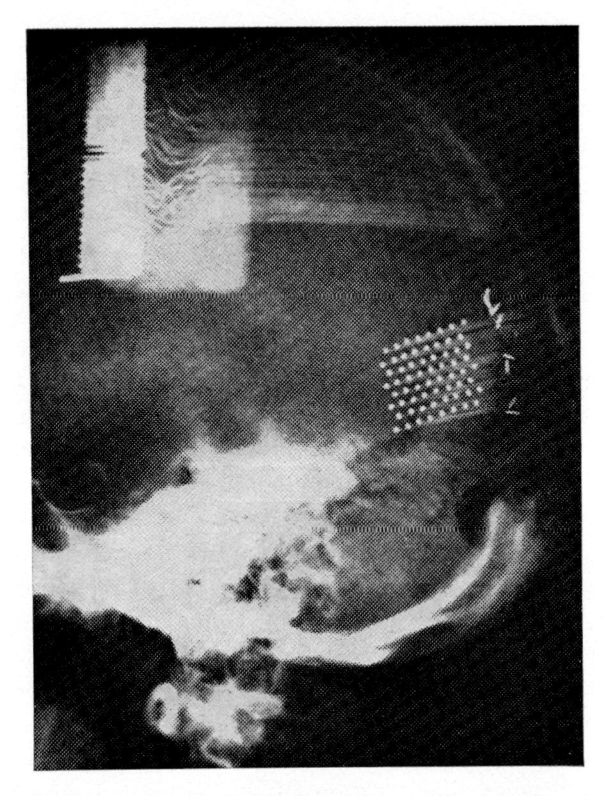

Figure 118. Electrode array of visual prosthesis, located in subdural space, as seen in a lateral roentgenogram. The external portion of the ribbon cable and the connector are likewise shown (from DOBELLE *et al.*, 1974).

cortices and their surroundings are, however, electrically excitable insofar as their stimulation, performed upon conscious Human patients, is known to evoke sensations (cf. section 10 of chapter XIII in volume 5/I, and further below in the present context).

With regard to the sensory projection cortices, the localized projection of retinal neighborhoods, with emphasis on the macula (or fovea centralis) has been well substantiated, particularly also concerning the Human area striata, and was dealt with in section 10 of chapter XIII in volume 5/I (cf. also Fig. 145 A–C, vol. 5/I, chapter XII).

Since electrical stimulation of the occipital cortex can evoke localized, discrete phosphenes, some preliminary attempts have been made to develop artificial vision for the blind by means of an electronic visual prosthesis[115] (DOBELLE *et al.*, 1974; STERLING *et al.*, 1971; BRINDLEY and LEWIN, 1968; VAN DEN BOSCH, 1959). Such artificial vision involves the transmission of signals from a television device or similar sensor to the surface of the visual cortex. This necessitates the implantation of an electrode array molded to fit the occipital lobe (cf. Fig. 118). The various problems presented by the development of such visual prostheses are evidently considerable and far from being successfully solved. Nevertheless, in a temporary experiment of this type by DOBELLE *et al.* (1974), two totally blind volunteers reported localized photic sensations, and the stimulation by multiple electrodes allowed one of these patients to recognize simple patterns, including letters. It seems thus possible that, despite the numerous substantial difficulties, a to some degree functional visual prosthesis could perhaps be developed in the course of technological progress. This was also to some extent implied in a report by UEMATSU *et al.* (1974), concerning experiments with electrical stimulation by sterotaxically implanted electrodes into striate cortex and geniculocalcarine tract of patients undergoing thalamotomy for intractable pain. The phosphenes, as described by the patients *qua* location, shape, and brightness, were either simple white or multicolored. The experiments usually began on the first postoperative day and were completed within 7–13 days. Of 17 cases, 7 reported phosphenes, and in one of several cases that came to autopsy, a detailed identification of the stimulated structures was made. This, and histological studies on some of the other autopsied brains suggested that the phosphenes were more often evoked from the striate cortex than from the geniculocalcarine fibers. The authors do not elaborate on their statement that only 'seven of the 17 cases reported phosphenes upon electrical stimulation of striate cortex and geniculocalcarine tract'. Diverse particular details of organization in the visual cortex of Cats and Monkeys have been investigated with the so-called single unit microelectrode recording technique in a series of studies by HUBEL (1958, 1959, 1960), and HUBEL and WIESEL (1962, 1965, 1968, 1969, 1972, 1974a, b, c). A columnar organization was shown to be superimposed upon the topographic representation of the

---

[115] Such visual prosthesis differs from the various efforts made to develop aids for the blind through the conversion of the optical image obtained by a sensor device to either auditory or tactile stimuli (e.g. vibrations by an 'Optacon' opto-electronic camera). So far, the limited performance of such conversion devices, and the difficulty in training patients to interpret the converted signals have been an obstacle to widespread practical application.

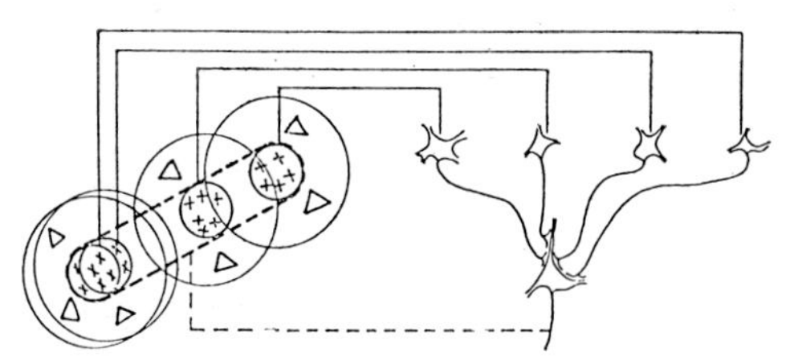

Figure 119 A. Possible scheme explaining the organization of a simple receptive field in the visual cortex (after HUBEL and WIESEL, 1962, and BRAIN, 1965). A large number of lateral geniculate neurons, of which four are depicted at right, have receptive fields with 'on centers' arranged in a straight line on the retina. All these geniculate cells project upon a single cortical cell through excitatory synapses. The receptive field of the cortical cell will then have an elongated 'on center' indicated by the broken line in the receptive field diagram at the figure's left (crosses: on; triangles: off).

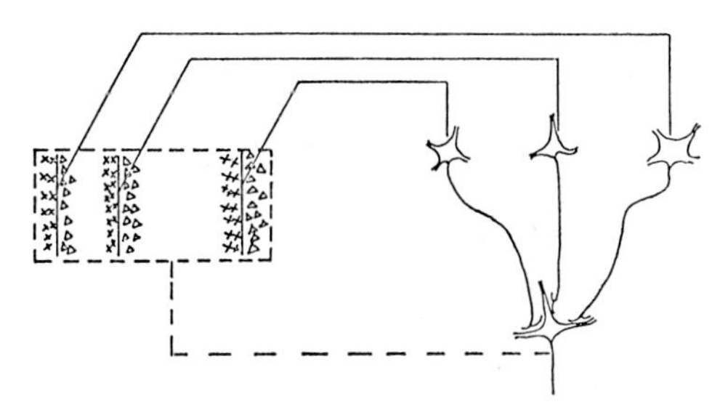

Figure 119 B. Possible scheme explaining the organization of complex receptive fields (after HUBEL and WIESEL, 1962, and BRAIN, 1963). A number of cells with simple fields, of which three are shown, project upon a single complex element. Each simple element has a receptive field as shown at left, with an excitatory region at left, and an inhibitory one at right of vertical trace line boundary. The boundaries of the fields are staggered within an area outlined by the broken line. Any vertical-edge stimulus falling across that rectangle, regardless of its position, will excite some simple field cells, leading to excitation of the complex element.

retina. Several sorts or 'families' of cortical columns seem to obtain, such as for 'receptive field axis orientation' (*'orientation columns'*), and for 'eye preference' (*'ocular dominance columns'*). The former sort includes as many types of columns as there are recognizable differences in orientation, while the cell responsivity of 'eye preference' columns depend upon whether a highly localized stimulus was presented to a corresponding region of homolateral or contralateral retina. The 'ocular dominance' columns described by HUBEL and WIESEL on the basis of electrophysiological data were also demonstrated by KENNEDY *et al.* (1975) in experiments using an autoradiographic technique.

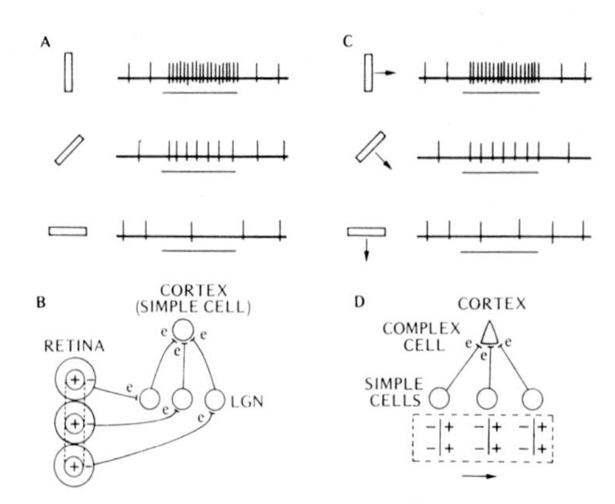

Figure 120 A–D. Responses of neurons in the visual cortex to stimulation of the retina with light (after HUBEL and WIESEL, 1962, from SHEPHERD, 1974). A: Extracellular recordings of unit spikes. Responses of a 'simple cell' are shown to stimulation of the retina with a bar of light with three orientations; duration of stimulus shown by line beneath recording trace. B: Schema of possible circuitry mediating response in A (cf. also Fig. 119 A). C: Responses of a 'complex cell' to bar of light moving as indicated by arrow. D: Schema illustrating possible circuitry mediating responses in C (cf. also Fig. 119B). e means excitatory synapse.

It will be recalled that, in the retina, 'on-off' receptive fields with annular 'surround region' are presumed to obtain, and that this organization also holds for the relay through the lateral geniculate griseum.[116]

In the primary visual projection cortex (area striata), however, the simplest responses are to 'bars' or 'lines' of light with a specific inclination (e.g. vertical, horizontal, oblique). Elements displaying this response are designated as '*simple cells*'. More complex responses are to a bar or edge moving over the retina in a particular direction, being manifested by '*complex cells*'. Still more complex responses, by '*hypercomplex cells*', are to bars and edges, moving with particular orientations, with various critical dimensions, and with antagonistic regions within their peripheral fields.

The sequence within a column of the primary visual cortex is presumably mediated by vertical circuits providing for excitatory-inhibitory interactions. The 'simple cells' may be stellate elements in layer IV, and the 'complex' respectively 'hypercomplex' cells may be

[116] Cf. volume 5/I, chapter XII, section 1B, p. 87, concerning retinal receptive fields, and volume 5/1, chapter XII, section 9, p. 357 concerning the dorsal lateral geniculate griseum.

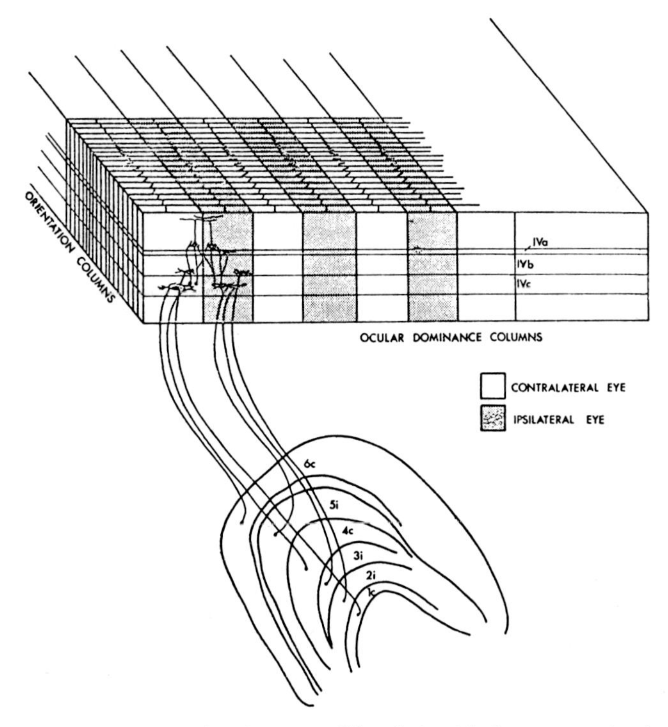

Figure 120 E. Diagram showing a possible relationship between ocular-dominance columns and orientation columns. The width of the orientation slabs is presumed to be much less than that of the ocular-dominance columns. A complex cell in an upper layer is shown receiving input from two neighboring ocular-dominance columns, but from the same orientation column (from HUBEL and WIESEL, 1972).

mostly pyramidal neurons in the more superficial or deeper layers. The diagrams of Figures 119 A, B and 120 A–E illustrate some aspects of organization in the area striata as inferred from the investigations by HUBEL and WIESEL.

As regards the neuronal sequence from retinal sensory cells to visual cortex, showing, except for the pathway from retinal ganglion cell to lateral geniculate griseum, a convergence of output from one functional class upon another functional class, Figure 120F illustrates a simplified diagram based on the data recorded by the investigations of KUFFLER (1953) and HUBEL and WIESEL (1965 and other publications). BURNS (1968) points out that the visual system seems, at least in this respect, to operate on the principle of a so-called 'classification machine' as outlined by UTTLEY (1954) and briefly referred to below on p. 316.

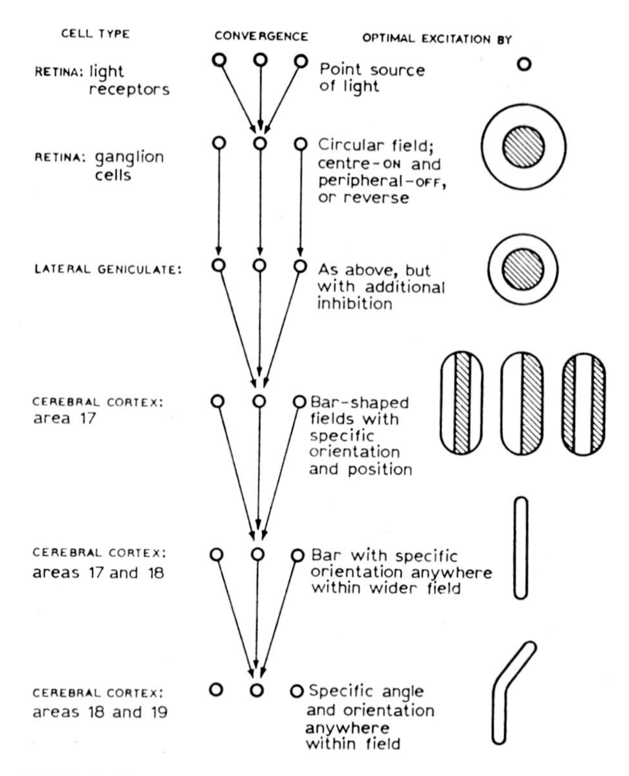

Figure 120 F. A diagrammatic summary of the functional classes of neurons that have been described at various levels of the visual system (adapted after HUBEL and WIESEL, 1965, from BURNS, 1968).

Simple cells are said to be present only (but not excluding other types) in the striate cortex receiving direct input from lateral geniculate griseum, while the columns in parastriate and peristriate areas seem to contain only complex and hypercomplex elements, the hypercomplex ones being of various hierarchical order.

Extracellular microelectrodes have the advantage over *intracellular* ones insofar as they can be maintained within recording distance of one unit for relatively long periods (1 h or more). In order to retain a satisfactory signal-to-noise ratio, the tip of the extracellular electrode must be kept within about 15 μ of the unit's surface membrane (cf. e.g. BURNS, 1968). The technique has certain obvious limitations and only provides reliable information concerning the precise times of neuronal discharge. Quite apart from the fact that the electrode may pick up sig-

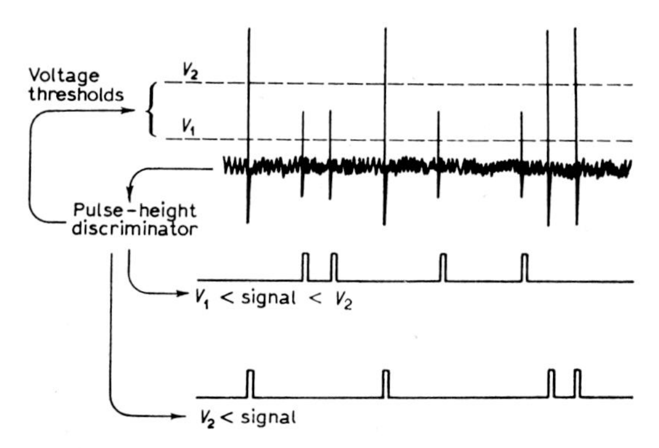

Figure 121 A. Spike activities of two neighboring neuronal elements recorded with one extracellular microelectrode and distinguished by means of a pulse-height discriminator (from BURNS, 1968). V_1, V_2: differing voltage thresholds of records from the two cells. In addition to the recorded spikes, the background potentials below those of the two discharging cells are shown.

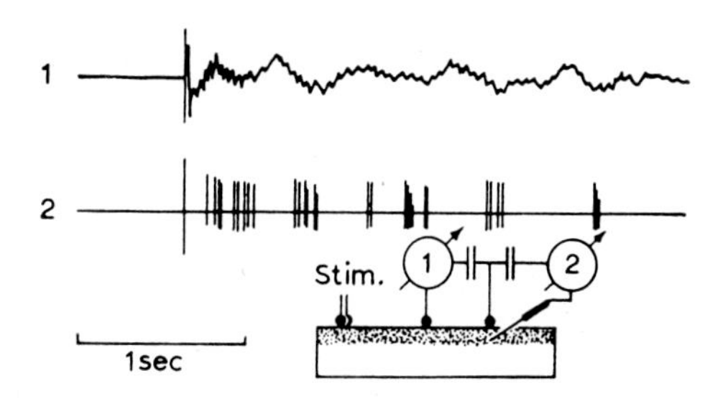

Figure 121 B. Records from isolated cortex upon a single stimulus, simultaneously obtained by surface electrode (1) and by extracellular microelectrode (2) (from BURNS, 1968).

nals from more than one unit, the records do not indicate why a unit fired, since presynaptic events are not usually recorded, nor are any postsynaptic events recorded that may precede discharge. Intracellular electrodes would here provide better information, but require a degree of mechanical stability not attainable for a prolonged registration of a single unit's activity. Moreover, records from a single neuron, whether extracellular or intracellular, provide only indirect information about the spatial spread of excitation among a population of neurons. Figure 121 A illustrates details of a record obtained by an extracellular microelectrode registering activities of two neighboring cells, and Figure 121 B shows the simultaneous records obtained by a 'gross' electrode on the cortical surface and by an extracellular microelectrode. Figure 121 A, moreover, shows the random background noise which presumably includes activities of the numerous synaptic endings on the neuron 'contacted' or 'touched' by the extracellular microelectrode.[117]

It should also be pointed out that designations such as 'simple', 'complex', 'hypercomplex', or 'opponent-color' cells, although to some extent justified, are semantically awkward, since they unduly stress the functional significance of a particular cell within a neuronal network. It is the particular type of event occurring in the network rather than the particular cell which is functionally relevant, the cell as such merely representing, by its discharge, a many-one switch turned on in an activated circuit processing R-events encoded by N-events.

The peristriate and parastriate cortices appear to receive optic input from the area striata, and represent, as it were, second and third optic cortices. In the parastriate area, through callosal fibers from contralateral area striata, there is a representation of the entire visual field of each eye, while the area striata includes only half of the visual field of both eyes.

In recent years, many attempts have been made to elucidate, with the diverse contemporary technical methods, further details of fiber connections related to the visual cortex and its subdivisions (e.g. area striata, 17; area parastriata, 18; and area peristriata, 19), as well to other grisea of the optic system, supplementing the basic data obtained by the earlier investigators. Some of the obtained results are not altogether convincing nor conclusive, quite apart from the probability of non-negligible taxonomic differences, individual variations, and intrinsic limitations of the methods. It will here be sufficient to refer the interested reader to the papers by Gilbert and Kelly (1975), Lund *et al.* (1975), Niimi *et al.* (1974), Updyke (1975), and Wagor *et al.* (1975) which also contain pertinent bibliographies relevant to this topic.

With regard to the '*generalization of patterns*' or to the *abstraction of invariants* performed by the cortical mechanisms of vision, Pollen *et al.* (1971) assume that the striate cortex transforms the topographic representation of retinal visual space in the lateral geniculate griseum into a *Fourier transform* or frequency representation at the complex cell level *via* the intermediary simple cell stage of 'strip integration'. Each of these three stages contains essentially the same amount of information,

[117] Cf. the comments on p. 616, section 8, chapter V, of volume 3/I.

thereby expressing a 'principle of information conservation'. However, the form of information is changed. In the transform domain, invariant registration of visual input can be derived to serve as the basic sets required for pattern recognition and subsequent storage.

With respect to the obtaining complex circuitry, the assumption that the essential patterns of neuronal connectivity are established under genetic control, and thereby to a large extent prefunctional, seems supported by strong evidence (cf. e.g. HUBEL and WIESEL, 1963b, 1974c). On the other hand, there is no doubt that, in the maturing brain, and particularly at early critical periods, environmental stimuli strongly influence the development of functional capacities as well as of neuronal connectivity. LOCKE's concept of a '*tabula rasa*', while to some extent valid, must be qualified by the statement that the receptivity or capacity of said '*tabula*' is not initially or suddenly given, but develops concomitantly with its incipient use.

BERGER (1909) showed that the visual cortex of Dogs and Cats, deprived of visual experience since birth by suturing of the eyelids, remained, at the adult stage, in an incomplete state of development substantially differing from the visual cortex of concomitantly reared normal adults. The pyramidal and other neuronal elements of the operated animals were characterized by a conspicuous deficiency in the outgrowth of dendritic processes. Much later, comparable experiments were performed by others. Thus, WIESEL and HUBEL (1963, 1965a, b) reported that absence of pattern vision in one eye of the kitten leads, after a few months, to a functional disconnection of that retina from neurons in the cerebral cortex. Three months of monocular deprivation in the adult animal, by lid-closure, dit not, however, result in detectable physiologic abnormalities. HIRSCH and SPINELLI (1970) found that newborn kittens exposed only to a visual environment of either horizontal or vertical stripes and thus deprived of other visual experience exhibit a preponderance of cells in the visual cortex which are selective for the orientation seen, and an absence of units sensitive to orthogonal or oblique fields. A number of comparable observations have been reported by other authors. It seems, however, that, with respect to this overall topic, differences, *qua* early deprivation effects, may obtain in taxonomically differing Mammalian forms (cf. e.g. also HUBEL and WIESEL, 1974c, reporting on short-term experiments in the Monkey).

Concerning the functional organization of other sensory cortices, there is some evidence that, in the primary somatosensory area, there are neurons activated by only one of several types of stimuli, such as e.g. movement of hairs, pressure upon the skin, or mechanical deformation of deep tissues (cf. e.g. MOUNTCASTLE, 1957). The type-specific elements may be related to a given peripheral receptive field, and the cortical columns or cylinders may respond to, or further process, signals related to only one type of stimulus, in a manner roughly comparable to the mechanisms obtaining in the visual cortex.

The sinus hairs (vibrissae) of the mouse were found to project upon '*barrels*' (cf. above, section 3, p. 234) in layer IV of the contralateral primary somatosensory cortex. Injury to selected individual vibrissae in the newborn mouse was reported to result in subsequent absence of the corresponding barrels (cf. e.g. WOOLSEY and VAN DER LOOS, 1970; VAN DER LOOS and WOOLSEY, 1973).

A stimulus moving across the skin is said to elicit no response in the cells of layer IV of the primary somatosensory cortex (S_1) but selectively to elicit neurons in layers III and V (WHITSEL *et al.*, 1972). Some sort of analogy to so-called complex cell responses in the visual cortex has been suggested. Again, single cells in the secondary somatosensory area (s_2)[118] are said to respond upon multiple joint movements in one or more limbs, in contrast with the spatially more restricted and type-specific cell columns in S_2 (DUFFY and BURCHFIEL, 1971). Both excitatory and inhibitory convergence seem to obtain, suggesting an organizational hierarchy within the somatosensory system likewise roughly analogous to lower-order hypercomplex neuronal behavior in the visual system.

As regards another aspect of somatosensory signal processing, regional localization with two point discrimination, Figure 121 C illustrates a hypothetical pattern of input distribution through neuronal channels from skin to cortex. The resulting cortical so-called modal excitation fields are presumed to provide the information for the perception of two closely adjacent points as two, despite the multiple factors resulting in a merging of impulse streams originating from the adjacent stimulated points.[119]

---

[118] As regards the assumed locations of S_1 and S_2 in the Mammalian neopallium cf. e.g. Figures 241 D, E of chapter XIII, volume 5/I.

[119] It is of interest that the minimal distance between two stimulated spots required for two point discrimination in the esthesiometer (or compasses) test not only differs with respect to the regions of the body surface, but also varies, within certain limits, in accordance with diverse physiologic and 'psychologic' states of the tested individual (cf. footnote 155, p. 368 of section 6, chapter V of volume 3/1).

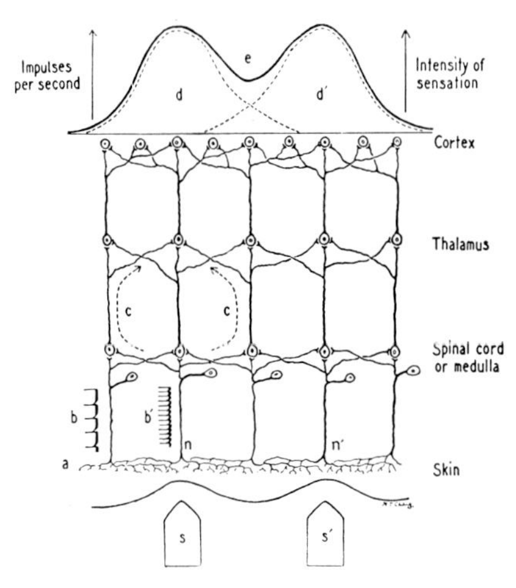

Figure 121 C. Diagram illustrating possible excitation spread resulting in localization and two-point discrimination (combined after concepts of WEBER and other authors, from RUCH, 1961). a: plexus of interlocking sensory nerve terminals; b, b′: rate of discharge of neuron stimulated at periphery respectively center of its terminal arborization field; c: tendency of excitation to spread at first synaptic level; d, d′: frequency of corticopetal impulses arriving at cortical cells when due to s and s′, respectively; e: summed activity pattern (d, d′, e are termed 'modal excitation fields'); n: peripheral (afferent) neurite of spinal ganglion cell; s, s′: points of esthesiometer (compass) used in determining the two-point threshold.

It is of importance to realize that, in Man, spatial configuration of external events is only registered by two sensory cortical regions with topological projection of the peripheral sensory fields, namely by the visual areas and the somatosensory areas, and such spatial configuration seems also encoded in the somatomotor regions.

This configurated spatial representation is, as it were, contralaterally projected upon the cortex by the decussating afferent channels, correlated with a decussation of the efferent one (pyramidal tract). The still poorly understood principle of decussating tracts was dealt with on pp. 283–285, section 12, chapter IV of volume 2.

In addition to this right-left transposition, the body as well as the retinal image are projected upside down. Thus, the inverted retinal picture remains inverted in its cortical representation, corresponding to that of the somatosensory and somatomotor projection. It is of interest that, by a rotation of 180° around a midsagittal body axis, the projection of the external world upon the cortex would again be brought into correspondence with the original orientation of body and its environment.

The cortex, presumably in cooperation with subcortical grisea, including the diencephalon, doubtless registers a scheme of body and environment, in accordance with HEAD (1926) who assumed a relatively plastic three-dimensional map of body and referred external events molded by tactile, postural and visual data (cf. also SCHILDER, 1923, 1935). This registration can be assumed to involve 'ordinative' local sign-encoding values, as e.g. discussed, with reference to some previous theories, by RENSCH (1968).

The plasticity of this body and environmental 'scheme' is evidenced by the well-known experiment of STRATTON (1897, 1902), EWERT (1957), and SNYDER and PRONKO (1952). These experiments disclosed that effective adjustment and familiarity with the environment, despite inverted visual fields, could be achieved. Temporary confusion resulted upon restoration of normal vision, but this soon subsided (cf. also SMITH, 1960).

With regard to the registration of visual spatial configuration, the *saccadic eye movements*, causing a continuous shifting of the excitation pattern within the visual cortex, seem to be of particular importance. In tactile registration, however, although motion (e.g. in palpation by the hands) enhances the registration of spatial configuration, this latter is also quite clearly registered by the immobilized body or hand in contact with external objects.

With respect to the neural mechanisms in other primary sensory areas (auditory, visceral including taste, and olfactory) there are less detailed and significant data available than for visual and somesthetic cortices. An important 'organizing parameter' in the auditory system is doubtless frequency, and a 'tonotopic localization' in the first and second auditory area seems well established, as pointed out in section 10 of chapter XIII in volume 5/I (cf. also Figs. 243A, B, vol. 5/I). Yet, EVANS *et al.* (1965), using extracellular microelectrodes, failed to find convincing evidence of tonotopic organization in the auditory cortex of the unrestrained, unanesthetized Cat.

Although the auditory projection area receives bilateral input, the contralateral projection may predominate. Differences in the bilateral input are presumably of significance for the spatial localization of sounds (e.g. right, left, anterior or posterior to body and, more vaguely, distance).

As regards the auditory cortex of a Bat, SUGA and JEN (1976) recently reported that a disproportionally large part of the auditory cortex is occupied by neurons processing the predominant components in the orientation signal and *Doppler-shifted* echoes. The orientation signals of that Bat, a Pteronotus species, is said to consist of long constant-frequency and short frequency-modulated components.

Concerning the special visceral (taste), general visceral, and olfactory primary (and secondary) cortices, significant data beyond those suggesting the general location of these areas, dealt with in chapter XIII of volume 5/I, are not available.[120]

Burns (1968) summarizes the relevant problems pertaining to the mechanisms of primary sensory cortices as follows:

(a) What sort of input to cortical neurons most readily excites them?

(b) How is the external world represented in sensory cortex? By what sorts of codes of neural behavior?

(c) What is transmitted from the neurons of sensory cortex, for use by the rest of the brain?

It seems evident that events in the postulated physical world, that is to say R-events (reduced events) are transduced by sense organs or sensory endings of nerves, and thereby transformed into N-events, namely coded signals superimposed upon the biological R-events of neuronal elements. The signals transmitted by nerve fibers consist of discrete pulses (or bursts) of action potentials transmitted at variable speeds essentially related to degree of myelinization respectively fiber diameter. The all-or-nothing relationship between magnitude of input and neuronal discharge provides a barrier to the weak stimulus. The strength of a supraliminal stimulus is apparently encoded in terms of pulse frequency. In addition, sensory structures or sensory endings are particularly 'adapted' by a low threshold to 'specific' or 'adequate' stimuli, although also responding to unspecific stimuli above a certain strength. Afferent channels concerned with particular types of R-events differ in respect to their central connections.

If all signals consist of discrete pulses and are thus essentially alike as regards their structure, then apart from the specific adaptation of the receptors, all specificities would be reduced to the spatial arrangement respectively pattern of the network, but differences in frequency, pat-

[120] With regard to the Human brain, a short discussion of these inconclusive data is also included in Daly's (1975) chapter on cerebral localization of Baker's Clinical Neurology.

terns of sequences, and differences in conduction rate could account for coding possibilities within a given spatially arranged system. Differences in magnitude or amplitude of the signal pulse might likewise obtain between diverse conducting fibers of a channel and would complicate an assumed simple discrete signal pattern.[121] To this, however, must be added biochemical differences of excitatory and inhibitory transmitter substances at the synaptic connections, and biochemical differences between neuronal elements.

The total picture is thus by no means sufficiently clarified. BURNS (1968) justly points out that because so little is reliably known of cortical function, generalizations are hard to formulate and have no greater value than working hypotheses. He stresses that the neural codes by which information is transmitted are very poorly understood. The difficulties of breaking the code, or rather codes, obtaining e.g. in the visual system are the same whether the retinal, the cortical level, or the transmission of signals from the visual cortex to the rest of the brain are investigated. In addition, as BURNS likewise points out, there appear to be relevant differences of information processing in different Vertebrate (respectively also Invertebrate) species.

ADRIAN (1947) nevertheless suggested that the nerve net carries out its work on a simple and uniform plan; he assumed that the activity of the brain from moment to moment should be capable of definition in terms of rapidly fluctuating spatio-temporal excitation patterns, built up everywhere of essentially identical elements. As v. BONIN (1960) remarked, electrophysiologists who failed to detect any qualitative differences in the impulse transmission tend to think of nervous impulses being merely '*pips*', the differences between various neurons concerning nothing but the size (and sequences) of the '*pip*', the amount of current necessary to evoke it, and the time necessary for a restitution of the cell. Although, according to v. BONIN, it is not quite clear in what else the cells actually differ, it appears conceivable that differences in neuronal cell physiology may represent relevant parameters whose numerical or code values still remain obscure. Among important differences are doubtless the diverse types of transmitter substances and their actions upon the postsynaptic membranes of different neurons.

Be that as it may, and regardless of the normally relevant specific

---

[121] Cf. the comments in the author's monograph 'Brain and Consciousness' (K., 1957, pp. 211–212 *et passim*).

nature of the receptor (e.g. eye, ear, etc.), the type of sensation experienced doubtless depends on the connections which an afferent channel makes in the brain.

Reverting now from the localizatory aspect of the cerebral cortex, as particularly displayed by its motor and sensory projection areas, to the '*holistic*' viewpoint considering the cortex as a population of interconnected neurons representing a continuum throughout which waves of activity are spreading, it seems appropriate to mention the views elaborated by BURNS (1958, 1968) on the basis of his experimental studies.

BURNS emphasizes the stochastic and 'indeterminate' behavior[122] of nerve cells and random nerve nets. Not even in the investigation of 'simple reflexes' can the relation between stimulus and response be predicted with absolute certainty.

Figure 122A illustrates what BURNS conceives as regular and random nerve nets. A nerve net is defined as a population of similar neurons which are functionally interconnected so that, when fully excitable, activity among some of them always spreads to invade the remainder. Random networks with positive feedback provide for excitation conduction in all directions, sustained by said positive feedback which include self-reexciting circuits permitting reverberation. A nerve net of this type has thus the following properties:

1. a wave front of excitation can travel in any direction through the system;
2. it will spread with the same velocity in all directions;
3. positive feedback loops are available within the system for self-reexcitation.

This formulation by BURNS must, however, be qualified by keeping in mind the restrictions imposed by inhibiting activities, e.g. by negative, inhibitory feedback, and by differences in conduction velocity displayed by different sorts of fibers.

The experiments by BURNS seem to indicate that the Mammalian cerebral cortex displays many of the properties that can be theoretically predicted for the activity of random networks (cf. e.g. Fig. 121B. This author particularly emphasizes sustained activity manifested by repeti-

---

[122] 'Indeterminate' should here, in my opinion, of course mean 'unpredictable' respectively 'probabilistic' or 'statistically determinate' or 'statistically predictable' behavior. There is no valid reason to equate unpredictability with true 'indeterminacy', that is, with non-causal events (cf. e.g. vol. 4, chapter VIII, section 2, pp. 27–30, 49–50).

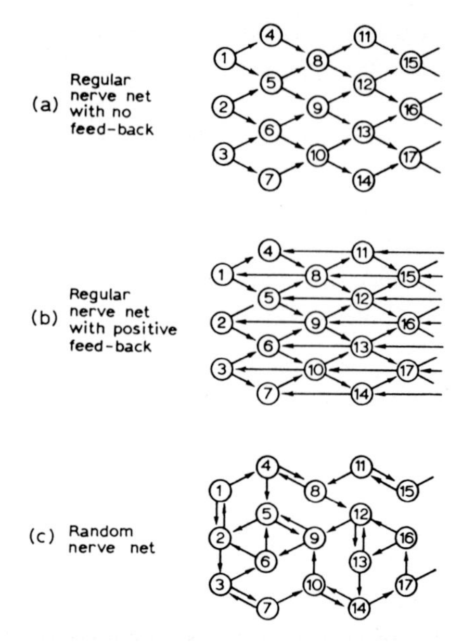

Figure 122 A. Neural networks as defined by BURNS (from BURNS, 1968). The difference in feedback between net (b) and net (c) should be noted.

tive discharges within the net in response to a single surface stimulus in an isolated slab of cortex. Two mechanisms providing for such afterdischarges are considered, namely delay paths and self-reexciting chains.[123] The former are believed to be insufficient because requiring excessive length of the thereby postulated neuronal chains for long duration afterdischarges, while self-reexciting chains of interneurons afford a satisfactory explanation. It should, however, be added that both sorts of mechanisms are not mutually exclusive and can be as-

[123] BURNS (1958, 1968) credits FORBES (1929) with the concept of self-reexciting circuits which, however, I formulated in 1927 on the basis of observable synaptic connections, particularly in the olfactory bulb. The history of the concept of closed circuits is discussed on pp. 11–13, section 3, chapter I of volume 1, where a previous rather different concept of FORBES (1922) related to reciprocal innervation was mentioned. FORBES' 1929 concept of true reverberating circuits in arrangements of functional interneurons had then not come to my attention. Cf. also volume 4, chapter VIII, section 2, p. 26, Figure 5 concerning the allegedly first elaboration of said concept by RANSON and HINSEY in 1930. The question of so-called '*autapses*' formed by recurrent axon collaterals to soma or dendrites of their own neuronal entity will be taken up further below in the discussion of logical mechanisms.

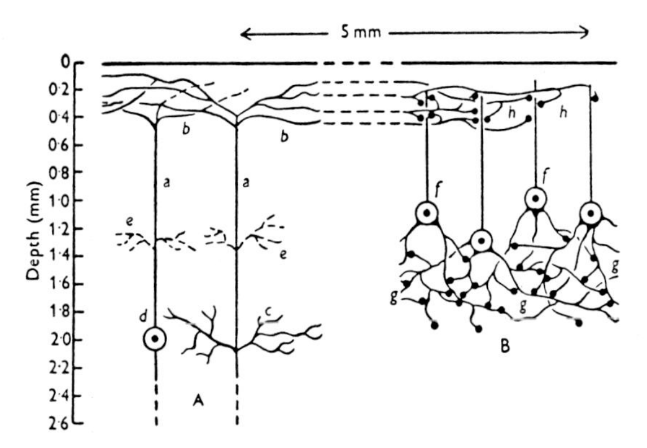

Figure 122 B. Diagrammatic summary of the conclusions reached by BURNS and GRAFSTEIN (1952) about the size and shape of neurons giving the surface negative (A) and the surface positive (B) responses of isolated cortex in the Cat to single stimuli (from BURNS, 1958). a: radial axon; b: long branching extension of a; c: 'second family of branches'; d: cell bodies at this depth; e: 'third set of branches'; f, g: cell bodies and their dendrites; h: synapses linking A with B elements. This Figure should be compared with Figure 403 E, p. 927 of volume 4, in which the spread of impulses through the Avian cortex tecti mesencephali is illustrated.

sumed to obtain. There ist, furthermore, the possibility of persistent humoral transmitters.

In addition, the spontaneous activity of cortical neurons must be considered. Spontaneous activity can be defined as the capability of a nerve cell to discharge without being triggered by synaptic events at its surface. Such spontaneous activity, either rhythmic, e.g. of pacemaker type, or random and unpredictable, can evidently be evaluated as strictly determinate, being caused by metabolic events within the neuronal elements. Events of this type, nervertheless, also depend upon external parameters relevant to the cell's biologic activities (cf. vol. 1, chapter I, section 3, pp. 17–18). Although there is convincing evidence for the presence of spontaneously active neurons in the Mammalian brain, presumably including the cerebral cortex, BURNS (1968) found that spontaneous activity is absent in slabs of neurologically isolated cerebral cortex. BURNS and GRAFSTEIN (1952) investigated the burst response spread to single stimuli in isolated cortex slabs of the Cat, considered to represent a 'random network of cells and interconnecting synapses, the network being distributed as a tangential sheath about 1 mm beneath the cortical surface'. Figure 122B illustrates in diagrammatic manner their conclusions about the site and shape of neurons giving the surface negative and the surface positive responses. As regards the rapid dispersion of activity through a population of neurons it will be recalled that this was already recognized by CAJAL (1911) and subsumed under his concept of *avalanche conduction*.[124]

In addition to the above-mentioned localizatory aspects, and to the 'random aspect' of cortical nerve nets stressed by BURNS, as dealt with in the preceding paragraphs, events based on so-called neuronal local circuits within grisea have been emphasized by various recent authors (cf. e.g. SCHMITT *et al.*, 1976). While the transmission of signal between grisea is generally assumed to be effected through axonic discharges accompanied by all-or-none spike-potentials, the 'local circuits' are supposed to involve 'graded slow potentials', transmitted without spike action potentials through peculiar 'high-sensitivity' synapses related to short axon or axonless neurons and dendritic networks, which latter include 'dendrodendritic synapses' or junctions, respectively 'reciprocal synapses'. Thus, small graded changes in potential at a dendritic locus might 'synaptically influence' electrical activity in other neurons, e.g. by 'electrotonic coupling' through 'gap junctions'. Intraneuronal and interneuronal transport of diverse substances could be functionally interwoven with the bioelectrical parameters. These concepts, suggesting that the interactions of neuronal local circuits with their presumed distinctive properties may play an important role in 'higher brain function' can be regarded as still inconclusive. Some reference to the here relevant data were given in volume 3/I, chapter V (sect. 4, p. 277: 'dendrodendritic junctions'; sect. 8, p. 639: 'slow potentials'; p. 640; 'electrotonic spread').

As regards the neuronal signal transmission by axonic discharges with spike potentials and through typical (or 'classical') synapses, it seems most likely that several particular types of patterned cortical cir-

---

[124] It should here be pointed out that, in CAJAL's concept of '*avalanche conduction*', the simile is based on a popular misconception concerning the nature of an avalanche, supposed to be started by a small rolling snow mass which gathers more and more additions in its downward motion. Any competent mountaineer knows that this is not the case. Avalanches start as a disruption within a snow cover on a steep slope, such that an entire massive snow layer breaks off *in toto* and slides downward, gathering momentum, and usually blowing up in an explosive and destructive dense snow cloud. People caught in the sliding snow layer are carried along and finally buried in the terminal pile-up. The momentum of an avalanche, apart from the steepness of the slope and the length of the fall, essentially depends on the initial size (area and volume, i.e. mass) of the snow layer which has broken off and slides downward.

cuits obtain which are highly relevant for the *logical mechanisms* of the cerebral cortex.

In his treatise on the laws of thought, George Boole (1854) outlined a system of algebra operating with non-numerical relations, properties, classes, statements, and following certain rules of logic. General properties of relations and statements, such as e.g. truth values, and grounds for reasoning can be formulated by means of an appropriate notation. Many subsequent authors, including Couturat, Frege, Hilbert, Jevons, Peano, Russell and Whitehead, Schröder, Tarski, Venn, Zermelo and others, have elaborated this system into propositional calculus and present-day symbolic logic or '*logistics*'.

Boolean algebra became of particular significance for communication engineering after Cl. E. Shannon showed in 1938 that this algebra applied to electrical circuit problems. Thus, in Shannon's symbolic relay analysis, any circuit is represented by a set of equations, the terms of the equations corresponding to the various relays and switches in the circuit. Circuit diagrams referring to on-off circuits can be expressed and even to a certain extent entirely replaced by the notations of *Boolean algebra*, and this application of symbolic logic is commonly referred to as *circuit algebra* or *calculus*.

Boolean algebra is especially suited for operations with variables of only two values, such as on-off, yes-no, true-false, with the binary digits 0 and 1, and with step-functions. Various combinations of properties and rules governing the selections from the combinations are given; the combination determined by a particular rule can be easily expressed.

Some of the fundamental operations, expressed by notations and operational terms are: 'and', 'or' (inclusive disjunction), 'or' ('or else', 'or, but not both' (exclusive disjunction), 'not', 'if ... then', 'if ..., and only if ... then', 'except', 'unless', and various others. In other words, this algebra formalizes the use of logical connectives. The notations are, unfortunately, not entirely standardized, and vary with different authors. However, some of the fundamental notations are widely adopted. In order to discuss a few theoretical models of neural networks from the viewpoint of circuit algebra and communication engineering, some elementary examples of *Boolean algebra* in simplified notation may be illustrated.

By juxtaposition or by a point, the conjunction *and* is expressed, thus ab or a · b denotes a and b. The sign v denotes inclusive disjunction: avb to mean a or b or both. The sign + denotes exclusive disjunction: a + b is thus a or b but not both. The sign ′(prime) or preferably ~ denotes negation, thus a′ or ~ a means 'not a', e.g. (a′)′ = a, or ~ (~ a) would signify 'the denial of the denial is the statement itself'. The sign ⊃ is used to express the conditional 'if ... then', thus a ⊃ b means: if a, then b. This notation is also occasionally read to mean 'a implies b'. Again, in the algebra of sets, this notation means 'b is a subset of a', or 'a contains b'. The notation ≡ expresses the biconditional 'if ..., and only if ... then', thus a ≡ b signifies a if, and only if b.

Numerous different rules can be formulated and expressed as theorems, such as associative, commutative, distributive laws, absorption, combination, relation of order, contradiction, laws of opposites, and others more.

The time-element is not necessarily significant. However, in circuit activities, it becomes essential to operate with events changing as time changes. The expressions begin, finish, in the event that, before, after, during, change, happen and others assume intrinsic importance. Such applications of *Boolean algebra* are sometimes referred to as *Boolean calculus.*

An interesting application of circuit algebra to neural networks was presented by McCulloch and Pitts (1943). However, while circuit algebra can be directly and implicitly put to use for problems concerning electric relay or switching circuits, including electronic tubes and transistors, the transmission of impulses through neural networks cannot at the present time be adequately analyzed by means of *Boolean algebra.*

McCulloch and Pitts are therefore compelled to formulate a set of postulates or assumptions which amount to a high degree of arbitrary simplification. Nevertheless, their procedure provides interesting theoretical models.

These postulates, applying to the diagrammatic networks of Figure 123 A and with minor alterations of the authors' original wording, can be enumerated as follows:

(1) Activity of the neuron follows the all or none principle.

(2) A minimal number of synaptic knobs must be excited within the period of latent addition in order to obtain a response. This minimum number is assumed to be two in the diagram, and this number is assumed to be independent of previous activity and of position of the neuron. In the diagram, a single stimulating synaptic knob is indicated by a dot.

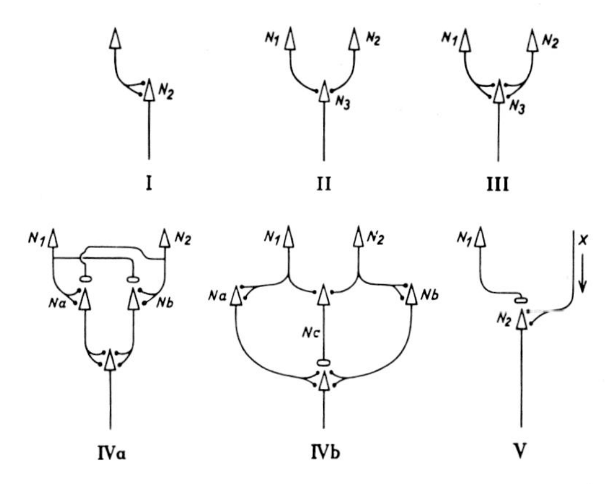

Figure 123 A. Diagrams of elementary neuronal networks expressing fundamental logical operations of *Boolean algebra* as expressed by SHANNON's adaptation to circuit algebra (adapted from McCULLOCH and PITTS, 1943, from K., 1957, 1973). Explanation in text. Two versions of circuit IV are given. Upper neuron in I is N_1.

(3) The only significant delay within the theoretical network is the synaptic delay, assumed to be uniform throughout.

(4) The activity of any inhibiting synapse absolutely prevents excitation of the neuron at that time. In the diagrams, an inhibiting synapse is indicated by a small circle or oval.

Disregarding the intricate and complex mathematical formulations, including the various theorems proffered by McCULLOCH and PITTS, a few elementary circuits, represented in Figure 123A, may be discussed.

McCULLOCH and PITTS have introduced the time element in their notation. The discharge of a neuron x at time t is equivalent to the discharge of a neuron y at time t–1 or time t–2 etc., if time is counted in terms of corresponding synaptic delays. If, instead of using *Boolean calculus* in the terminology of some authors, *Boolean algebra* is used, t may be omitted.

Circuit I is expressed and $N_2(t) \equiv N_1(t-1)$; in *Boolean algebra* $N_2 \equiv N_1$, or N_2 (discharges) if, and only if N_1 (discharges).

Circuit II is expressed as $N_3(t) \equiv N_1\ (t-1) \cdot N_2(t-1)$; simplified $N_3 \equiv N_1 \cdot N_2$, that is N_3 if N_1 and N_2. In communication engineering, this arrangement is designated as an 'and gate' with the equation $r = p \cdot q$.

Circuit III is expressed a $N_3(t) \equiv N_1(t-1) \vee N_2(t-1)$, simplified $N_3 \equiv N_1 \vee N_2$, that is N_3 if N_1 or N_2 or both. In communication engi-

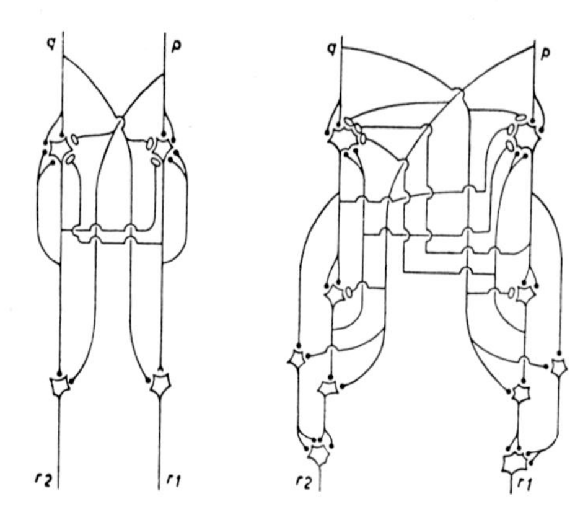

Figure 123 B. Diagrams of neuronal models of so-called flip-flop elements used in communication engineering (from K., 1957, 1973).

neering, this arrangement is known as an 'or gate', with the equation $r = p \vee q$.

Circuit IV can be expressed as $N_3(t) \equiv N_1(t-2) + N_2(t-2)$, simplified $N_3 \equiv N_1 + N_2$, that is N_3 if N_1 or else N_2, but not both.

Circuit V is expressed as $N_2(t) \sim N_1(t-1)$, assuming that N_2 is continuously fired by another source (x); simplified $N_2 \sim N_1$. This circuit and the preceding circuit have certain features in common. In communication engineering, these features are combined in so-called 'except gates' with the general equation $r = p \cdot q'$.

In the circuit of Figure 123B, I have suggested a theoretical neural model of a network corresponding to a *flip-flop element* of communication engineering (K., 1957). This type of element consists of a '*black box*' with two inputs p and q, and two outputs, r_1 and r_2. The circuit element reports on its two output lines which input line was the last one to receive a pulse, according to the formulation $r_1 = S\ (p, q)$; $r_2 = S\ (q, p)$. In other words, according to the algebra of relations, it reports temporally spaced pulses as an ordered pair. In the proposed (highly theoretical and artificial) neural solution there are two additional postulates: the mechanism must not respond if the input pulses are simultaneous, and it must be shut off and be ready for the next operation as soon as signal function has been accomplished. Moreover, the time difference between the two input pulses must not be less than one cycle, that is t to t + 1. The first impulse must be remembered by a short-term memory mechanism until the second pulse arrives at time t + x

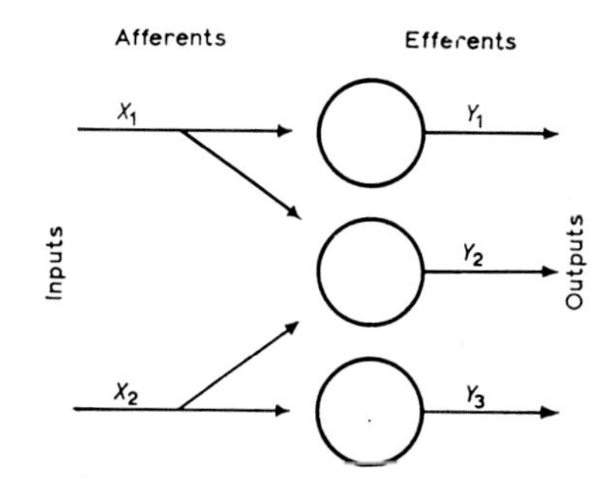

	AFFERENT PATTERNS		
	I	*II*	*III*
Behaviour of X_1	1	0	1
Behaviour of X_2	0	1	1
Causing efferent signals from . . .	Y_1 Y_2	Y_2 Y_3	Y_1, Y_2 Y_3

Figure 123 C. Diagram illustrating UTTLEY's (1954) 'classification machine' (from BURNS, 1968).

of the cycle, and the second pulse must not be registered by the memory mechanism.

As regards the postulated synaptic connection in the left diagram of Figure 123B, provided by an axon collateral running back to its cell of origin, such arrangements which could be interpreted as single neuron self-reexciting or self-inhibiting circuits were suggested to me by *Golgi impregnations* of diverse sorts of neuronal elements, including pyramidal and Purkinje cells. In his first theoretical diagrams McCULLOCH assumed single neuron self-reexciting circuits, while RASHEWSKY denied that possibility, claiming that such excitation would fall within the neuron's own refractory phase. I nevertheless still considered such single neuron mechanisms to be possible.[125] More recently VAN DER LOOS and GLAZER (1972) also found evidence for such connection, which they designated as *autapse*, a term apparently now widely accepted (cf. e.g. SHEPHERD (1974).

Another problem concerning the theoretical networks of Figures 123A and B is the question of *Dale's law or principle*,[125a] which postu-

[125] Cf. my discussion remark to a paper presented by SHOLL (1960). In his reply, SHOLL (p. 27) remarked: 'I think a single neuron self-re-exciting circuit perfectly possible. I have never been able to assure myself that this occurs, much as I would like to do'.

[125a] Cf. volume 3/I, pp. 631 f., and volume 4, p. 663.

lates that a given neuron can only release the same transmitter substance at all its synapses, thus implying some sort of biochemical specificity of the nerve cell. This is commonly interpreted to mean that a given neuron can only be excitatory or inhibitory, but not both.

Although there is evidence that this interpretation holds for some neurons, there is no definite proof that all neurons can only produce the same transmitter substance at all of their (pre)synaptic terminals. *Dale's principle*, moreover, does not apply to the effect the transmitter may exert at the postsynaptic membrane of the target cell. Thus, as mentioned on p. 632 of section 8, chapter V of volume 3/I, one and the same transmitter (acetylcholine) produced by one neuron in Aplysia, mediates excitation of one target cell, and inhibition of another one.

Still less substantiated is a so-called structural corollary of *Dale's law*, claiming that all synaptic endings made by a given neuron on other nerve cells are of the same structural type.

UTTLEY (1954) described the logical requirements of a neuronal machine designed for the classification of input signals, as illustrated by the theoretical circuit of Figure 123C. The output indicates the type of spatial patterns composed of simultaneous inputs in the manner shown by the following tabulation:

Input		Output		
X_1	X_2	Y_1	Y_2	Y_3
0	0	0	0	0
1	0	1	1	0
0	1	0	1	1
1	1	1	1	1

LANDAHL *et al.* (1943), and LANDAHL (1945) have, by abstract notations, derived a formal method for converting logical relations among the actions of neurons in a net into statistical relations among the frequencies of their impulses. In a further elaboration, LANDAHL (1945) suggests a theoretical neural mechanism which reacts to differences in different modalities of a stimulus pattern. Another such mechanism reacts to similarity, which is measured by the number of neurons in common to the two-stimulus patterns.

RASHEVSKY (1945, 1946) has presented a further development of LANDAHL's equations and discusses mechanisms providing for abstraction and for some forms of logical thinking. A mechanism for logical inferences is suggested, and equations are derived which give the probability of making an error in a reasoning consisting of a chain of syllogisms, as well as the probability of being unable to complete the chain of reasoning at all. In addition, a theory of such neural circuits is developed which provides for formal logical thinking. Moreover, a neural mechanism is indicated which provides for the conception of ordinary numbers. Included in these elaborations is a quantitative theory of the probability of erroneous reasoning and of the speed of reasoning in its relations to other psychological phenomena. The circuits postulated by RASHEVSKY perform operations corresponding to the categorical syllogisms of traditional *Aristotelian logic*, such as the figure or mode 'Darii' (All A's are B's, some C's are A's ∴ some C's are B's), and others.

The eight circuits discussed above represent models of partial processes or functions. The circuits and equations elaborated by RASHEVSKY, and performing a complete logical operation are too complex to be reviewed here.

Nevertheless, it is possible to devise more elementary theoretical network models demonstrating a complete operation. E. C. BERKELEY (1949), in his first version of '*Simon*', has originated a miniature model computer designed to demonstrate the working principles of such machines. A set of only four binary numbers from zero to three (00, 01, 10, 11), referring to the counterclockwise rotation of a line through four right angles, is handled. With these four numbers, four different operations can be performed, namely $c = a \cdot b$ (addition), $c = -a$ (negation), $p = T(a > b)$ or discrimination of 'greater than', and $c = a \cdot p + b\ (1-p)$ or selection. A second somewhat more complicated and advanced version of '*Simon*' has been subsequently evolved by that author.

Adapting the postulates of McCULLOCH and PITTS for a simplified neuronal network, and E. C. BERKELEY's elementary set of four numbers handled by '*Simon*', a model neuronal circuit may be constructed, presumed to represent crudely simplified and diagrammatic cortical mechanisms. Figure 124B shows a circuit performing the operation of negation, if, and only if, the operation is properly programmed by a corresponding program circuit. The coding for the program or routine input is likewise given in binary numbers, assumed to be 00 for addition, 01 for negation, 10 for discrimination of 'greater than', and 11 for selection. Only negation is worked out in the model; negation, in

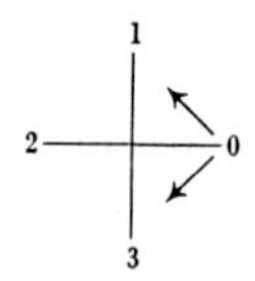

	Conventional Procedure Counterclockwise Rotation						Simonoid Negation Clockwise Rotation	
None	No R	0	00	=	00	0	No R	None
One	One R	1	01	=	11	3	One R	One
Two	Two R	2	10	=	10	2	Two R	Two
Three	Three R	3	11	=	01	1	Three R	Three

Figure 124 A. Binary numerical sequence in E. C. BERKELEY's *Simonoid negation* explaining operations performed by model circuit of Figure 124 B (from K., 1957, 1973).

relation to what might be considered a matrix of the given four numbers (Fig. 124 A), has a specific significance, and differs from negation in *Boolean algebra*. As regards the problem circuit, p and q represent the digital binary signal input, and r_1 and r_2 the output, signalling the computed information. The model thus demonstrates a complete operation depending on an *problem circuit* and on a *program or routine circuit*. These two circuit types are fundamental for many aspects of communication engineering, and likewise for presumably comparable neural mechanisms. The model cortical network of Figure 124B shows that circuits of this type could easily be provided by the structural arrangement of the neocortex (Figs. 124 C, D) and its columnar organization discussed above in section 3, p. 236 of the present chapter.

Discounting the specific details, it will also be seen that the general circuitry obtaining in various computers and control devices, as depicted in Figure 125 A, shows conspicuous similarities with certain pattern of cortical networks as e.g. displayed by *Golgi impregnations* and shown in Figure 125 C. It is evident that neuronal synapses can, to some extent be compared with automatic switches, with electronic vacuum tubes, or with transistors.

There are numerous publications concerning the similarities in the overall working principles applying to the performance of computers and related artificial devices and to the functions of the brain, respectively of the nervous system of Metazoa in general. It will here be suffi-

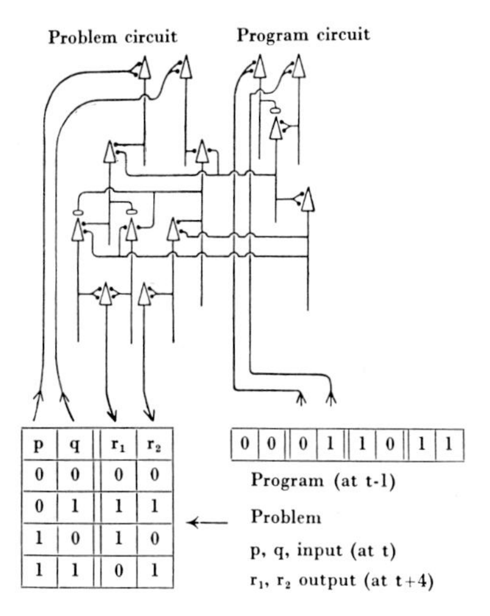

Figure 124 B. Model cortical network illustrating problem and program circuit in *Simonoid negation* explained in Figure 124 A and in text (from K., 1957, 1973).

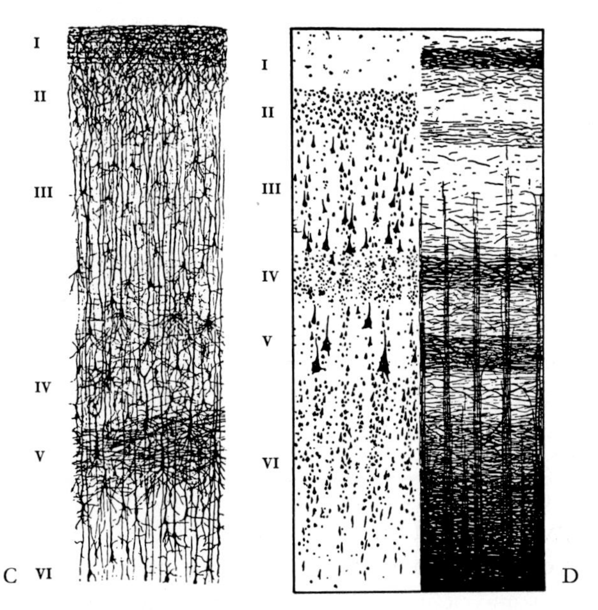

Figure 124 C, D. C: Structure of Human neocortex as revealed by *Golgi method* (postcentral gyrus of newborn, after CAJAL, layers in *Brodmann's notation)*, for comparison with preceding model network. D: Cyto- and myeloarchitecture of Human neocortex (after BRODMANN and C. and O. VOGT, for comparison with preceding model network. The columnar arrangement of cells and the corresponding radial fiber bundles, which the *Golgi picture* does not display, are clearly shown (from K., 1957, 1973).

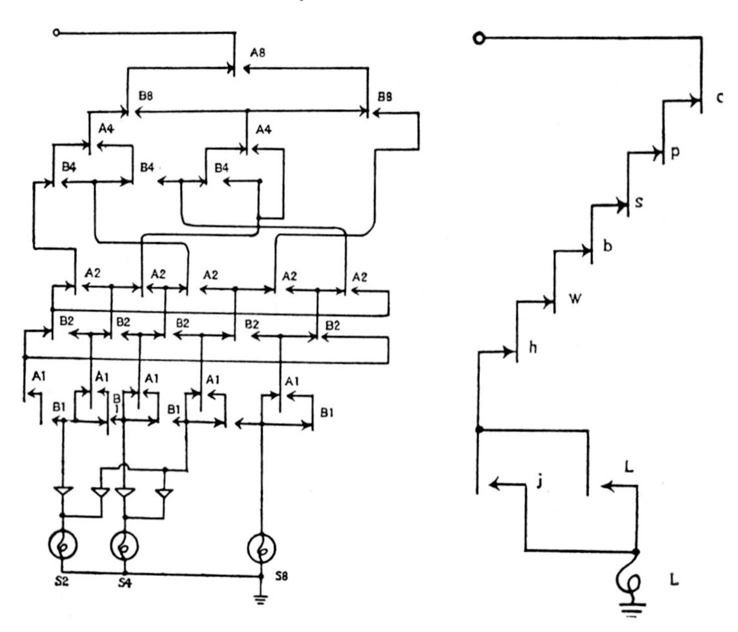

Figure 125 A. Examples of circuit patterns in computers and control mechanisms displaying similarities to cortical circuitry. The cascade-like sequence of *Boolean* 'if – then' (implication) connections should be compared with that obtaining in Figure 125 B. The details indicated by the lettering are not relevant in the present context (after E. C. BERKELEY, 1959, from K., 1965a).

cient to point out the contributions by E. C. BERKELEY (1949, 1959), CRAIK (1943), GEORGE (1962), KÜPFMÜLLER (1958), v. NEUMANN (1958), STEINBUCH (1961), and WIENER (1948). This latter author re-introduced the old term '*cybernetics*' for control and communication in the animal and the machine. Further comments on this topic can also be found in several writings by the present author (K., 1957, 1965a, b, 1966, 1973).

On the other hand, it also seems evident that the oversimplified theoretical networks based on the postulates of MCCULLOCH and PITTS represent very crude models disregarding significant differences between electric and neuronal circuits. These latter are complicated by numerous biological parameters, such e.g. as diverse conduction velocities, diversity of synaptic transmission by a variety of transmitter substances, and the very large number of excitatory and inhibitory synaptic endings on a single nerve cell.

It seems, however, not improbable that circuit processes compara-

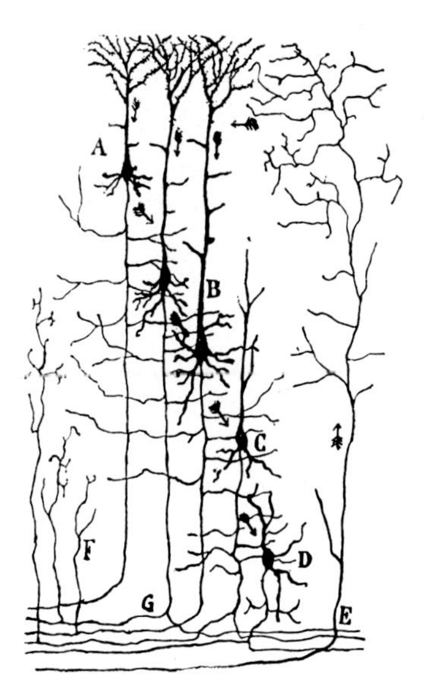

Figure 125 B. Semidiagrammatic drawing, by CAJAL, of neocortical structure based on Golgi pictures. The oblique arrows indicate a sequence of 'if – then' circuits in BOOLE's and SHANNON's sense, similar to those shown in Figure 125 A, but not identified as such by CAJAL (after CAJAL, from JOHNSTON, 1906). A: small pyramidal cell; B: large pyramidal cell; C, D: polymorph cells; E: afferent fiber reaching upper cortical layers; F: collaterals of afferent fibers; G: axon of pyramidal cell bifurcating in white substance.

ble to those performed by the theoretical networks may be carried out, in a probabilistic manner, by the complex and highly redundant cortical network, representing a very large system of connecting lines between 'unreliable' elements. In the rigid theoretical networks, a single minor defect or disturbance would render the circuit inoperative. In the '*uncertain*' nervous system, however, the statistical orderliness displayed by the very large number of components in compound stable and ultrastable systems could provide the basis for fairly efficient operations despite unreliable components and considerable disturbing interference.[126]

Reverting once more to the reflex concept which represents the unit reaction in nervous integration (SHERRINGTON, 1906) and can be extended to include any end-effect reached through the mediation of a

[126] Cf. also the comments in sections 4 and 5 *et passim* of chapter I, volume 1.

nerve fiber following an initiating neural reaction, the term reflex[127] may be applied to any neuronal circuit activity, and can thus subsume behavior of central, e.g. cortical grisea, even if their neural discharges should be initiated by intrinsic or so-called 'spontaneous' neural activity.

It is here of interest that v. MONAKOW (1911) considered 'reflexes' as the basis of all nervous functions, such that the more complex activities differ from the simpler ones by their much more complicated temporal structure or by their rhythm. The richer a nervous function is, in the more extended manner polymorph components enter into its structure, and the larger a circle of nervous discharges has to be assumed. The cited author particularly emphasized that the cortex participates in reflexes and furnishes for them particular specialized components. GOLTZ (1888), referring to his experimental work on the Mammalian cerebral hemispheres, stated that he had ceased to make a distinction between 'voluntary' and reflex movements. The psychiatrists BERGER (1921) and KRETSCHMER (1926) used the designations '*psychoreflex*' respectively '*psychical reflex*' for certain types of conscious motor reactions. CAJAL (1909), with reference to concepts of FOREL, justly considered activities involving triggered engrams as reflexes of higher order.[128] Quite evidently, there obtains an hierarchical order of neural activities, all of which are based on the reflex principle, and can be conceived as strictly determinate and 'mechanistic' in the wider sense (i.e. causal), despite uncertainty and unpredictability. For clinical purposes, however, as pointed out on p. 25 of volume 4, it becomes convenient to distinguish automatic (CAJAL) or involuntary reflexes from 'voluntary' activities.

Among authors disagreeing with this generalized evaluation of reflexes, one might quote ELLIOTT (1969), who states: 'Reflex responses have certain definite qualities: They are built into the nervous system as racial memories, not learned – that is, developed by individual experience; they can vary in strength and direction as needed, unlike a mechanical toy, but the range of variation is narrow; and they are com-

---

[127] Cf. section 2, chapter I of volume 1, and section 2, pp. 18–49, chapter VIII of volume 4. Although UNZER (1771) was, as mentioned on p. 18 of volume 4, one of the first to use the term reflex in neurology, THOMAS WILLIS (1621–1675) should be credited, according to SHERRINGTON (1951), with the designation of 'automatic nervous actions' as being 'reflex'.

[128] *C'est donc bien un réflexe, mais fort compliqué et à longue échéance, ayant son substratum probable dans l'ensemble des neurones d'association de l'écorce cérébrale.'* CAJAL (1909, p. 19) designates this as '*réflexe, d'un ordre plus élevé que le réflexe automatique à qui il commande.*'

pletely involuntary though subject to suppression by voluntary brain levels.' The neurons of a reflex are said to be always confined to the segmental nervous system, namely spinal cord and brain stem. 'Any path that runs through higher levels of the brain is not a reflex arc.'

The cited author also objects to the fact that 'transferred response and engram-building may seem to reduce the mind to mechanics' and claims that transferred responses have been misleadingly labelled conditioned reflexes. The inference that, since reflexes are mechanical, cortical thought is mechanical too, ELLIOTT dismisses with the comment that 'this is pure belief, an opinion held without proof, since calling cortical thought a reflex does not make it so, and calling reflexes mechanical is equally presumptuous'.

Although ELLIOTT's (1969) publication, which brings an interesting generalized account of the evolution of the Human brain, contains numerous shrewd remarks, I cannot agree with the quoted comments nor with various other views of that author.

Quite naive, moreover, are the views of ECCLES (1970), who assumes '*free will*' residing as a little ghost in the cerebral cortex, directing, like a railroad-switchman, the flow of neuronal activity.[129] ECCLES claims that 'the denial of free will and the advocacy of a universal determinism have been asserted within the scientific framework both of a primitive type of reflexology as an epitome of brain performance and of the now-discredited nineteenth-century deterministic physics'.[130]

ECCLES refers to what he calls the extraordinary doctrine that all behavior is determined solely by inheritance and conditioning. He then adds: 'It is unfortunate that the advocates of this doctrine have been blind to the fact that its logical consequences make its assertion meaningless; for their act of assertion should be recognized by them as

---

[129] Cf. chapter V, section 3, pp. 231–232 of volume 3/I.

[130] It will be recalled that the two main founders of 20th century physics, MAX PLANCK *(quantum)* and ALBERT EINSTEIN *(relativity)* fully upheld determinism and causation. Causality (determinism) is merely one aspect of the principle of sufficient reason which is a necessity of logical thought but remains unprovable since it represents the basis of proof. One cannot formulate a proof for the requirement of a proof. Thus, the 'logical analyses' that even deterministic physics would 'not render untenable our belief in freedom of the will', as cited by ECCLES (1970), cannot be taken seriously. Quite evidently, one may hold on to any kind of belief supported by emotional metalogical reasoning. The problem of determinism and causality is dealt with *in extenso* by the present author in the monograph 'Mind and Matter. An appraisal of their significance for neurologic theory' (K., 1961).

merely the result of a prior conditioning thus signaling merely the effectiveness of this conditioning!' The logical weakness of ECCLES' reasoning in failing to detect a *non sequitur* is evident: the effectiveness of a conditioning has no bearing whatsoever on truth or falsity respectively probability or improbability of the determined or conditioned assertion. One may be effectively conditioned to understand and recite the multiplication table and to work with logarithms, or, on the other hand, to believe in werewolves, goblins, or religious mythology. Thus, a computer, working on the strictly determinate mechanistic application of *Boolean algebra*, can indicate the logical truth value, i.e. truth or falsity of a proposition.

SHERRINGTON (1951, p. 182) justly stated that 'If "free-will" means a series of events in which at some point the succeeding is not conditioned by reaction with the preceding, such an anomaly in the brain's series of events is scientifically unthinkable'. From the practical standpoint of course, as SHERRINGTON (1951, p. 163) rightly comments, 'the important thing is less that man's will should be free than that man should think that he is free'.

The introspective, that is to say mental (conscious) experience of 'free will' respectively 'will' in general, as pointed out in volume 4, p. 20 of this series, pertains to the consciousness modality of affectivity and emotion, whose neural correlate seems to be provided by the mechanisms of limbic and prefrontal lobe circuits. Concerning the still poorly elucidated functional problems and detailed aspects of circuitry thereby involved, no further comments can here be given which would add to the overall description and evaluation of these circuits in section 10 of chapter XIII in volume 5/I (Figs. 255 A–E, vol. 5/I). It could, nevertheless, be said that, although still rather vague, the concept of these modulating circuits is far better supported by evidence than LUGARO's (1899) theory on the localization of 'intellectual and emotional processes', sceptically discussed by CAJAL (1911). LUGARO assumed that the 'affective phenomenon' was elaborated within the nerve cells, while the 'intellectual phenomenon' is produced between them, namely at the level of the articulation between afferent fiber endings and dendrites or body of pyramidal cells.

With respect to the cortical mechanisms involved in sleep and dreaming (paradoxical sleep) there is some recent evidence that biochemical factors, such as concentration of dopamine, e.g. in the hippocampus, may play a certain role. KOVAČEVIČ and RADULOVAČKI (1976) reported increase in the metabolism of 5-hydroxytryptamine (serotonin) and increase of dopamine concentration in the hippocampus of Cats during slow-wave sleep, possibly related to the subsequent appearance of paradoxical sleep. The metabolism of dopamine, on the other hand, is said to decrease in striatum and thalamus during slow-wave sleep. The hippocampal circuit presumably affects, in a still poorly understood manner, the sleep-wakefulness cycle, which also seems to depend on activities of the centrencephalic grisea and the reticular formation (cf. e.g. K., 1972).

Memory, which involves storage and retrieval of information, can be considered to represent a highly important function of the Mammalian cerebral cortex. General aspects of memory, as formulated in the basic elaborations by HERING (1870, 1905) and SEMON (1904, 1909, 1920) were dealt with in section 6, chapter I of volume 1. It will be recalled that SEMON coined the now firmly established term *engram* for stored information in living structures. In his phraseology, *engraphy* designates the process of storage, and *ecphory* the engram's retrieval. SEMON used the designation *mneme* for the capacity of living protoplasm to form engrams resulting in ecphory. He distinguished, in particular, genetic memory, and central nervous memory.

As regards specific neuronal aspects of association and memory, CAJAL (1911) refers to the hypothesis of TANZI (1893) who assumed that a nervous stream which passes more frequently over a sequence of neurons will provoke in this pathway a more active nutrition and therefore an hypertrophy just as in well exercised muscles. This hypertrophy will lead to a lengthening of cellular processes, reducing the distance which separates the articulated surfaces. Therefore, exercise which essentially tends to decrease the intervals of articulation, will increase the functional power of neurons. CAJAL (1911), who quotes TANZI's hypothesis with approval, supplements it with his own, which postulates the creation of new connections by growth of more dendritic and axonal ramifications. This is said to be the primordial condition, although not the only one, for cerebral capacity and organic memory. As additional conditions, CAJAL mentions number of neurons and undefined 'other factors'.

GOLDSCHEIDER (1906), who likewise was concerned with the cortical mechanisms of association and memory, assumed molecular changes to occur at nodal interconnections *(Knotenpunktlinien)* between simultaneously activated nerve cells, facilitating subsequent activation within the network (Fig. 125 C). He admitted, however, in the discussion following the presentation of his paper, that nothing certain

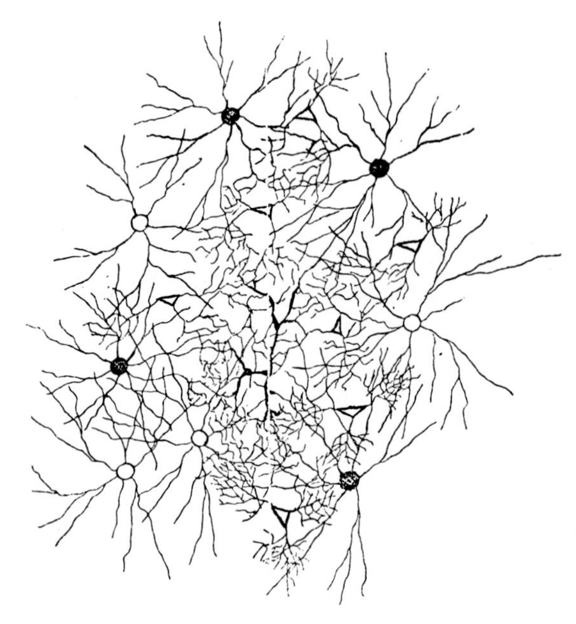

Figure 125 C. So-called 'nodal line' *('Knotenpunktlinie')*, indicated by thickening, between four nerve cells (darkened circles), stimulated by input from periphery (from GOLDSCHEIDER, 1906).

could be said about the type of interconnections between nerve fibers, and he did not refer to the concept of synaptic junctions.[131]

HERRICK (1926), in discussing the conditioned reflexes discovered and studied by PAVLOV (1923, 1927), presented an instructive diagram (Fig. 126 A) illustrating possible neuronal connections, within a 'correlation center', between the reflex arcs of an auditory and of a salivatory reflex, providing a plausible explanation for the conditioned salivatory reflex upon the ringing of a bell.[132]

---

[131] This concept was introduced by SHERRINGTON about 1897 in a textbook of Physiology in cooperation with FOSTER. It can be seen that it was unknown to TANZI in 1893. CAJAL (1909, 1911), who, in his important treatise, only perfunctorily mentions SHERRINGTON, does not use the the term 'synapse' for his description of interneuronal connections, nor did he conceive their significance in SHERRINGTON's sense, since he adhered to his own theory of 'dynamic polarization', discussed on pp. 513–514, section 7, chapter V of volume 3/I. GOLDSCHEIDER (1906) does not refer to SEMON, TANZI, nor CAJAL, but stresses views expressed by EBBINGHAUS, HERING, and VERWORN.

[132] Cf. pp. 31–33 in section 6, chapter I of volume 1.

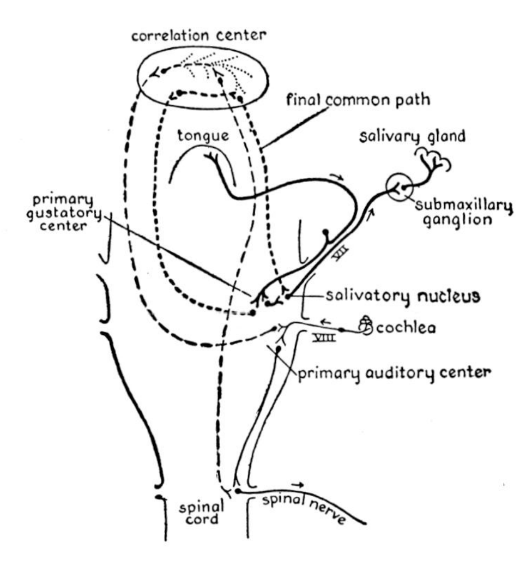

Figure 126 A. Simplified diagram explaining the conditioned salivatory reflex upon ringing of bell (from HERRICK, 1926). Primary unconditioned salivatory reflex in heavy continuous lines; primary auditory head-turning reflex in continuous lines; secondary loops to correlation center and within this latter in broken lines.

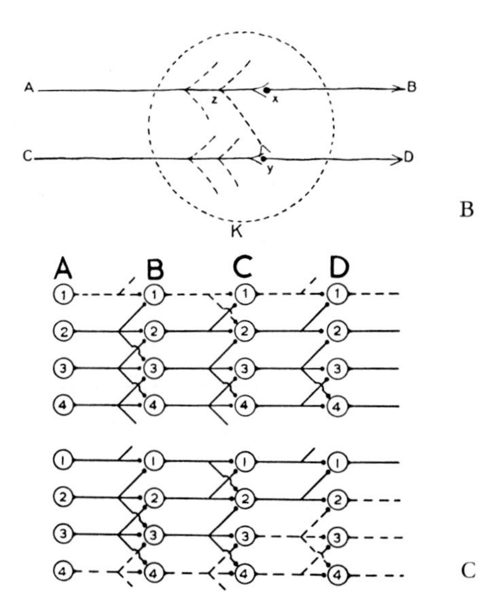

Figure 126 B, C. B: Simplified diagram indicating circuitry related to conditioned reflex and engraphy in general (adapted after Fig. 126 A of HERRICK, from K., 1927). C: ECCLES' diagram of 'the same neuronal network to show different response to a changed input' (from ECCLES, 1953). K in B: 'correlation center'.

Since the conditioned reflex can be regarded as a relatively simple manifestation of association, learning, and memory, I attempted to expand the concepts of PAVLOV and HERRICK into a generalized theory of long-term[133] memory (K., 1927, pp. 333–338).

It was therein assumed that an engram represented a patterned association of neurons, distributed as a network throughout an extensive region, and produced by an enduring modification of synapses, facilitating transmission. The discharge of such a patterned net of neurons could be triggered by an appropriate fractional input into its circuitry.

This was illustrated by the diagram of Figure 126B, adapted from HERRICK, to whom I gave due credit. Thus, if the circuit A–B is activated, one of the numerous collateral impulses reaches a synapse at y in the circuit C–D, but is here not sufficient to activate that circuit. If, however, both circuits are simultaneously activated, the synaptic threshold for the collateral zy becomes lowered, and an impulse transmission zy results. After repeated simultaneous activations an enduring change of the synapse zy results, such that an impulse merely coming from A becomes now sufficient for a discharge over D. This corresponds to SEMON's postulate that '*alle gleichzeitigen Erregungen innerhalb eines Organismus bilden einen zusammenhängenden Erregungskomplex*, *der als solcher engraphisch wirkt*'.

It seems likely that the neuronal elements in the central nervous system, respectively their synaptic connections, display considerable differences with respect to their plasticity or capacity to produce engraphic changes. Although some simple types of conditional reflexes have been demonstrated in the spinal cord (FRANZISKET, 1951, and other authors), and relevant engrams seem to be stored in the tectum opticum of Fishes (RENSCH and RAHMANN, 1966; RENSCH *et al.*, 1968), it is likely that certain neuronal elements of the Mammalian cerebral cortex possess the highest degree of engraphic capacity (K., 1927, p. 336). It appears probable that the ecphory of a single engram represents the simultaneous activity of a whole system of interconnected neurons, which are distributed upon different and distant *(auseinanderliegende)* regions of the cerebral cortex (K., 1927, p. 324).

In higher thought processes there obtains apparently the activity of complexly combined neuronal groups, distributed upon the secondary regions adjacent to several separate sensory regions, such that the

[133] As regards memory, relatively enduring long-term memory must be distinguished from rapidly vanishing or transitory short-term memory, to be dealt with further below.

fibers of these groups extend in net-like fashion over wide portions of cerebral cortex and subcortical white matter (K., 1927, p. 337).

There are thus combined, e.g. optic-acoustic-tactile engrams, resulting in further patterned combinations of the acquired engram stock, such that subsequent ecphories produce new engram complexes, as particularly stressed by FOREL (1922), who elaborated on SEMON's concepts, but did not deal with the specific neuronal mechanisms of engraphy. FOREL, however, emphasized the intrinsic relationship of engraphy to the psychologic process of association, that is to say, to the correlation of thoughts and other mental events.

As regards my own interpretation of the aforementioned patterned, respectively encoding neuronal network activities, I stated that '*diese Vorgänge stellen die höheren Gedächtnisfunktionen, Begriffsbildung, Sprachfunktionen und die höheren Denkvorgänge dar. Wenn auch hier wieder angenommen werden darf, dass jedem Begriff, überhaupt jeder speziellen Gedächtnis- oder Denkleistung die Tätigkeit von irgendwie gesetzmässig lokalisierten Neuronengruppen zugrunde liegen muss, so ist doch diese Lokalisation nicht so aufzufassen, als ob die betreffenden Leistungen an umschriebene Rindenfelder gebunden wären*' (K., 1927, p. 337).[134]

This would well agree with the fact that extensive cortical damage may not eliminate memory functions: so long as the essential pattern of a given combination is allowed to remain, a substantial number of nerve cells and fibers may be destroyed without affecting its significant code value. Thus, a given configuration shot through with sieve-like holes may still keep its recognizable features. This resistance to damage, moreover, is enhanced by the doubtless obtaining considerable redundancy.

It is also most likely that one and the same neuron may participate, in accordance with the different synaptic activations, as link in a variety of different engram patterns. In other words, what remembers is not the individual neuron, but the patterned network. Individual neurons, however, may play an important biochemical role *qua* synaptic changes.

The very large number of neurons in the Human neocortex and parahippocampal cortex, the much greater number of possible synaptic

[134] It is of interest that, in his glossary, ELLIOTT (1969) defines engram as 'a complex system of cortical neurons, developed by experience, and corresponding to a concept or idea'. The cited author, however, claims that this definition contains his original ideas, for which he takes responsibility.

connections between these neurons, and the still much greater number of different states which the system may display at a given time, approximately 2^x, where x is the number of neuronal elements (cf. vol. 3/I, pp. 711–712), provide a number of combinations and permutations exceeding all known astronomical orders of magnitude. Such undefined number is more than sufficient to encode a practically unlimited variety of memories and thoughts.

It will be recalled that the number of neurons in the Human cerebral cortex is estimated at between 10^9 and 10^{10}. PAKKENBERG (1966), whose estimate amounts to 2.6×10^9, brings an interesting review of the various estimates by previous authors, e.g. DONALDSON 1.2×10^9, THOMPSON 9.3×10^9, BERGER 5.5×10^9, v. ECONOMO and KOSKINAS 14×10^9.

Although the concept of enduring synaptic changes leading to specific excitatory efficiency represents the most plausible explanation for memory and association, definite experimental proofs are most difficult to obtain and have not yet been provided.[135]

As regards the postulated synaptic changes, one may wonder whether they are presynaptic or postsynaptic. The former could involve local, facilitating factors involving the regular transmitter substances by the addition of chemical groups, or membrane changes of the synaptic vesicles, or of the presynaptic membrane, or the formation of substances within the cytoplasm of the synaptic knobs. Postsynaptic changes would involve the postsynaptic membrane pertaining to surface of dendrites or perikaryon.

In the generalized memory theory elaborated by the present author, engrams were assumed to result from still unknown chemical alterations or deposits of specific substances at numerous synapses along the pathway of discharges through a complex and extended circuit of a particular pattern, in adaptation of concepts propounded by ROBERTSON (1912, 1913, 1914), who had published an interesting series of papers on the chemical dynamics of the central nervous system. ROBERTSON reached the conclusion that certain chemical processes related to

[135] The present author's comment of about 50 years ago remains still valid. '*Die durchaus hypothetische Natur aller auf den bisherigen Forschungsergebnissen beruhenden Auffassungen über das Wesen der Engraphie sei zum Schluss noch einmal besonders betont. Immerhin hat es aber den Anschein, als ob auch bei diesen Vorgängen, wie beim gesamten Verhalten des Neurons, die Synaps eine sehr wichtige, um nicht zu sagen die wichtigste Rolle spielt*' (K., 1927, p. 337). ELLIOTT (1969, p. 35) remarks: 'Memory is perhaps the most interesting gift of the synapse'.

central nervous activities are of *autocatalytic* type, whereby one of the resulting chemical products catalyzes the process. He assumed furthermore that the fading of a long-term memory trace is due to the disappearance of a substance laid down within a colloidal medium. This substance forming the memory trace is supposed to be washed out by the circulating fluids of the nervous parenchyma. The process of forgetting is believed to be not particulate but continuous. It is, of course, exceedingly difficult to obtain numerical data on processes of this type. Nevertheless, it appears possible that the resulting curves are closely related to the *die-away curve*, in which the rate of decrease remains proportional to the magnitude of what is decreasing, as expressed by the *die-away factor* e^{-at}.

Subsequently to the generalized theory of memory presented in the '*Vorlesungen*' (K., 1927) which assumed net-like patterned engrams to result from the enhanced synaptic functions of their neuronal linkages, other authors (e.g. ECCLES, 1953, 1970; HEBB, 1949; LASHLEY, 1950), without reference to that already long ago proposed theory, have elaborated the same concept, often in almost identical words.[136] Thus, ECCLES (1970, p. 41) states 'that a long term memory must depend on enduring increase in the synaptic efficiency that has been built up in a specific neuronal pattern'. He also stresses, as in my old theory, 'the important property that any one nerve cell can participate in a large number of separate neuronal patterns. Its response can belong to this or that pattern according to the ensemble of neurones that are activated with it' (ECCLES, 1970, p. 21). With respect to the general property the conditioned reflex, ECCLES (1953, Fig. 76, p. 222) brings a diagram, here reproduced as Figure 126C, which, discounting a slight modification by additional parallel and serial connections, is closely similar to that of Figure 126B from the 1927 *Vorlesungen*. HEBB (1949, 1959) designated my net-like engrams respectively neuronal cell arrays of 1927 as 'cell assemblies'. Exactly as in my old theory, 'an elementary

[136] An alternate theory of memory was proposed by BOK (1959), who regarded vacuoles between crossing cortical nerve fibers ('interfibral vacuoles' with two stable states) as specific organs of memory in analogy to magnetic artificial memory devices used in electronic engineering. The fiber crossings are interpreted as 'interfibrous synapses'. BOK described his 'vacuoles' in phase contrast microscopic preparations of the Mammalian cerebral cortex. Regardless whether these vacuoles are or are not artefacts, BOK's theory does not seem very convincing. Yet, even here, an engram would remain a patterned network, in which, however, the vacuoles rather than the synapses are conceived as the significant functional structures.

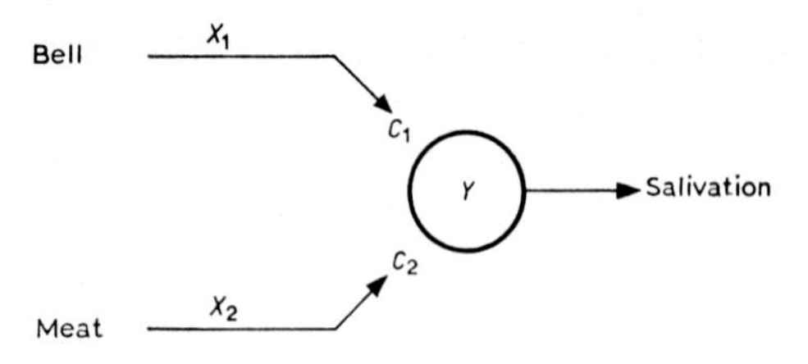

Figure 126 D. 'The classes of a neural pathway involved in formation of a conditioned reflex' (from BURNS, 1968).

idea consists of activity in a complex closed loop called 'cell assembly', developed as the result of repeated stimulation during childhood'. 'One or more of these, simultaneously active, can excite another, in a series celled a "phase sequence" – that is, a train of thought.' LASHLEY (1950) assumed that, as in my 1927 theory, memory traces are not localized in any one cortical district, but must be diffuse, involving a large number of neurons distributed all over the cortex, operating as a functional unit, but not located as a unit within the coordinates of gross anatomy (cf. also BURNS, 1958). Again, Figure 126D, representing the elements of a neural 'conditional probability machine necessary for learning', as discussed by UTTLEY (1956) and BURNS (1968), closely corresponds to my 1927 diagram (Fig. 126B) illustrating the presumable synaptic connection pattern obtaining in conditioned reflex, learning, and memory in general.

As regards the biochemical aspect of memory, apparently first formulated about 1912 by ROBERTSON, and which I had included in the memory theory of 1927, before the significance of the nucleic acids DNA and RNA were understood,[137] the subsequent studies of HYDÉN (1959, 1964, 1973, and other publications) have provided strong support for the assumption that specific macromolecular events are involved in the formation of engrams. HYDÉN assumes that, in a learning situation, neuronal and perhaps glial[137a] DNA may produce uniquely

[137] It may be recalled that K. VOIT and myself, on the basis of our studies on DNA in cell nuclei, were apparently the first to point out, against strong opposition from an expert on cytology and 'living substance', the significance of the DNA components as representing the active nuclear structures concerned with the transmission of variety (Anat. Anz. *75:* Erg. Bd., p. 71, 1932). Cf. pp. 100–109, particularly footnote 32, p. 103, of section 2, chapter V, volume 3/I. We recognized that DNA was the essential component of chromatin respectively chromosomes.

[137a] As regards the participation of neuroglia, cf. pp. 233–234, section 3 and pp. 626–627, section 8, chapter V, volume 3/I.

specific RNA which synthesizes specific proteins changing the biochemical functions of the synapses activated during the learning process, or, in other words during the process of engraphy. The preferred pathways thus developed in the neuronal nets presumably provide the biological basis for information storage in the central nervous system (cf. e.g. MATHIES, 1973). First, network discharge patterns become transduced into molecular structures facilitating synaptic transmission ('recognition molecules'), and second, such molecular markers, as substances encoding acquired information, become 'signposts' directing the flow of neuronal impulses. Whether these molecular markers consist of RNA, or, as RENSCH (1971) and others suggest, of proteins, remains an open question, likewise whether these biochemical traces involve the synaptic knobs or the postsynaptic membrane. The formation of specific RNA and of specific protein molecules is by no means mutually exclusive, and could also be interpreted in accordance with ROBERTSON's original theory of autocatalytic processes. Again, the relationship of these postulated processes to immunological specificites has been considered (SZILARD, 1964; cf. also the discussion remarks by GILLISSEN to the paper of RENSCH, 1971). RENSCH and his collaborators (e.g. RAHMANN, 1973) used the method of autoradiographic tracers (^{3}H-histidine) to demonstrate locally increased protein synthesis presumably related to engram formation.

There is also the question whether a *transfer* of 'memory' or 'learning' can be effected by injection or ingestion of substances from trained into untrained animals, that is by chemical transfer of specific behavior inducing substances. Besides the experiments with conditioned Planarians, discussed on pp. 42–44 of volume 2, experiments in other animals have been undertaken (cf. UNGAR and OCEGUERA-NAVARRO, 1965; UNGAR *et al.* (1968). Thus, conditioned 'fear of light' is claimed to have been induced by transfer of a 'scotophobin' from conditioned to unconditioned Fishes. As stated in the cited comments of volume 2, such transfer seems to concern merely the tendency toward a certain type of reaction rather than the 'learning' of a specific task. RENSCH (1971) likewise assumes that merely the 'steering' of a mechanism, but not the transfer of '*Gedächtnis-Inhalten*' appears to be involved.

ECCLES (1970) discounts, in a rather cavalier fashion, the theories of 'molecular memory' and of 'immunological' analogies, although he fails to account, in any convincing or intelligible manner, for the assumed engraphic facilitation of synapses.

With regard to *facilitation* of synapses it should also be kept in mind that for formation and ecphory of an engram by triggered, sustained, or reverberating activity in an engraphic network, not only synaptic activation but *synaptic inhibition* or '*repression*', closing the channels not pertaining to the encoding network pattern, must be presumed.

In addition to the *long-term memory* dealt with in the preceding discussion, there is a *short-term memory* lasting for seconds or minutes and not followed by long-term storage (K., 1957, p. 21). Short-term memory is e.g. evidently required in the execution of complex reflexes involving time intervals of some duration, or in the transitory retention of number sequences during calculations. Such short-time memory can be plausibly explained by the activity of self-reexciting, reverberating circuits pointed out in the 'Vorlesungen' of 1927.[138]

Yet, it seems probable that in the formation of long-term memory the aforementioned mechanism of short-time memory also may play an initial role.[139] Quite evidently, the formation of a stable memory trace requires a certain amount of time, necessary for repetitive activation of the particular network and formation of the biochemical 'marker'. This may be provided by the self-reexciting, reverberating circuits. Furthermore, the relatively slow development of stable memory traces might be concomitant with undefined additional synaptosomal regulations by conformational changes in enzymes, or correlated with changes in membrane proteins. Processes of this kind might represent one or several types of *intermediate memory* (cf. e.g. MATHIES, 1973).

Again, the selection of information patterns for engraphy, and the diverse components of the mechanism for recall, that is, for the ecphory of engrams, pose numerous, still poorly understood questions. It is generally assumed that the limbic system, whose efferent cortical griseum seems to be the caudal (postcommissural) hippocampal formation, may here play an important role. Various authors (e.g. SCOVILLE

---

[138] It was stated that through such circuits '*ein die ursprüngliche Erregung verlängernder bzw. verstärkender Erregungskreislauf entsteht*' (K., 1927, pp. 57, 191, 249). This was qualified by the comment that '*Verstärkung*' would require heterobolic systems. Yet, a '*Verstärkung*' could also be achieved in isobolic systems by increased discharge frequency (cf. also K., 1928).

[139] Thus, RENSCH (1971) assumes that '*das Kurzzeitgedächtnis kann aber in das Langzeitgedächtnis übergehen*'. RENSCH adds that '*bei höheren Tieren und beim Menschen ist das normale Erlöschen des Kurzzeitgedächtnisses ein wichtiger Vorgang, weil das Gehirn dadurch vor der Überflutung von unzähligen, im praktischen Leben völlig unwichtigen Eindrücken bewahrt wird*'.

and MILNER, 1957), have reported memory impairment following hippocampal lesions.

The *Papez circuit* in its original concept and in the diverse modifications[140] by subsequent authors appears, on rather convincing grounds, to be concerned with affectivity and correlated visceral activities, in general agreement with a cautious interpretation of the *James-Lange theory*. Since 'will' and 'volition' are conscious aspects of affectivity, the role of the limbic system and of its main efferent griseum's for memory fixation and recall does not seem implausible. The great flexibility of the hippocampal cortex as an 'energizing', 'facilitating' or 'activating' griseum within the limbic circuit is well supported by findings suggesting that excitatory synapses reach the dendrites, and inhibitory synapses the soma of the pyramidal cells in the cornu Ammonis. Thus, the inhibitory synapses may suppress all discharges that arrive from the dendrites on their way toward the axon, and could account for a high capability for differentiated response. Such flexibility of response might be the basis for the versatility of the limbic systems performance (cf. e.g. ANDERSEN, 1966).[141]

Attentiveness, short-term memory, long-term memory and its recall, as well as temporal organization of 'experience' could possibly depend upon activating and repressing or scanning discharge patterns of the limbic circuit. However, because of the doubtless obtaining considerable redundancy, other circuits, such as those related to brain stem reticular formation, hypothalamus, and prefrontal cortex may also play an important role in the just mentioned multifactorial events,[142] whereby information reaching the cortex is selected and transformed into either short- or long-term memory, and engrams of this latter may be recalled. The observations of PENFIELD (1955), implicating the temporal lobe cortex in memory recall and hallucinatory experiences upon electrical stimulation can be explained, in a very general way, by the

---

[140] It should be recalled that the telencephalic grisea of the limbic system *sensu latiori* include septum (paraterminal grisea) and amygdaloid complex.

[141] Summarizing discussions on the various theories concerning functions of the limbic system are included in STEPHAN's (1975) treatise on the allocortex.

[142] It should also be added that CAJAL's theory, postulating further growth of new neuronal connections, appears quite acceptable and may represent one of the many factors involved. Cf. also K. (1927, pp. 337–338), referring not to new pathways, but to the myelinization of previously non-medullated ones, which was assumed on the basis of findings by KAES (1907). This, however, could also be interpreted in accordance with CAJAL's hypothesis.

triggering of engram complexes whose networks are partly 'anchored' in the temporal association area,[143] part of which, moreover, borders on the limbic cortex with its hippocampal formation.

There is, moreover, as regards the formation and ecphory of engrams, an interesting general analogy with the principle of *holography*, discovered by GABOR (1972) about 1948. *Holography* is a new and ingenious method for recording a three-dimensional image of a scene. It records not only the light waves issuing from the scene, but a second set of coherent waves called the reference beam. The two sets of waves interfere with each other, and this interference pattern is recorded. Holography acts as a complicated diffraction device, causing a beam of coherent laser light to be diffracted in such a manner as to 'reconstruct' a three-dimension image of the original scene. In other words, holography is an interferometric wavefront reconstruction (cf. e.g. FARHAT, 1975).

It will be recalled that neuronal events spreading through the cortex are likewise conceived as representing a wavefront of excitation (BURNS, 1968; cf. above p. 307). This wavefront can be considered to result from the interference of the input wave with the 'reference beam' provided by the modulating (e.g. limbic) circuit, the 'reconstruction' being subsequently effected by another 'modulating neural wave'.

With regard to the diffuse localization of the engram, whose ecphory is not prevented by considerable cortical damage nor by removal of substantial cortical mass, the following similarity with the hologram is of interest. If this latter is cut in half, each half produces the entire image, and this is also the case with further tearing apart. However, when the pieces become very small, the image becomes more and more fuzzy, until from the smallest pieces only a blur is obtained. Again, a hole punched in the middle of a hologram will hardly affect the produced image. That this property of the hologram is similar to that of the distributed memory storage in the brain was clearly recognized by GABOR (1972), who, in reviewing the development of holography, also gave credit to BRAGG, ZERNICKE and others for contributions of significance for his own work. Thus, BRAGG's 'X-ray microscope', an optical *Fourier transformer* device, could, under certain special cases, produce a projection of the electron densities.[144]

---

[143] Cf. also RENSCH (1971).

[144] Cf. e.g. also Figure 243 A, p. 389 of volume 3/I, showing a composite x-ray diffraction photograph displaying a benzene ring formed by six carbon atoms.

In concluding the present comments on engraphy and ecphory, it should again be stated that, concomitantly with an increase in relevant recorded data concerning basic neural mechanisms, some rather plausible general concepts about memory can be formulated, but that, *qua* significant details, little can be said with any degree of certainty. Formation, localization, and recall of memory traces are still not sufficiently well understood despite very numerous investigations and a multitudinous literature dealing with this topic. It will here be sufficient to mention the publications by ADAM (1971), KIMBLE (1965), MARK (1974), RENSCH (1973), YOUNG (1966) and ZIPPEL (1973) which contain a diversity of bibliographic references. Of interest is also the treatise by ANDERSON and BOWER (1974), which deals with memory from the linguistic and semantic viewpoint. This publication, elaborates, *inter alia*, also on the highly sophisticated but rather unconvincing various formulations of 'generative' respectively 'transformational grammar', highly vocally propounded by CHOMSKY. It is evident that such attempts must remain inconclusive until a better understanding of the neural coding systems and of the relevant brain mechanisms is reached. In particular, although GABOR and others recognized the relationship of holography to the neuronal engraphy process, no detailed model of the mechanism of ecphory based on a similar 'reconstruction process' by the known aspects of brain circuitry can be elaborated at the present time.

It is, however, important to realize that engraphy by a neural network implies abstraction of invariants and the formation of 'concepts', which, in turn, can be 'handled' by the logical mechanisms discussed above, as well as by the mechanisms of 'affectivity'. Human concepts differ from animal concepts in an important manner by being usually combined with words. Yet, if by 'thought' the processing of concepts is meant, identical 'laws of thoughts' obtain for Man and all animals provided with fairly complex central nervous mechanisms. The indeed very relevant difference between Man and animals can be assessed as a difference in degree, and not as a difference in essence or quiddity (K., 1957). Generally speaking, even in the 'higher' animals, concepts remain averbal. But even in Man, a large amount of conscious thought and unconscious cerebration remains averbal (*nicht sprachgebundenes begriffliches Denken*, cf. e.g. WEIGL and METZE, 1968). It seems well established (RENSCH, 1968, 1971) that animals may already possess 'self-awareness', respectively a limited concept of the 'self'. Much the same could be said about 'death-awareness', e.g. manifested by fear of death

(Todesangst). Social behavior displayed by some animals likewise suggest occurrence of 'axiologic values' which, in Man, become related to 'conscience' and 'morality'.[144a] As regards some present-day concepts about various psychological aspects of Human memory, the interested reader may be referred to the publications by MURDOCK (1974), and KLATZKY (1975).

In the wider sense, memory includes the *genetic engrams* encoded in the chromosomes, which originate and maintain cell structure as well as structural and configurational features of the organism. In contradistinction to cortical, acquired individual memory, the genetic memory, modified by mutations, involves a quite different mechanism of engraphy, which appears either to preclude, or at least severely to limit, the inheritance of acquired characters.[145]

As regards abstraction, generalization, recognition, and thought processes in general, there is little doubt that the present-day developments of information theory and 'cybernetics' by ASHBY (1952, 1956, 1957, 1960), E. C. BERKELEY (1962), BRILLOUIN (1956), CHERRY (1968), KÜPFMÜLLER (1958), MACKAY (1954), v. NEUMANN (1958), PIERCE (1961), SCHAEFER (1960), SHANNON and WEAVER (1949), SHANNON (1950), STEINBUCH (1961), TURING (1936, 1937, 1950), WIENER (1948), and others have substantially contributed to a better approach and to an overall understanding of the far more complicated neural mechanisms which involve additional biological events not obtaining in hardware contraptions.

Nevertheless, circuit processes in artificial devices (cf. e.g. Figs. 127A, B) and in neural network seem to proceed according to a comparable mathematical and logical orderliness (K., 1957, 1961,

---

[144a] *Sir John (facing Reality)* ECCLES, in his 'philosophical adventures by a brain scientist' (1970), vainly attempts to support the notion assuming the transcendent state of Man, differing 'radically in kind' from other animals, and exclusively possessing a '*soul*'. This latter is supposed to be a component of the so-called 'World 2', which is 'non-material and hence not subject in death to the disintegration that effects all components of the individual in World 1 – both the body and the brain' (ECCLES, 1970, p. 174). In contradiction to that author, RENSCH (1968) elaborates a sober and well supported survey of Man's status as a result of phylogenetic evolution. Many of RENSCH's conclusions agree with my own views as expressed in the section 'Uniformity of Nature' (pp. 96–109) of the monograph 'Brain and Consciousness' (1957). However, to another and somewhat more enthusiastic one of his prolific publications (Homo sapiens, 1959), RENSCH added the subtitle '*Vom Tier zum Halbgott*'. As regards this 'demigod', one is inclined to quote the quip from GOETHE's *Faust* (I): '*Dir wird gewiss einmal bei deiner Gottähnlichkeit bange!*'.

[145] Cf. section 7, chapter II of volume 1.

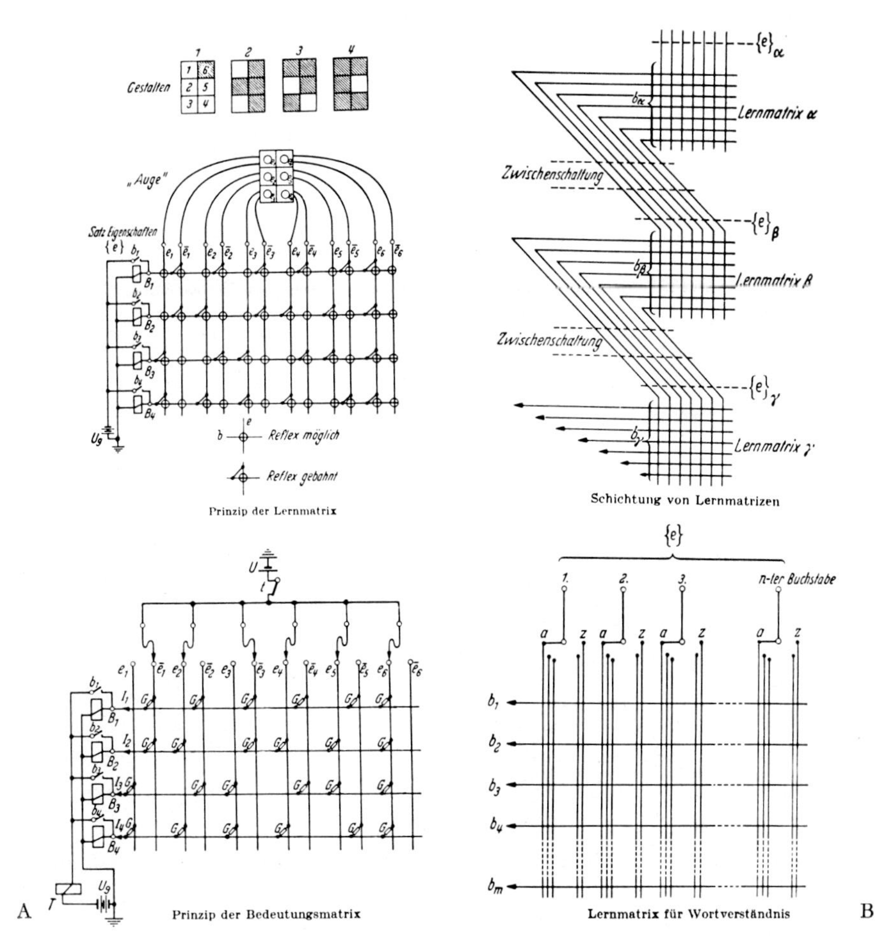

Figure 127 A. Two artificial network patterns, above indicating pattern recognition matrix (*'Lernmatrix'*, 'perceptron'), and below the corresponding significance matrix (*'Bedeutungsmatrix'*) as devised in 'cybernetic' engineering (from STEINBUCH, 1961).

Figure 127 B. Serial arrangement of recognition matrices (*'Lernmatrizen'*) above, and general arrangement of *'Lernmatrix'* for word recognition below (from STEINBUCH, 1961). Omitting the details, the diagrams are merely depicted to illustrate the capacities for pattern recognition and logical processes inherent in artificial electric, or in neuronal circuitry.

1965a, 1966, 1973). Keeping in mind the considerable differences between organic, living systems, and inorganic, non-living, artificial ones, it seems rather irrelevant that the pattern recognition, the chess play, the language translation, or the 'imitation game' by computers are rather mediocre. The important fact remains that, however unsatisfactorily, such activities can be performed by mechanisms in accordance with general rules presumably also applying to the brain, and apparently substantiating what could be called, for want of a better wording, the 'uniformity of nature'. Yet, as is evident from the attempt by SMYTHIES and ADEY (1970) to summarize contemporary views on 'brain mechanisms and behavior', the various controversial and hazy theories hitherto propounded on that topic have, so far, remained at a quite unsatisfactory state. However, it seems at least in principle possible to relate neuronal activities of the brain to so-called higher psychological processes such as pattern registration and identification, memory, learning, and thought, if these processes are conceived in terms of behaviorism. This statement means that it should be possible to explain these processes as resulting from the physiocochemical and mathematical properties of neuronal circuit activities, that is to say, as the outcome of strictly causal and purely mechanistic events. Although, as just mentioned, it cannot be said that this goal already has been reached, some fairly promising progress in that direction has been achieved.

On the other hand it should be emphasized that any attempt of 'explaining' *consciousness* in terms of neuronal events is, *ab initio*, intrinsically and absolutely impossible. Neuronal activities cannot be related to conscious events in accordance with concepts of natural sciences implying causal physicochemical events respectively energy transformations. Moreover, neuronal events cannot 'generate' or 'produce' consciousness, since neuronal events already are 'products' of human consciousness. Physical events, 'matter', organisms, and brains have no independent spatio-temporal existence except as phenomena of consciousness. At best, they can be interpreted as manifestations within a given consciousness-manifold (private perceptual space-time system) of an orderliness x, which is devoid of space-time attributes and remains beyond the possibilities of Human understanding. This intrinsic incomprehensibility is expressed by the *brain paradox* (K., 1957, 1959).

The semantic universe of discourse pertaining to physics and chemistry is based on the concept of orderly events, assuming transformations, respectively interactions, of mass and energy in a public four-dimensional space-time system.[146] These events, as public data, can be

---

[146] In his treatise on the principle of sufficient reason, SCHOPENHAUER (*Werke, editio* GRISEBACH, vol. 3) stressed that 'empiric reality' (which I prefer to designate as the

registered by the sensory endings (tranducers) of the nervous system or by instruments, which, again, can be 'read' by the senses of an observer. Although phenomena of consciousness are experienced by the awake observer, this latter cannot, by means of his senses detect consciousness phenomena extraneous to his own, nor can such phenomena be registered by instruments. Moreover, consciousness has no logical existence in the universe of discourse pertaining to physics and chemistry. Much the same could be said about most aspects of biology, concerned with the study of organic life, insofar as life represents a physical and biochemical phenomenon displayed by 'matter' (i.e. mass and energy) under particular conditions.

ASHBY (1952 in his 'Design for a Brain', although recognizing consciousness as 'the most fundamental fact of all', justly ignores this phenomenon in dealing with his topic: 'Vivid though consciousness may be to its possessor, there is as yet no method known by which he can demonstrate his experience to another. And until such method, or its equivalent, is found, the facts of consciousness cannot be used in scientific method.'

This view, as here applied to brain function, corresponds, of course, to WATSON's (1925, 1929) concept of *behaviorism*. This author, while originally not denying that consciousness exists, attempted to deal with behavior exclusively in terms of physiological and neurological states which can be recorded by an external observer or by instruments.[147] In this respect, consciousness is indeed neither a definable nor a usable concept. Thus, 'psychology as a science of consciousness has no community of data' (WATSON, 1929).

Quite evidently, a presumably conscious living Human or Animal being cannot 'demonstrate' its state of consciousness to an observer.

'actuality' of conscious experience) involves the *unification of space and time*. SCHOPENHAUER postulated '*Vereinigung des Raumes und der Zeit zur Gesamtvorstellung der empirischen Realität*' (p. 44); '*sogar ist eine innige Vereinigung beider die Bedingung der Realität*' (p. 42).

Concerning the space and time of physics, MINKOWSKI (1908) elaborated, in mathematical terms, the four-dimensional space-time concept which proved to be of fundamental significance for EINSTEIN's theory of relativity. MINKOWSKI, in his lecture at the Congress of German natural scientists and physicians, September 1908, stated: 'The views of space and time which I wish to lay before you have sprung from the soil of experimental physics, and therein lies their strength. They are radical. Henceforth, space by itself, and time by itself, are doomed to fade away into mere shadows, and only a kind of union of the two will preserve an independent reality' (LORENTZ *et al.*, n.d., p. 75).

[147] Cf. also pp. 70–74, section 1A, chapter VI of volume 3/II.

This latter, or for that matter, a suitable instrument, can only register the observed being's physical behavior (including, e.g. verbal communication) from which that being's consciousness can merely be inferred *per analogiam*. Such unverifiable inference, although in many cases quite 'certain' or rather convincing, is not an actual observation. No conceivable method for such observation can be described in terms of exact sciences, its only equivalent (cf. above ASHBY's remark) being an inference from observable behavior.

Thus, from the viewpoint of 'objective science', methodologic behaviorism appears fully justified. Extreme or dogmatic behaviorism assumes the ludicrous viewpoint of denying the 'existence' of consciousness. Thus, HEBB (1954) regarded 'consciousness' to be a 'venerable hypothesis' (cf. K., 1957, p. 93). According to LASHLEY (1923), WATSON, in some of his later writings, likewise seems to have taken the position that consciousness does not exist. He is quoted, by LASHLEY (1923, p. 240) to have regarded as a serious mistake the statement that a behaviorist does not deny the existence of mental states, but merely prefers to ignore them. According to WATSON, 'he "ignores" them in the same sense that chemistry ignores alchemy, astronomy horoscopy, and psychologists telepathy and psychic manifestations. The behaviorist does not concern himself with them because as the stream of his science broadens and deepens, such older concepts are sucked under, never to reappear.'

LASHLEY (1923) does not share this extreme view and makes a determined effort to struggle with the problem of consciousness in relation to behaviorism. He attempts to analyze the 'organization of consciousness' under various headings, such as limitations of contents, unity, consciousness of self, and self-arrangement. He also attempts to define 'awareness' and 'contents'. He fails, however, to realize that consciousness is the given private space-time system in which the perceptual brain is located, and thus becomes unable to detect the therefrom resulting *brain paradox*.

LASHLEY believes that as complete an account of the attributes of consciousness can be given in behavioristic terms as can be given in subjective terms as a result of introspective study. According to LASHLEY, 'the statement "I am conscious" does not mean anything more than the statement that "such and such physiological processes are going on within me".' It is of interest that LASHLEY in elaborating on the problem of a 'conscious machine', anticipates some of TURING's (1950) views concerning the so-called 'imitation game' (cf. K., 1966).

In my own attempts at correlating behavioristically conceived brain events with the given phenomenon of consciousness, I found it appropriate, as repeatedly stated (e.g. pp. 180–183, section 9, chapter IV, vol. 2) to distinguish all physicochemical events as R-events (reduced events), and all events encoded as signals by the nervous system as N-events. In other words, R-events become, at the receptor endings, transduced into N-events, and N-events, in turn, become transduced, at the effector level, into R-events.

Roughly speaking, one could also say that all events externally to the nervous system are R-events, and that N-events are all neural signal activities. This, however, requires the qualification that the nervous system itself consists of biological R-events, upon which the N-events are imposed, or which 'carry' the N-events as 'signals' encoding information (variety, negentropy, theoretically expressible in *bits*). Some, but by no means all N-events appear correlated with events of consciousness, that is, with P-events (parallel events), and such physical N-events could be designated as Np-events. It seems likely that, at least in the Human brain, Np-events represent essentially cortical, or corticothalamic events.[148]

As regards *consciousness*, or, in other words, P-events, the claim, by WATSON and other authors,[149] that consciousness is undefinable, indeed holds if its denotational definition requires reference to one or more other known 'entities'. This claim of undefinability, however, does not hold for a structural definition of consciousness, obtained by enumeration of the significant known elements and relations constituting a 'structure'. In this respect, *consciousness can easily be defined, in a satisfactory and unequivocal manner, as the designation for any private perceptual space-time system* (K., 1927, 1957, 1961, 1973).

---

[148] This does not exclude the possibility or probability that, particularly in 'lower Mammals', in Anamnia and Invertebrates, Np-events may occur in other central nervous grisea.

[149] Thus, v. BONIN in his review of 'Problems of Consciousness. Third Conference, ABRAMSON, E.H., ed.' points out that 'people tried (in vain, of course) to define consciousness over many pages'. VON BONIN then adds that 'consciousness is one of the "ur-Erlebnisse" which cannot be defined in simpler terms' (J. comp. Neurol. *98:* 554, 1953). To this, I would reply that consciousness is not '*one* of the "*ur-Erlebnisse*"', but rather *the* one and only given comprehensive *Urphänomen*, which, as elaborated further below in the text, indeed can be defined in relative simple descriptive terms. It should also be emphasized that, in the here adopted terminology, consciousness is considered as synonymous with mind, mentation, soul, or psyche. HUGHLINGS JACKSON (1931/32) and KRETSCHMER (1926) had likewise adopted this connotational usage.

Such system, designated as a *consciousness-manifold*, is given by its interrelated contents, which may be called *percepts*, and thus represents a plenum. 'Pure consciousness', that is to say consciousness without contents, is non-existent, except as abstraction manifested by a conscious thought. All percepts are *localized* and occur *in time*, that is, 'here and now'. *Sensory percepts*, pertaining to the conventional five and additional modalities dealt with in section 1, chapter VII of volume 3/II, can be distinguished from *extrasensory percepts*, represented by thoughts and certain aspects of emotion (affectivity).

Only two sorts of Human vivid sensory percepts, namely optic ones and some body sensations (e.g. tactile and proprioceptive percepts) are, in addition to being well localized, also *spatially configurated* (i.e. implying limiting surfaces with outlines and shapes). Acoustic percepts may be vivid, but are less distinctly localized, and, at least in Man, are shapeless, that is, lack spatial configuration.

Olfactory and gustatory modalities may be vivid but are quite vaguely localized with emphasis on nasal cavity respectively tongue, and also lack spatial configuration. Vagueness of localization also occurs with regard to the manifestations of referred pain, which are familiar to the clinician.

Extrasensory percepts, namely, non-eidetic memory images, abstractions, thoughts,[150] as well as emotions and vague feelings are poorly localized and likewise without spatial configuration. Extrasensory percepts frequently, but not always, may display a low degree of vividity. Nevertheless, all extrasensory percepts are localized, thoughts, e.g. being generally referred to a region corresponding to the interior of the body, dorsally to the visual space. Not infrequently, sensory and extrasensory percepts may be closely linked together by superposition. If accompanied by vague body sensations, extrasensory percepts may be localized in chest or abdomen. Thus, as e.g. implied by the *James-Lange theory*, the hypothesis of a localization of 'mind' in heart or diaphragm (phrenes), as expressed by some views of antiquity, had at least a modicum of justification. However, as justly stressed by authors such as FLECHSIG (1896) and FOREL (1922), we do not localize our thoughts in our feet or externally to the body.

There is, moreover, a third type of percepts, which could be called *pseudosensory percepts*, being intermediate between the sensory and the

[150] A definition and discussion of thought is given in the author's contribution to the *Helen Adolf Festschrift* (K., 1968).

extrasensory ones. Such pseudosensory percepts, occurring in dreams or as hallucinations, may display, even quite vividly, the aspects of sensory modalities, without however being mediated by concomitant R-events transduced into N-events by the sense organs. It can be assumed that said pseudosensory events are P-events concomitant with Np-events related to, or triggered by, engraphic complexes pertaining to previously experienced sensory events.

The consciousness-manifold, as manifested by normal adult Human beings, tends to display two distinctive groupings or neighborhoods, namely the '*external world*' and the '*internal world*' or *self sensu latiori*. This latter is represented by the perceptual body containing the sensory and motor, respectively efferent, nerve endings related to the likewise included brain which generates that consciousness (or, of which said consciousness is a dependent variable). The zones of consciousness corresponding to 'internal' and 'external' world complexes have, however, ill-defined boundaries and overlap with wide, fluctuating border zones (KRETSCHMER, 1926). The body-complex, experienced as 'ego' or 'self', and the 'external world' manifest interaction, such that the latter acts upon the former, and vice-versa. Some activities of the body are experienced as involuntary and others as voluntary, namely accompanied by a percept designated as '*will*' (*conation*), which can be subsumed under the phenomena of affectivity. It is absolutely impossible to define the experience 'will', except by its relationship, just as it is absolutely impossible to define the sensation 'red'. Both pertain to what POINCARÉ (1909) appropriately called '*qualités pures*', which are intransmissible by words and utterly incommunicable, except for their relationships (cf. K., 1961, p. 241). '*Will*', moreover, is an exceedingly vague and ambiguous term with a variety of diverse connotations and denotations.

The distinctly localized and spatially configurated sensory optic, tactile (in the wider sense) and proprioceptive percepts, as well as the indistinctly localized and not spatially configurated acoustic, olfactory and visceral percepts are localized externally to the brain. This latter, enclosed in the perceivable head, is likewise perceived *qua* weight through proprioception.[151]

Although all conscious percepts (a tautology) seem to depend upon

---

[151] There is, in the aspect here under consideration, little difference between perceiving, with closed eyes, a dead brain held in one's hand, and perceiving one's own living brain enclosed in the head.

the function of the perceptual (i.e. perceived) brain, consciousness is thus not localized within this latter, but said brain, contrariwise, becomes localized within the space-time system of the consciousness-manifold.

It follows that the assumed physical space, in which the physical brain is presumed to be located, cannot be identical with the observer's perceptual space, in which his percepts are located.[152] It becomes therefore necessary to distinguish a perceptual and a physical brain. Both cannot be identical, unless it is assumed that, by an entirely undescribable and thus incomprehensile process, percepts are 'projected' from such 'identical' brain into its 'external world', but such 'projection' can be dismissed as highly unbelievable.

One might also ask how the perceptual brain, upon whose activity consciousness depends, can 'produce' consciousness, since this brain itself is a manifestation of consciousness. This question, as repeatedly stated, leads to the *brain paradox* (K., 1959, 1961).

There is, moreover, no describable relationship between the coded circuit events in the physical brain and the correlated percepts of consciousness. No intelligible explanation can be given for the assumption that physical brain events, involving energy and mass, respectively biochemical and electric processes in physical space-time, become transformed into the corresponding conscious percepts, and no *energy transfer* from physical space-time to perceptual space-time appears conceivable.[153]

Although events in perceptual space-time appear correlated with brain events in physical space-time, this correlation cannot be con-

---

[152] Since, as pointed out above in footnote 146, space and its dimensions cannot be separated from time as a fourth dimension (MINKOWSKI, 1908), the assumption of both a perceptual and a physical space-time system would become necessary. The distinction of private perceptual spaces and public physical space has been recognized by a number of previous authors (cf. the comments in chapter III, p. 59, footnote 2 of Mind and Matter; K., 1961). However, to the best of my knowledge, nobody seems to have drawn the self-evident logical conclusion that, therefore, consciousness represents, and must be defined as, a private perceptual space-time system.

[153] In other words, no event whatsoever, along a given path and in a given time, leads e.g. from a circuit activity in the visual cortex to the percept of a colored object. In physical, causal interactions all events occur along given pathways and with given velocities. As ZIEHEN (1924) stated: '*Vorgänge, die auf bestimmten Wegen mit bestimmter Geschwindigkeit ablaufen.*' '*Von den Rindenzellen der Sehsphäre führt kein "Weg" zu der Empfindung rot, und daher ist auch von keiner Geschwindigkeit die Rede, mit der etwa die Empfindung rot auf die Sehsphärenerregung folgté.*' Cf. also Fig. 128.

ceived to involve interaction. For want of a better word, the term '*parallel events*' seems therefore appropriate for said relationship. HUGHLINGS JACKSON (1931, 1932), concluding that consciousness 'attends' brain function, used the term 'concomitance'. As regards the activities of the physical brain, consciousness has also been considered an incomprehensible 'epiphenomenon'.

The meaningless 'philosophical' so-called psychoneural identity hypothesis, claiming that both sets of events are 'identical',[154] merely slurs over the crux at issue and conceals the gist of the problem by means of a mere word. How can a cortical discharge pattern be 'identical' with a concomitant pattern of entirely different events?

It is, furthermore, important to keep in mind that only some, but by no means all brain events (N-events) are correlated, as Np-events, with consciousness, i.e. with parallel events (P-events). There is little doubt that complex logical and paralogical, evaluating operations, in all respects comparable to thought-processes, and to those manifested by affectivity except for the occurrence of consciousness, can be performed by unconscious cortical N-events. If thought and affectivity are defined as involving consciousness, then the term '*unconscious cerebration*' can be applied to such sequences of N-events not correlated with consciousness, i.e. with P-events. Operations similar to those by 'unconscious cerebration' can, moreover, also be performed by artificial mechanisms (computers).

The procedures of so-called *psychophysics* merely compare relatively accurate measurements concerning a stimulus, and inaccurate measurements ('estimates') made by the operation of physical N-events, i.e. by the central nervous system, regardless of occurrence or non-occurrence of parallel mental events correlated with some physical N-events. The *modus operandi* of synaptic and encoding neuronal events appears fully consistent with relationships between stimulus and response expressible by either the logarithmic *Weber-Fechner law*, or by *Stevens' power law*, as the case may be. The relevant problems of psychophysics with respect to physical or mental events have been discussed elsewhere (K., 1971) and need not to be dealt with in the present context.[155]

---

[154] Cf. e.g. MOUNTCASTLE (1966).

[155] The *Weber-Fechner law* is also discussed in the monograph on brain and consciousness (K., 1957) and, together with *Steven's power law*, in the subsequent German edition (K., 1973).

Again, in observing another Human (or comparable Animal) being, possessor of an awake, functioning brain, which is located in the perceptual space of the observer, this latter cannot observe, nor register by instruments, the corresponding conscious percepts 'produced' by the brain of the observed and presumably likewise located externally to said brain. It follows that the observer's perceptual space is not identical with the assumed perceptual space of the observed. The different perceptual spaces, respectively space-time systems (in this instance of observer and observed) are thus strictly '*private*', while the assumed physical space-time system can be conceived as '*public*'.

In the preceding discussion, and in agreement with the universe of discourse pertaining to the physical sciences, including the relevant aspects of biology, the 'existence' or occurrence of events in a public physical space-time system was taken for granted, and the philosophical, epistemologic aspects were ignored.

In order to understand the significance of consciousness, however, it becomes necessary to consider the problem of *reality in the first sense*,[156] that is to say the 'existence', 'occurrence' or 'persistence' of events when they are not perceived, respectively when they do not represent contents of a private perceptual space-time system (i.e. of a consciousness-manifold).

JOHN LOCKE (1689, 1690) and other previous or contemporary authors, assuming physical 'matter' to be real in the first sense, distinguished *primary qualities*, pertaining to that matter, and *secondary qualities*, pertaining to the sensations, i.e. to the consciousness or 'mind' 'perceiving' that matter. The primary qualities were described as extension, figure (configuration), motion, rest, solidity or impenetrability, and number, the secondary qualities being color, darkness and light, sounds, odors, tastes, cold, heat, softness, hardness, etc.[156a]

GEORGE BERKELEY (1710, 1713, 1725), however, convincingly argued that all primary qualities, such as extension, figure, motion,

[156] A 'thing' is *real in the first sense* when it persists or exists at times when it is not perceived. A 'thing' is *real in the second sense* when it is correlated with other things in a way which experience (respectively the application of the principle of sufficient reason) has led us to expect. This distinction of reality already expressed in the writings of SCHOPENHAUER was also given by BERTRAND RUSSELL.

[156a] It seems evident that the so-called 'primary qualities' correspond to what I prefer to call R-events (including the N-events superimposed upon a particular sort of biologic R-events). R-events, in turn, correspond to 'public data'.

etc., are likewise only percepts in consciousness,[157] whose existence consists in being perceived.[158]

In BERKELEY's words: 'Some truths there are so near and obvious to the mind, that a man need only open his eyes to see them. Such I take this important one to be, to wit, that all the choir of heaven and furniture of the earth, in a word all those bodies which compose the mighty frame of the world, have not any subsistence without a mind, that their being (esse) is to be perceived or known; that consequently so long as they are not actually perceived by me, or do not exist, in my mind or that of any created spirit, they must either have no existence at all, or else subsist in the mind of some eternal spirit: it being perfectly unintelligible and involving all the absurdity of abstraction, to attribute to any single part of them an existence independent of a spirit.'

DAVID HUME (1739–1740, 1748) likewise recognized the cogency of BERKELEY's argumentation denying the intrinsic difference between secondary and primary qualities. He stated: 'Tis universally allowed by modern enquirers, that all the sensible qualities of objects, such as hard, soft, hot, cold, white, black, etc., are merely secondary and exist not in the objects themselves, but are perceptions of the mind, without any external archetype or model which they represent. If this be allowed, with regard to secondary qualities, it must also follow with regard to the supposed primary qualities of extension and solidity; nor can the latter be any more entitled to that denomination than the former. The idea of extension is entirely acquired from the senses of sight and feeling; and if all the qualities, perceived by the senses, be in the mind not in the object, the same conclusion must reach the idea of extension, which is wholly dependent on the sensible ideas or the ideas of secondary qualities.'

HUME, moreover, also recognized the weakness or *non sequitur* in BERKELEY's important argumentation, namely that things (or events) not perceived must have no existence at all, or else subsist in the mind of some eternal spirit. Quite evidently, since the so-called primary qualities merely represent the attributes of space and time, there is a third possibility, namely that an *unknown 'orderliness', devoid of space and*

---

[157] BERKELEY uses the wording 'ideas existing in the mind', 'ideas' thus representing percepts, and 'mind' being synonymous with consciousness.

[158] BERKELEY states: 'Their esse is percipi, nor is it possible they should have any existence, out of the mind or thinking which perceive them.' 'In truth the object and the sensation are the same thing, and cannot therefore be abstracted from each other.'

time, represents reality in the first sense, and becomes manifested in consciousness. Since our thinking involves extrasensory percepts within a private perceptual space-time system, our thoughts are obviously bound to the symbolism of space and time, and the nature of reality in the first sense remains unintelligible.[159] Yet, in accordance with the valid aspect of BERKELEY's argumentation, the space-time of science is thus absolutely swept away, and the 'external world' of physics has no existence, except in our thoughts.

With regard to this problem, HUME remarked: 'Thus the first philosophical objection to the evidence of sense or to the opinion of external existence is this, that such an opinion, if rested on natural instinct, is contrary to reason, and if referred to reason, is contrary to instinct, and carries no rational evidence with it, to convince an impartial enquirer. The second objection goes farther, and represents this opinion as contrary to reason: at least, if it be a principle of reason, that all sensible qualities are in the mind, not in the object. Bereave matter of all its intelligible qualities, you in a manner annihilate it, and leave only a certain unknown, inexplicable something, as the cause of our perceptions; a notice so imperfect, that no sceptic will think it worth while to contend against it.'

HUME furthermore comments as follows on the views of BERKELEY, whom he greatly admires: 'and indeed most of the writings of that very ingenious author form the best lessons of scepticism, which are to be found either among the ancient or modern philosophers, BAYLE not excepted. He professes, however, in his title page (and undoubtedly with great truth) to have composed his book against the sceptics as well as against the atheists and free-thinkers. But that all his arguments, though otherwise intended, are, in reality, merely sceptical, appears from this, that they admit of no answer and produce no conviction. Their only effect is to cause that momentary amazement and irresolution and confusion, which is the result of scepticism.'

The concept of *transcendental neutralism*, as elaborated by myself (K., 1957, 1973 *et passim)* assumes that the phenomenon of consciousness, displayed by private perceptual space-time systems, which are indirect-

---

[159] An amusing demonstration of an unintelligible relationship is elaborated in the amiable romance '*Flatland*' (1952 reprint) by E.A. ABBOTT (1838–1926) in which two-dimensional sentient beings are unable to conceive a three-dimensional world. Conversely, our spatio-temporal mind is unable to conceive, in intelligible terms, a reality devoid of space-time attributes.

ly interrelated through '*public data*'[160] manifested by perceptual 'matter', is dependent[161] upon an orderliness x which is neither mental nor physical, being devoid of space-time characteristics. This formulation represents a synthesis of the valid epistomologic viewpoints contained in the writings of LOCKE (1632–1704), GEORGE BERKELEY (1684–1750), HUME (1711–1776), KANT (1724–1804), SCHOPENHAUER (1788–1860), and VAIHINGER (1852–1933). Reduced to the simplest terms, it could be said that LOCKE firmly established empiricism (or positivism *sensu strictiori)*, BERKELEY showed that there is no physical 'matter', HUME showed that there is no 'mind', as a substantial 'self' or 'spirit', thus independently rediscovering the *anâtman principle* of Buddhism, and KANT emphasized the distinction between phenomena (consciousness, perception) and independent reality in the first sense. SCHOPENHAUER recognized that our world of consciousness is a *brain phenomenon*, and VAIHINGER elaborated a notable concept of *fictionalism*, suitable for coping with the intrinsically unsolvable *brain paradox*.

As regards some further details, LOCKE, besides his concise discussion of primary and secondary qualities,[162] convincingly argued that there are no innate ideas, and that all our knowledge derives from experience. He conceived mind, at birth, to be a *tabula rasa*. With certain qualifications, the gist of this view can be upheld. The neuraxis is evidently not a tabula rasa at birth, since, besides the developmental processes which continue to proceed after birth, prenatal neural activities have doubtless occurred concomitantly with development. However,

---

[160] It is also of interest that SHERRINGTON (1951) expresses what could be interpreted as a somewhat similar concept. He states (p. 206): 'Mind, always as we know it, finite and individual, is individually insulated and devoid of direct liaison with other minds. These latter too, are individual and each one finite and insulated. By means of the brain, liaison as it is between mind and energy, the finite mind obtains indirect liaison with finite minds around it.'

Since mass and energy, including the brain, are, however, already mental, i.e. in BERKELEY's terminology, mere ideas, mind cannot be a product of one of its contents. SHERRINGTON (1951, p. 229) seems to apprehend this difficulty. He expresses the standpoint of natural sciences by stating: 'Thoughts, feelings and so on are not amenable to the energy (matter) concept. They lie outside it. Therefore they lie outside Natural Science.'

[161] Since, in SCHOPENHAUER's formulation of the principle of sufficient reason, which I have adopted with minor modifications, a 'cause' implies dynamic 'material' interaction within a given space-time system, and thereby 'continuity', I prefer here to use the term 'dependent upon' instead of 'the cause of' (cf. above DAVID HUME's formulation) for the inexplicable 'relationship' between orderliness x and consciousness.

[162] Cf. e.g. footnote 43c, p. 74, section 1A, chapter VI in volume 3/II of the present series.

substantial aspects of neural circuitry appear genetically determined and provide, as it were, a pre-functional repertory of performances. These latter, however, need not be conceived as based on 'knowledge'. The not at once given, but gradually developing '*tabula rasa*' may be considered as provided by the higher cerebral registering and processing mechanism, and the 'knowledge' as its subsequently registered and processed content. Thus LOCKE's dictum *nihil est in intellectu quod non ante fuerit in sensu* may be qualified by the addition *nisi intellectus ipse*.[163]

Although BERKELEY's hypostatization of a 'spirit', 'soul' or 'self' as that 'indivisible, unextended thing, which thinks, acts, and perceives' cannot be upheld, that outstanding philosopher almost reached the proper formulation. One can evidently state that the private consciousness-manifold, namely mind *sensu completo*, represents that indivisible (unified, gestalt-like), extended four-dimensional space-time system, which is a 'whole' consisting of extrasensory percepts (thoughts, volition, emotions) and of sensory percepts.

This transitory consciousness-manifold can be designated as the 'self' or the '*subject*' in the widest sense, and one may thus, without contradiction, reintroduce the eliminated concept of a 'substantial self' in the guise of an auxiliary fictional construction. That manifold, moreover, might be likened, in some of the aspects under consideration, to GIORDANO BRUNO's conception of a *monad*, later adopted, with some modifications, by LEIBNIZ.

The fact that a percept must be perceived by a perceiver, i.e. by a mind, or that a perceived object requires a perceiving subject, merely means that, in order to be actualized, a percept must be the content of a consciousness-manifold, that is, occur as intrinsic component of a private perceptual space-time system related to its perceptual brain. BERKELEY justly stated that 'in truth the object and the sensation are the same thing'. He nevertheless, presumably because of his ecclesiastical status, could not draw the inevitable conclusion, and felt compelled to retain the concept of a substantial soul, spirit, mind or 'myself', which perceived the perception.

The percepts (or perceptions) represent, however, irreducible given phenomena[164] *(Urphänomene)* which constitute a private conscious-

---

[163] This qualification is attributed to LEIBNIZ and can be interpreted as referring to the given neural mechanism (cf. K., 1961, pp. 186–189).

[164] The '*Gignomene*' of ZIEHEN (1922).

ness-manifold whose sustained 'identity', despite disruptions by periods of unconsciousness, is, as experience evinces, correlated with an awake perceptual brain, pertaining to a given perceptual body. Said percepts, all of which together represent a transitory self *sensu latiori* with a relatively persistent vague self *sensu strictiori*, based on aspects of memory, are as they are and not otherwise.[165] To quote a pithy remark, which the earthy late President HARRY S. TRUMAN was wont to use in a different context: '*the buck stops here*'. Anything going beyond this 'given' is an extrapolation into the unknown, but remaining within that consciousness-manifold as a thought percept in terms of said manifold's spatio-temporal symbolism. This 'unknown' remains, therefore, intrinsically unknowable, thus justifying the statement: *ignorabimus*.

As regards BERKELEY's extrapolation, the following statement of this author is of interest: 'But whatever power I may have over my own thoughts, I find the ideas actually perceived by sense have not a like dependence on my will. When in broad daylight I open my eyes, it is not in my power to choose whether I shall see or no, or to determine what particular objects shall present themselves to my view, and so likewise as to the hearing and other senses, the ideas imprinted on them are not creatures of my will. There is therefore some other will or spirit that produces them.' My own extrapolation, as discussed further above, postulates the unknown orderliness x, which, in some respects, is comparable to KANT's '*Ding an sich*'. The term 'thing', however, does not seem very appropriate. Neither BERKELEY nor KANT, moreover, sufficiently considered the significance of the perceptual *brain*.

With respect to 'reality', BERKELEY designated the 'ideas imprinted on the senses by the author of nature' as the 'real things', and those excited in the imagination as the images or ideas 'of things'. Both 'real

---

[165] HIRST (1966), who evidently disliked what he called my 'showy but unsatisfactory article' (K., 1965b) in the publication 'Brain and Mind' edited by SMYTHIES, completely failed to grasp this relevant point in his critique and naively asked 'how can a bundle or manifold be aware of itself?' Consciousness is obviously awareness and represents a manifold or 'bundle' of contents, which are as they are, thus being 'itself'. My admission that the existence of consciousness cannot be explained (since it includes the basis for all explanations) is merely an honest reply to an unanswerable question which I did not evade, but over which others slur with meaningless platitudes, unless they choose to invoke the by now almost defunct Deity. I doubt that Mr. HIRST, who stressed my admission of unanswerability, is capable to provide an intelligible answer to the question which I pointed out, since I am not afraid 'to face the music'. Mr. HIRST, however, to quote an old German saying '*geht wie die Katze um den heissen Brei*'.

things' and ideas of our own framing, imaginations or chimeras 'equally exist in the mind, and in that sense are like ideas'.

In BERKELEY's 'Third Dialogue between *Hylas* and *Philonous*', *Hylas* asks: 'What think you therefore of retaining the name matter, and applying it to sensible things.' In the guise of *Philonous*, BERKELEY replies: 'With all my heart: retain the word matter, and apply it to the objects of sense, if you please, provided you do not attribute them to any subsistence distinct from their being perceived. I shall never quarrel with you for an expression.' BERKELEY, however, then again repeats his strong denial of matter or material substance as concepts 'introduced by philosophers' and used by them to imply a 'sort of independency, or a subsistence distinct from their being perceived'.

In *Pars II*, *Propositio VII* of his *Ethica, ordine geometrico demonstrata*, SPINOZA formulated his well-known dictum: '*substantia cogitans et substantia extensa una eademque est substantia, quae iam sub hoc, iam sub illo attributo comprehenditur*'.

In full agreement with BERKELEY's and with my own view, SPINOZA's statement easily can be modified as follows: *Substantia extensa et substantia cogitans duo attributa sunt unae eiusdemque substantiae, quae conscientia nominatur.*

ERNST MACH (1922), in a formulation essentially similar to basic aspects of BERKELEY's and HUME's views, elaborated the thesis that our world is built up of sensations and perceptions *(Empfindungen)* which he interprets as elements. Matter is not conceived as anything given, but merely as a 'though-symbol' for a relatively stable complex of sensory elements. He suggests that the laws of physics may be formulated as functional relations between perceptions such as warm, hard, color sensations, and others; he includes among these elements space and time in terms of space perceptions and time perceptions: '*Farben, Töne, Wärme, Drücke, Räume, Zeiten usw. sind in mannigfaltiger Weise miteinander verknüpft und an dieselben sind Stimmungen, Gefühle und Willen gebunden.*' He emphasizes, however, that, since the world must be interpreted in terms of functional relations between these elements, it cannot be regarded as a mere sum of these sensations. MACH, moreover, in conformity with HUME and the *Buddhist anâtman concept*, eliminated the hypostatization of an 'ego' or 'self'.

The distinction of what is 'outside' and 'inside myself' depends, according to MACH, on the amount of immediate sense-impression. Figure 128 shows a sketch by MACH, in which he depicts his left monocular optic space: 'In a frame formed by the ridge of my eyebrow, by my

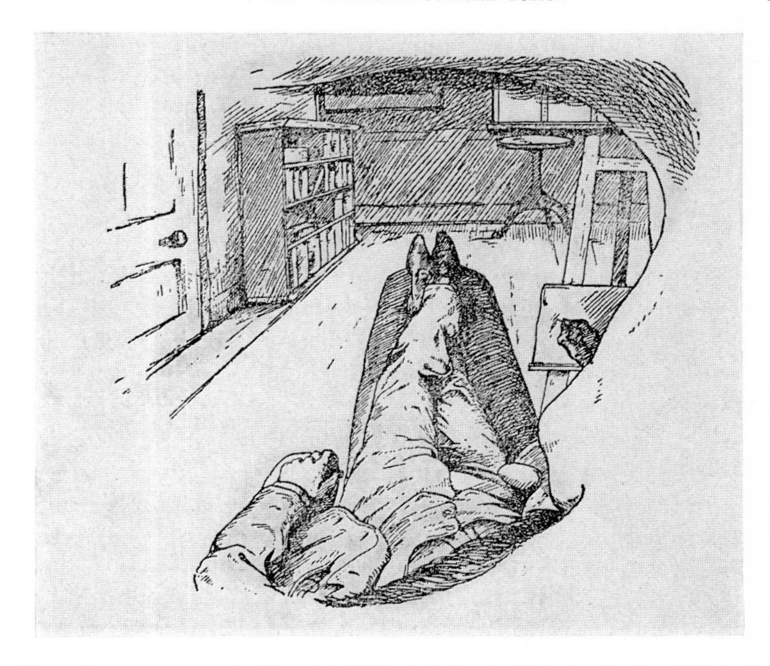

Figure 128. Sketch illustrating monocular visual space of left eye (from MACH, 1922). This, of course, represented a component of MACH's consciousness-manifold, that is of his private perceptual space-time system in the terminology which I have adopted. MACH's perceptual brain was located, in his consciousness, dorsally to the visual space depicted at the time of his sketch, and included in his proprioceptive and tactile space (head). Visual, proprioceptive, and tactile spaces however, were integrated into the common private perceptual space-time system representing his consciousness.

nose, and my moustache, appears part of my body so far as it is visible, and also the things and space about it.' MACH then adds that, 'If I observe an element, A, within my field of vision, and investigate its connection with another element, B, within the same field, I go out of the domain of physics into that of physiology and psychology if B, to use the apposite expression that a friend of mine employed upon seeing this drawing, passes through my skin.'

The perceptual visual space here depicted by MACH is located, within the common, unified space-time system of his consciousness-manifold, externally and rostrally to the tactile and proprioceptive space occupied by his brain. It is utterly impossible to 'explain', on the basis of physicochemical and physiologic processes or events assumed by the 'natural sciences' (including the 'neurosciences') this commonplace mental (i.e. conscious) event depicted by MACH. With reference to these sciences, said event is indeed truly 'miraculous'.

As regards the problems of space, time, and causality in the physical model of the phenomenal world, I can recommend, to the interested reader, the delightful series of essays entitled '*Drei Dialoge über Raum, Zeit und Kausalität*' by my esteemed old friend GEORGE JAFFÉ (1954). The late professor JAFFÉ (1880–1965) was one of the few theoretical physicists with deep insight into the epistemologic problems at issue, and who had completely understood the significance as well as the importance of GEORGE BERKELEY's fundamental thesis.

MACH intended to use a terminology suitable for a unified language of science, characterized by 'economy of thought'. All metaphysical propositions were to be eliminated, and he regarded both materialism and idealism as metaphysical propositions which cannot be stated in terms of a genuine scientific problem. These tendencies and opinions of MACH have been further developed in the voluminous writings of authors belonging to the so-called *Vienna circle* and associated groups.

Concerning the *Unity of Science Movement* which is closely related to the *Vienna circle* and to logical positivism, CRAIK (1943) has advanced the following criticism: 'There is a good deal of evidence that the precise definitions of the symbolic logicians and the logical positivists do not, in fact, cover very much of the range even of sensory experience. As HALDANE (1938) remarks, logistics "will only work for material that has certain highly abstract properties, which are rather less frequently and much less completely exemplified in the real world than logicians would like us to believe".'[166]

CRAIK (1943) also justly comments that 'admittedly the phenomenalists give a somewhat more sophisticated statement of their position, and evade all questions as to where sense data exist, what status they have, or how they are related to each other and to us'.

This very interesting remark points out the fallacy that sensations respectively percepts could independently exist as combining and recombining 'elements', without definite spatio-temporal relationships to a given consciousness-manifold pertaining to a perceptual brain, said manifold being a 'gestalt-property' of the manifested correlated

---

[166] This is, of course, quite evident since the logical systems, such e.g. as that of CARNAP, predominantly use empty syntax language instead of meaningful referential language (cf. K., 1968).

percepts.[167] The fallacy of 'floating' and 'status-less' percepts is the opposite to that which postulates a perceiving soul, spirit, self or subject as the perceiver (cf. above, footnote 165).

Reverting to the inescapable conclusion that the inner world of thoughts and feelings, as well as the so-called external world of material objects are private manifestations of consciousness, it seems evident that the public world of physics and natural sciences becomes a mere fiction. Nevertheless, it appears that only on the basis of this fiction a rational explanation of the experienced phenomena becomes possible. The *fiction* of an orderly 'world', independent of consciousness, and represented by a public physical space-time system, must therefore be considered an indispensable *postulate of practical reason.*

As already stated in the present author's old '*Vorlesungen*': '*Sowohl die Gedanken und Gefühle, also die innere Welt, geistige oder Geisteswelt des gewöhnlichen Sprachgebrauches wie die körperlich in Raum und Zeit ausgebreitete äussere (materielle) Welt sind Bewusstseinszustände (Vorstellung, Erscheinung). Das naturwissenschaftliche Weltbild mit seiner unabhängig von einem vorstellenden Bewusstsein vorhandenen objektiven, räumlichen, zeitlichen, kausalen Welt ist eine Fiktion. Aber nur mit Hilfe dieser Fiktion ist ein Verständnis der Erscheinungsreihen möglich*' (K., 1927, p. 2).

RIESE (1954) remarks that 'the history of medicine lists no physician daring enough to adopt the idealistic thesis that man's brain is but an idea, if not an illusion'. It is evident that my esteemed colleague was not acquainted with my views (1927, 1957, 1961, and other publications), which unequivocally conceived the perceptual brain as an 'idea' in BERKELEY's sense. There exists, in my interpretation, no physical brain, except as an operatively valid fiction, since the unknowable orderliness x must be assumed devoid of spatio-temporal relationships.

[167] The statements 'no percept without a perceiver' or 'no object without a subject' are only valid if referring to percepts as components of a consciousness-manifold respectively private perceptual space-time system. This latter, as such, has no existence or actuality, except for its correlated and organized contents. The childishly naive theory of ECCLES (1953, 1970) postulating a 'little ghost' inhabiting the brain, and switching neural activities by 'psychokinesis' while in turn receiving information from the encoded neural circuit processes, can hardly be taken seriously.

Other, likewise rather naive views of hylozoistic type assume that living cells, or even physical particles are endowed with 'mental properties'. SHERRINGTON appropriately remarked that: 'Most life, I imagine, is mindless.' This author also stated 'That the brain derives its mind additively from a cumulative mental property of the individual cells composing it, has therefore no support from any facts of its cell-structure' (SHERRINGTON, 1951).

The indeed *real (in the second sense)* perceptual brain represents merely, translated into the space-time symbolism of our consciousness, relevant aspects of said independent orderliness. Much the same can be said about the concept of molecules, atoms, particles, quanta, or about the four main sorts of physical interactions (gravity, electromagnetism, weak and strong interactions etc. pertaining to particles).

It should also be emphasized that the idealistic viewpoint, properly understood, does not conflict with the contemporary theories of cosmogony which assume the universe as arisen from the primordial explosion of a superdense mass perhaps about more or less 10^{10} years ago.[168] Within a given time the present types of atoms were formed and, in a hierarchy of condensations, galaxies resulted, some stars of which may include planetary systems. In this expanding universe with its population of galaxies and diverse types of evolving and dying stars, organic life seems to be a rather rare phenomenon. Yet such life can reasonably be assumed as having evolved or still evolving in fairly numerous, widely scattered regions of the universe. Quite evidently, organic life originated much later than the preceding cosmic events, and again consciousness, presumed to be correlated with activities of a nervous system, arose much later than the origin of organic life.

Regardless of unsettled details, these theories appear rather plausible, and pose the problem of their proper formulation in the terminology of the idealistic viewpoint, since it must be presumed that the occurrence of consciousness, only in which spatio-temporal 'material events' become manifested, was preceded by the necessary 'material' evolutionary events. These latter, assumed to be unconscious, were thus not manifested in a consciousness of their own, which would presuppose an adequate nervous system, nor in that of conscious beings, since these did not yet exist. Thus, these 'preceding' states were not spatio-temporal events, but merely unknowable aspects of the unknowable orderliness x. However, if translated into the spatio-temporal symbolism of an intelligent conscious being endowed with reason, these aspects can be conceived as having occurred in the manner postulated by the theories.[169] More naively expressed, if a con-

---

[168] Despite various uncertainties, it does not seem improbable that this explosion followed upon the collapse of a preceding universe. Likewise, the present expansion, indicated by HUBBLE's red shift, might finally slow down, and change into a collapse, followed by another primordial explosion. Thus, a 'pulsating' or 'oscillating' universe could be assumed.

[169] Cf. SCHOPENHAUER *(editio* GRISEBACH, vol. 5, '*Zur Philosophie und Wissenschaft der*

scious observer, possessor of a perceptual brain comparable to that of an intelligent Human being, could have been there during these presumed 'periods', said events would have appeared to him as so occurring.

From the viewpoint of the operationally valid fiction postulating the public spatio-temporal world of physicalisms and natural sciences, including biology, and thereby the existence of a physical brain characterized by physiologic neuronal circuit events, it becomes necessary to formulate, in appropriate terms, the correlation of consciousness (P-events) with some of the 'physical' neural N-events, namely with Np-events.

Since this correlation or concomitance cannot be assumed to be causal, involving energy transformations from public physical space-time into private perceptual space-time, the term psychophysical parallelism seems rather appropriate, but it is quite irrelevant if objectors to the term 'parallelism' prefer to use, instead, the words 'correlation', or 'concomitance', or 'epiphenomenon'. This latter term, however, implies an accessory or accidental significance of consciousness as manifested by a 'real' physical brain, in complete reversal of, and contradiction to, the actually obtaining phenomenology. There is, of course, no psychophysical parallelism, because there are no spatio-temporal physical events including excitation or activation of neuronal circuits, since the independent and real orderliness x lacks both the so-called secondary and primary qualities.[170]

In accordance with ASHBY's (1956, 1957) elementary formulation of

---

Natur', pp. 154–155). Also 'Brain and Consciousness' (K., 1957, pp. 41–42, 89). In other words, and from the viewpoint of VAIHINGER's fictionalism, it is '*as if*' the cosmogonic events had taken place or occurred in a public physical space-time system.

[170] *Dieser Erregungszustand ist jedoch nicht als die Ursache der Empfindung oder des sonstigen Bewusstseinszustandes anzusprechen (materialistische Auffassung).*'

'*Auch der primitive Dualismus von Materie und Geist, welcher in der bekannten Hypothese vom psycho-physischen Parallelismus eine wissenschaftliche Fassung gefunden hat, entbehrt jeglicher Grundlage. Gegeben sind nur gesetzmässig miteinander verbundene Bewusstseinszustände, welche künstlich in "Materie" und "Geist" eingeteilt werden. Insbesondere mache man sich klar, dass dem Begriff "Ganglienzelle" keine reale (d.h. an sich bestehende) materielle Wesenheit zukommt. Der Begriff der Ganglienzelle ist lediglich ein Symbol für unter bestimmten Bedingungen auftretende optische Bewusstseinsinhalte, farbige oder schwarze Flecke im Gesichtsfeld beim Mikroskopieren*' (K., 1927, p. 336). Yet, despite its purely fictional aspect and its obvious shortcomings, psychophysical parallelism or its equivalents (correlation, concomitance, epiphenomena) remain the inevitable formulation, if the unavoidable practical postulate of a public spatio-temporal world involving physicalisms is adopted.

group and transformation theory,[171] incorporating the concepts of closure and identity, one may distinguish *closed transformations*, which create no new elements of a given set or system, and *open transformations*, in which the result of the operation creates an entirely new element, not included within the original set.

As regards the events in the physical brain, the sequence from R-events to transduced N-events, the further sequences of these latter in the synaptic neural network, and the transduction, at the effector level, from N-events to R-events represent a series of closed transformations, involving energy and mass in physical space-time. In other words, there is a set of operands, transformed by the neurosensory, neural, and effector mechanisms, representing an operator, the results of whose operations remain, as physical respectively physicochemical events, elements of the given system or original set.

At a certain level, however, some N-events, namely Np-events, undergo a two-valued, or double transformation. One of these transformations remains closed in the adopted sense. The other transformation becomes open, creating an entirely new element, namely events in private perceptual space-time.

This transition, from physical N-events to parallel P-events is a purely logical, imaginary one, not involving interaction or 'causality' as defined by an analysis of the principle of sufficient reason. Since imaginary physical events and experienced mental events occur in different space-time systems which have no dimensions in common, no path, course, direction, vector, trajectory, velocity or motion could lead, in a given time, from one system into the other.

The kinematic graph of Figure 129 B displays these series of events in an oversimplified manner, which nevertheless illustrates the relevant aspect of these transformations.

The tabulation of Figure 129 A expresses the transformations with respect to their parameters. T designates the transformation; T's subscript x represents the parameter 'physical space-time', and y stands for the parameter 'perceptual space-time'. There are, of course, a number of successive transformations Tx, such as Tx_1, Tx_2, ... etc., as indicated in the legends to Figure 129.[172]

Finally, since not all N-events but only some of these latter (the Np-events) are correlated with states of consciousness, one might ask

[171] Cf. volume 1, pp. 18 and 166.

[172] For further details, cf. K. (1958).

↓	r_1	r_2	r_3	r_4
Tx	n_1	n_2	n_3	n_4
Ty	P_1	P_2	P_3	P_4

Figure 129 A. Tabulation illustrating parameters of closed and open transformations involving R-, N-, and P-events; (from K., 1958). n_1 etc.: N-events; P_1 etc.: P-events; r_1 etc.: R-events; Tx: parameter public physical space-time; Ty: parameter private perceptual space-time (i.e. consciousness). There are, of course, a number of successive transformations Tx such as Tx_1, Tx_2, ... etc., which have been omitted for the sake of simplification; horizontal single line indicates closed transformation, and double line stands for open transformation.

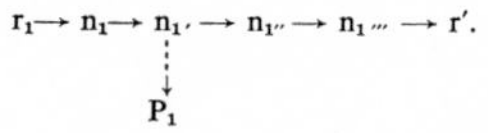

$$r_1 \rightarrow n_1 \rightarrow n_{1'} \rightarrow n_{1''} \rightarrow n_{1'''} \rightarrow r'.$$
$$\downarrow$$
$$P_1$$

Figure 129 B. Kinematic graph illustrating a sequence of R-events and N-events by closed transformation and a concomitant P-event related to an N-event by open transformation (from K., 1958). r_1: R-event becoming transduced into N-event; r': R-event resulting from transduced N-event; $n_{1'}$ represents here a Np-event, correlated, by open transformation, with a P-event. Solid arrows: closed transformations; dotted arrow: open transformation. An undefined series of additional closed transformations between r_1, n_1, $n_{1'}$ etc. has been omitted.

(1) whether Np-events occur only in certain grisea of the neuraxis, and (2) what are the characteristic features distinguishing N-events from Np-events. As regards the first question, it seems likely that, at least in Man and perhaps Mammals in general, Np-events require particular but still undefinable cortical respectively corticothalamic activities, although these latter may, in turn, perhaps depend upon 'energizing' parameters provided by the 'centrencephalic grisea'. In lower Vertebrates, it seems not impossible that Np-events might occur in the tectum mesencephali. Again, Np-events might possibly occur in undefined centra of the Invertebrate nervous system. Yet, no conclusive evidence whatsoever seems here obtainable. With respect to Man, DANDY (1946) noted that consciousness was lost after necrosis of the anterior portion of the corpus striatum following damage to the anterior cerebral artery. He thus assumed the corpus striatum to be the 'conscious center' in the Human brain. It was, however, subsequently shown that such damage to the anterior cerebral artery is correlated with widespread effects on other branches of the internal carotid artery and on the cerebral circulation in general, thereby resulting in deficient blood supply to all or most of the cerebral cortex.

Assuming consciousness to be a dependent variable of corticothalamic Np-events involving antimeric structures, namely two hemispheres with their adjoining diencephalic grisea, one could conclude that the unified perceptual space time manifold of consciousness, that is to say the experienced unity of consciousness, depends on the 'physical' continuity of the entire neural network whose activities become correlated with consciousness. This question was considered by FECHNER (1860) and various other authors. FECHNER believed that, if the two halves of the brain were divided in the midline, a condition alike to a doubling of the consciousness would result. This can be interpreted to mean that two disjoint perceptual space-time systems would thereby ensue. VON HARTMANN (1869, 1931), in his 'philosophy of the unconscious' also considered the experienced unity of consciousness to be based on the continuity of nervous conduction. He even believed that, if the brains of two men could be effectively joined by a 'bridge of nervous matter', the two men would have a single, common consciousness.

Among contemporary authors dealing with this topic ZANGWILL (1974) has recently reviewed the questions at issue, particularly with reference to the data provided by split-brain operations such as discussed in section 10 of chapter XIII in volume 5/I. ZANGWILL is compelled to admit that these operations, as performed in Man, fail 'to impair appreciably the unity of consciousness as apprehended by introspection'. He emphasizes, nevertheless, that this unity 'masks a genuine duality of consciousness'. Yet, he qualifies his statement by adding that the assumption of such 'duality as a necessary consequence of duplex brain organization remains controversial'.

Reverting to the well substantiated retained 'unity of consciousness' after transection of most prosencephalic commissures except the spared commissura supraoptica, supramammillaris, and, at the diencephalo-mesencephalic boundary, the commissura posterior, it would appear that the interconnections through these spared commissures suffice to maintain that unity which, moreover, might also in part depend on biochemical parameters specific for a given individual as pointed out further below in dealing with question (2). It should also be added that conditions figuratively interpreted as 'double consciousness', such e.g. as 'split personality', and performances of post-hypnotic suggestions, or, in other words, manifestations not unlike to those observed in Human split-brain patients, can be noted to occur in persons whose brains, with complete, intact commissural systems, otherwise, and most of the time, provide normal behavior.

On the other hand, the strange speculation elaborated by v. HARTMANN and cited above implies procedures whose realization seems highly improbable and whose results must be left to the imagination. Yet, cases of diprosopia with doubling of the prosencephalon, such as dealt with in chapter VI, section 1 C, pp. 241–243 of volume 3/II pose an interesting question. Since four hemispheres and two diencephala, joined to a single brain stem and spinal cord obtain, would such diprosopus manifest two disjoint consciousness-manifolds? If terata of this kind survived to a stage with fairly normal mental development and acquisition of speech, this question could perhaps be answered by inference from verbal behavior, but not otherwise. Only two cases in which 'fusion of two heads', classifiable as diprosopia, occurred, and which survived to an undefined period of childhood, have come to my attention, but the rather sketchy description of their behavior (GOULD and PYLE, 1896, p. 187) does not permit relevant conclusions concerning the question under consideration.

In the case of the double-headed boy *Tocci*, depicted and briefly referred to on pp. 239–240 of volume 3/II, and in the very similar *Ritta-Christina* case, two vertebral canals with spinal cord extended as far as a double *(Tocci)* or fused *(Ritta-Christina)* sacrum. 'All their sensations and emotions were distinctly individual and independent' (GOULD and PYLE, 1896, pp. 184–186), except for sensations along the fused midline strip and the common abdomen including common indication for and performance of urination and defecation. This was presumably due to overlapping of the peripheral innervation.

Sensations in and performance with each pair of the upper extremities were distinct, and also as regards the right and left inferior extremity. Since each head merely 'had power over the corresponding leg on his side, but not over the other one', walking was impossible and support was needed for standing.[173]

[173] Cf. GOULD and PYLE (1896) and the popularizing publication by F. DRIMMER 'Very special people' (Amjon, New York 1973) which includes data on the *Tocci* boy. The double-headed girl *Ritta-Christina* (born in Sardinia, 1829) died during infancy. A postmortem was performed and a skeleton was preserved. The double-headed boy *Tocci (Giovanni-Batista*, right, and *Giacomo*, left) was born 1877 (according to other sources 1875) near Turin. Until 1891 the two-headed boy was exhibited in Europe and the United States. After the *Tocci* 'twins' returned to Italy, their earnings permitted them to take up residence in a villa in Venice, and they refused to exhibit themselves again. No record has become available to me concerning their further fate.

As regards their recorded behavior, the independence of their moods and actions was noted. One head might be wide awake, and the other sleeping, one might be crying, and

Reverting now to the question (2) as formulated above, and concerning the *characteristic neurological features of Np-events*, no satisfactory answer can be given at this time. One might, nevertheless, conceive consciousness to be correlated with events in neuronal networks containing nerve cells of specific, not yet ascertained biochemical characteristics. Moreover, in order to be correlated with consciousness, these events might need to occur above a certain threshold, theoretically involving numerical parameters (either relative number of discharging elements, or frequencies of discharge bursts), or details of biochemical synaptic events. There would, thus, obtain a time lag between the closed transformation of N-events into Np-events, related to, but not identical with the 'reaction time' of psychology, which likewise involves conduction velocities and synaptic delays.[174]

The assumed, and theoretically perhaps ascertainable physical or physicochemical characteristics of Np-events would, of course, not provide an *ab initio* impossible causal explanation of consciousness, but could merely indicate the 'symbolic' or 'code value' for the Np-events correlated with consciousness. The assumed physical brain events, moreover, cannot be correlated with conscious events, because there exists no physical brain, except as a fiction in our conscious experience, this fiction being based upon the actual existence of a perceptual brain, which is already located in, and a mere aspect of, consciousness.

Public data, represented by perceptual matter, including body and brain, are merely mental phenomena, manifesting, in the space-time symbolism of conscious events, aspects of the unknown and unknowable reality in the first sense, i.e. of the 'orderliness x'.

---

the other in good humor. Although they could use their own arms freely, the inner ones were very crowded. They readily learned to write and draw, one using his left hand, and the other his right. In addition to their native Italian, they learned to speak some French and German in the course of their travels. MARK TWAIN, in his novel '*Pudd'nhead Wilson*', made use of an adaptation of the *Tocci-twin's* story.

[174] The complex and difficult problems related to 'simultaneity' have been discussed in § 20, pp. 79–88 of the author's monograph 'Mind and Matter' (K., 1961), to which the interested reader is referred. As regards 'the time required for the elaboration of the neuronal substrate of a very simple conscious experience' (LIBET, 1966; ECCLES, 1970) or for the 'so-called threshold of conscious experience', respectively for 'perceptual delay', the experiments of LIBET performed by means of electrical stimulation on the somesthetic cortex of conscious patients during brain operations can hardly be taken seriously. LIBET and ECCLES fail here to take into consideration the entirely 'unphysiological' character of electric stimuli applied to the cortex. Both cited authors, moreover, seem to be quite insufficiently acquainted with the relevant problems of time relationships.

7. References to Chapter XV

ABBIE, A.A.: Cortical lamination in Monotremata. J. comp. Neurol. *72:* 429–467 (1940a).

ABBIE, A.A.: The excitable cortex in Perameles, Sarcophilus, Dasyurus, Trichosurus and Wallabia (Macropus). J. comp. Neurol. *72:* 469–487 (1940b).

ABBIE, A.A.: Cortical lamination in a polyprotodont marsupial, Perameles nasuta. J. comp. Neurol. *76:* 509–536 (1942).

ABBOTT, E.A.: Flatland. A romance of many dimensions (Dover, New York 1952).

ADAM, G. (ed.): The biology of memory (Plenum Press, New York 1971).

ADRIAN, E.D.: The physical background of perception (Clarendon Press, Oxford 1947).

ALTSCHUL, R.: Die Blutgefässverteilung im Ammonshorn. Z. ges. Neurol. Psychiat. *163:* 634–642 (1938).

ANDERSEN, P.O.: Correlation of structural design with function in the archicortex; in ECCLES. Brain and conscious experience, pp. 59–84 (Springer, Berlin 1966).

ANDERSON, J.R. and BOWER, G.H.: Human associative memory (Wiley, New York 1974).

ARANTIUS, J.C.: De humano foetu etc. (Venetiis, 1587; quoted after LEWIS 1923).

ARNOLD, J. C.: Handbuch der Anatomie des Menschen, vol. 2 (Herder, Freiburg 1851).

ASANUMA, H. and ROSEN, I.: Spread of mono- and polysynaptic connections within cat's motor cortex. Exp. Brain Res. *16:* 507–520 (1973).

ASHBY, W.R.: Design for a brain (Wiley, New York 1952; 2nd ed. 1960).

ASHBY, W.R.: An introduction to cybernetics (Wiley, New York 1956, 1957).

BAILEY, P. and BONIN, G. VON: The isocortex of Man (University of Illinois Press, Urbana 1951).

BAILEY, P.; BONIN, G. VON, and MCCULLOCH, W.S.: The isocortex of the Chimpanzee (University of Illinois Press, Urbana 1950).

BAILLARGER, J.G.F.: Recherches sur la structure de la couche corticale des circonvolutions du cerveau. Mém. Acad. roy. Méd., Paris *8:* 149–183 (1840).

BARR, M.L.: The human nervous system; 2nd ed. (Harper & Row, Hagerstown 1974).

BECCARI, N.: Neurologia comparata anatomo-funzionale dei vertebrati, compreso l'uomo (Sanzoni, Firenze 1943).

BECHTEREW, W. VON: Die Leitungsbahnen in Gehirn und Rückenmark; 2. Aufl. (Georgi, Leipzig 1899).

BECK, E.: Zur Exaktheit der myeloarchitektonischen Felderung des Cortex cerebri. J. Psychol. Neurol. *36:* 281–288 (1925).

BERGER, H.: Experimentell anatomische Studien über die durch den Mangel optischer Reize veranlassten Entwicklungshemmungen im Occipitallappen des Hundes und der Katze. Arch. Psychiat. *33:* 521–567 (1909).

BERGER, H.: Psychophysiologie in 12 Vorlesungen (Fischer, Jena 1921).

BERKE, J.J.: The claustrum, the external capsule and the extreme capsule of Macaca mulatta. J. comp. Neurol. *115:* 297–331 (1960).

BERKELEY G.: A treatise concerning the principles of human knowledge (1710; Everyman's Library, Dent, London 1946).

BERKELEY, G.: Three dialogues between Hylas and Philonous, in opposition to sceptics and atheists (1713, 1725; Everyman's Library, Dent, London 1946).

BERKELEY, E.C: Giant brains or machines that think (Wiley, New York 1949).

BERKELEY, E.C.: The construction of living robots (Berkeley and Associates, New York 1952).

BERKELEY, E.C.: Symbolic logic and intelligent machines (Reinhold, New York 1959).

BERKELEY, E.C.: The computer revolution (Doubleday, Garden City 1962).

BERRY, M. and ROGERS, A.W.: Histogenesis of mammalian neocortex. In HASSLER and STEPHAN Evolution of the forebrain, pp. 197–205 (Thieme, Stuttgart 1966).

BETZ, W.: Anatomischer Nachweis zweier Gehirncentra. Centralbl. med. Wiss. *12:* 578–580, 595–599 (1874).

BETZ, W.: Über die feinere Structur der Gehirnrinde des Menschen. Centralbl. med. Wiss. *19:* 193–195, 209–213, 231–234 (1881).

BOK, S.T.: Der Einfluss der in den Furchen und Windungen auftretenden Krümmungen der Grosshirnrinde auf die Rindenarchitektur. Z. ges. Neurol. Psychiat. *121:* 682–750 (1929).

BOK, S.T.: Histonomy of the cerebral cortex (Elsevier, Amsterdam 1959).

BONIN, G. VON: The striate area of Primates. J. comp. Neurol. *77:* 405–429 (1942).

BONIN, G. VON: The cortex of Galago. Illinois Monogr. med. Sci. V/3 (University of Illinois Press, Urbana 1945).

BONIN, G. VON: Essay on the cerebral cortex (Thomas, Springfield 1950).

BONIN, G. VON: Some papers on the cerebral cortex. Translated from the French and German (Thomas, Springfield 1960).

BONIN, G. VON and BAILEY, P.: The neocortex of Macaca mulatta. Illinois Monogr. med. Sci. V/4 (University of Illinois Press, Urbana 1947).

BONIN, G. VON; GAROL, H.W., and MCCULLOCH, W.S.: The functional organization of the occipital lobe. Biol. Symp. *7:* 165–192 (1942).

BONIN, G. VON and MEHLER, W.R.: On columnar arrangement of nerve cells in cerebral cortex. Brain Res. *27:* 1–9 (1971).

BOOLE, G.: An investigation of the laws of thought, on which are founded the mathematical theories of logic and probabilities; 1854 (Dover, New York n.d.).

BOSCH, F.J.G. VAN DEN: Extending vision with electronics (Centre pour Electro-Microscopie, Antwerp 1959).

BRAAK, H.: Zur Pigmentarchitektonik der Grosshirnrinde des Menschen. I. Regio entorhinalis. II. Subiculum. Z. Zellforsch. *127:* 407–438; *131:* 235–254 (1972).

BRAIN, W.R.: Some reflections on brain and mind. Brain. *86:* 381–402 (1963).

BRAIN, W.R.: Perception: a trialogue. Brain *88:* 697–710 (1965).

BRAIN, W.R. and WALTON, J.W.: Brain's diseases of the nervous system; 7th ed. (Oxford University Press, London 1967).

BRILLOUIN, L.: Science and information theory (Academic Press, New York 1956).

BRINDLEY, G. and LEWIN, W.: The sensations produced by electrical stimulation of the visual cortex. J. Physiol., Lond. *196:* 479–493 (1968).

BROCA, P.: Remarques sur le siège de la faculté du langage articulé; suivies d'une observation d'aphémie. Bull. Soc. Anat. Paris, 2ème Ser. *6:* 330–357 (1861).

BRODMANN, K.: Vergleichende Lokalisationslehre der Grosshirnrinde in ihren Prinzipien dargestellt auf Grund des Zellenbaues (Barth, Leipzig 1909).

BROWN, J.W.: Some aspects of the early development of the hippocampal formation in certain insectivore bats; in HASSLER and STEPHAN Evolution of the forebrain, pp. 92–103 (Thieme, Leipzig 1966).

BURNS, B.D.: The mammalian cerebral cortex (Arnold, London 1958).

BURNS, B.D.: The uncertain nervous system (Arnold, London 1968).

BURNS, B.D. and GRAFSTEIN, B.: The function and structure of some neurones in the cat's cerebral cortex. J. Physiol., Lond. *118:* 412–433 (1952).

BURR, H.S.: The neural basis of human behavior (Thomas, Springfield 1960).
BURTON, H. and JONES, E.G.: The posterior thalamic region and its projection in New World and Old World monkeys. J. comp. Neurol. *168:* 249–301 (1976).
CAJAL, S.R. Y: Studien über die Hirnrinde des Menschen (übersetzt von Dr. J. BRESLER). I. Die Sehrinde. II. Die Bewegungsrinde. III. Die Hörrinde. IV. Die Riechrinde beim Menschen und Säugetier. V. Vergleichende Strukturbeschreibung etc. (Barth, Leipzig 1900a, 1900b, 1902, 1903, 1906).
CAJAL, S.R. Y: Histologie du système nerveux de l'homme et des vertébrés. 2 vols. (Maloine, Paris 1909, 1911; Instituto Ramon y Cajal, Madrid 1952, 1955).
CAJAL, S.R. Y: Textura de la corteza visual del gato. Trab. Lab. Invest. biol. *19:* 113–144 (1921).
CAJAL, S.R. Y: Estudios sobre la fina estructura de la corteza regional de los roedores. Trab. Lab. Invest. biol. *20:* 1–30 (1922).
CAJAL, S.R. Y: Recuerdos de mi vida; ed. (Pueyo, Madrid 1923).
CAJAL, S.R. Y: Histology. Transl. by M. FERNAN-NUÑEZ (Wood, Baltimore 1933).
CALLEJA, C.: La región olfatoria del cerebro (Moya, Madrid 1893).
CAMPBELL, A.W.: Histological studies on the localisation of cerebral function (Cambridge University Press, Cambridge 1905).
CAMPBELL, B.: Re-interpretation of the structure of the cerebral cortex (Abstract). Anat. Rec. *109:* 277 (1951).
CAMPBELL, B.: The organization of the cerebral cortex. I. Introduction and methodology. J. Neuropath. exp. Neurol. *13:* 407–416 (1954).
CAVINESS, V.S., JR.: Architectonic map of neocortex of the normal mouse. J. comp. Neurol. *164:* 247–263 (1975).
CHERRY, C.: On human communication (Science Editions, New York 1968).
CHUSID, J.G.: Black-and-white supplement for the color brain map of Bailey and von Bonin. Neurology *14:* 154–157 (1964).
CLARK, W.E. LE GROS: A note on cortical cytoarchitectonics. Brain *75:* 96–104 (1952).
COLONNIER, M.L.: The structural design of the neocortex; in ECCLES Brain and conscious experience, pp. 1–23 (Springer, Berlin 1966).
CONEL, J.L.: The postnatal development of the human cerebral cortex, vol. 1–7 (Harvard University Press, Cambridge 1939, 1941, 1947, 1951, 1955, 1959, 1963).
CRAIGIE, E.H.: An introduction to the finer anatomy of the central nervous system based upon that of the albino rat (Blakiston, Philadelphia 1925).
CRAIK, K.J.W.: The nature of explanation (Cambridge University Press, Cambridge 1943).
CRINIS, M. DE: Die Cytodendrogenese des menschlichen Grosshirns. Ein Beitrag zur Kenntnis der morphologischen Grundlagen der Ganglienzellfunktion. Proc. Kon. Akad. Wetensch. Amsterdam *35:* 197–204 (1932).
CROSBY, E. C.: The forebrain of Alligator mississipiensis. J. comp. Neurol. *27:* 325–402 (1917).
CROSBY, E.C.; HUMPHREY, T., and LAUER, E.W.: Correlative anatomy of the nervous system (Macmillan, New York 1962).
CUÉNOD, M.: Split-brain studies. Functional interaction between bilateral central nervous structures; in BOURNE The structure and function of nervous tissue, vol. V, chapter 8, pp. 455–506 (Academic Press, New York 1972).
CURTIS, H.J.: Intercortical connections of the corpus callosum as indicated by evoked potentials. J. Neurophysiol. *3:* 407–413 (1940).

DALY, D.D.: Cerebral localization; in BAKER Clinical neurology; 3rd ed., vol. 1, chapter 7, pp. 1–42 (Harper & Row, New York 1975).

DANDY, W.E.: The location of the consciousness center in the brain – the corpus striatum. Bull. Johns Hopk. Hosp. *79:* 34–58 (1946).

DIMOND, S.J. and BEAUMONT, J.G. (eds.): Hemisphere functions of the human brain (Wiley, New York 1974).

DOBELLE, W.H.; MLADEJOWSKY, M.G., and GIRVIN, J.P.: Artificial vision for the blind: electrical stimulation of visual cortex offers hope for a functional prosthesis. Science *183:* 440–444 (1974).

DUFFY, F.H. and BURCHFIEL, J.L.: Somatosensory system organizational hierarchy from single units in monkey area 5. Science *172:* 273–275 (1971).

ECCLES, J.C.: The neurophysiological basis of mind (Clarendon Press, Oxford 1953).

ECCLES, J.C. (ed.): Brain and conscious experience (Springer, Berlin 1966).

ECCLES, J.C.: Facing reality. Philosophical adventures by a brain scientist (Springer, New York 1970).

ECONOMO, C. VON: Zellaufbau der Grosshirnrinde des Menschen. Zehn Vorlesungen (Springer, Berlin 1927).

ECONOMO, C. VON: Nochmals zur Frage der arealen Grenzen in der Hirnrinde (Antwort auf die Vogtschen Darstellungen) Z. ges. Neurol. Psychiat. *124:* 309–316 (1930).

ECONOMO, C. VON und KOSKINAS, G.N.: Die Cytoarchitektonik der Hirnrinde des erwachsenen Menschen. Textband und Atlas (Springer, Berlin 1925).

EDINGER, L.: Vorlesungen über den Bau der nervösen Zentralorgane des Menschen und der Tiere. I. Das Zentralnervensystem des Menschen und der Säugetiere; 8. Aufl., 1911. II. Vergleichende Anatomie des Gehirns; 7. Aufl., 1908 (Vogel, Leipzig 1908–1911).

EDINGER, L.: Einführung in die Lehre vom Bau und den Verrichtungen des Nervensystems (Vogel, Leipzig 1912).

ELLIOTT, H.C.: The shape of intelligence. The evolution of the human brain (Scribner's Sons, New York 1969).

ETTLINGER, E.G.: Functions of the corpus callosum (Little, Brown, Boston 1965).

EULER, C. VON: On the significance of the high zinc content in the hippocampal formation; in PASSONANT Physiologie de l'hippocampe, pp. 135–145 (Centre natl. Rech. scient., Paris 1961).

EVANS, E.F.; ROSS, H.F., and WHITFIELD, I.C.: The spatial distribution of unit characteristic frequency in the primary auditory cortex of the cat. J. Physiol. *179:* 238–247 (1965).

EWERT, P.H.: A study of the effects of inverted retinal stimulation upon spatially co-ordinated behavior. Genet. Psychol. Monogr. *7* (1957).

FARHAT, N.H.: Advances in holography, vol. 1 (Dekker, New York 1975).

FECHNER, G.T.: Elemente der Psychophysik (Breitkopf & Härtel, Leipzig 1860).

FERRIER, D.: The functions of the brain (Smith-Elder, London 1876).

FILIMONOFF, I.N.: A rational subdivision of the cerebral cortex. Arch. Neurol. Psychiat. *58:* 296–311 (1947).

FILIMONOFF, I.N.: Homologies of the cerebral formations of mammals and reptiles. J. Hirnforsch. *7:* 229–251 (1964).

FILIMONOFF, I.N.: On the so-called Rhinencephalon in the dolphin. J. Hirnforsch. *8:* 1–23 (1965).

FLECHSIG, P.: Die Leitungsbahnen im Gehirn und Rückenmark des Menschen auf Grund entwicklungsgeschichtlicher Untersuchungen (Engelmann, Leipzig 1876).

FLECHSIG, P.: Gehirn und Seele; 2. Aufl. (Veit, Leipzig 1896).

FLECHSIG, P.: Gehirnphysiologie und Willenstheorien. Proc. 5th Int. Psychol. Congress, Rome 1905, pp. 73–89 (1905).

FLECHSIG, P.: Anatomie des menschlichen Gehirns und Rückenmarks auf myelogenetischer Grundlage (Thieme, Leipzig 1920).

FLECHSIG, P.: Meine myelogenetische Hirnlehre (Springer, Berlin 1927).

FLEISCHHAUER, K. und HORSTMANN, E.: Intravitale Dithizonfärbung homologer Felder der Ammonsformation von Säugern. Z. Zellforsch. *46:* 598–609 (1957).

FLOURENS, N.J.P: Recherches expérimentales sur les propriétés et les fonctions du système nerveux dans les animaux vertébrés (Crevot, Paris 1824).

FORBES, A.: The interpretation of spinal reflexes in terms of present knowledge of nerve conduction. Physiol. Rev. *2:* 361–414 (1922).

FORBES, A.: The mechanism of reaction; in MURCHISON The foundations of experimental psychology, pp. 128–168 (Clark University Press, Worcester 1929).

FOREL, A.: Gesammelte hirnanatomische Abhandlungen (Reinhardt, München 1907).

FOREL, A.: Gehirn und Seele; 13. Aufl. (Kröner, Leipzig 1922).

FRANZISKET, L.: Gewohnheitsbildung und bedingte Reflexe bei Rückenmarksfröschen. Z. vergl. Physiol. *32:* 142–178 (1951).

FRIEDE, R.L.: Topographic brain chemistry (Academic Press, New York 1966a).

FRIEDE, R.L.: The histochemical architecture of the ammon's horn as related to its selective vulnerability. Acta neuropath. *6:* 1–13 (1966b).

FRITSCH, G. und HITZIG, E.: Über die elektrische Erregbarkeit des Grosshirns. Arch. Anat. Physiol. wiss. Med. *37:* 300–332 (1870).

FUJITA, S.: Application of light and electron microscopic autoradiography to the study of cytogenesis in the forebrain; in HASSLER and STEPHAN Evolution of the forebrain, pp. 180–196 (Thieme, Stuttgart 1966).

FULTON, J.F.: Physiology of the nervous system; 3rd ed. (Oxford University Press, New York 1949).

GABOR, D.: Holography, 1948–1971. Science *177:* 299–313 (1972).

GANSER, S.: Vergleichend-anatomische Studien über das Gehirn des Maulwurfs. Morph. Jb. *7:* 591–725 (1882).

GASTAUT, H. et LAMMERS, H.J.: Anatomie du rhinencéphale; in ALAJOUANINE Les grandes activities du rhinencéphale, pp. 1–166 (Masson, Paris 1961).

GEORGE, F.H.: The brain as a computer (Pergamon Press, Oxford 1962).

GILBERT, C.D. and KELLY, J.P.: The projection of cells in different layers of the cat's visual cortex. J. comp. Neurol. *163:* 81–106 (1975).

GOLDSCHEIDER, A.: Über die materiellen Veränderungen bei der Assoziationsbildung. Neurol. Centralbl. *25:* 146–157; Discussion of paper on pp. 286–287 (1906).

GOLTZ, F.C.: Über die Verrichtungen des Grosshirns (Straus, Bonn 1881).

GOLTZ, F.: Über die Verrichtungen des Grosshirns. Pflügers Arch. ges. Physiol. *42:* 419–467 (1888).

GOULD, G.M. and PYLE, W.L.: Anomalies and curiosities of medicine (Saunders, Philadelphia 1896).

GRAY, P.A.: The cortical lamination pattern of the opossum, Didelphys virginiana. J. comp. Neurol. *37:* 221–263 (1924).

GUREWITSCH, M. und BYCHOWSKY, G.: Zur Architektonik der Hirnrinde (Isocortex) des Hundes. J. Psychol. Neurol. *35:* 283–300 (1928).

GUREWITSCH, M. und CHATSCHATURIAN, A.: Zur Zytoarchitektonik der Grosshirnrinde der Feliden. Z. Anat. Entw.-Gesch. *87:* 100–138 (1928).

HALDANE, J.B. SANDERSON: The Marxist philosophy and the sciences (Allen & Unwin, London 1938).

HALDANE, J. SCOTT: The philosophical basis of biology (Allen & Unwin, London 1931).

HALLER, B.: Die phyletische Entfaltung der Grosshirnrinde. Arch. mikr. Anat. *71:* 350–466 (1908).

HALLER, B.: Zur Ontogenie der Grosshirnrinde der Säugetiere. Anat. Anz. *37:* 282–293 (1910).

HALLER, GRAF V.: Über die Morphologie des Hippocampus, der Gyrus dentatus, Gyrus fasciolaris, Gyrus callosus und des Uncus. Anat. Anz. *66:* Ergänzungsheft, pp. 197–211 (1928).

HAMMARBERG, C.: Studien über Klinik und Pathologie der Idiotie, nebst Untersuchungen über die normale Anatomie der Hirnrinde (Berling, Upsala 1895).

HARTMANN, E. VON: Philosophie der Unbewussten. 3 vols. (Duncker, Berlin 1869; numerous subsequent editions).

HARTMANN, E. VON: Philosophy of the unconscious: speculative results according to the inductive method of physical science (Kegan Paul, Trench & Trubner, London 1931).

HASSLER, R.: Die Entwicklung der Architektonik seit Brodmann und ihre Bedeutung für die moderne Hirnforschung. Dtsch. med. Wschr. *87:* 1180–1185 (1962).

HASSLER, R.: Funktionelle Neuroanatomie und Physiologie; in GRUHLE *et al.* Psychiatrie der Gegenwart, vol. I/1a, pp. 152–285 (Springer, Berlin 1967).

HASSLER, R. und MUHS-CLEMENT, K.: Architektonischer Aufbau des sensorimotorischen und parietalen Cortex der Katze. J. Hirnforsch. *6:* 377–480 (1963/64).

HAUG, F. M. S.: Electron microscopical localization of zinc in hippocampal mossy fibre synapses by a modified sulfide silver procedure. Histochemie *8:* 355–368 (1967).

HAUG, F. M. S.: Heavy metals in the brain. A light microscope study of the rat with Timm's sulfide silver method. Ergebn. Anat. EntwGesch. *47/4:* 1–71 (1973).

HAYMAKER, W.: *Bing's* local diagnosis in neurological diseases; 14th ed., 15th ed. (Mosby, Saint Louis 1956, 1969).

HEAD, H.: Studies in neurology. 2 vols (Cambridge University Press, Cambridge 1926).

HEBB, D.O.: The organization of behavior (Wiley, New York 1949, 1961).

HEBB, D.O.: The problem of consciousness and introspection; in DELAFRESNAYE Brain mechanisms and consciousness, pp. 402–422 (Blackwell, Oxford 1954).

HEBB, D.O.: Intelligence, brain function and the theory of mind. Brain *82:* 260–275 (1959).

HEIMAN, M.: Über Gefässstudien am aufgehellten Hirn. Schweiz. Arch. Neurol. Neurochir. Psychiat. *40:* 277–301 (1937/38).

HERING, E.: Das Gedächtnis als eine allgemeine Funktion der organisierten Materie. Vortrag, Akad. Wien 1870 (Engelmann, Leipzig 1905).

HERRICK, C. J.: Brains of rats and men (University of Chicago Press, Chicago 1926).

HIRSCH, H.V.B. and SPINELLI, D.N.: Visual experience modifies distribution of horizontally and vertically oriented receptive fields in cats. Science *168:* 869–871 (1970).

HIRST, R. J.: Modern concepts of mind (Book review). Brain *89:* 391–396 (1966).

HJORTH-SIMONSEN, A.: Hippocampal efferents to the ipsilateral entorhinal area: an experimental study in the rat. J. compl. Neurol. *142:* 417–437 (1971).

HJORTH-SIMONSEN, A.: Some intrinsic connections of the hippocampus in the rat: an experimental analysis. J. comp. Neurol. *147:* 145–161 (1973).

Hochstetter, F.: Beiträge zur Entwicklungsgeschichte des menschlichen Gehirns. I. Teil (Deuticke, Leipzig 1919).

Holmes, R.L. and Berry, M.: Electron-microscopic studies on developing foetal cerebral cortex of the rat; in Hassler and Stephan Evolution of the forebrain, pp. 206–216 (Thieme, Stuttgart 1966).

Hoog, E.G. van't: Über Tiefenlokalisation in der Grosshirnrinde. Psychiat. neurol. Bladen *1918* (Feestbundel Winkler): 281–298 (1918).

Hoog, E.G. van't: On deep localization in the cerebral cortex. J. nerv. ment. Dis. *51:* 313–329 (1920).

Hopf, A.: Über eine Methode zur objektiven Registrierung der Myeloarchitektonik. J. Hirnforsch. *8:* 301–313 (1966).

Hubel, D.H.: Cortical unit responses to visual stimuli in nonanesthetized cats. Amer. J. Ophthal. *46:* 110–122 (1958).

Hubel, D.H.: Single unit activity in striate cortex of unrestrained cats. J. Physiol., Lond. *147:* 226–238 (1959).

Hubel, D.H.: Single unit activity in lateral geniculate body and optic tract of unrestrained cats. J. Physiol., Lond. *150:* 91–104 (1960).

Hubel, D.H. and Wiesel, T.N.: Receptive fields of single neurons of the cat's striate cortex. J. Physiol., Lond. *148:* 574–591 (1959).

Hubel, D.H. and Wiesel, T.N.: Receptive fields of optic nerve fibers in the spider monkey. J. Physiol., Lond. *154:* 572–580 (1960).

Hubel, D.H. and Wiesel, T.N.: Receptive fields, binocular interaction and functional architecture in the cat's visual cortex. J. Physiol., Lond. *160:* 106–154 (1962).

Hubel, D.H. and Wiesel, T.N.: Shape and arrangement of columns in cat's striate cortex. J. Physiol., Lond. *165:* 559–568 (1963a).

Hubel, D.H. and Wiesel, T.N.: Receptive fields of cells in striate cortex of very young, visually inexperienced kittens. J. Neurophysiol. *26:* 994–1002 (1963b).

Hubel, D.H and Wiesel, T.N.: Receptive fields and functional achitecture in two nonstriate visual areas (18 and 19) of the cat. J. Neurophysiol. *28:* 229–289 (1965).

Hubel, D.H. and Wiesel, T.N.: Receptive fields and functional architecture of monkey striate cortex. J. Physiol., Lond. *195:* 215–243 (1968).

Hubel, D. H. and Wiesel, T. N.: Anatomical demonstration of columns in the monkey striate cortex. Nature, Lond. *221:* 747–750 (1969).

Hubel, D.H. and Wiesel, T.N.: Laminar and columnar distribution of geniculo-cortical fibers in the macaque monkey. J. comp. Neurol. *158:* 295–306 (1972).

Hubel, D.H. and Wiesel, T.N.: Sequence regularity and geometry of orientation columns in the monkey striate cortex. J. comp. Neurol. *158:* 267–294 (1974a).

Hubel, D.H. and Wiesel, T.N.: Uniformity of monkey striate cortex: a parallel relationship between field size, scatter, and magnification factor. J. comp. Neurol. *158:* 295–306 (1974b).

Hubel, D.H. and Wiesel, T.N.: Ordered arrangement of orientation columns in monkeys lacking visual experience. J. comp. Neurol. *158:* 307–318 (1974c).

Hume, D.: A treatise of human nature. 2 vols. (1739–1740; Everyman's Library, Dent, London 1911).

Hume, D.: An enquiry concerning human understanding (1750; Oxford University Press, Oxford 1902).

Humphrey, T.: The development of the human hippocampal formation correlated with some aspects of its phylogenetic history; in HASSLER and STEPHAN Evolution of the forebrain, pp. 104–116 (Thieme, Leipzig 1966).

HUMPHREY, T.: The development of the human hippocampal fissure. J. Anat. *101:* 655–676 (1967).

HYDÉN, H.: Biochemical changes in glial cells and nerve cells at varying activity; in HOFFMAN-OSTERHOF Proc. 4th Int. Congr. Biochem., vol. 3, pp. 64–68 (Pergamon Press, London 1959).

HYDÉN, H.: Introductory remarks to the session on memory processes. Neurosci. Res. Bull., pp. 23–38 (MIT Press, Cambridge 1964).

HYDÉN, H.: Neuronal plasticity, protein conformation and behavior; in ZIPPEL Memory and transfer of information, pp. 511–529 (Plenum Press, New York 1973).

ISAACSON, R.L. and PRIBRAM, K.H. (eds.): The hippocampus. I. Structure and development. II. Neurophysiology and behavior. 2 vols. (Plenum Press, New York 1975).

JACKSON, J. HUGHLINGS: Selected writings. 2 vols. (Hodder & Stoughton, London 1931, 1932).

JACOBS, M.S.; MORGANE, P.J., and MCFARLAND, W.L.: The anatomy of the brain of the bottlenose dolphin (Tursiops truncatus). Rhinic lobe (Rhinencephalon. I. The palaeocortex. J. comp. Neurol. *141:* 205–271 (1971).

JACOBSON, S.: Sequence of myelinization in the brain of the albino rat. A. Cerebral cortex, thalamus and related structures. J. comp. Neurol. *121:* 5–29 (1963).

JAFFÉ, G.: Drei Dialoge über Raum, Zeit und Kausalität (Springer, Berlin 1954).

JAKOB, C. und ONELLI, C.: Vom Tierhirn zum Menschenhirn (Lehmann, München 1911).

JANSEN, J. and JANSEN, J.K.S.: The nervous system of Cetacea; in ANDERSEN The biology of marine Mammals, chapter 7, pp. 175–252 (Academic Press New, York 1968).

JOHNSTON, J.B.: The nervous system of vertebrates (Blakiston, Philadelphia 1906).

JONES, E.G. and BURTON, H.: Areal differences in the laminar distribution of thalamic afferents in cortical fields of the insular, parietal, and temporal regions of primates. J. comp. Neurol. *168:* 197–247 (1976).

KAES, T.: Die Grosshirnrinde des Menschen in ihren Massen und ihrem Fasergehalt (Fischer, Jena 1907).

KAHLE, W.: Studien über die Matrixphasen und die örtlichen Reifungsunterschiede im embryonalen menschlichen Hirn. Dtsch. Z. Nervenheilk. *166:* 273–304 (1951).

KAHLE, W.: Zur ontogenetischen Entwicklung der Brodmannschen Rindenfelder; in HASSLER and STEPHAN Evolution of the forebrain, pp. 305–315 (Thieme, Leipzig 1966).

KAHLE, W.: Die Entwicklung der menschlichen Grosshirnhemisphäre. Neurology series, vol. 1 (Springer, Berlin 1969).

KANT, I.: Kritik der reinen Vernunft; 1781, 1783 (Reclam, Leipzig, n.d.).

KANT, I.: Prolegomena zu einer jeden Metaphysik, die als Wissenschaft wird auftreten können; 1783 (Reclam, Leipzig n.d.).

KANT, I.: Kritik der praktischen Vernunft; 1788 (Reclam, Leipzig n.d.).

KAPPERS, C.U.A.: Die vergleichende Anatomie des Nervensystems der Wirbeltiere und des Menschen, vol. 2 (Bohn, Haarlem 1921).

KAPPERS, C.U.A.: The development of the cortex and the functions of its different layers. Acta psychiat. *3:* 115–132 (1928).

KAPPERS, C.U.A.: Anatomie comparée du système nerveux, particulièrement de celui des mammifères et de l'homme. Avec la collaboration de E.H. STRASBURGER (Bohn, Haarlem/Masson, Paris 1947).

KAPPERS, C.U.A.; HUBER, G.C., and CROSBY, E.C.: The comparative anatomy of the nervous system of vertebrates, vol. 2 (Macmillan, New York 1936).

KAWATA, A.: Zur Myeloarchitektonik der menschlichen Hirnrinde. Arb. neurol. Inst. Wiener Uni. *29:* 191–225 (1927).

KENNEDY, C.; DES ROSIERS, M.H.; JEHLE, J.W.; REIVICH, M.; SHARPE, F., and SOKOLOFF, L.: Mapping of functional neural pathways by autoradiographic survey of local metabolic rate with ^{14}C-desoxy-glucose. Science *187:* 850–853 (1975).

KIMBLE, D.P. (ed.): The anatomy of memory (Science and Behavior Books, Palo Alto 1965).

KIRSCHE, W.: Die Entwicklung des Telencephalons der Reptilien und deren Beziehung zur Hirn-Bauplanlehre. Nova Acta Leopoldina *37/2:* 9–78 (1972).

KIRSCHE, W.: Zur vergleichenden funktionsbezogenen Morphologie der Hirnrinde der Wirbeltiere auf der Grundlage embryologischer und neurohistologischer Untersuchungen. Z. mikr.-anat. Forsch. *88:* 21–57 (1974).

KLATZKY, R.L.: Human memory. Structures and processes (Freeman, San Francisco 1975).

KNOOK, H.L.: The fiber connections of the forebrain (van Gorcum, Assen 1965).

KOELLIKER, A. VON: Handbuch der Gewebelehre des Menschen, vol. 2: Nervensystem des Menschen und der Tiere; 6. Aufl. (Engelmann, Leipzig 1896).

KOIKEGAMI, H.; IGARASI, K. und ATUMI, H.: Zur Zytoarchitektonik der Grosshirnrinde des Elefanten, insbesondere des Area gigantopyramidalis. Okajimas Fol. anat. japon. *22:* 95–110 (1943).

KORNMÜLLER, A.E.: Die bioelektrischen Erscheinungen der Hirnrindenfelder (Thieme, Leipzig 1937).

KOVAČEVIČ, R. and RADULOVAČKI, M.: Monoamine changes in the brain of cats during slow-wave sleep. Science *193:* 1025–1027 (1976).

KRETSCHMER, E.: Medizinische Psychologie; 3. Aufl. (Thieme, Leipzig 1926).

KRIEG, W.J.S.: Connections of the cerebral cortex. I. The albino rat. A. Topography of the cortical areas. B. Structure of the cortical areas. C. Extrinsic connections. J. comp. Neurol. *84:* 221–275, 277–323 (1946); *86:* 267–394 (1947).

KRIEG, W.J.S.: Connections of the frontal cortex in the monkey (Thomas, Springfield 1954).

KRIEG, W.J.S.: Connections of the cerebral cortex (Brain Books, Evanston 1963).

KRIEG, W.J.S.: Architectonics of human cerebral fiber systems (Brain Books, Evanston 1973).

KUFFLER, S.W.: Discharge patterns and functional organization of mammalian retina. J. Neurophysiol. *16:* 37–68 (1953).

KUHLENBECK, H.: Über den Ursprung der Grosshirnrinde. Eine phylogenetische und neurobiotaktische Studie. Anat. Anz. *55:* 330–339 (1922).

KUHLENBECK, H.: Über den Ursprung der Basalganglien des Grosshirns. Anat. Anz. *58:* 49–74 (1924a).

KUHLENBECK, H.: Über den Ursprung der Grosshirnrinde. Neurologia, Tokyo *24:* 1–18 (1924b).

KUHLENBECK, H.: Über die Homologien der Zellmassen im Hemisphärenhirn der Wirbeltiere. Folia anat. japon. *2:* 325–364 (1924c).

KUHLENBECK, H.: Weitere Mitteilungen zur Genese der Basalganglien. Über die sogenannten Ganglienhügel. Anat. Anz. *60:* 35–40 (1925–26).

KUHLENBECK, H.: Vorlesungen über das Zentralnervensystem der Wirbeltiere (Fischer, Jena 1927).

KUHLENBECK, H.: Über die anatomischen Grundlagen nervöser Mechanismen. Psychiat.-neurol. Wschr. *30:* 336–338 (1928).

KUHLENBECK, H.: Die Grundbestandteile des Endhirns im Lichte der Bauplanlehre. Anat. Anz. *67:* 1–51 (1929).

KUHLENBECK, H.: The ontogenetic development and phylogenetic significance of the cortex telencephali in the chick. J. comp. Neurol. *69:* 273–301 (1938).

KUHLENBECK, H.: The human diencephalon. A summary of development, structure, function, and pathology (Karger, Basel 1954).

KUHLENBECK, H.: Brain and consciousness. Some prolegomena to an approach of the problem (Karger, Basel 1957).

KUHLENBECK, H.: The meaning of 'postulational psycho-physical parallelism'. Brain *81:* 588–603 (1958).

KUHLENBECK, H.: Further remarks on brain and consciousness: the brain-paradox and the meanings of consciousness. Confin. neurol. *19:* 462–485 (1959).

KUHLENBECK, H.: Mind and matter. An appraisal of their significance for neurologic theory (Karger, Basel 1961).

KUHLENBECK, H.: Gehirn und Intelligenz. Confin. Neurol. *25:* 35–62 (1965a).

KUHLENBECK, H.: The concept of consciousness in neurological epistemology; in SMYTHIES Brain and mind – modern concepts of the nature of mind, pp. 137–161 (Routledge & Kegan Paul, London 1965b).

KUHLENBECK, H.: Weitere Bemerkungen zur Maschinentheorie des Gehirns. Confin. neurol. *27:* 295–328 (1966).

KUHLENBECK, H.: Some comments on words, language, thought and definition; in BUEHNE *et al.* Helen Adolf Festschrift, pp. 9–29 (Ungar, New York 1968).

KUHLENBECK, H.: Some comments on psychophysics. Confin. neurol. *33:* 245–257 (1971).

KUHLENBECK, H.: Schopenhauers Satz 'Die Welt ist meine Vorstellung' und das Traumerlebnis. Schopenhauer Jahrb. *53* (Festschrift Hübscher): 376–392 (1972).

KUHLENBECK, H.: Gehirn und Bewusstsein. Erfahrung und Denken. Schriften zur Förderung der Beziehungen zwischen Philosophie und Einzelwissenschaften, vol. 39. Transl. by Prof. J. GERLACH and Dr. U. PROTZER (Duncker & Humblot, Berlin 1973).

KUHLENBECK, H. und DOMARUS, E. VON: Zur Ontogenese des menschlichen Grosshirns. Anat. Anz. *53:* 316–320 (1920).

KUHLENBECK, H.; SZEKELY, E.G., and SPULER, H.: Some remarks on the zonal pattern of Mammalian cortex cerebri as manifested in the rabbit: its relationship with certain electrocorticographic findings. Conf. neurol. *20:* 407–423 (1960).

KÜPFMÜLLER, K.: Informationsverarbeitung durch den Menschen. Nachrichtentechn. Z. *12:* 68–74 (1958).

LAISSUE, J.: Die histogenetische Gliederung der Rindenanlage des Endhirns. Acta anat. *53:* 158–185 (1963).

LANDAHL, H.D.: Neural mechanisms for the concepts of difference and similarity. Bull. math. Biophys. *7:* 83–88 (1945).

LANDAHL, H.D.; MCCULLOCH, W.S., and PITTS, W.: A statistical consequence of the logical calculus of nervous nets. Bull. math. Biophys. *5:* 135–137 (1943).

LANDGREN, S.; PHILLIPS, C.G., and PORTER, R.: Cortical fields of origin of the monosynaptic pyramidal pathways to some alpha motoneurones of the baboon's hand and forearm. J. Physiol., Lond. *161:* 112–125 (1962).

LANGWORTHY, O.R.: A description of the central nervous system of the porpoise (Tursiops truncatus). J. comp. Neurol. *32:* 437–499 (1932).

LASHLEY, K.S.: The behavioristic interpretation of consciousness. Psychol. Rev. *30:* 237–272, 329–353 (1923).

LASHLEY, K.S.: Thalamo-cortical connections in the rat's brain. J. comp. Neurol. *75:* 67–121 (1941).

LASHLEY, K.S.: In search of the engram. Symp. Soc. exp. Biol., physiol. mechanisms in animal behavior, pp. 454–482 (Cambridge Universit. Press, Cambridge 1950).

LASHLEY, K.S. and CLARK, G.: The cytoarchitecture of the cerebral cortex of Ateles: a critical examination of architectonic studies. J. comp. Neurol. *85:* 223–305 (1946).

LEWIS, F.T.: The significance of the term Hippocampus. J. comp. Neurol. *35:* 213–230 (1923).

LEWIS, W.BEVAN: On the comparative structure of the cortex cerebri. Brain *1:* 79–96 (1878).

LEWIS, W.BEVAN: On the comparative structure of the brain in rodents. Philos. Trans. roy. Soc. *173, I:* 699–749 (1882).

LEYTON, A.S.F. and SHERRINGTON, C.S.: Observations on the excitable cortex of the chimpanzee, orangutan and gorilla. Quart. J. exp. Physiol. *11:* 135–222 (1917).

LIBET, B.: Brain stimulation and the threshold of conscious experience; in ECCLES Brain and conscious experience, pp. 165–181 (Springer, Berlin 1966).

LILLY, J.C.: Man and dolphin (Doubleday, New York 1963).

LILLY, J.C.: Lilly on dolphins. Humans of the sea (Anchor-Doubleday, Garden City 1975).

LILLY, J.C. and MILLER, A.M.: Vocal exchanges between Dolphins. Science *134:* 1873–1876 (1961).

LINDENBERG, R.: Compression of brain arteries as pathogenetic factor for tissue necroses and their areas of predilection. J. Neuropath. exp. Neurol. *14:* 223–243 (1955).

LOCKE, J.: An essay concerning human understanding; 1689, 1690 (Oxford University Press, Oxford 1894).

LOO, Y.T.: On formation of human cerebral cortex. An ontogenetic study with a discussion of the function of different cortical layers. Anat. Anz. *68:* 305–324 (1929).

LOOS, H. VAN DER and GLAZER, E.M.: Autapses in neocortex cerebri: synapses between a pyramidal cell and its own dendrites. Brain Res. *48:* 355–360 (1972).

LOOS, H. VAN DER and WOOLSEY, T.A.: Somatosensory cortex: structural alterations following early injury to sense organs. Science *179:* 395–398 (1973).

LORENTE DE NÓ, R.: La corteza cerebral del ratón. Trab. Invest. biol. *20:* 41–78 (1922/23).

LORENTE DE NÓ, R.: Studies on the structure of the cerebral cortex. II. Continuation of the study of the ammonic system. J. Psychol. Neurol. *46:* 113–177 (1934).

LORENTE DE NÓ, R.: Cerebral cortex; in FULTON Physiology of the nervous system; 3rd ed., chapter XV, pp. 288–312 (Oxford University Press, New York 1949).

LORENTZ, H.A.; EINSTEIN, A.; MINKOWSKI, H., and WEYL, H.: The principle of relativity (Dover, New York, n.d.).

LUDWIG, E. and KLINGLER, J.: Atlas cerebri humani (Karger, Basel 1956).

LUGARO, E.: I recenti progressi dell'anatomia del sistema nervoso in rapporto alla psicologia ed alla pischiatria. Riv. Patol. nerv. mentale 4 (1899). Quoted after CAJAL (1911).

LUND, J.S.; LUND, R.D.; HENDRICKSON, A.H., and FUCHS, A.F.: The origin of efferent pathways from the primary visual cortex, area 17, of the macaque monkey, as shown by retrograde transport of horseradish peroxidase. J. comp. Neurol. 164: 287–303 (1975).

MACH, E.: Die Analyse der Empfindungen und das Verhältnis des Physischen zum Psychischen; 9. Aufl. (Fischer, Jena 1922).

MACKAY, D. M.: Comparing the brain with machines. Amer. Scient. *42:* 261–268 (1954).

MARK, R.: Memory and nerve cell connections. Criticisms and contributions from developmental neurophysiology (Clarendon Press, Oxford 1974).

MARTINOTTI, C.: Beiträge zum Studium der Hirnrinde und dem Centralursprung der Nerven. Int. Mschr. Anat. Physiol. *7:* 69–90 (1890).

MASKE, H.: Über den topochemischen Nachweis von Zink im Ammonshorn verschiedener Säugetiere. Naturwissenschaften *42:* 424 (1955).

MATHIES, H.: Biochemical regulation of synaptic activity; in ZIPPEL Memory and transfer of information, pp. 531–548 (Plenum Press, New York 1973).

MCCULLOCH, W.S. and GAROL, H.W.: Cortical origin and distribution of corpus callosum and anterior commissure in the monkey (Macaca mulatta). J. Neurophysiol. *4:* 555–563 (1941).

MCCULLOCH, W.S. and PITTS, W.: A logical calculus of the ideas immanent in nervous activity. Bull. math. Biophys. *5:* 115–133 (1943).

MCLARDY, T.: Neurosyncytial aspects of the hippocampal mossy fiber system. Confin. neurol. *20:* 1–17 (1960).

MCLARDY, T.: Zinc enzymes and the hippocampal mossy fiber system. Nature, Lond. *194:* 300–302 (1962).

MCLARDY, T.: Some cell and fibre peculiarities of uncal hippocampus. Progr. Brain Res. *3:* 71–88 (1963).

MCLARDY, T.: Second hippocampal zinc-rich synaptic system. Nature, Lond. *201:* 92–93 (1964).

MCLARDY, T.: Hilum hippocampi: some primate and subprimate histologic features relevant to modeling of habituation circuitry IRCS (Research on: Anatomy and Human Biology; Neurobiology and Neurophysiology; Neurology and Neurosurgery) *2:* 1442 (1974); as quoted by STEPHAN (1975).

MEYNERT, T.: Der Bau der Grosshirnrinde und seine örtlichen Verschiedenheiten, nebst einem pathologisch-anatomischen Corollarium. Vjschr. Psychiat. *1:* 77–93 (1867); *2:* 88–113 (1868). Also (Heuser, Neuwied 1868).

MEYNERT, T.: Vom Gehirn der Säugethiere; in STRICKER Handbuch der Lehre von den Geweben des Menschen und der Thiere, vol. 2, Kapitel XXXI, pp. 694–808 (Engelmann, Leipzig 1872).

MINKOWSKI, H.: Space and time (1908); reprinted in LORENTZ *et al.* The principle of relativity, pp. 75–91 (Dover, New York n.d.).

MONAKOW, C. VON: Lokalisation der Hirnfunktionen. J. Psychol. Neurol. *17:* 185–200 (1911).

MOORE, R.Y. and HALARIS, A.E.: Hippocampal innervation by serotonin neurons of the midbrain raphe in the rat. J. comp. Neurol. *164:* 171–183 (1975).

MOTT, W.W.: The progressive evolution of the structure and function of the visual cortex. Arch. Neurol. (London County Asylums) *3:* 1–48 (1907).

MOUNTCASTLE, V.B.: Modality and topographic properties of single neurons of cat's somatic sensory cortex. J. Neurophysiol. *20:* 408–434 (1957).

MOUNTCASTLE, V.B.: The neural replication of sensory events in the somatic afferent system; in ECCLES Brain and conscious experience, pp. 85–115 (Springer, Berlin 1966).

MUNK, H.: Über die Funktionen der Grosshirnrinde (Hirschwald, Berlin 1881).

MURDOCK, B.B.: Human memory. Theory and data (Wiley, New York 1974).

NAUTA, J.W.: Terminal distribution of some afferent fiber systems in the cortex (Abstract). Anat. Rec. *118:* 333 (1954).

NEUMANN, J. VON: Probabilistic logics (California Institute of Technology, Pasadena 1952).

NEUMANN, J. VON: The computer and the brain (Yale University Press, New Haven 1958).

NGOWYANG, G.: Beschreibung einer Art von Spezialzellen in der Inselrinde, zugleich Bemerkungen über die v. Economoschen Spezialzellen. J. Psychol. Neurol. *44:* 671–674 (1932).

NGOWYANG, G.: A further contribution to the morphology of the 'fork-cells' (Chinese, with English summary). Science, Nanking *18:* 216–221 (1934).

NGOWYANG, G.: Neuere Befunde über die Gabelzellen. Z. Zellforsch. mikr. Anat. *25:* 236–239 (1936).

NGOWYANG, G.: Structural variations of the visual cortex in primates. J. comp. Neurol. *67:* 89–107 (1937).

NIIMI, K.; KADOTA, M., and MATSUSHITA, Y.: Cortical projections of the pulvinar nuclear group in the cat. Brain Behav. Evol. *9:* 422–457 (1974).

NIKLOWITZ, W.: Elektronenmikroskopische Untersuchungen an den Pyramidenzellen des Ammonshorns. Nervenarzt *35:* 463–469 (1964).

NIKLOWITZ, W.: Elektronenmikroskopische Untersuchungen am Ammonshorn. II, III. Z. Zellforsch. *70:* 220–239; *75:* 485–500 (1966).

NIKLOWITZ, W. und BAK, I.J.: Elektronenmikroskopische Untersuchungen am Ammonshorn. I. Die normale Substruktur der Pyramidenzellen. Z. Zellforsch. *66:* 529–547 (1965).

NOBLE, G.K.: The biology of the Amphibia (McGraw-Hill, New York 1931/Dover, New York 1954).

OBERSTEINER, H.: Anleitung beim Studium des nervösen Zentralorgans im gesunden und kranken Zustande; 5. Aufl. (Deuticke, Leipzig 1912).

O'LEARY, J.L.: Structure of the area striata in the cat. J. comp. Neurol. *75:* 131–164 (1941).

PAKKENBERG, H.: The number of nerve cells in the cerebral cortex of man. J. comp. Neurol. *128:* 17–20 (1966).

PAPEZ, J.W.: A proposed mechanism of emotion. Arch. Neurol. Psychiat. *38:* 725–743 (1937).

PASTERNAK, J.F. and WOOLSEY, T.A.: The number, size and spatial distribution of neurons in lamina IV of the mouse Sm1 cortex. J. comp. Neurol. *160:* 291–306 (1975).

PAVLOV, I.: Die normale Tätigkeit und allgemeine Konstitution der Grosshirnrinde. Skand. Arch. Physiol. *44:* 32–41 (1923).

PAVLOV, I.: Conditioned reflexes (Oxford University Press, Oxford 1927).

PENFIELD, W.: The role of the temporal lobe in certain psychical phenomena. J. ment. Sci. *101:* 451–465 (1955).

PFEIFER, R.A.: Grundlegende Untersuchungen für die Angioarchitektonik des menschlichen Gehirnes (Springer, Berlin 1930).

PFEIFER, R.A.: Die angioarchitektonische Gliederung der Grosshirnrinde (Thieme, Leipzig 1940).

PIERCE, J.F.: Symbols, signals, and noise. The nature and process of communication (Harper & Bros., New York 1961).

PILLERI, G.: Die Grosshirnrinde des kanadischen Bibers (Castor canadensis). Z. Hirnforsch. *6:* 1–63 (1963).

PILLERI, G.: (1962, 1964, 1965/66a, b) cf. References to Chapter XIV.

POINCARÉ, H.: La valeur de la science (Flammarion, Paris 1909).

POINCARÉ H.: The foundation of science. Transl. by G.B. HALSTEAD (Science Press, New York 1913).

POLIAK, S.: The main afferent fiber systems of the cerebral cortex in primates (University of California Press, Berkeley 1932).

POLIAKOV, G.I.: Some results of research into the development of the neuronal structure of the cortical ends of the analyzers in man. J. comp. Neurol. *117:* 197–212 (1961).

POLLEN, D.A.; LEE, J.R., and TAYLOR, J.H.: How does the striate cortex begin the reconstruction of the visual world? Science *173:* 74–77 (1971).

PURPURA, D.P. (ed.): Methodological approaches to the study of brain maturation and its abnormalities (University Park Press, Baltimore 1974).

RAHMANN, E.: Radioactive studies of changes in protein metabolism by adequate and inadequate stimulation of the optic tectum of teleosts; in ZIPPEL Memory and transfer of information, pp. 547–570 (Plenum Press, New York 1973).

RAISMAN, G.; COWAN, H.M., and POWELL, T.P.S.: An experimental analysis of the efferent projection of the hippocampus. Brain *89:* 83–108 (1966).

RANSON, S.W.: The anatomy of the nervous system; 7th ed. (Saunders, Philadelphia 1943).

RASHEWSKY, N.: Some remarks on the Boolean algebra of nervous nets in mathematical biophysics. Bull. math. Biophys. *7:* 203–211 (1945).

RASHEWSKY, N.: The neural mathematics of logical thinking. Bull. math. Biophys. *8:* 29–40 (1946).

REBLET, C.: La encrucijada ventricular cerebral alterada experimentalmente y vías de degeneración Walleriana consecuente (referencias a las conexiones del hipocampo). Anal. Anat., Zaragoza *25:* 407–435 (1976).

REMANE, A.: Phylogenetische Entwicklungsregeln von Organen; in HASSLER and STEPHAN Evolution of the forebrain, pp. 1–8 (Thieme, Stuttgart 1966).

RENSCH, B.: Homo sapiens. Vom Tier zum Halbgott (Vandenhoeck et Rupprecht, Göttingen 1959).

RENSCH, B.: Biophilosophie auf erkenntnistheoretischer Grundlage (Fischer, Stuttgart 1968).

RENSCH, B.: Probleme der Gedächtnisspuren. Rheinisch-Westfälische Akad. Wiss. Abt. H. *211:* 7–67 (1971).

RENSCH, B.: Gedächtnis, Begriffsbildung und Planhandlung bei Tieren (Parey, Berlin 1973).

RENSCH, B. und RAHMANN, H.: Autoradiographische Untersuchungen über visuelle 'Engramm'-Bildung bei Zahnkarpfen. I. Pflügers Arch. ges. Physiol. *290:* 158–166 (1966).

RENSCH, B.; RAHMANN, H. und SKRZIPEK, K.: Autoradiographische Untersuchungen über visuelle 'Engramm'-Bildung bei Fischen. II. Pflügers Arch. ges. Physiol. *304:* 242–252 (1968).

RIESE, W.: Formprobleme des Gehirns. II. Über die Hirnrinde der Wale. J. Psychol. Neurol. *31:* 275–280 (1925).

RIESE, W.: Konvergenzerscheinungen am Gehirn, nebst Bemerkungen zu der Arbeit von ROSE: 'Der Grundplan der Cortexarchitektonik beim Delphin'. J. Psychol. Neurol. *33:* 84–96 (1927).

RIESE, W.: The cellular structure of the marsupial cortex. Natural. Canad. *70:* 139–144 (1943).

RIESE, W.: Hughlings Jackson's doctrine of consciousness. Sources, versions and elaborations. J. nerv. ment. Dis. *120:* 330–337 (1954).

RITTER, J.; MEYER, U. und WENK, H.: Zur Chemodifferenzierung der Hippocampusformation in der postnatalen Entwicklung der Albinoratte. II. Transmitterenzyme. J. Hirnforsch. *13:* 255–278 (1971/72).

ROBERTSON, B.T.: Further studies in the chemical dynamics of the central nervous system. 1. The time-relations of simple voluntary movements. 2. On the physiological conditions underlying phenomena of heightened suggestibility, hypnosis, multiple personality, sleep, etc. 3. On the nature of the process of forgetting. Folia neurobiol. *6:* 553–578 (1912); *7:* 309–337 (1913); *8:* 485–506 (1914).

ROSE, M.: Histologische Lokalisation der Grosshirnrinde bei kleinen Säugetieren (Rodentia, Insectivora, Chiroptera). J. Psychol. Neurol. *19:* 391–479 (1912).

ROSE, M.: Über das histogenetische Prinzip der Einteilung der Grosshirnrinde. J. Psychol. Neurol. *32:* 97–160 (1926a).

ROSE, M.: Der Grundplan der Cortextektonik beim Delphin. J. Psychol. Neurol. *32:* 161–169 (1926b).

ROSE, M.: Der Allocortex bei Tier und Mensch. I. Teil. J. Psychol. Neurol. *34:* 1–111 (1927a).

ROSE, M.: Die sogenannte Riechrinde beim Menschen und beim Affen. II. Teil des 'Allocortex bei Tier und Mensch'. J. Psychol. Neurol. *34:* 261–401 (1927b).

ROSE, M.: Die Ontogenie der Inselrinde. Zugleich ein Beitrag zur histogenetischen Rindeneinteilung. J. Psychol. Neurol. *32:* 182–209 (1928).

ROSE, M.: Die Inselrinde des Menschen und der Tiere. J. Psychol. Neurol. *37:* 467–624 (1929a).

ROSE, M.: Cytoarchitektonischer Atlas der Grosshirnrinde der Maus. J. Psychol. Neurol. *40:* 1–51 (1929b).

ROSE, M.: Cytoarchitektonischer Atlas der Grosshirnrinde des Kaninchens. J. Psychol. Neurol. *43:* 353–440 (1931).

ROYCE, G.J.; MARTIN G.F., and DOM, R.M.: Functional localization and cortical architecture in the nine-banded Armadillo (Dasypus novemcinctus mexicanus). J. comp. Neurol. *1975:* 495–521.

RUCH, T.C.: Somatic sensation; in RUCH, PATTON, WOODBURY and TOWE Neurophysiology, chapter 13, pp. 300–322 (Saunders, Philadelphia 1961).

RYZEN, M. and CAMPBELL, B.: Organization of the cerebral cortex. III. Cortex of Sorex pacificus. J. comp. Neurol. *102:* 365–424 (1955).

SANIDES, F.: Die Architektonik des menschlichen Stirnhirns. Zugleich eine Darstellung der Prinzipien seiner Gestaltung als Spiegel der stammesgeschichtlichen Differenzierung der Grosshirnrinde. Monogr. Gesamtgeb. Neurol. Psychiat., vol. 98 (Springer, Berlin 1962).

SANIDES, F.: Representation in the cerebral cortex and its areal lamination patterns; in BOURNE The structure and function of nervous tissue, vol. V, pp. 329–453 (Academic Press, New York 1972).

SARKISOV, S.A.; POPOVA, E.N., and BOGOLEPOV, N.N.: Structure of synapses in evolutionary aspect (optic and electron-microscopic studies); in HASSLER and STEPHAN Evolution of the forebrain, pp. 225–236 (Thieme, Stuttgart 1966).

SCHAEFER, E.: Das menschliche Gedächtnis als Informationsspeicher. Elektron. Rdsch. *14:* 79–84 (1960).

SCHAFFER, K.: Beitrag zur Histologie der Ammonsformation. Arch. mikr. Anat. *39:* 611–632 (1892).

SCHARRER, E.: Vascularization and vulnerability of the cornu Ammonis in the opossum. Arch. Neurol. Psychiat. *44:* 483–506 (1940).

SCHILDER, P.: Das Körperschema (Springer, Berlin 1923).

SCHILDER, P.: The image and the appearance of the human body; studies in constructive energies of the psyche (Kegan, London 1935).

SCHMITT, F.O.; DEV, P., and SMITH, B.H.: Electrotonic processing of information by brain cells. Science *193:* 114–120 (1976).

SCHOPENHAUER, A.: Sämmtliche Werke (1813–1831), 6 vols. Handschriftlicher Nachlass, 4 vols. (ed. *Grisebach*, Reclam, Leipzig n.d.; ed. *Hübscher*, 7 vols., Brockhaus, Wiesbaden 1946–1950; Nachlass, 5 vols. Kramer, Frankfurt 1966–1975).

SCHWALBE, G.: Lehrbuch der Neurologie (Besold, Erlangen 1881).

SCOVILLE, W.B. and MILNER, B.: Loss of recent memory after bilateral hippocampal lesions. J. Neurol. Neurosurg. Psychiat. *20:* 11–21 (1957).

SEMON, R.: Die Mneme als erhaltendes Prinzip im Wechsel des organischen Geschehens (Engelmann, Leipzig 1904).

SEMON, R.: Die mnemischen Empfindungen in ihren Beziehungen zu den Originalempfindungen (Engelmann, Leipzig 1909).

SEMON, R.: Bewusstseinsvorgang und Gehirnprozess (Bergmann, Wiesbaden 1920).

SHANNON, C.E.: A symbolic analysis of relay and switching circuits. Trans. amer. Inst. electr. Engin. *57:* 713–723 (1938).

SHANNON, C.E.: Programming a computer for playing chess. Philos. Mag. *41:* 256–275 (1950).

SHANNON, C.E.: and WEAVER, W.: The mathematical theory of communication (University of Illinois Press, Urbana 1949).

SHELLSHEAR, J.L.: A contribution to our knowledge of the arterial supply of the cerebral cortex in man. Brain *50:* 236–253 (1927).

SHELLSHEAR, J.L.: The arterial supply of the cerebral cortex. Proc. kon. nederl. Akad. Wet. *36:* 700–710 (1933).

SHEPHERD, G.U.: The synaptic organization of the brain. An introduction (Oxford University Press, New York 1974).

SHERRINGTON, C.S.: The integrative action of the nervous system (Yale University Press, Hew Haven 1906).

SHERRINGTON, C.: Man on his nature; 2nd ed. (Cambridge University Press, Cambridge 1951).

SHOLL, D.A.: The organization of the cerebral cortex (Hafner, New York 1956, 1967).

SHOLL, D.A.: Anatomical heterogeneity in the cerebral cortex; in TOWER and SCHADÉ Structure and function of the cerebral cortex, pp. 21–27 (Elsevier, Amsterdam 1960).

SMITH, G.ELLIOT: The brain in the edentata. Trans. Linnean Soc. Lond. 2. ser. Zool. *7.:* 277–394 (1898).

SMITH, G.ELLIOT: Notes upon the natural subdivision of the cerebral hemispheres. J. Anat. *35:* 431–454 (1901).

SMITH, G.ELLIOT: A new topographical survey of the human cerebral cortex. J. Anat. *41:* 237–254 (1907).

Smith, G. Elliot: Some problems relating to the evolution of the brain. The Arris and Gale lectures. Lancet *i:* 1–6, 147–153, 221–227 (1910a).
Smith, G. Elliot: The term 'archipallium' – a disclaimer. Anat. Anz. *35:* 429–430 (1910b).
Smith, G. Elliot: A preliminary note on the morphology of the corpus striatum and the origin of the neopallium. J. Anat. *53:* 271–291 (1919).
Smith, F. V.: Explanation of human behaviour; 2nd ed. (Constable, London 1960).
Smythies, J.R. and Adey, W.R.: Brain mechanisms and behaviour; 2nd ed. (Academic Press, New York 1970).
Snyder, F.W. and Pronko, N.H.: Vision with spatial inversion (University Press, Wichita 1952).
Sommer, A.: Erkrankungen des Ammonshorns als etiologisches Moment der Epilepsie. Arch. Psychiat. *10:* 631–675 (1880).
Spatz, H.: Gehirnentwicklung (Introversion-Promination) und Endocranialausguss; in Hassler and Stephan Evolution of the forebrain, pp. 136–152 (Thieme, Stuttgart 1966).
Spielmeyer, W.: Zur Pathogenese örtlich elektiver Gehirnveränderungen. Z. ges. Neurol. Psychiat. *99:* 756–766 (1925).
Steinbuch, K.: Automat und Mensch. Über menschliche und maschinelle Intelligenz (Springer, Berlin 1961).
Stephan, H.: Vergleichend-anatomische Untersuchungen an Insektivorengehirnen. In Hirnform, palaeo-neocortikale Grenze und relative Zusammensetzung der Cortexoberfläche. Morph. Jb. *97:* 123–142 (1956).
Stephan, H.: Die quantitative Zusammensetzung der Oberflächen des Allocortex bei Insektivoren und Primaten; in Tower and Schadé Structure and function of the cerebral cortex, pp. 51–58 (Elsevier, Amsterdam 1960).
Stephan, H.: Allocortex; in von Möllendorff and Bargmann Handbuch der mikroskopischen Anatomie des Menschen, vol. 4, Teil 9 (Springer, Berlin 1975).
Sterling, T.D.; Bering, E.A., jr.; Pollak, S.V., and Vaughan, H.G., jr.: Visual prosthesis. The interdisciplinary dialogue. Proc. 2nd Conf. Visual Prosthesis (Academic Press, New York 1971).
Steward, O.: Topographic organization of the projection from the entorhinal area to the hippocampal formation in the rat. J. comp. Neurol. *167:* 285–314 (1976a).
Steward, O.: Reinervation of dentate gyrus by homologous afferents following entorhinal cortical lesions in adult rats. Science *194:* 426–428 (1976b).
Stratton, G.M.: Vision without inversion of the retinal image. Psychol. Rev. *4:* 341–360, 463–481 (1897).
Stratton, G. M.: Eye-movements and the aesthetics of visual form. Wundts philos. Studien *20:* 336–359 (1902).
Suga, N. and Jen, P.H.S.: Disproportionate tonotopic representation for processing CF-FM sonar signals in the mustache bat auditory cortex. Science *194:* 542–544 (1976).
Sugita, N.: Comparative studies on the growth of the cerebral cortex. I–VIII. J. comp. Neurol. *28:* 495–510, 511–591 (1917); *29:* 1–39, 61–117, 119–162, 177–240, 241–278 (1918).
Szilard, L.: On memory and recall. Proc. nat. Acad. Sci., Wash. *51:* 1092–1099 (1964).
Tanzi, E.: I fatti e le induzioni nell'odierna istologia del sistema nervoso. Rassegna critica. Riv. sper. Fren. Med. leg. *19:* 419–472 (1893).
Timm, F.: Zur Histochemie des Ammonshorngebiets. Z. Zellforsch. *18:* 548–555 (1958).

TOWER, D. B.: Structural and functional organization of mammalian cerebral cortex. The correlation of neurone density with brain size. Cortical density in the fin whale (Balaenoptera physalus L.) with a note on the cortical density in the Indian elephant. J. comp. Neurol. *101:* 19–51 (1954).

TURING, A. M.: On computable numbers with an application to the Entscheidungsproblem. Proc. London math. Soc. Ser. 2 *42:* 230–265; *43:* 544–546 (1936, 1937).

TURING, A. M.: Computing machinery and intelligence. Mind *59:* 433–460 (1950).

UCHIMURA, J.: Über die Gefässversorgung des Ammonshorns. Z. ges. Neurol. Psychiat. *112:* 1–19 (1928).

UEMATSU, S.; CHAPANIS, N.; GUCER, G.; KÖNIGSMARK, B., and WALKER, A. E.: Electrical stimulation of the cerebral visual system in man. Confin. neurol. *36:* 113–124 (1974).

UNGAR, G.; GALVAN, L., and CLARK, R. H.: Chemical transfer of learned fear. Nature, Lond. *217:* 1259–1261 (1968).

UNGAR, G.: Evidence for molecular coding of neural information; in ZIPPEL Memory and transfer of information, pp. 319–341 (Plenum Press, New York 1973).

UNGAR, G. and OCEGUERA-NAVARRO, G.: Transfer of habituation by material extracted from brain. Nature, Lond. *207:* 301–302 (1965).

UPDYKE, B. V.: The patterns of projection of cortical areas 17, 18, and 29 onto the laminae of the dorsal lateral geniculate nucleus in the cat. J. comp. Neurol. *163:* 377–395 (1975).

UTTLEY, A. M.: The classification of signals in the nervous system. Electroceph. clin. Neurophysiol. *6:* 479–494 (1954).

UTTLEY, A. M.: A theory on the mechanism of learning based on the computation of conditional probabilities. Proc. 1st Int. Congr. Cybernetics (1956). Quoted by BURNS (1968).

VAIHINGER, H.: Die Philosophie des Als Ob. System der theoretischen, praktischen und religiösen Fiktionen der Menschheit auf Grund eines idealistischen Positivismus. 1. und 2. Aufl. (Reuther & Reichard, Berlin 1911, 1913).

VILLIGER E.: Gehirn und Rückenmark; 7. Aufl. (Engelmann, Leipzig 1920).

VOGT, O.: Zur anatomischen Gliederung des Cortex cerebri. J. Psychol. Neurol. *2:* 160–180 (1903).

VOGT, O.: Architektonik der menschlichen Hirnrinde. Allg. Z. Psychiat. *86:* 247–266 (1927).

VOGT, C. and VOGT, O.: Allgemeinere Ergebnisse unserer Hirnforschung. J. Psychol. Neurol. *25:* Erg. H. 1, pp. 279–462 (1919).

VOGT, C. und VOGT, O.: Über die Neuheit und über den Wert des Pathoklisenbegriffes. J. Psychol. Neurol. *38:* 147–154 (1929).

VOLKMANN, R. VON: Vergleichende Untersuchungen an der Rinde der 'motorischen' und der 'Sehregion' von Nagetieren. Verh. anat. Ges. Erg. H. Anat. Anz. *61:* 234–243 (1926).

VOLKMANN, R. VON: Vergleichende Cytoarchitektonik der Regio occipitalis kleiner Nager und ihre Beziehung zur Sehleistung. Z. Anat. EntwGesch. *85:* 561–657 (1928a).

VOLKMANN, R. VON: Sehrinde und Binokularsehen. Klin. Wschr. *7:* 1320–1323 (1928b).

VOLKMANN, R. VON: Sehrinde und Binokularsehen (Abstract). Münch. med. Wschr. *1928:* 758 (1928c).

WÄCHTLER, K.: Vergleichend-histochemische Untersuchungen zur Acetylcholinesteraseverteilung im Telencephalon von Wirbeltieren; Habilitationsschrift, Tierärztliche Hochschule Hannover (1973).

WAGOR, C.; LIN, S., and KAAS, J. H.: Some cortical projections of the dorsomedial visual area (DM) of association cortex in the owl monkey, Aotus trivirgatus. J. comp. Neurol. *163:* 227–250 (1975).

Walker, A. E.: The primate thalamus (University of Chicago Press, Chicago 1938).

Walshe, F. M. R.: On the mode of representation of movements in the motor cortex, with special reference to 'convulsions beginning unilaterally' (Jackson). Brain *66:* 104–139 (1943).

Walshe, F. M. R.: On the role of the pyramidal system in willed movements. Brain *70:* 329–334 (1947).

Watson, J. B.: Behaviourism (Kegan Paul, London 1925).

Watson, J. B.: Psychology from the standpoint of a behaviorist; 3rd ed. (Lippincott, Philadelphia 1929).

Weigl, E. und Metze, E.: Experimentelle Untersuchungen zum Problem des nicht sprachgebundenen begrifflichen Denkens. Schweiz. Z. Psychol. *27:* 1–17 (1968).

Welker, C.: Receptive fields of barrels in the somatosensory cortex of the rat. J. comp. Neurol. *166:* 173–189 (1976).

Whitsel, B. L.; Roppolo, J. R., and Werner G.: Cortical information processing of stimulus motion on primate skin. J. Neurophysiol. *35:* 691–719 (1972).

Wiener, N.: Cybernetics or control and communication in the animal and the machine (Wiley, New York 1948).

Wiesel, T. N. and Hubel, D. H.: Single cell responses in striate cortex of kittens deprived of vision in one eye. J. Neurophysiol. *26:* 1003–1017 (1963).

Wiesel, T. N. and Hubel, D. H.: Comparison of the effects of unilateral and bilateral eye closure on cortical unit responses in kittens. J. Neurophysiol. *28:* 1029–1040 (1965a).

Wiesel, T. N. and Hubel, D. H.: Extent of recovery from the effects of visual deprivation in kittens. J. Neurophysiol. *28:* 1060–1072 (1965b).

Woolsey, T. A. and Loos, H. van der: The structural organization of layer IV in the somatosensory region (S1) of mouse cerebral cortex. The description of a cortical field composed of discrete cytoarchitectonic units. Brain Res. *17:* 205–242 (1970).

Yorke, C. H. and Caviness, V. S., jr.: Interhemispheric neocortical connections of the corpus callosum in the normal mouse: a study based on anterograde and retrograde methods. J. comp. Neurol. *194:* 233–245 (1975).

Young, J. Z.: The memory system of the brain (University of California Press, Berkeley 1966).

Zangwill, O. L.: Consciousness and the cerebral hemispheres; in Dimond and Beaumont Hemisphere function in the human brain, pp. 264–278 (Wiley, New York 1974).

Zeman, W. and Innes, J. R. M.: *Craigie's* neuroanatomy of the rat revised and expanded (Academic Press, New York 1963).

Ziehen, T.: Grundlagen der Naturphilosophie (Quelle & Meyer, Leipzig 1922).

Ziehen, T.: Das Leib-Seele Problem. Deutsch. med. Wschr. *30:* 1267–1269 (1924).

Zippel, H. P. (ed.): Memory and transfer of information (Plenum Press, New York 1973).

XVI. The Central Nervous System as a Whole: Typologic Features in the Vertebrate Series

1. General Remarks

The Vertebrate neuraxis displays a tubular pattern characterized by distinctive regional, dorsoventral and rostrocaudal differentiations of the tube's wall. The thereby resulting diverse components represent *open topologic neighborhoods* arranged in longitudinal strips or zones whose extent, distribution, and caudorostral regional variance determine the configurational features of the major and easily recognizable subdivisions of the central nervous system, namely, in caudorostral sequence, spinal cord, deuterencephalon (rhombencephalon and mesencephalon), and archencephalon (prosencephalon: diencephalon and telencephalon).

The *primary zonal system*, extending through *spinal cord* and *deuterencephalon*, consists of unpaired *roof plate*, paired *alar plate*, paired *basal plate*, and unpaired *floor plate*. The floor plate seems to disappear in the mesencephalon. The prosencephalon essentially displays only alar plate and roof plate, the basal plate ending as tegmental cell cord at the mesencephalo-prosencephalic, respectively mesencephalo-diencephalic boundary.

In the course of ontogenetic development, three different *secondary zonal systems* become superimposed upon the primary zonal system and are characteristic for (a) spinal cord and deuterencephalon, (b) diencephalon, and (c) telencephalon.

In the *spinal cord and deuterencephalon*, the secondary zonal system displays (1) *ventral zone*, (2) *intermedioventral zone*, (3) *intermediodorsal zone*, and (4) *dorsal zone*. The first two pertain to basal plate, the two last ones pertain to alar plate.

In the *diencephalon*, the secondary zonal system consists of *epithalamus*, *thalamus dorsalis*, *thalamus ventralis*, and *hypothalamus*, all of which derive from the primary alar plate. Both deuterencephalon and diencephalon retain, topologically, the feature of a single (unpaired) tube with bilaterally symmetric wall neighborhoods.

The *telencephalon* arises at early ontogenetic stages by unpaired rostral evagination, from which subsequently the paired rostral evaginations of the olfactory bulbs originate. Depending on the different vertebrate groups the formal aspects of ontogenetic shaping events, involving growth and displacement of organic mass, assume the following sorts of tubular wall moulding: (1) unpaired evagination (exclusively rostral, (2) paired rostral evagination, (3) paired caudal evagination, (4) inversion, (5) eversion, and (6) overall bending.

The *secondary longitudinal zonal pattern* displayed by the telencephalic wall derived from alar plate are the *basal zones* B_1, B_2, B_3, B_4, and the *dorsal* or *pallial zones* D_1, D_2, D_3. Zone D_1, pallial in the telencephalon of Anamnia, becomes secondarily basal in Amniota. The topologic telencephalic transformations can be expressed by means of pattern formulae *(bauplanformeln)* comparable to those used by botanists for short-hand formulations of flower diagrams.

It is of interest that the topologic neighborhoods represented by the longitudinal zones and some of their further differentiations may become outlined by *folds* respectively *sulci* of the neural tube's inner surface (lumen) as well as, to some extent, of its external surface. Such folds and sulci can be interpreted as resulting from interactions between morphogenetic fields intrinsic to the topologic neighborhoods. These sulci do not seem related to, and appear to be independent of, the variable histogenetic processes such as KAHLE's (1951) matrix phases or 'time of neuron origin' stressed by CREPS (1974) both of whom failed to apprehend the topologic aspect of brain configuration.

The reciprocal spatial relationships of distinctive configurational units were designated by JACOBSHAGEN (1925) as characterizing a morphologic *bauplan*. This latter, however, can be conceived as a simplified three-dimensional *topologic space* representing an arcwise connected set whose subsets are partially and hierarchically ordered neighborhoods corresponding to cellular aggregates and their derivatives. Regardless of their shape and size, different morphologic configurations display the same bauplan if they are topologically identical, that is *homeomorph*. The comparison of two or more morphologic configurations, indicating whether such identity does or does not obtain, is based upon an operation or 'transformation' called *mapping*, whereby the identifiable neighborhoods of a given configuration can be brought in one-to-one correspondence with those of other configurations such that the basic *relationships of connectedness* are preserved. If to each element of a space A there corresponds a unique element f (x) of a space B, then there is said to be a mapping or a map of the set A in the set B, and the element (x) is said to be the image of the element x. A mapping is said to be continuous, if neighborhoods map on neighborhoods. In order to be homeomorph, a one-to-one mapping must be bicontinuous, that is, continuous both ways, whether configuration A is mapped on configuration B or vice versa. Under certain rules, the mapping of a set A correlating points of A with points of B is said to be a mapping of A *into* B. It is a mapping of A *onto* B if each point of B is the image of at least one point A. With respect to this distinction it seems appropriate to remark that, in the simplified topologic space of morphology, in which the smallest elements are spots of finite smallness, mathematical points have no existence. Because of this and additional complications, the application of the rigorous topological concepts 'into' and 'onto' as referring to mappings is here beset with certain difficulties and ambiguities. It seems therefore preferable to use the noncommittal terms *on* or *upon* in connection with mappings in the simplified topology of morphologic space.

This nonrigorous definition of topologic mapping should be sufficient in the aspect under consideration. More rigorously, a mapping T–T′ is continuous if the original image of every open set of T′ is open in T (*'eine Abbildung T nach T′ ist stetig, wenn das Urbild jeder offenen Menge in T′ offen in T ist'*).

The *topologization of morphology*, as indicated in the preceding paragraphs, represents a new approach to comparative anatomy, which the present author elaborated in volumes 1 and 3/II of this series, after recognizing (K., 1954, p. 63) that JACOBSHAGEN's (1925, 1927, 1928) fundamental formulations implied concepts of topology. Previously (K., 1927, 1929a, b, 1931, 1936), in following JACOBSHAGEN's procedure with respect to the neuraxis, I had referred to *gestalt theory* which, in a much less rigorous manner, also stresses invariants under transformation.

For the purpose of a simplified morphologic topology, bicontinuous one-to-one homeomorphism may be considered as synonymous with *isomorphism sensu strictiori*. Since morphologic patterns, however, despite fundamental invariance with respect to their basic components (*'Grundbestandteile'*), can display considerable differences with regard to the differentiations of further subdivisions of these components into subsets (form-elements, *'Formbestandteile'*), the application of one-many and many-one mappings becomes necessary. It is here convenient to use the term *homomorphism* in agreement with ASHBY's procedure. Homomorphism (as distinguished from homeomorphism or isomorphism *sensu stricto*) obtains, when a many-one transformation, applied to the more complex domain, can reduce it to a form that is isomorphic with the simpler. Again, it is evident that, while an isomorphic *(sensu stricto*, i.e. homeomorphic) mapping is necessarily one-to-one, a homomorphic mapping is single-valued only in one direction. Thus, two different organisms, or given organs in such organisms, may each display a bauplan which is topologically identical, i.e. homeomorph with respect to *grundbestandteile*, but involves greater or lesser complexity with respect to further subdivisions of these basic components.

It is then possible to state that all formed elements or sets of formed elements which, although occurring in different taxonomic groups of organisms, can be mapped by an isomorphic or homomorphic topologic transformation upon a given component (set or subset) of a common morphologic pattern, are the '*same*' or, in other words, *homologous*. If the mapping is isomorphic, that is one-one, *special orthohomology* obtains. If the mapping is homomorphic by one-many or many-one transformations, *special kathomology* obtains, *augmentative* in the first case, and *defective* in the second. If the homomorphism involves repetitive sequences along the main body axis (metamerism), an *allomeric type* of *special kathomology* obtains.

In contradistinction to special homology, *general morphologic homology* refers to grouped as well as randomly distributed repetitive elements within an extended domain for which a morphologic pattern can be formulated, and applies to multiple configurations such as bony and horny scales, teeth, feathers, hairs, claws, nails and hoofs.

A third major class of homology is represented by *promorphologic homology*, namely by isomorphism or homomorphism *within* promorphologic entities of a single organism such as metameres, antimeres, and parameres. Thus, in the metameric extremities of one and the same vertebrate organism, the humerus and the femur provide an example of promorphologic orthohomology (JACOBSHAGEN). Generally speaking, some authors consider homology as only referring to interindividual morphologic relationships, while the intraindividual correspondences representing promorphologic homology are then subsumed under the term homotypy.

In addition to the qualifying terms as defined above, namely ortho- and kat- (kata, implying a downgrading), augmentative, defective, and allomeric (including 'mixed forms'), the supplemental terms homo- (iso-), and hetero- can be used, as shown in the following tabulation, to indicate whether homologous configurations do or do not display the same shape, structure (texture, grain), function, and anatomical (as contrasted with topological) position, or whether configurations which are homologous in their definitive (adult) connectedness, are derived from nonhomologous ontogenetic matrices (e.g. certain muscles of Mammalia in comparison with those of Anamnia and vice versa):

Shape	homo- and heteroeidetic homology
Structure	homo- and heterotypic homology
Function	homo- and heteropractic homology
	iso- and heteropractic homology
Position	homo- and heterotopic homology
Matrix	homo- and heteromeric homology

Different organic configurations which cannot be compared with each other by means of isomorphic or homomorphic mappings although displaying some sorts of invariants concerning structure or function, or both, do not represent, or do not pertain to, the same bauplan, and therefore do not manifest homology. The obtaining invariance or similarity is then expressed by the term (morphologic, biologic) *analogy*, of which several definitions can be given.

For practival purposes, a morphologic pattern or bauplan, which is an elementary three-dimensional topologic space, can conveniently be depicted by two-dimensional diagrams in a plane to which one of the three conventional sorts of orthogonal body axes (e.g. rostro-caudal, dorso-ventral, right-left) is perpendicular. The cutting of a configurated space in the course of a deformation, which is here provided by the mapping upon a figure or diagram, becomes a permissible procedure. It is admitted that if a figure is cut during a deformation, and the edges cut are joined after the deformation in exactly the same way as before, the process still defines a topological transformation of the original figure. In morphology, mental reconstruction by the scanning of serial sections, or actual physical reconstructions by means of graphic or plastic models provide the required joining. Such cutting-up or '*tearing*' of topologic spaces representing complex interconnected systems greatly simplifies their analysis, and methods of '*tearing*' have been applied to the practical solution of certain topologic problems.

Again, as pointed out above with respect to the definition of the bauplan as a topologic space, the formulation of the homology concept in terms of topology represents an entirely new approach. The hitherto elaborated attempts at a definition of morphologic homology will disclose, to any competent and sceptical reader, the multiplicity of fluctuating and ambiguous terms propounded by the numerous authors who discussed the homology problem, for which, in my opinion, simplified concepts of topology, devoid of phylogenetic implications, provide the only reasonably satisfactory formulation. This latter, of course, by no means precludes or prevents further plausible phylogenetic speculations based on the observed morphologic relationships.

Because of the evident differences in shape, size, and relative position displayed by comparable form-entities of living organisms, relevant organic forms cannot be adequately described, and still less defined, in the aspect here under consideration, by *quantitative* metric procedures.

The 'relaxed morphologic topology', however, applies just to those 'distortions' and transformations which still completely defy quantitative metric methods in view of their complexity involving unknown parameters. It places into the given manifold a vaguely localized but combinatorially exactly determined framework. Being thus a higher generalization of 'deformation', 'distortion', or transformation than the metric methods, it is therefore, in the above-mentioned aspects under consideration, not only more powerful, but much more accurate.

As regards the one-many and the many-one transformations obtaining in phylogenetic evolution, the following types should be pointed out. One-many transformations can result (a) by increase in the number of distinctive, definable, but *conjoined* open sub-neighborhoods within a given neighborhood (e.g. lamination respectively stratification of cerebral cortex, or increase in the number of architectural grisea within diencephalon and brain stem, etc.).

The second main type (b) of one-many transformations results by increase in the number of *disjoined* distinctive components, e.g. by cleavage, splitting, or branching (e.g. increase in the number of vertebrae).

The corresponding sorts of many-one transformations are the following. Type (a) by *fusion* of distinctive components, such that the number of definable open neighborhoods remains unchanged (e.g. fusion of epibasal D_1 and basal B_1 and B_2 components in Mammalian corpus striatum, cf. below p. 438). Type (b) is characterized by *disappearance* of distinctive, discrete, partly or wholly disconnected components (e.g. in the evolution of the Equine extremities as shown in Figure 153B).

One could also say that one-many transformations result either (a) by *intrinsic* ('intussusceptive') differentiation or (b) by *extrinsic* addition respectively apposition, while many-one transformations are either (a) *inclusive*, by fusion, or (b) *exclusive*, by extrinsic diminution-respectively subtraction.

In all these instances, the given set remains identical under the transformation, but, in accordance with the required or arbitrarily adopted definitions, the number of its describable elements changes in (a) by conjoined, and in (b) by disjoined increase respectively decrease. It is here, moreover, of importance to realize that the relevant topologic neighborhoods can be classified as represented by two hierarchically different categories, namely *grundbestandteile* essential to a *bauplan*, and *formbestandteile* as subsets within a *grundbestandteil*. On the basis of these concepts, we proceeded with a rigorous analysis of homologies in the Vertebrate neuraxis (K., 1929a, b, 1931, 1936, 1954; K. and NIIMI, 1969; GERLACH, 1933, 1947; MIURA, 1933).

With respect to the term *bauplan* it might be recalled that HAECKEL, in his fundamental work '*Generelle Morphologie der Organismen*' (1866) expressed misgivings about the use of said term because some of its connotations could imply purpose, respectively teleologic or theologic concepts. This, however, is not necessarily the case, since *bauplan* can merely denote orderliness in an arrangement displaying relations of parts, that is to say *pattern (Muster, Grundriss)* or 'design' *qua* 'natural' make-up and composition without implication of a 'purpose'. JACOBSHAGEN, a disciple of HAECKEL, used the term bauplan in this sense, and I have retained that semantic procedure. HERRICK (1933), who strongly opposed my views, regarded *baupläne* as abstractions 'endowed with some occult dynamic power'. To this it can be replied that *baupläne* are perfectly accurate descriptions of actual topologic orderliness. This orderliness is doubtless encoded in the genome and results from physico-chemical causal interactions during ontogenesis. The genome, in turn, is causally related to the organism's phylogenetic evolution, and no purpose is here implied. Thus, if, by 'occult dynamic power', HERRICK refers to supernatural effects, he would have completely misunderstood our use of the bauplan concept. On the other hand, since nothing is known about the details of the presumed encoding, nor of the obtaining formative mechanisms, the 'dynamic power', namely the mechanism resulting in a given configuration still remains unknown, and is, in this particular sense, indeed 'occult'. It should be added that HERRICK, in a personal communication of 1954, finally admitted the justification of the formanalytic morphologic approach, and stated 'I think we are now not far apart' (cf. vol. 1, chapter III, p. 295, footnote 83).

It should also be recalled that, about 150 years ago, CARL ERNST VON BAER (1828) had clearly recognized the significance of typologic configuration. He stated: '*Der Grad der Ausbildung des thierischen Körpers besteht in der grösseren histologischen und morphologischen Sonderung (Differenzierung). Je gleichmässiger die ganze Masse des Leibes ist, desto geringer die Stufe der Ausbildung. Je verschiedener sie ist, desto entwickelter das thierische Leben in seinen verschiedenen Richtungen. Der Typus dagegen ist das Lagerungsverhältniss der organischen Elemente und der Organe. Dieses Lagerungsverhältniss ist der Ausdruck von gewissen Grundverhältnissen in der Richtung der einzelnen Beziehungen des Lebens. Der Typus ist von der Stufe der Ausbildung durchaus verschieden, so dass derselbe Typus in mehreren Stufen der Ausbildung bestehen kann, und umgekehrt dieselbe Stufe der Ausbildung in mehreren Typen erreicht wird.*'

On the other hand, the famed physiologist CARL LUDWIG (1816–1895), whose merits as one of the founders of modern experimental physiology are uncontestable, had no understanding whatsoever for morphological problems. He considered morphologic configurations to be entirely irrelevant, and claimed that morphology was not a topic for scientific investigation. He stated *verbatim* that '*die Morphologie ohne alle wissenschaftliche Berechtigung, höchstens eine künstlerische Spielerei ist*'.[1] LUDWIG's attitude was severely criticized by the noted zoologist RUDOLF LEUCKART (1823–1898) and others,[1a] including CARL GEGENBAUR. Yet, depreciation of morphology by biologists concerned with functional problems was still noticeable at the time when I began my own studies[1b] and this undervaluation, due to lack of proper understanding, is even at the present time not infrequently quite evident.

Since, as can be seen from recent discussions concerning the mor-

[1] Quoted after HEBERER (1968, p. 320). Seen from a higher viewpoint however, the concept of '*game*' is of course a rather interesting one, which includes some connotations with rather profound meaning. Thus, it may be recalled that the noted mathematician HILBERT defined pure mathematics as 'a meaningless formal game' (cf. § 64, p. 289 in K., 1961). On the other hand, LUDWIG did not use the term '*Spiel*' for game, but the diminutive and pejorative form '*Spielerei*', which could perhaps be rendered by 'gamelet' or 'gamekin'. As regards 'heuristic' or 'practical' values in science, cf. also the remarks on p. 640, volume 3/II with footnote 273, which contains a comment by CAJAL.

[1a] LEUCKART, a pioneer in animal ecology and parasitology (discoverer of the life cycle of Taenia echinococcus) had published, in 1848, a paper dealing with classification and characteristics of animal forms, with special reference to Invertebrates. He was then sharply attacked by LUDWIG in *Schmidts Jahrbücher der Medizin* (vol. 62, 1849, pp. 341f.). and responded with a strong retort in *Z. wiss. Zool.* (vol. 2, 1850, pp. 271f.) under the title: '*Ist die Morphologie denn wirklich so ganz unberechtigt?*'.

[1b] Cf. the comments on p. 297, section 8, chapter III of volume 1.

phological homology concept, ambiguity about its significance continues to obtain, a few additional critical comments on this topic seem appropriate.

Intuitively, everybody senses the '*sameness*' displayed by homologous morphologic configurations such e.g. as, in diverse Vertebrates, by humerus, by femur, by ribs, or by pelvis, which represent very simple cases. The topologic transformations appear here trivial and obvious to the superficial observer, who does not attempt to formulate said transformations, respectively the relevant topologic propositions, into rigorous and precise terms.

The questions to be answered are, in particular: (1) what is the topologic meaning of morphological 'sameness', and (2) how can such 'sameness' rigorously be established. The alleged obviousness (or '*Selbstverständlichkeit*') is deceiving, and all definitions of homology based on phylogenetic derivation or on 'identical' developmental mechanisms (whose detailed causal aspects, resulting from physicochemical interactions, are entirely unknown) involve several gross logical errors,[2] namely *petitio principii* and *ignoratio elenchi*, combined with the fallacy of ὕστερον πρότερον.

STARCK (1950), in his paper on the homology concept, distinguishes the following sorts:

'(a) *typologische Homologie. Homolog sind Teile, die im Bauplan die gleichen Lagebeziehungen besitzen, unabhängig von Form und Abstammung.*

(b) *typologisch-ontogenetische Homologie. Die Embryonalentwicklung wird zur Feststellung der typologischen Beziehung herangezogen.*

(c) *phylogenetische Homologie (= Homogenie* RAY LANKESTER), *Homolog sind Organe, die gleiche Abstammung besitzen (von gleichen Vorfahren ererbt).*

(d) *entwicklungsphysiologische Homologie* (V. BERTALANFFY). *Homolog sind Organe, die unter vergleichbaren organisierenden Beziehungen entstanden sind. Dieser Begriff schliesst zwangsläufig a, b, c in sich ein.*

(e) *erbbiologische Homologie. Homolog ist, was durch homologe Mutation homologer Gene entstanden ist* (J. HUXLEY). *Blutsverwandtschaft kann also*

---

[2] *Petitio principii* (or begging the question) is the fallacy in which a premiss is assumed to be true without proof, or, in which that which is to be proved is implicitly taken for granted. The closely related fallacy of ὕστερον πρότερον is, in general, the explanation of a thing by that which presupposes it, thereby inverting the natural order of reasoning in logic. In rhetoric, this results in an arrangement reversing the natural sequence, e.g.: *valet ergo vivit* (he is healthy and therefore he is alive). *Ignoratio elenchi* is the failure to grasp the essential point at issue.

fehlen. Hiervon praktisch meist kaum abtrennbar sind Fälle, in denen nicht homologe Mutationen ähnliche Entwicklungsabläufe verursachen oder Ähnlichkeiten im Phaenotypus durch verschiedenartige Faktoren verursacht werden. Die Homomorphien von NOVIKOFF *sind zum Teil durch den erbbiologischen Homologiebegriff erfassbar.*

Aus dem Gesagten ergibt sich, dass die Gruppen a, b und c von d mit umfasst werden. Trotzdem verzichten wir nicht auf die engeren Definitionen, da sie zur Kennzeichnung entsprechender Tatbestände praktischen Wert besitzen. Der erbbiologische Homologiebegriff hingegen umschreibt etwas grundsätzlich Neues. Er ist unentbehrlich, da in praxi oft nicht unterscheidbar sein wird, ob c, d oder e vorliegt.

Die Ähnlichkeit feinbaulicher Strukturen stellt einen Sonderfall der Analogie dar. Es wird vorgeschlagen zur Erfassung feinbaulicher Ähnlichkeiten den Begriff der "anatomischen Äquivalenz", den C. und O. Vogt in die Hirnanatomie eingeführt haben, allgemein zu übernehmen.'

It seems evident that STARCK's typologic and typologic-ontogenetic homologies both represent the topologic homology as defined by myself. In the course of ontogeny, transitory stages may reveal a topologic connectedness which becomes subsequently blurred in the definitive adult stage (e.g. that of the diencephalic and telencephalic grisea of Amniota, as dealt with in volume 3/II of this series).

STARCK's formulation in a), namely '*unabhängig von Form und Abstammung*' contains, moreover, a logical weakness, namely a ὕστερον πρότερον. In the universe of discourse pertaining to 'typologic', that is to say topologic form-analysis, '*Abstammung*' respectively phylogenesis have no logical existence whatsoever, and must thus be excluded from the pertinent definitions. Omitting '*Abstammung*', STARCK's wording should be changed to: '*unabhängig von Form und Struktur*'.

Once the topologic respectively morphologic homologies are established, then, of course, their significance as circumstantial evidence for phylogeny can evidently be stressed. Yet, although in many instances homologous configurations indeed strongly support the assumption of common ancestry, no conclusive proof obtains that this must be the case in all instances.

The phylogenetic homology concept, postulating common ancestry, is merely a judicious inference from topologic homology, since the phylogenetic sequences, whose actual course cannot be directly observed, are merely suggested by the directly observable and definable topologic homologies. If homologies are defined on the basis of common ancestry, we have a *petitio principii* and ὕστερον πρότερον, since the common ancestry is deduced from the obtaining (topologic) homologies. These latter, with respect to their invariance under one-one, one-many, and many-one transformations involving changes of outline or size, are arranged in sequences suggesting plausible phylogenetic lines of descent or of 'radiations'.

The so-called homologies of 'developmental physiology' and 'genetic biology' can be entirely discounted, since their relevant causal mechanisms still remain entirely unknown, thus precluding any intelligible formulation of such homologies. To invoke the assumed factors resulting in a given configuration merely reduces *obscurum to obscurius.*

The elaborations on homology by SIMPSON (1967), BOYDEN (1969), CAMPBELL and HODOS (1970), and HODOS (1974) retain the *petitio principii* and *circulus vitiosus* of the phylogenetic homology definition. Thus, HODOS (1974) defines homology by stating that 'structures or other entities are homologous when they could, in principle, be traced back through a genealogical series to a stipulated common ancestral precursor, irrespective of morphological similarity'. He seems to forget that the common ancestral precursor is 'stipulated' on the basis of homology. Moreover, the cited author is rather vague about what represents 'morphological similarity'. In addition, he and other writers on this subject matter fail to realize that not 'structure' but topological connectedness of a morphologically definable neighborhood *(grundbestandteil, formbestandteil)* is the only rigorous criterion of homology. Thus, structurally quite different neighborhoods can be homologous (e.g. primordium hippocampi of Anamnia and hippocampal formation of Mammals).[2a]

It is also of particular interest that CHARLES DARWIN, in the glossary which he appended to his 'Origin of Species', defined homology in strictly configurational terms, including ontogenetic derivation, but omitting any reference to descent from a common ancestor. This definition, prepared by W. S. DALLAS in collaboration with DARWIN, reads as follows: homology is 'that relation between parts which results from their development from corresponding embryonic parts, either in different animals, as in the case of the arm of man, the foreleg of a

---

[2a] As regards further comments on the homology problem, reference should be made to chapter III of volume 1. Homology based on fiber tracts and their connections is critically considered on p. 209 and 294 of volume 1, and footnote 215, p. 531 of volume 3/II. The genetic concept of homology, and the lack of any sufficiently detailed knowledge concerning the relevant aspects of developmental mechanisms were pointed out on pp. 29–30, 57, and 73 of volume 3/II.

quadruped, and the wing of a bird; or in the same individual, as in the case of fore and hind legs in quadrupeds, and the segments or rings and their appendages of which the body of a worm, a centipede, etc., is composed. The latter is called serial homology. The parts which stand in such relation to each other are said to be homologous, and one such part or organ is called the homologue of the other. In different plants the parts of the flower are homologous, and in general these parts are regarded as homologous with leaves.' DARWIN's homology concept thus included promorphologic homology, of which his 'serial homology' (metameric promorphologic homology) is an instance, the two others being antimeric and parameric homologies. Analogy was defined by DARWIN as 'that resemblance of structures which depends upon similarity of function, as in the wings of insects and birds. Such structures are said to be analogous and to be analogues of each other.'

Turning to recent publications dealing with mathematical, physical and engineering approaches to problems of organic configuration, the publication by OXNARD (1973) may be mentioned. This author attempts to extend the analysis of form and pattern as suggested by D'ARCY THOMPSON, J. HUXLEY, and R.A. FISHER, briefly pointed out in chapters II and III of volume 1. OXNARD, *inter alia*, emphasizes multivariate statistical analysis, clustering techniques, and experimental stress analysis. As regards the significance of these methods and their relevant differences with respect to the topologic-morphologic approach, reference to the comments on pp. 186–194 of volume 1, and pp. 66–67 of volume 3/II will here be sufficient.

Since the differentiation and topologic connectedness of morphologic entities (neighborhoods) can be assumed to depend on macromolecular aspects of the genome as well as of that of enzymes and protein structures resulting from the genome's activities, morphologic homology is evidently related to invariants pertaining to these macromolecular structures and events. Thus, the term 'homology' could, in theory, obviously be applied to proteins etc., but would then, of course, have a quite different denotation (cf. WINTER *et al.*, 1968; MARGOLIASH, 1969; *Ciba Symposium 29*, 1974). It should, however, again be kept in mind that, until now, nothing whatsoever of significance is known about the actual details of the mechanisms of interaction whereby the observable ontogenetic and adult morphologic patterns are produced and maintained (cf. above, p. 394). There obtain, nevertheless, already at this time, some data suggesting evidence of 'biochemical phylogeny'. These latter, however, have not yet substantially advanced the original phylogenetic concepts based on ontogenetic and adult morphology combined with the paleontological record. Brief comments on 'evolutionary biochemistry', including serologic respectively immunological data, which require here no further elaboration, were given in chapter II (section 9, p. 141) of volume 1, and chapter IV (section 14, p. 343) of volume 2.

ELLIOTT (1969), in attempting to give a generalized account of the evolution of the Vertebrate central nervous system, emphasized the particular significance of its tubular configuration, which indeed represents a 'radical feature' distinguishing the Chordates from Invertebrates with ganglionic central nervous systems. The dorsal Vertebrate neuraxis, ventralward separated from other organs by the notochord and its derivatives, as well as by the cranial basis, could evolve without interference from the evolution of other organs. Every brain feature is related to the expansion of the primitive neural tube which retains its unity. Thus, ELLIOTT (1969) considers the Vertebrate phylogenetic ascent culminating in the Human brain and in the emergence of superior 'intelligence' as intrinsically resulting from the development of a neural tube, whose potentialities, in his opinion, are superior to those inherent in an Invertebrate ganglionated central nervous system. Be this as it may, the course of higher Metazoan evolution on the surface of our planet at least *appears* to corroborate ELLIOTT's surmise.

Within the Vertebrate tubular neuraxis, the neuronal elements pertaining to the longitudinal zones provide distinctive arrays (grisea) with neuropils containing excitatory and inhibitory synapses. These synapses can be roughly compared to on and off switches. The grisea, whose constituent elements are interrelated by an intrinsic circuitry, are also connected with other grisea. They can be conceived as '*black boxes*' with input and output channels.

Grisea in spinal cord and brain stem are related to the processing of segmental peripheral input and output. Moreover, in brain stem and prosencephalon, grisea are concerned with the processing of input from special sensory organs such as gustatory, lateral line, otic, optic, and olfactory receptors.

The promorphologic serial homology of dorsal and ventral spinal and deuterencephalic nerve roots, and the various morphologic or functional classifications of nerve components were dealt with in sections 3 and 4, chapter III of volume 1, as well as in sections 2, 3, and 4, chapter VII of volume 3/II. In this latter chapter and on p. 930, section 8, chapter XI of volume 4, the still not sufficiently elucidated problem of afferent (proprioceptive) fibers and even nerve cells in some spinal respectively cranial ventral roots was pointed out.

In addition to primary input and output grisea with their intrasegmental, intersegmental, and antimeric fiber connections, there are relay grisea and processing grisea of higher order. Among these latter, the brain stem reticular formation, the hypothalamus, the thalamus, the telencephalic basal ganglia, and the suprasegmental grisea represented by cerebellar cortex, tectum mesencephali and cerebral cortex are of particular import.

The regional configuration of the neuraxis as well as the structure, the main communication channels, and the functional significance of the various grisea were dealt with in volume 4, 5/I and in the preceding chapters of the present volume.

In the subsequent sections of this final chapter a few short summarizing comments on some relevant typologic features of the central neuraxis in the diverse craniote Vertebrate classes shall be given. The essentially conservative character of the neuraxis throughout the entire Vertebrate series is evident. Although specific progressive differentiations, such as those of basal ganglia of Sauropsida, and of cerebral cortex of Mammals are evident in the 'higher' classes, no additional, 'new' fundamental topologic component has been added, in the course of phylogenetic evolution, to the basic griseal bauplan of telencephalon, diencephalon, mesencephalon, cerebellum, oblongata and spinal cord, characteristic for all Vertebrates. All progressive differentiations have resulted from one-many transformations of the given topologic neighborhoods pertaining to the fundamental longitudinal zonal systems. A few special cases of differentiation involve secondary many-one transformations.[3]

Mutatis mutandis, this uniformity or conformity within the entire Vertebrate series applies also to the main ascending and descending communication channels of the central neuraxis. Ascending systems from spinal cord reach reticular formation, cerebellum, tectum mesencephali and diencephalon. From these two latter subdivisions, especially from thalamic grisea, the input into spinal cord finally reaches the telencephalon already in Anamnia. This is also the case with respect to the brain stem's sensory input. Thus, there are the spino-cerebellar and the spino-bulbo-tecto-thalamic channels, of which the latter one rep-

[3] E.g. the Mammalian striatum derived from a fusion of D_1, B_1, and B_2 components or, in the telencephalon of some Teleosts, fusions of D or of B components.

resents a general lemniscus system, which also includes the vestibulo-lateral-acoustic channel to mesencephalon and diencephalon, representing the lateral lemniscus system.

Optic input reaches tectum mesencephali, pretectal grisea, reticular formation, hypothalamus and thalamus, especially the lateral geniculate grisea, which latter, in particular, provide a relay to telencephalon.

Olfactory input directly reaches the telencephalon and is relayed to diencephalon and deuterencephalon by grisea and channels retaining rather constant features throughout the Vertebrate series.

As regards descending fiber tracts, respectively 'motor' or output channels, the reticulospinal system (including fasciculus longitudinalis medialis) the vestibulospinal system, the tectotegmental system, the cerebellotegmental system, and descending pretectal, hypothalamic and telencephalic channels are constant features. The ascending and, descending systems connecting telencephalon with more caudal levels are gathered, at the hemispheric stalk, into medial and lateral forebrain bundles.

Direct descending channels from telencephalon to spinal cord seem to be lacking in Anamnia, but are provided by the corticospinal system of Mammals. An analogous descending channel from the Avian epibasal and basal complex, reaching spinal cord levels, is perhaps present.

The main *commissural systems*, namely commissura pallii, commissura anterior, supraoptic and postoptic commissures, habenular commissure, posterior commissure, tegmental commissures, and lemniscal decussations are present, in essentially identical manner, although with considerable differences in details of connections and circuitry, throughout the Vertebrate series.

As regards the *autonomic* or '*vegetative*' components, which partly pertain to the central neuraxis, and partly to the peripheral nervous system, the overall organization of this 'system', dealt with in section 6, chapter VII of volume 3/II, is likewise identical in the entire Vertebrate series, but displays a progressive development in the 'higher' forms, in which various autonomic grisea and connections are by no means 'phylogenetically old' but represent rather 'new' differentiations. Thus, the distinction between an essentially 'autonomic' or 'vegetative' periventricular central gray and the more peripheral 'nuclei'[4]

[4] Some of the 'nuclei' externally to the 'central gray' may, however, also be related to 'vegetative' or 'autonomic' functions.

e.g. in diencephalon and mesencephalon is characteristic for Amniota rather than for Anamnia.

A peculiar feature of most of the main communication channels is their crossing or *decussation in the midline*, as dealt with in the apposite chapters concerned with the diverse subdivisions of the neuraxis. No satisfactory 'explanations' for this contralateral input (sensory) and output (motor) projection from or toward the opposite side, which is also manifested to a significant degree by Invertebrates, can be given. The various, rather unsatisfactory and unconvincing hypotheses elaborated, *inter alia*, by CAJAL, SPITZER, JACOBSON-LASK, and MISKOLCZY, are discussed on pp. 283–285, section 12, chapter IV of volume 2, and require no further comments in the present context.[5] It will here suffice to point out that, in Man, the two sensory modalities characterized not only by localization but by spatial configuration, that is to say by definite shapes or outlines, namely body sense (touch in the wider sense as well as proprioception), and vision, involve, as regards the main projection[6] of their afferent channels, an inverted (upside-down) distorted but topologically coherent mapping representing body respectively visual fields, upon contralateral postcentral gyrus and occipital lobe (area striata). An efferent (motor) mapping, roughly corresponding to the sensory body mapping, is outlined upon the precentral region and correlated with the sensory mapping upon postcentral cortex.[7] In all Vertebrate classes, a topological mapping of the contralateral visual fields, respectively of the retinal quadrants upon the cortex of the mesencephalic lobi optici likewise obtains.

---

[5] Some comments on the theory of decussations can also be found in a recent paper by SCHMATOLLA (1974).

[6] It should be recalled that, in addition to the contralateral projections, a variable amount of homolateral ones obtains.

[7] In Man, whose private conscious experiences are generally well known since they can be communicated and compared by means of their verbal formulation, the configurated spatial relationships within the individual consciousness-manifold seem, in some manner, related to this topologic mapping upon the cortical surface, as well as to the engram-complexes representing a 'body image'. The open transformations thereby involved imply, as it were, a reduction of the topologic distortion, such that spatial expansion (e.g. macular area, thumb, fingers etc.) may be transformed into acuity or vividity. The spatial arrangement of N-, respectively Np-events doubtless represents an important factor in the still rather unsatisfactory local sign theory proposed by various authors and discussed by RENSCH (1968). The spatio temporal organization of the private consciousness-manifold can be assumed to depend on a not yet adequately understood neural coding system implying a multiplicity of relevant factors. The neural coding of 'external

The connections of the central neuraxis with the various regions and structures of the body are provided by the cranial and spinal nerves which, *in toto*, display a striking instance of homology that holds throughout the Vertebrate series. In addition to the peculiar archencephalic (olfactory, terminal, optic, and epiphysial) nerves, there are the serially arranged deuterencephalic and spinal dorsal and ventral nerve roots, which, moreover, manifest promorphologic, metameric homology. In the deuterencephalon, there are the branchial nerves V, VII, IX, X, including the octavus system (VIII) derived from VII, and the cranial accessory (XI) derived from X. All these represent dorsal roots, while the eye muscle nerves III and VI, as well as the hypoglossus (XII, derived from spino-occipital segments), are ventral roots. The relevant morphologic aspects of cranial nerves and of dorsal and ventral spinal roots were considered in sections 2, 3, and 4 of chapter VII of volume 3/II. Three still poorly elucidated problems concerning the deuterencephalic cranial nerves may here be recalled, namely (1) the phylogenetic derivation of nervus trochlearis (IV) as either an originally ventral or dorsal root, (2) the so-called accessorius problem, related to the evolution of the joint cranial and spinal accessorius complex, and (3) the possible or probable presence of afferent (proprioceptive) fibers and even of thereto related ganglion cells (NICHOLSON, 1924, and other authors) in the cranial nerves III and VI, which are doubtless ventral roots.

With regard to an understanding of configuration, structure, communication channels, and functions of the Vertebrate neuraxis, considerable progress indeed has been achieved in the course of the past hundred years, since e.g. the summary by MEYNERT in STRICKER's *Handbuch der Gewebelehre* (1872). Yet, very numerous uncertainties as well as ambiguities remain, and the obtained knowledge can still be considered unsatisfactory, although an understanding of the relevant functions and structural arrangements of the central nervous system has been obtained as far as the overall outlines are concerned, i.e. '*in groben Umrissen*'.

With respect to the main communication channels or fiber tracts and various details of circuitry, rather convincing and generally ac-

---

world' and body seems to display an inverted and contralateral mapping as would result from a rotation of 180° around an assumed midsagittal promorphological body axis. The morphological aspects of the several body axes were dealt with on pp. 213–220, section 4, chapter III of volume 1, and pp. 111–116, section 1A, chapter VI of volume 3/II.

cepted data are available, based on the classical methods of myelin stains, myelogenesis, and metallic impregnations, supplemented by the experimental techniques involving degeneration effects (tigrolysis, *Marchi method, Glees, Nauta, and Fink-Heimer terminal degeneration methods)*, combined with pathological and clinical data, with electrophysiological techniques, the investigation of reflexes, and other procedures of neurophysiology.

More recent developments concomitant with progress in neurochemistry[8] have led to methods bases on fluorescence histochemistry, on the chemical identification and the behavior of transmitter substances, on the antegrade and retrograde neuronal (or axonal) flow as particularly demonstrated by the investigations of P. WEISS,[9] and on autoradiographic techniques. With regard to some publications concerning these methods, the reader may be referred to those by NAUTA *et al.* (1974), GILBERT and KELLY (1975), and SNOW *et al.* (1976), dealing with the horseradish peroxidase technique, and to the papers by COWAN *et al.* (1972) and CONRAD and PFAFF (1975) on autoradiographic demonstration of axonal connections. A symposium concerned with various studies of neuronal connectivity by use of the axonal (or neuronal) flow was recently edited by COWAN and CUÉNOD (1975). Another recent publication dealing with the various methods in brain research, with emphasis on the contemporary ones, has been edited by BRADLEY (1975). The data obtained by electron microscopy, in combination with other methods, have likewise provided some evidence for detailed interpretations of synaptic connections.[10]

Although various uncertainties of interpretation, and perhaps still incompletely understood sources of error, may impose limits to the efficiency and accuracy of the new methods, as in the case of the older ones, it can be expected that many new details of circuitry will be elucidated. While the fundamental arrangement of the main communica-

---

[8] Methods related to chemical stimulation of the brain, and a discussion of the thereby obtained results are surveyed in an extensive publication by MYERS (1974). A textbook of neurochemistry has been published by ALBERS *et al.* (1972), and a collective review of one hundred years neurochemistry (1875–1975) was presented by TOWER (1977). The current status of research on neurotransmitter amino acids has been summarized and reviewed by DAVIDSON (1976).

[9] Cf. the discussion of 'neuronal flow' on pp. 640–643, section 8, chapter V in volume 3/I.

[10] A concise review and discussion of the relevant technical methods is given in STEPHAN's (1975) handbook volume.

tion channels, demonstrated by means of the old techniques, may be considered rather well understood, it seems probable that numerous tentative and provisory interpretations concerning details of connection will require diverse revisions.[10a]

The period following the work of the classical authors in neurobiology such as BECCARI, BRODMANN, CAJAL, EDINGER, ELLIOT SMITH, FLECHSIG, FOREL, VAN GEHUCHTEN, HERRICK, HIS, JOHNSTON, KAPPERS, LANGLEY, MEYNERT, PAPEZ, SHERRINGTON and others up to the first third of the 20th century has been characterized by an almost exponential 'increase in available details without any corresponding clarification of the principles involved that would make the details more meaningful'.[11]

The uncertainty and flimsiness typical for many contemporary aspects of 'neurosciences' is e.g. strikingly displayed to the competent reader in a recent critical review by NATHAN (1976) concerning the 'gate-control theory' of pain.[12] This lengthy paper discusses a large number of more or less plausible but unsatisfactory hypotheses relevant to registration, processing and further transmission of an important type of sensory input. It considers, *inter alia*, the still very poorly understood significance and origin of the antidromic dorsal root potential reported by MATTHEWS (1934) and BARRON and MATTHEWS (1935, 1938), long neglected by electrophysiologists, until re-investigated by numerous authors after about 1955. NATHAN's (1976) paper also deals with 'use and misuse of electrical stimulation' and with assumed aspects of presynaptic inhibition,[13] whereby some axon collaterals respectively groups of end-arborizations may be prevented from transmitting impulses channeled by the stem neurite.

Summarizing the reasonably valid and generally accepted views concerning significance of the Vertebrate neuraxis as a control, signal-

---

[10a] Thus, a recent report by RUSTIONI (1977) claims, on the basis of retrograde horseradish peroxidase transport evidence, that dorsal and ventral horn cells of the spinal cord project to ipsilateral dorsal column nuclei (nn. gracilis et cuneatus). Although the interpretation by the cited author does appear possible, his actual findings seem inconclusive, since transsynaptic transport through the numerous collaterals of the dorsal root fibers cannot be excluded as another contingency.

[11] Cf. the review of ROSENBLUETH's tretaise on nerve impulse transmission by D.H. BARRON in *Journal of comparative Neurology*, vol. 94, pp. 513–514 (1951).

[12] Cf. also pp. 801–802, section 1, chapter VII of volume 3/II, and p. 229, section 11, chapter VII of volume 4.

[13] Cf. p. 279, section 4, and p. 468, section 6, chapter V of volume 3/I.

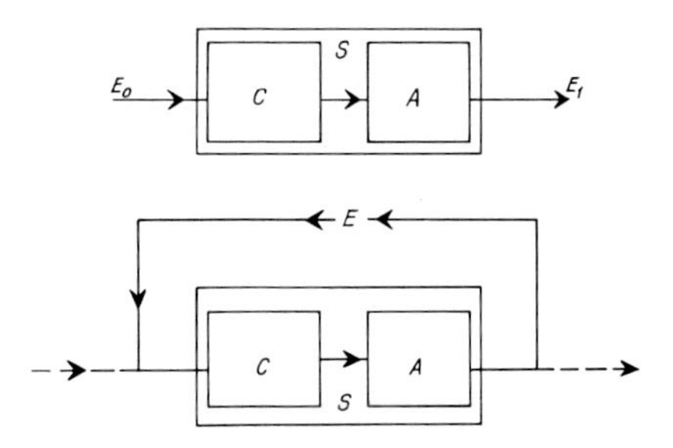

Figure 130. Diagrams illustrating the neuraxis as conceived in terms of open cycle control mechanisms (above) and of control mechanisms with feedback (below). The systems with their subsystems are represented as '*black boxes*' whose internal arrangements remain undefined. A: actuating part of system, providing output; C: part of system providing control; E: environment; E_0: state of environment acting on open cycle control mechanism; E_1: state of environment resulting from action of open cycle control mechanism; S: system as distinguished from environment.

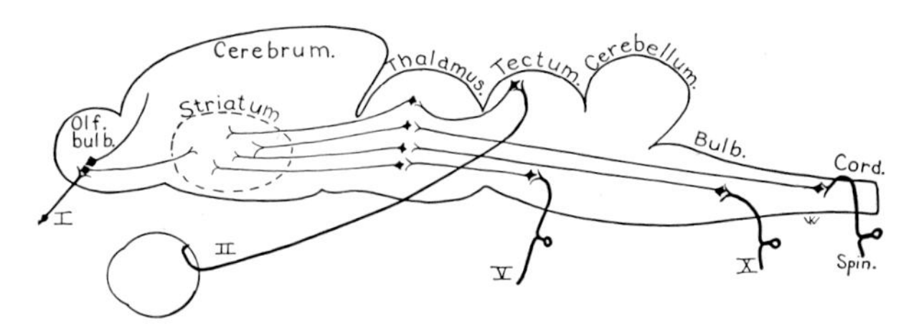

Figure 131. Diagram of simple Vertebrate brain showing how a telencephalic griseum may receive input from diverse sensory systems (from Krieg, 1966).

transmitting and signal processing mechanism with its communication channels, it is perhaps appropriate to consider five generalized models representing successive first order approximations with increasing amounts of complexity, as illustrated by the self-explanatory Figures 130, 131, and 132A, B. The two diagrams of Figure 130 depict the neuraxis as *black boxes* with input and output, without, and with feedback effect. Figure 131 indicates the assumed convergence of diverse sensory input upon a generalized Vertebrate telencephalic griseum. Figure 132A takes into consideration some simplified overall aspects of circuitry in the main subdivisons of the Mammalian neuraxis, and Figure 132B, likewise based on the Mammalian central nervous system, gives a diagrammatic representation of some main channels with emphasis on their functional significance. Comments on the clas-

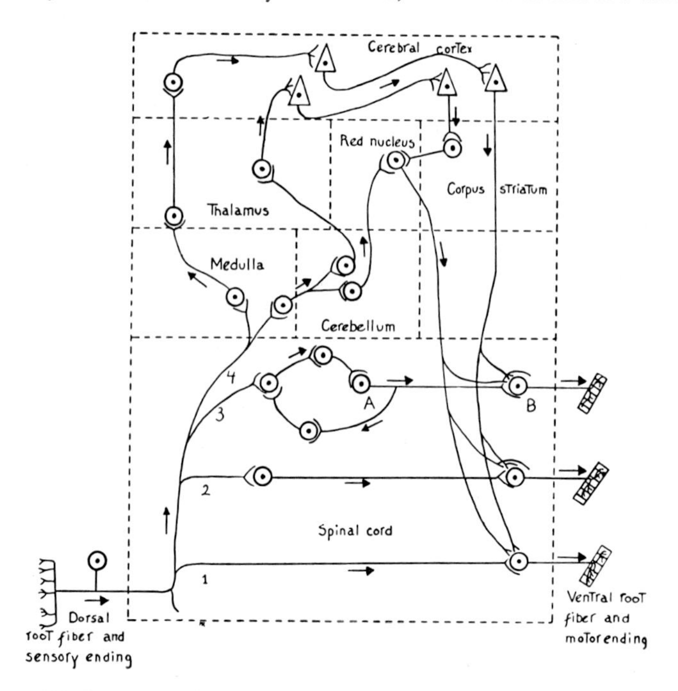

Figure 132 A. Diagram illustrating essential arrangements of the Mammalian neuraxis (modified after BAYLISS, from RANSON, 1943).

sification of the central nervous system's peripheral input und output, and on the classification of the 'senses' were given in section 1, chapter VII of volume 3/II.

Any evaluation of the Vertebrate 'senses' in an order of 'importance' is highly arbitrary. Quite evidently, the general body senses with their 'direct contact receptors' are basic. As far as 'distance receptors' are concerned, there are electric receptors in some Fishes, perhaps receptors to magnetic fields in certain Vertebrates, and the olfactory, auditory, and optic receptors of Vertebrates in general. The lateral line system of Anamnia assumes an intermediate status, being, to some extent, a distance and even perhaps 'auditory' as well as proprioceptive receptor related to the vestibular apparatus.

In some Vertebrates, the olfactory system predominates, and in others the optic system. This latter, however, can become essentially eliminated in various 'blind' forms, and the former in a few entirely anosmatic ones. Again, in certain Fishes, the chemical receptor system, designated as 'gustatory', which differs from the likewise chemical ol-

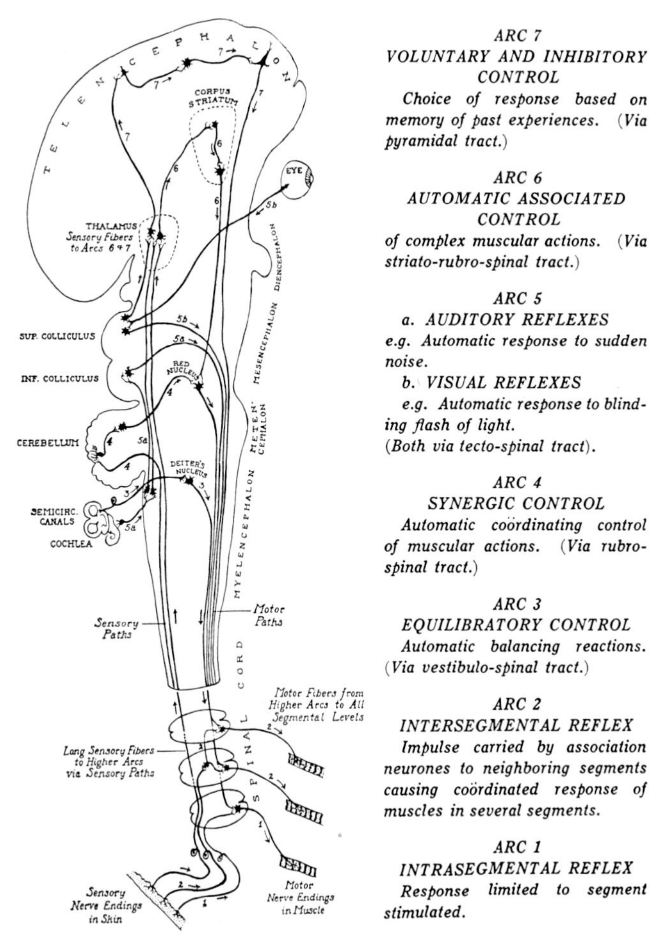

Figure 132B. Schematic drawing showing main communication channels in the Mammalian neuraxis and including a functional classification of so-called 'arcs' (from PATTEN, 1931).

factory one, can assume the function of a cutaneous 'distance receptor'.

As regards Man, and in addition to the basic 'body senses' including proprioception, it could be said that the optic system assumes primary importance, followed by the auditory system, while olfactory and gustatory systems are relegated to a minor role. In cases of blindness, which is a particularly severe disability for '*oculocentric Man*', audition and touch vicariously replace functions of the optic system, and in very special cases, such as that of HELEN KELLER (1954), touch assumes a preponderant multiple role.

With respect to the typologic features of the neuraxis as a whole in the diverse forms displayed by the Vertebrate series, the latest extant (post World War I) standard texts of comparative neurology, by KAPPERS (1920, 1921), KAPPERS *et al.* (1936) and by BECCARI (1943), as well as my '*Vorlesungen*' of 1927, and the present treatise, deal in topographic sequence with the major subdivisions of the neuraxis in the entire series. In his posthumous treatise of 1947, KAPPERS, however, after four preliminary chapters on nervous structures, on Invertebrates, Amphioxus, and general principles concerning the Vertebrate nervous system, deals separately, in 6 chapters, with the entire neuraxis as a whole of specific classes from Cyclostomes (Petromyzon) to Birds. The Mammalian and Human central nervous system is then, with greater details, taken up in 8 chapters, of which the last one is entirely devoted to anthropologic aspects.

PAPEZ (1929) adopts a comparable arrangement, although with inverse order, presenting a detailed survey of the Mammalian neuraxis, in 14 chapters on gross morphology, beginning with the cerebrum, and in 19 chapters on microscopic structure, from peripheral nervous system to cerebral cortex. The nervous system of the submammalian forms is dealt with as a whole but more briefly in 6 chapters (Reptiles, Birds, Amphibians, Fishes, Petromyzon, Amphioxus), and a final short chapter concerns the evolution of the forebrain.[13a]

There are, moreover, numerous papers and treatises devoted to either the entire brain or the entire neuraxis as a whole in single Vertebrate forms or groups. It will here suffice to mention the following. On Cyclostomes there are the papers by JOHNSTON (1902), HEIER (1948) and SAITO (1930) on Petromyzonts, and those by EDINGER (1906) and JANSEN (1930) on Myxine. As regards Plagiostomes, the publications by KAPPERS (1906), KAPPERS and CARPENTER (1911), STERZI (1909), GERLACH (1947), and K. and NIIMI (1969) may be

---

[13a] After the printing of the present volume was completed, a book notice in Science 196 (May 20, 1977) announced the publication of a treatise 'The vertebrate brain' by R. and L. PEARSON (Academic Press, New York 1976). Another book notice in Science 196 (April 23, 1977) reported two multi-authored publications edited by MASTERTON *et al.*, one being entitled 'Evolution, brain and behavior. Persistent problems' (Halsted/Wiley, New York 1976) and the other 'Evolution of brain and behavior in vertebrates. Papers from a conference, Tallahassee, Fla., Feb. 1973 (Halsted/Wiley, New York 1976). I was unable to procure these three publications in time for comments in the present context.

pointed out. Ganoids and Teleosts were dealt with by JOHNSTON (1901), KAPPERS (1906, 1907), BURR (1928), and HOOGENBOOM (1929). The peculiar neuraxis of Latimeria was described as a whole by MILLOT and ANTHONY (1965). With respect to Dipnoans, there are the studies on Ceratodus by HOLMGREN and VAN DER HORST (1925), and on Protopterus by GERLACH (1933).

The Urodele Amphibian neuraxis as a whole was dealt with in considerable detail in HERRICK's (1948) monograph. A rather detailed description of the entire Anuran neuraxis (Frog) was given by GAUPP (1899). Shorter publications, on the neuraxis respectively the brain as a whole, are those by KAPPERS and HAMMER (1918) on a species of Anurans, and by myself (K., 1922) concerning Gymnophiona. An entire Reptilian brain (Lacerta) was described in the short but well illustrated and valuable monograph by FREDERIKSE (1931). A detailed treatise on the Avian brain, with extensive references, and including functional aspects, was prepared by R. PEARSON (1972). An entire 'lower' Mammalian neuraxis (Rat) is dealt with in CRAIGIE's (1925) monograph, revised and expanded by ZEMAN and INNES (1963). The Elephant's neuraxis is described in the paper by DEXLER (1907) and the brain of CETACEANS is dealt with by JANSEN and JANSEN (1968).

Treatises on the Human neuraxis as a whole are, for obvious reasons, very numerous. In addition to the relevant chapters included in the standard texts of Human Anatomy, significant publications are, *inter alia*, those by DEJERINE (1895, 1901), EDINGER (1912), OBERSTEINER (1912), VILLIGER (1920), RANSON (1943, 1961), CLARA (1959), PEELE (1961), CROSBY *et al.* (1962), ELLIOTT (1963), and KRIEG (1966). A more recent and concise introductory summary, stressing essentials, is that by BARR (1974).[13b]

A number of *atlases* have been published, some of which deal with the neuraxis of a single form and are either prepared for general orientation or more specifically for providing stereotaxic coordinates.

Among useful atlases depicting overall aspects are those by WINKLER and POTTER (1911, 1914) for the brains of the Rabbit and of the Cat. KRIEG (1975) has published an 'interpretative atlas' of a Monkey brain (Macaque). For the Human brain, the large atlas of myelin-stained sections prepared by JELGERSMA (n.d.), and the atlases by MARBURG (1927) and, of myelin-stained sagittal sections, by SINGER and

---

[13b] Of considerable interest, among the older publications, are also the elaboration on the history of neurology by SOURY (1899) and the treatise by BARKER (1901).

YAKOVLEV (1954) may be mentioned. Atlases of the Human neuraxis based on other technical approaches are those by FLECHSIG (1920), concerning myelogeny, and by LUDWIG and KLINGLER (1956), depicting macroscopic preparation of the brain obtained by dissection of fiber tracts (*Abfaserungsmethode* or *Zerfaserung*, HULTKRANTZ, 1929). A smaller, but interesting and unusual atlas of this type was also published by GÜNERAL (1972). An atlas of schematic figures illustrating the generally recognized principal pathways in the Human neuraxis was prepared by RASMUSSEN (1941).

Stereotaxic atlases concerning particular single Vertebrate forms are rather numerous, and some of these concern only selected regions of the neuraxis. Among those depicting the entire brain or at least most of it, are the following. For Avian forms TIENHOVEN and JUHASZ (1962) have prepared an atlas of the Chick's telencephalon, diencephalon, and mesencephalon; KARTEN and HODOS (1967) have published a more detailed atlas of the Pigeon's brain with transverse as well as sagittal sections, containing unlabelled low-power photomicrographs supplemented by the outline drawings indicating their interpretation by the cited authors. Among stereotaxic atlases of Mammalian forms are those by KÖNIG and KLIPPEL (1963) of the Rat's brain, by SNIDER and NIEMER (1962) of the Cat's brain, by LIM *et al.* (1960) of the Dog's brain, by SNIDER and LEE (1962) of the Monkey's brain (Macaca mulatta), and by DE LUCCHI *et al.* (1965) of the Chimpanzee's brain. Following their introduction of stereotaxic surgery for clinical purposes in Man, SPIEGEL and WYCIS (1952) have prepared a first stereotaxic atlas of the Human brain, which was followed by those of TALAIRACH *et al.* (1957) and of subsequent other authors.

Finally, with reference to the increasingly important data obtained by investigations of neurochemistry, the topographic atlas by JACOBOWITZ and PALCOVITZ (1974), of catecholamine and acetylcholesterase-containing grisea in the Rat brain may be mentioned.

2. Cyclostomes

Typologic features of the *Petromyzont* neuraxis, depicted in Figures 133, 134 and 140A, can roughly be summarized as follows. The telencephalon is inverted, with rostral and caudal paired evagination, which latter results in the formation of a true polus posterior. The bulbus olfactorius is about as large as the lobus hemisphaericus. The common telencephalic ventricle, that is to say the pars impar telencephali is practically absent, being reduced, as in Amniota, to the space between the foramina opening into the two lateral ventricles dorsally to anterior commissure. In the lobus hemisphaericus, the D and B components do not display any sufficiently distinctive further subdivisions.

The diencephalon is characterized by the presence of pineal and parapineal organs, dealt with in chapter XII of volume 5/I. A saccus vasculosus is not present. The thalamus ventralis displays rostrally an extensive eminentia thalami. This latter was interpreted, by various authors, as primordium hippocampi.[14]

The mesencephalon displays an epithelial roof plate forming a vascular plexus. The cerebellum is poorly developed, its corpus being a simple transverse cerebellar plate. The tegmentum of mesencephalon and oblongata contains the large *reticular cells of Müller* which manifest a conspicuous instance of constant cell number. Discounting possible subjective differences in the identification of large reticular cells as *Müller cells*, their number (24 to 16) seems to vary with respect to different species.[15]

The nervus octavus pertaining to the rostral group of branchial nerves includes, as nervus vestibulolateralis, a lateralis component, which is characteristic for Cyclostome as well as Gnathostome fishes and aquatic Amphibians (cf. vol. 3/II, chapter VII, section 1, pp. 793–796, section 3, p. 854). The lateralis system, providing, as it were, a sense of 'water touch' of significance for swimming activities, may also, to some extent, register 'sound waves'. According to BUDDENBROCK (1958), the lateralis system might also be called a 'long-range tactile sense'. If well developed it can enable the animal 'not only to detect anything that moves in its vicinity, but even to detect its exact location' (BUDDENBROCK, 1958).

The spinal cord of Cyclostomes is flat and ribbon-like, containing the uncrossed 'giant fibers' represented by the neurites of the *Müller cells*. It should also be added that medullated nerve fibers, present in the nervous system of all Gnathostome Vertebrates, do not seem to be present in that of Cyclostomes. The arrangement of the spinal nerves,

---

[14] Details concerning the interpretation of the diencephalic bauplan of Cyclostomes are dealt with in volume 3/II.

[15] In Entosphenus japonicus 12 pairs, in Petromyzon fluviatilis and subspecies of Petromyzon marinus 10 pairs, in another subspecies of Petromyzon marinus 8 pairs. Cf. section 3, chapter IX of volume 4.

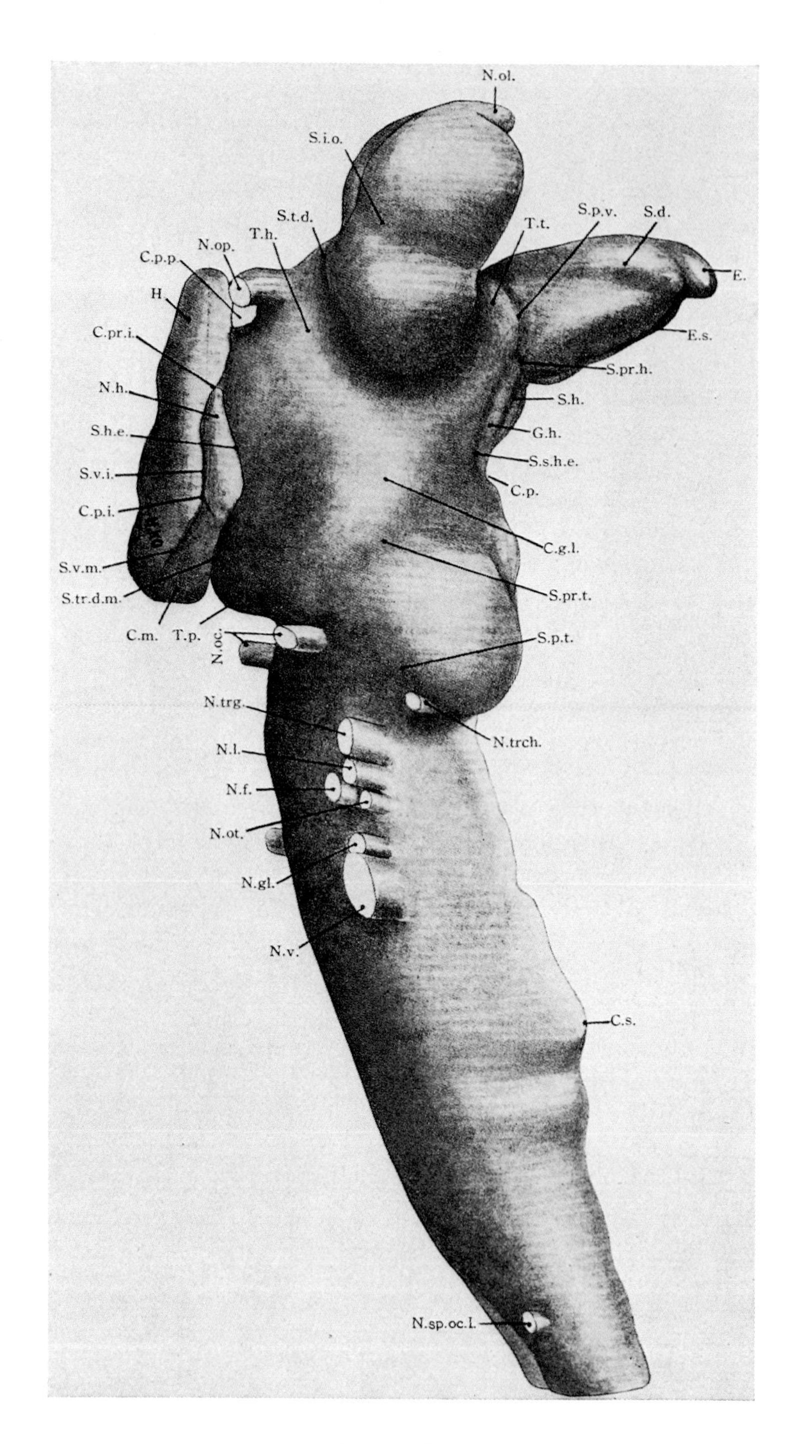

N.ol.
S.i.o.
S.t.d.
T.h.
N.op.
C.p.p.
H.
C.pr.i.
N.h.
S.h.e.
S.v.i.
C.p.i.
S.v.m.
S.tr.d.m.
C.m.
T.p.
N.oc.
T.t.
S.p.v.
S.d.
E.
E.s.
S.pr.h.
S.h.
G.h.
S.s.h.e.
C.p.
C.g.l.
S.pr.t.
S.p.t.
N.trg.
N.trch.
N.l.
N.f.
N.ot.
N.gl.
N.v.
C.s.
N.sp.oc.I.

Figure 133 A

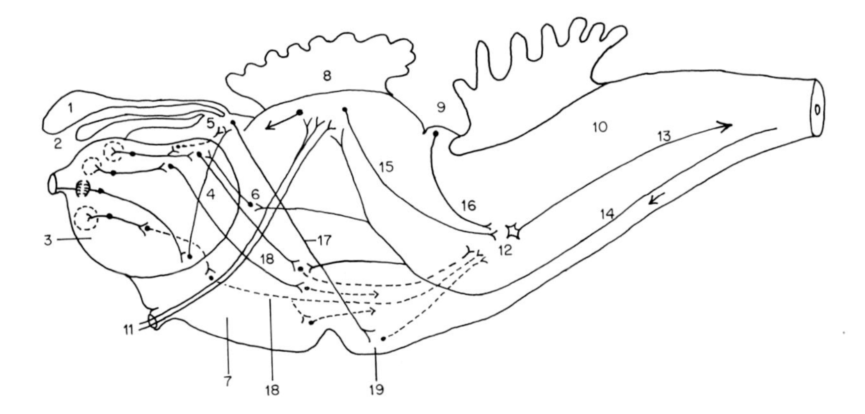

Figure 133 B. Schematic drawing showing in a generalized fashion some main communication channels in the neuraxis of a Petromyzont Cyclostome. 1: pineal organ; 2: parapineal organ; 3: olfactory bulb; 4: lobus hemisphaericus; 5: ganglion habenulae; 6: thalamus (dorsalis et ventralis); 7: hypothalamus; 8: tectum mesencephali (with choroid plexus); 9: cerebellum; 10: oblongata; 11: optic nerve; 12: nucleus reticularis tegmenti (one *Müller cell* is indicated); 13: tegmentobulbar and tegmentospinal channel (including *Müller fibers);* 14: ascending lemniscus channels; 15: tectobulbar (and perhaps tectospinal) channel; 16: efferent cerebellar (e.g. cerebellotegmental) channel; 17: fasciculus retroflexus (habenulopeduncular and related channels); 18: basal forebrain bundle system; 19: interpeduncular nucleus. The olfactory channels in telencephalon, and the olfacto-habenular channel (near 4) have not been specifically labelled.

Figure 133 A. Brain and transition to spinal cord in the Petromyzont Cyclostome Entosphenus japonicus, as displayed by a wax plate reconstruction (from Saito, 1930). C.g.l.: corpus geniculatum laterale; C.m.: 'corpus mammillare' (lobus inferior hypothalami); C.p.: commissura posterior; C.p.i. : commissura postinfundibularis; C.p.p.: commissura postoptica; C.pr.i.: commissura praeinfundibularis; C.s.: 'calamus scriptorius'; E: 'epiphysis' (pineal organ); E.s.: pineal organ stalk; G.h.: ganglion habenulae; H.: adenohypophysis; N.f.: nervus facialis; N.gl.: nervus glossopharyngeus; N.h.: neurohypophysis; N.l.: nervus lateralis; N.oc.: nervus oculomotorius; N.ol.: nervus olfactorius; N.op.: nervus opticus; N.sp.oc.I.: first spino-occipital nerve; N.trch.: nervus trochlearis; N.trg.: nervus trigeminus; N.v.: nervus vagus; S.d.: saccus dorsalis; S.h.e.: 'sulcus hypothalamicus externus'; S.i.o.: 'sulcus interolfactorius' (sulcus limitans bulbi olfactorii); S.pr.h.: sulcus praehabenularis; S.pr.t.: sulcus praetectalis; S.p.t.: sulcus post-tectalis; S.p.v.: 'sulcus parencephalicus ventralis'; S.s.h.e.: sulcus subhabenularis externus; S.t.d.: sulcus telodiencephalicus; S.tr.d.m.: 'sulcus transversalis dorsalis mammillaris' (S.tr.d. externus hypothalami posterioris); S.v.i.: sulcus ventralis infundibuli; S.v.m.: 'sulcus ventralis mammillaris' (s.v. externus hypothalami posterioris); T.h.: tuber hypothalami; T.p.: tuberculum posterius; T.t.: tuber (eminentiae) thalami.

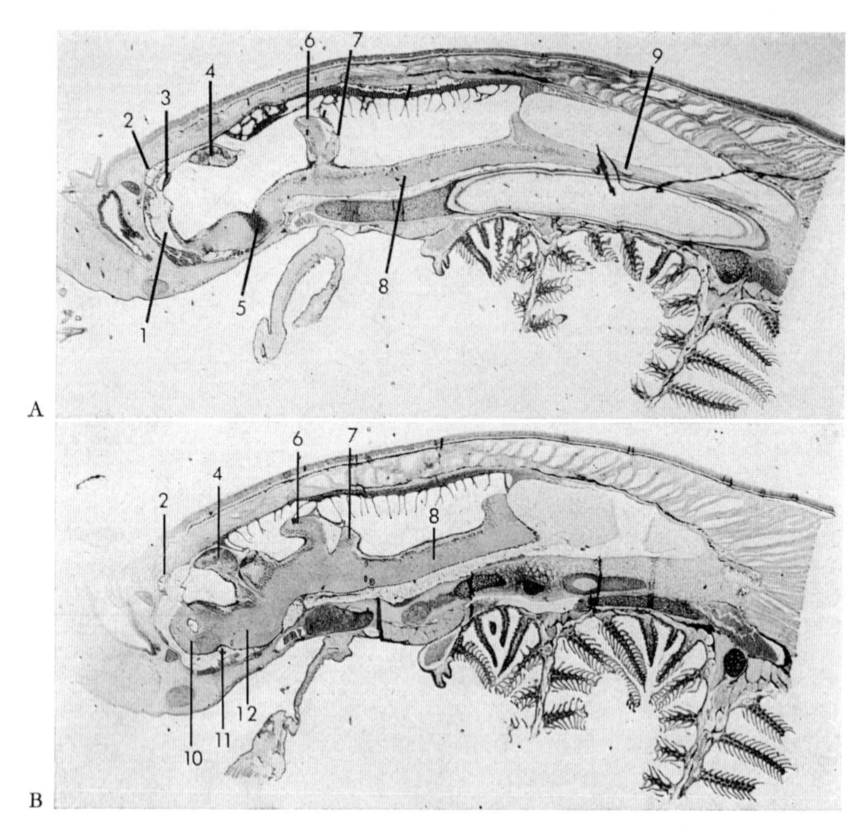

Figure 134 A, B. Sagittal sections through neuraxis of a Petromyzont Ammocoetes larva. B is more lateral than A. 1: lamina terminalis; 2: pineal organ; 3: parapineal organ; 4: ganglion habenulae; 5: posterior hypothalamus; 6: posterior part of tectum mesencephali; 7: cerebellum; 8: oblongata; 9: spinal cord; 10: telencephalon; 11: sulcus telodiencephalicus; 12: diencephalon.

whose dorsal and ventral roots remain separate was dealt with in section 2, chapter VII of volume 3/II.

The neuraxis of *Myxinoids* is characterized by what could be called highly aberrant features, such as the substantial obliteration of the ventricular system, the exaggerated 'telescoping' of brain subdivisions into each other, the lack of epiphysial differentiations, and the absence of unambiguously recognizable cerebellar development.[16] Pertinent as-

[16] As regards further details, not relevant to the aspect here under consideration, reference is made to pp. 192–195, section 1B, chapter VI of volume 3/II (ontogenetic

pects of their cranial and spinal nerves were discussed in the apposite chapters and sections of volumes 3/II and 4.

As regards the phylogenetic relationship of Cyclostomes to Acrania (e.g. Amphioxus) on one hand, and to the Gnathostome fishes on the other hand, several morphologic features in addition to those concerning the neuraxis and its peripheral nerves should be recalled.

In contradistinction to Gnathostomes, both Acrania and Cyclostomes lack paired appendages such as pectoral and pelvic fins displayed by Selachians, Ganoids, and Teleosts. These fins, in accordance with GEGENBAUR's archipterygium theory and other slightly modified ones, are assumed to have given origin to the Tetrapod extremities with their shoulder and pelvic girdles.

Concerning the *phylogenetic evolution of the mouth*, there is little doubt that the mouths of all Vertebrates are homologous, the sucking mouth of Cyclostomes being no exception. Less certain, however, is the homology of the mouth of Craniote Vertebrates with that of 'lower' Chordates, including Amphioxus. Yet, the considerable enlargement of the larval mouth of Amphioxus, as well as its asymmetry, can be easily interpreted as 'larval adaptations' (cf. e.g. NEAL and RAND, 1936).

The ectodermal hypophysial duct of Myxinoids opens into the pharynx (cf. vol. 5/I, chapter XII, section 1C, p. 94). These Cyclostomes are therefore classified as Hyperotreta in contradistinction to the Petromyzont Hyperoartia in which this unpaired duct ends blindly. The opening of the Myxinoid hypophysial duct into the pharynx has been interpreted in favor of the hypothesis that, in Vertebrate phylogenesis, an old *paleostoma*, of which said duct represents a remnant, preceded the formation of a new mouth or *neostoma* characteristic for Acrania and true Vertebrates. Said duct is assumed to be homologous with the mouth of Urochorda. The paleostoma-neostoma hypothesis, however, as e.g. proposed by BEARD and v. KUPFFER, is considered unconvincing by NEAL and RAND (1936) and some other morphologists.

With respect to the morphologically and phylogenetically highly important branchial structures the number of gill slits varies in different Cyclostomes, particularly in the Myxinoid genus Bdellostoma where up to 14 gill pouches may extend caudalward. In Myxine, by backward growth of the hyoid septum, the external gill slit apertures

---

development) and the sections on Cyclostomes in chapters VIII–XI of volume 4, and in chapters XII–XIII of volume 5/I.

become reduced to a single pair, a condition not unlike that seen in Osteichthyes.

Again, in the transformation from Ammocoetes to Petromyzon, the internal gill openings become separated from the esophagus by the caudal evagination of a blind sac as a branchial duct, which is not present in Myxine.

The *visceral skeleton* providing the gill arches (branchial arches), that is to say the bars within the total branchial structures lying between two successive gill slits, is represented in Cyclostomes by a complex cartilaginous gill-basket, in which cartilage rods surround the gill clefts. In Amphioxus, these 'gill arches' are slender rods of a material which, from its resemblance to cartilage, is called 'procartilage'. Typical sequences of branchial arches rostrally beginning with maxillo-mandibular arch, followed by hyoid, then by third and additional arches, make their appearance in Plagiostomes (cf. Fig. 424, p. 839, vol. 3/II).

3. Selachians

The neuraxis of Selachians, depicted in Figures 135A–C and 140B, displays the following general features. The rostrally evaginated and essentially inverted telencephalon consists of a rostral paired portion and of caudal pars impar telencephali with a common telencephalic ventricle. The paired portion includes, besides bulbus olfactorius, a variable part of lobus hemisphaericus. The olfactory bulb may develop lateralward and include an olfactory stalk. In Chimaeroids, the rostral olfactary bulb is directly adjacent to lobus hemisphaericus. Chimaeroids, moreover, are characterized by a moderate degree of pallial eversion in pars impar telencephali. Since no caudal paired evagination is manifested in Selachians, a true polus posterior with caudal expansion of lateral ventricle is lacking.

As regards the telencephalic grisea, the pallial ones become differentiated into D_1, D_2 and D_3 zones, and the basal ones into the zones B_1, B_2, B_3 and B_4. In addition to a periventricular griseum, a pallial and particularly a basal cortical or corticoid plate are generally present.

In the diencephalon, epithalamus and thalamus tend to recede caudalward from the rostral levels of hypothalamic preoptic recess, and caudal parts of thalamus become 'telescoped' into rostral mesencephalic levels. A saccus vasculosus is present.

The tectum mesencephali becomes, to a variable but substantial ex-

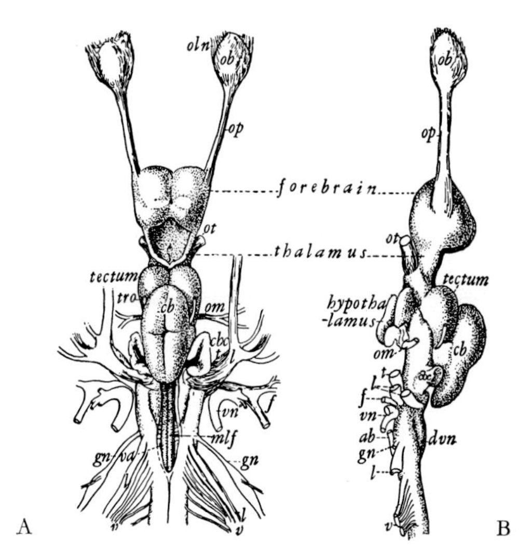

Figure 135 A, B. Brain of the Selachian Squalus acanthias (from Papez, 1929). ab: nervus abducens; cb: cerebellum; cbc: auricula cerebelli and cerebellar crest; dvn: vestibular area; gn: nervus glossopharyngeus; l: nervus lateralis; mlf: fasciculus longitudinalis medialis; ob: olfactory bulb; oln: olfactory nerve; om: nervus oculomotorius; op: olfactory peduncle; ot: nervus opticus; t: nervus trigeminus; tro: nervus trochlearis; v: nervus vagus; va: vagus area of oblongata; vn: nervus vestibularis; 'forebrain' should read: telencephalon, and 'thalamus': diencephalon.

tent, covered by the large corpus cerebelli. The cerebellar auricles and the cerebellar crest are likewise well developed. The oblongata,[17] as depicted in Figures 135 A and B, displays a rostral trigemino-facialis-octavus, and a caudal vagus group of dorsal nerve roots in an arrangement comparable to that obtaining in Cyclostomes (Fig. 133 A). The reticular formation, containing rather large multipolar elements, is well developed but does not include typical *Müller* nor a distinctively differentiated pair of *Mauthner cells*. The medulla spinalis assumes, in contradistinction to that of Cyclostomes, the rounded cord-like shape characteristic for all Gnathostome Vertebrates.

The peculiar, presumably neurosecretory *Dahlgren cells* in the caudal portion of the spinal cord of various Plagiostomes were dealt with in section 5, chapter VIII of volume 4. The electric organs of Torpedinidae and Rajidae were considered in section 5, chapter VII of vol-

[17] Concerning the grisea pertaining to the electric organs of some Selachians and Osteichthyes, the reader is referred to chapter IX of volume 4.

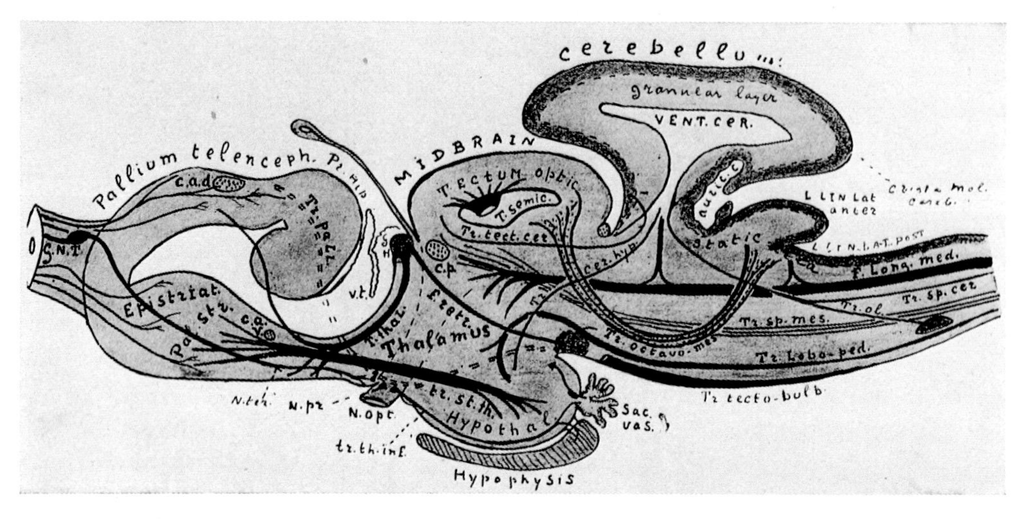

Figure 135C. Some main communication channels in the brain of the Selachian Spinax niger as interpreted by KAPPERS (from KAPPERS, 1929). c.a.d.: 'commissura anterior pars dorsalis' (commissura pallii); c.a.v.: 'commissura anterior, pars ventralis' (commissura anterior); c.p.: commissura posterior; Crista mol.cereb.: molecular layer of cerebellum; 'Epistriatum': nucleus basilateralis superior; f.retr.: fasciculus retroflexus; G.h.: ganglion habenulae; G.N.T.: ganglion nervi terminalis; N.pr.: nucleus praeopticus; N.ter.: nervus terminalis; Pal.str.: nucleus basilateralis inferior; Pr.Hip.: 'primordium hippocampi'; Tr.ol.: tractus olivocerebellaris to inferior olivary griseum; tr.st.th.: lateral forebrain bundle; tr.th.inf.: 'tractus thalamo-infundibularis'; Sac.vas.: saccus vasculosus; v.t.: velum transversum. Other designations self-explanatory.

ume 3/II. It will be recalled that the innervation of these organs is provided by branchial nerves in Torpedinidae, and by caudal spinal nerves in Rajidae. A few Selachians also display luminescent organs, likewise considered in the above-mentioned section of volume 3/II.

In contradistinction to the Cyclostomes (cf. section 2, p. 409), the nervous system of Selachians, as well as that of all other gnathostome Vertebrates, is provided with typical *medullated* nerve fibers in addition to non-medullated ones.

As regards origin and structure of *myelin sheaths* pertaining to peripheral and to central nerve fibers, reference may be made to volume 3/I, chapter V, section 4, pp. 238–261, and section 6, pp. 420–439. With respect to the Vertebrate *central* medullated nerve fibers, about which, since then, more detailed data have been recorded, it seems perhaps appropriate to add here three illustrations (Figs. 135D, E, F) depicting some by now well established essential features of their rela-

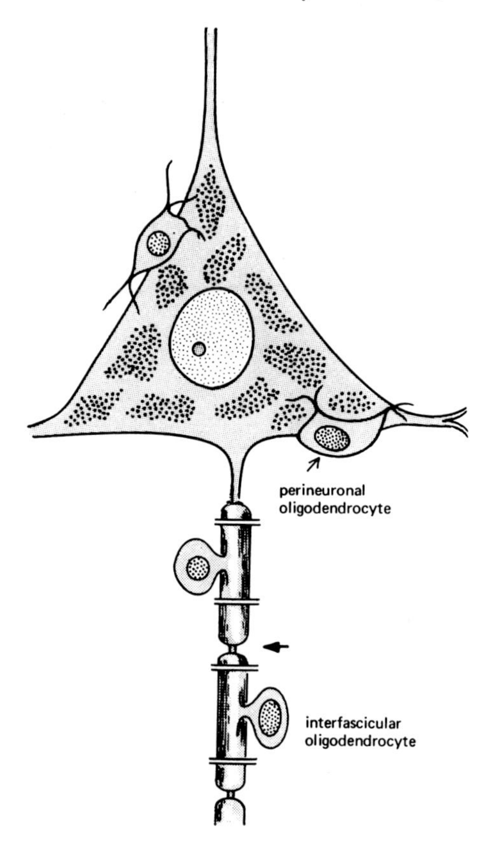

Figure 135 D. Simplified diagram illustrating a central medullated nerve fiber whose myelin sheath is formed by interfascicular oligodendroglia cells (from BARR, 1974). Added unlabelled arrow: central '*node of Ranvier*'. Although this and the following Figures E and F refer to findings in Man respectively Mammals, they can be regarded as illustrating the relevant relationships obtaining in the entire gnathostome Vertebrate series.

tionship to *oligodendroglia cells*. Although formed by these cells in a manner comparable to the formation of myelin sheaths by the likewise neuroectodermal *Schwann cells* of the peripheral nerves, certain peculiarities, already briefly pointed out in volume 3/I (loc.cit.) obtain with respect to said sheaths, which are wrapped around the neuronal axons.

The oligodendroglia cells extend tongue-like processes, which widen peripheralward to form the myelin membranes between two '*nodes of Ranvier*'. Again, in apparent contradistinction to *Schwann cells*, one single oligodendroglia cell may provide myelin sheaths for several

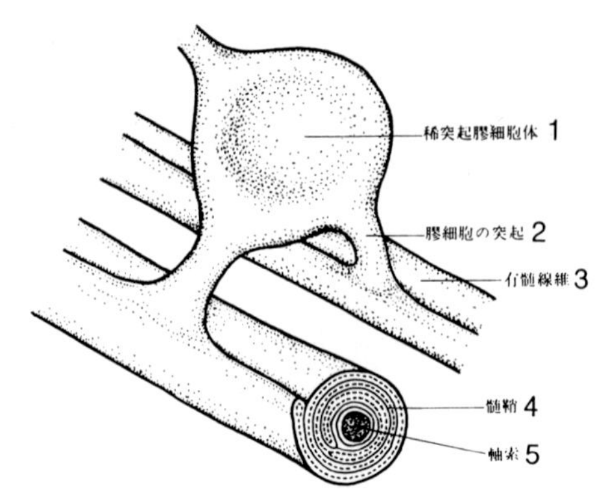

Figure 135E. Oligodendroglia cell with several tongue-like processes providing myelin sheaths for a corresponding number of neighboring nerve fibers (from NIIMI, 1976). 1: oligodendroglia cell body; 2: tongue-like extension of cell body, ending as expanded myelin-forming process; 3: medullated nerve fiber; 4: myelin sheath; 5: axon.

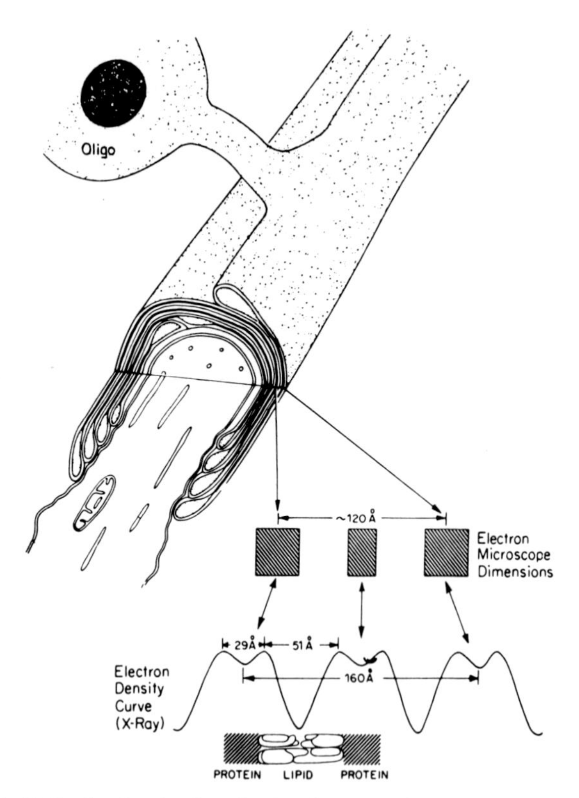

Figure 135F. Further details of myelin sheath in central nervous system (after NORTON, from TOWER, 1977).

neighboring nerve fibers (cf. Fig. 135E). Figures 135D–F should be compared with Figures 131 (p. 189), 170 (p. 241) and 257A (p. 433) of volume 3/I.

4. Ganoids and Teleosts

These Vertebrate forms represent together the class Osteichthyes, that is to say Fishes possessing a skeleton in part or wholly of bone. With respect to the telencephalon, this class differs from all other Vertebrates by eversion of the *entire* lobus hemisphaericus, which develops as a telencephalon medium *sive* impar formed by unpaired rostral evagination. In certain instances, the everted lobus hemisphaericus also displays some degree of paired caudal evagination with true polus posterior. The Osteichthyan bulbus olfactorius is, however, inverted and formed by paired rostral evagination. It may be directly contiguous to lobus hemisphaericus, or connected with this latter by an olfactory stalk. The everted pallium displays the D_1, D_2 and D_3 neighborhoods. In some forms these zones show additional differentiations, and in others some fusions of the basic neighborhoods obtain. The degree of eversion varies from moderate to extreme. The B zones display B_1, B_2, and B_3 neighborhoods which likewise may be subdivided by further differentiations. In conformity with the eversion of the entire lobus hemisphaericus, a B_4 zone is not present.

Telencephalic eversion, as manifested in Osteichthyes, tends to increase the ventricular surface and to decrease the external surface of the alar plate, concomitantly with an expansion of the roof plate which forms what was at one time interpreted as a 'membranous pallium'. The telencephalic grisea developing within the alar plate remain partly periventricular, and are partly crowded within the interior of the region corresponding to convexity of internal, and concavity of external surface, thereby precluding a development of a superficial pallial and basal cortical or corticoid layer present in Selachians and in Dipnoans.

The caudal part of the Osteichthyan diencephalon is partly 'telescoped' into rostral mesencephalic neighborhoods as is also the case in Selachians. The posterior hypothalamus, representing the so-called lobi inferiores, is commonly very large and complexly differentiated. A saccus vasculosus is generally present but may be missing in some forms.

As a rule, the tectum opticum is well developed and conspicuously large, displaying, in some forms, a torus longitudinalis. The cerebellum is likewise large and well developed. It is characterized by a valvula cerebelli protruding into the mesencephalic ventricle. In some instances, the antimeric portions of tectum mesencephali become thereby partly separated but remain interconnected by a lamina epithelialis derived from the roof plate.

In various Teleosts, the oblongata displays expanded facial or vagal lobes related to a highly developed 'gustatory' system. The reticular formation of Osteichthyes generally includes a typical pair of *Mauthner cells*. The large crossed neurites of these cells thus characterize the spinal cord of most Ganoids and Teleosts (cf. section 6, chapter VIII, vol. 4).

As in various Plagiostomes, neurosecretory cells occur in the caudal portion of the spinal cord in diverse Ganoids and Teleosts. In some of these latter, said cells provide a conspicuous *neurohypophysis caudalis* or *urohypophysis*, dealt with in section 6, chapter VIII of volume 4.

Except for the completely everted telencephalic lobus hemisphaericus, and the presence of *Mauthner cells*, the neuraxis of Ganoids is fairly similar to that of Selachians. The neuraxis of the very large group of Teleosts, however, with a great diversity of forms, displays numerous peculiarities of configuration distorting the fundamental Gnathostome Anamniote pattern, which, nevertheless, remains also here topologically invariant. The pronounced configurational and structural differences are correlated with the behavioral characteristics of these Osteichthyans. In the Teleostean neuraxis, the grisea are more circumscribed as 'nuclei' than in that of Selachians and many Ganoids. Likewise, the basic communication channels, on the whole quite diffuse in Selachians, tend to become more discrete, well circumscribed, and further differentiated in Teleosts. Figures 136, 137, and 140C depict some overall aspects of the Osteichthyan central nervous system.

As regards the electric organs occurring in some Teleosts, their arrangement and innervation were considered in section 5, chapter VII of volume 3/I. This section also deals briefly with the luminescent organs found in a variety of Teleosts, and no further comments are here required.

A particularly aberrant form is represented by the Crossopterygian Coelacanth Latimeria[18] repeatedly dealt with in the relevant chapters of

[18] The ethmoid region of Latimeria includes a large tubular, apparently sensory organ, innervated by the nervus ophthalmicus profundus trigemini. MILLOT and ANTHONY (1965), who designate it as '*organe rostral*', '*pour ne pas préjuger de sa fonction, encore mystérieuse*', state that it includes a structure '*inconnue chez les autres Vertébrés*'.

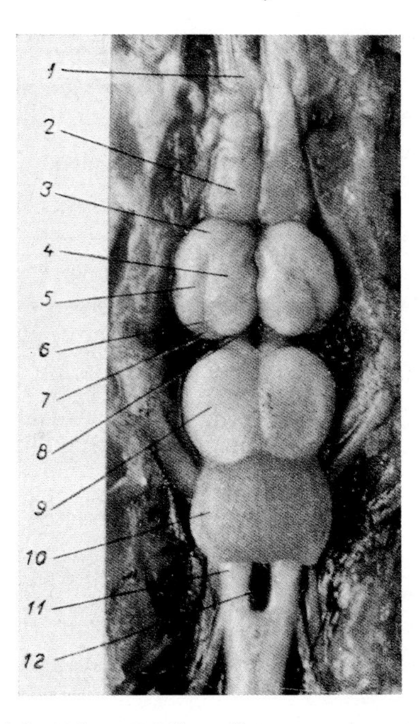

Figure 136. Brain of the Teleost Eel (Anguilla) as seen from the dorsal aspect in situ after removal of meninges and roof plate membranes (from BECCARI, 1943). 1: olfactory nerve; 2: olfactory bulb; 3: rostral part of lobus hemisphaericus; 4: pars basalis telencephali; 5: everted pallium telencephali; 6: sulcus ypsiliformis" (fd_1); 7: caudal extremity of telencephalon; 8: diencephalon; 9: tectum mesencephali; 10: cerebellum; 11: octavolateral area of oblongata *('tuberculo acustico');* 12: rhomboid fossa.

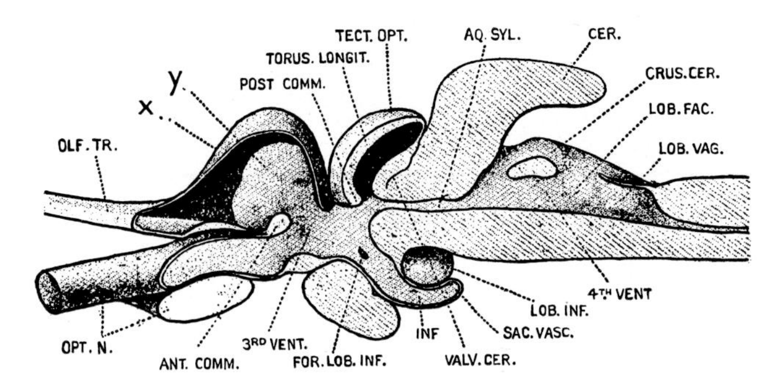

Figure 137. Brain of the Teleost Gadus morrhua (Codfish) as seen from the medial side of a midsagittal section (after ELLIOT SMITH, from PAPEZ, 1929). Crus.cer.: cerebellar crest; For.Lob.Inf., Inf., Lob.Inf.: recesses of the lobi inferiores hypothalami posterioris; Aq.Syl.: mesencephalic ventricle reduced by protruding valvula cerebelli, x: telencephalic roof plate; y: lobus hemisphaericus telencephali. Other abbreviations self-explanatory.

volumes 3/II, 4, and 5/I. Especially with regard to the morphology of its prosencephalic derivatives, I prefer to include this Coelacanth among the class Osteichthyes, and not in a class Choanichthyes together with the Dipnoans.[19]

5. Dipnoans

The typologic features of the Dipnoan neuraxis can be evaluated as representing a transition between those obtaining in Selachians and those characteristic for Amphibians.

The Dipnoan telencephalon is mainly formed by rostral paired evagination combined with inversion, but the caudal part of lobus hemisphaericus displays a telencephalon impar related to unpaired rostral evagination combined with a moderate degree of eversion. It manifests thus configurational aspects comparable with those obtaining in Plagiostome Chimaeroids. Again, as in Selachians, the Dipnoan telencephalon does not display a paired caudal evagination and thus lacks a

[19] The 'class' Choanichthyes was tentatively defined on p. 59, section 2, chapter II of volume 1.

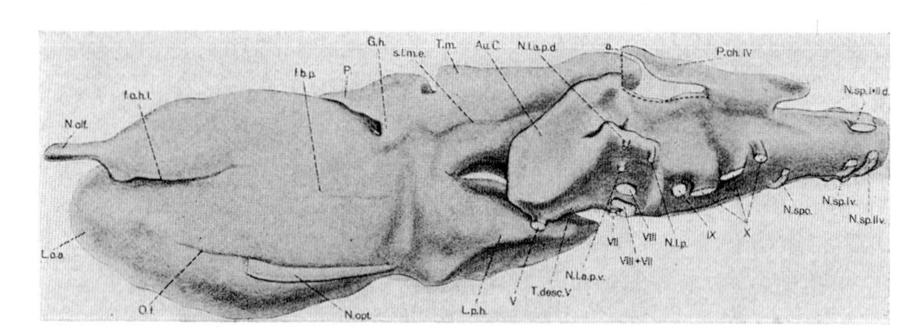

Figure 138. Lateral aspect of brain and rostral spinal cord in the Dipnoan Protopterus annectens, as seen in a wax-plate reconstruction (from Gerlach, 1933). a.: attachment line of choroid plexus; Au.C.: auricle of cerebellum; f.b.p.: external basopallial groove; f.o.h.l.: lateral olfactohemispheric groove; G.h.: ganglion habenulae; L.o.a.: 'anterior olfactory lobe'; L.p.h.: lobus posterior hypothalami; N.l.a.p.d. (p.v.): nervus lateralis anterior, dorsal and ventral part; N.l.p.: nervus lateralis posterior; N.sp.I, II d., I, II v.: dorsal and ventral roots of first two spinal nerves; N.spo.: spino-occipital nerve; P.: membranous parencephalon; O.f.: telencephalic groove corresponding to optic nerve; P.ch.IV: choroid plexus of fourth ventricle; s.l.m.e.: sulcus lateralis mesencephali externus; T.desc.V: bulge of radix descendens trigemini; T.m.: tectum mesencephali; V–X: cranial nerve roots.

true polus posterior. In the telencephalic wall, the longitudinal pallial zones D_1, D_2, D_3 and the basal zones B_1, B_2, B_3, B_4 are present. Bot pallium and basis include a rather conspicuous corticoid plate.

The diencephalon does not display the substantial *degree* of caudal 'telescoping' into mesencephalic neighborhoods manifested by Osteichthyans and Plagiostomes. In this and also in some other respects it is rather similar to that of Amphibians. A saccus vasculosus has not been identified with certainty and may be missing.

Mesencephalon and cerebellum are rather moderately developed. A valvula cerebelli comparable to that of Osteichthyans is lacking, but cerebellar auricles are present. The oblongata includes a pair of *Mauthner cells*. Figure 138 illustrates the neuraxis of the African Lungfish Protopterus. That of the South American Lepidosiren is quite similar, while the neuraxis of the Australian Ceratodus (or Neoceratodus) displays a few differences, which can be evaluated as minor in the aspect here under consideration.[20]

6. Amphibians

The intermediate position of the Tetrapod Amphibians, which generally develop as Fish-like forms in water and subsequently undergo a transformation into terrestrial animals, is of particular significance for a phylogenetic interpretation concerning the origin of Amniota, i.e. of 'higher Vertebrates'.

This intermediate position is likewise suggested by the configuration of the Amphibian neuraxis in all three extant orders (Urodeles, Anurans, Gymnophiones), although specific inferences concerning details of the phylogenetic transformations from Amphibians to Reptiles respectively to Mammals remain inconclusive.

The Amphibian telencephalon is entirely inverted. Discounting a minor degree of rostral unpaired evagination, resulting in a telencephalon medium *sive* impar (aula) of moderate size, rostral paired evagination provides not only olfactory bulb but the bulk of lobus hemisphaericus. In addition, there obtains a substantial paired caudal evagi-

---

[20] These differences were dealt with in the apposite sections of volumes 3/II and 4, as well as in chapters XII and XIII of volume 5/I. It should also be recalled that Ceratodus has an unpaired lung (Monopneumones), while Protopterus and Lepidosiren have a paired lung (Dipneumones).

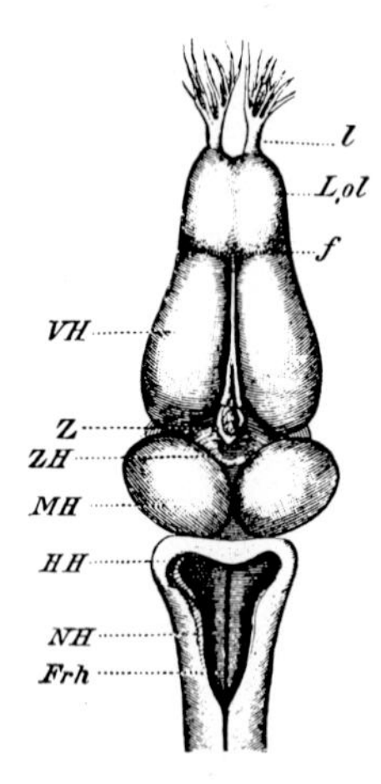

Figure 139 A. Dorsal aspect of the brain in the Anuran Amphibian Frog (from R. HERTWIG, 1912). f: groove between bulbus olfactorius and lobus hemisphaericus; Frh: rhomboid fossa; l: alfactory nerve; L.ol.: olfactory bulb; HH: cerebellum; MH: mesencephalon; NH: oblongata; VH: telencephalon; Z: epiphysial complex; ZH: diencephalon.

nation, resulting in a posterior recess of lateral ventricle caudally to interventricular foramen, combined with a true polus posterior of lobus hemisphaericus. The olfactory bulb, devoid of an olfactory stalk, remains contiguous to rostral portion of lobus hemisphaericus. Within this latter, the pallial D_1, D_2, D_3 zonal components, and the basal B_1, B_2, B_3, B_4 zones are well differentiated in all three orders. In contradistinction to Selachians and Ganoids, the grisea remain periventricular, and a cortex cerebri is not present in Urodeles and Anurans, but a very moderately developed basal cortex can be seen in Gymnophiona, in which, moreover, a cortex-like arrangement of caudal D_3 elements may become barely suggested.

The diencephalon, whose configuration displays most clearly the four longitudinal zones characteristic for the bauplan of that brain subdivision, is not, to any very significant degree, caudally 'telescoped' into the adjacent mesencephalic neighborhoods. A saccus vasculosus, even at aquatic stages, and in aquatic forms, is apparently lacking.

The mesencephalon, particularly the tectum opticum, is generally large and well developed in Anurans, moderately to various degrees in Urodeles, and poorly in Gymnophiona.

The cerebellum is far more poorly developed than in Dipnoans, being relatively best differentiated in Anurans, and less so in Urodeles. A corpus cerebelli is reduced to a nondescript vestigial edge in some of these latter (e.g. Proteus), and is likewise reduced or 'missing' in Gymnophiones, which, however, display fairly large, but modified cerebellar auriculae.

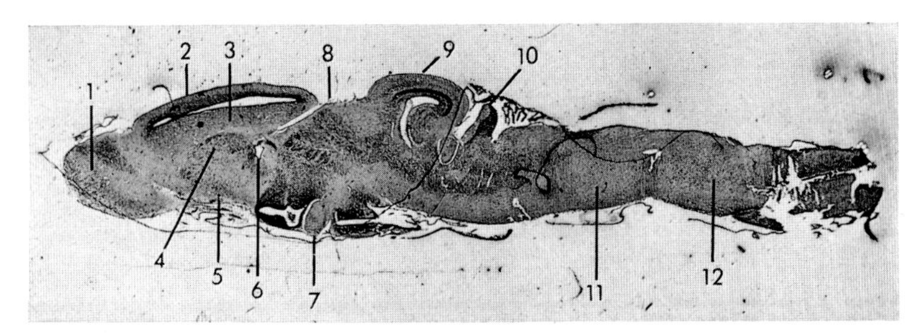

Figure 139 B. Sagittal section (cell stain) through brain and rostral spinal cord of a young Frog, shortly after metamorphosis. 1: olfactory bulb; 2: pallium (D_2); 3: pallium (D_3); 4: nucleus basimedialis superior; 5: nucleus basimedialis inferior; 6: interventricular foramen; 7: optic chiasma and supraoptic commissures (rostrally to 7 preoptic recess, and caudally to 7 ventricular space of posterior inferior hypothalamus); 8: diencephalon (the longitudinal zones of thalamus dorsalis, ventralis, and hypothalamus are clearly recognizable); 9: mesencephalon; 10: cerebellum; 11: oblongata; 12: spinal cord.

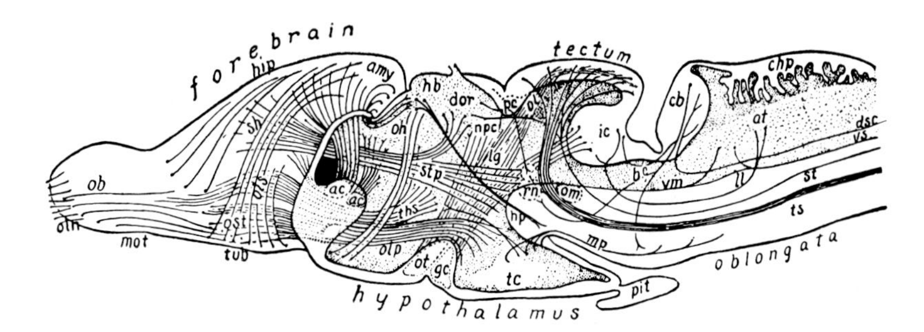

Figure 139 C. Schematic drawing showing main communication channels in the brain of a Frog as interpreted by PAPEZ (from PAPEZ, 1929). ac: anterior commissure; amy: posterior pole of hemisphere; at: cochlearis grisea; bc: 'brachium conjunctivum'; cb: cerebellum; chp: choroid plexus; dor: thalamus dorsalis; dsc: dorsal spinocerebellar tract; gc: postoptic commissure; hb: ganglion habenulae; hip: pallium telencephali; hp: habenulopeduncular tract; ic: torus semicircularis; lg: lateral geniculate grisea; ll: lateral lemniscus; mot: medial olfactory tract; mp: 'mammillary peduncle'; npc: nucleus of posterior commissure; ob: olfactory bulb; oh: olfactohabenular tract; oln: olfactory nerve; olp: medial forebrain bundle; ols: olfactoseptal fibers; om: oculomotor nerve; ost: basal ganglia; ot: optic tract; pc: posterior commissure; pit: hypophysial complex; rn: mesencephalic tegmentum; sh: 'septohippocampal fibers'; st: spinotectal tract; stp: lateral forebrain bundle; tc: posterior hypothalamus; ths: component of medial forebrain bundle; ts: tectobulbospinal tract; tub: 'olfactory tubercle' (basal telencephalic grisea); vm: vestibulomesencephalic tract; vs: vestibulospinal tract. Some of the original designations used by PAPEZ have been slightly modified.

As regards the oblongata, the lateralis system is present in aquatic Urodeles and generally in most Amphibian aquatic larvae. In Anurans, it commonly disappears, and a *primordial cochlearis system* develops. A pair of *Mauthner cells* is found in aquatic Urodeles as well as in Anuran larvae.[21] The occipito-spinal nerve roots, already variously present in Fishes, become, particularly in Anurans, more closely similar to the homologous Amniote nervus hypoglossus, and can, with some justification, be designated by this name. Said system is also present in Gymnophiona.

Figures 139 A and B illustrate typologic configurational aspects of the Amphibian brain, and Figures 139 C and 140 D depict, in two different versions, the arrangement of main communication channels.

As regards the variable degree of differentiation displayed by the Amphibian neuraxis in Urodeles, Anurans, and Gymnophiones, it is evident that, in diverse respects, said differentiation remains below the level obtaining in some Selachians, particularly in many Osteichthyes, moreover in at least some neuraxial subdivisions of Dipnoans.[22] Whether the more primitive and 'generalized' condition displayed by Amphibians, especially by diverse Urodeles, is a retained primary or a 'regressive' one, remains a moot question. Be that as it may, there is no doubt that the 'generalized' configuration and structure of the neuraxis in Urodeles, as stressed by LUDWIG EDINGER and subsequently intensively studied by HERRICK, represents a paradigm which indeed illustrates the most distinct manifestation of the fundamental morphologic pattern characteristic for the neuraxis of all Vertebrates. Thus, in particular, the Amphibian central nervous system provides a most suitable frame of reference for an understanding of both comparative anatomy and presumptive phylogenetic evolution of the Vertebrate neuraxis.

7. Reptiles

Reptiles, of which four or five recent orders[23] can be distinguished, are here of some importance since they represent the 'lowest' or 'most

---

[21] Details concerning the various development of lateralis respectively octavolateralis system, and the presence or absence of *Mauthner cells* in Amphibians are dealt with in section 7, chapter IX of volume 4.

[22] It will be recalled that in both groups of Dipnoans (cf. footnote 20), the telencephalon displays a conspicuous pallial and basal cortical (or 'corticoid') plate.

[23] These orders are: Lacertilia, Ophidia, Rhynchocephalia, Chelonia, and Crocodilia.

primitive' class of Amniota. Birds and Mammals may be conceived as Vertebrate forms derived, by a divergent phylogenetic evolution, from unknown Reptilian-like ancestors.[24] Recent Reptiles are more closely related to Birds than to Mammals. Reptiles, together with Birds, can

---

Lacertilia and Ophidia can also be subsumed, as suborders, under a single order Squamata. Cf. also pp. 62–63, section 2, chapter II of volume 1.

[24] Whether the Reptilian-like Mammalian ancestors were more close related to Amphibian forms than to the main Reptilian evolutionary line remains a moot question. This means, in other word, that a branching into Sauropsidan and Mammalian 'phylogenetic radiation' might already have occurred at the 'Amphibian level'.

Figures 140 and 141 see p. 428 and 429

Figure 140. Diagrams illustrating main communication channels in the brains of Anamniote Vertebrates (redrawn after PLATE, from NEAL and RAND, 1936). A: Petromyzont Cyclostome; B: Selachian Scymnus; C: Teleost Cyprinus carpio; D: Anuran Amphibian Rana. For designations of tracts, see legend of Figure 141.

Figure 141. Diagrams illustrating main communication channels in the brains of Amniota (redrawn after PLATE, from NEAL and RAND, 1936). E: Reptilian Lacerta; F: Avian Anas (Duck); G: generalized intermediate Mammal. 1: olfactory nerve; 2: tr. olfactohabenularis; 3: tr. olfactohypothalamicus; 4: tr. olfactocorticalis; 5: tr. olfacto-epibasalis; 6: tr. olfactopeduncularis; 7: tr. olfacto-ammonicus; 8: tr. olfactomammillaris; 9: tr. olfacto-amygdalinus; 10: tr. parolfactohabenularis; 11: tr. opticus; 12: tr. opticotectalis; 13: tr. isthmo-opticus; 14: tr. praeopticohabenularis; 15: tr. striohypothalamicus; 16: tr. striothalamicus; 17: tr. striomesencephalicus; 18: fasciculus retroflexus; 19: tr. pallialis; 20: tr. hippocampothalamicus; 21: tr. hippocampomammillaris (fornix); 22: tr. bulbothalamicus; 23: tr. bulbocorticalis; 24: tr. frontobulbaris; 25: tr. tectobulbaris; 26: tr. tectospinalis; 27: tr. tectothalamo-occipitalis; 28: tr. tectocerebellaris; 29: tr. frontothalamicus; 30: tr. frontoepibasalis; 31: tr. corticohabenularis; 32: tr. corticothalamicus; 33: tr. corticomammillaris (fornix); 34: tr. corticobulbaris; 35: tr. lobopedunculospinalis; 36: tr. habenulo-interpeduncularis; 37: tr. habenulocorticalis; 38: tr. mammillotegmentalis; 39: tr. mammillothalamicus; 40: tr. rubrothalamicus; 41: tr. rubrocerebellaris; 42: tr. rubrospinalis; 43: tr. thalamicospinalis; 44: tr. tegmentocerebellaris; 45: tr. trigeminocerebellaris; 46: tr. quintocerebellaris; 47: tr. septomesencephalicus; 48: tr. octavocerebellaris; 49: tr. olivocerebellaris; 50: tr. vestibulocerebellaris; 51: tr. diencephalocerebellaris; 52: tr. mesencephalocerebellaris; 53: tr. lateralicerebellaris; 55: tr. facialicerebellaris; 56: tr. spinocerebellaris; 57: tr. cerebellotegmentalis; 58: tr. cerebellodiencephalicus; 59: tr. spinothalamicus; 60: tr. spinohypothalamicus; 61: tr. octavomesencephalicus; 62: tr. mammillopeduncularis; 63: tr. frontobulbaris; 64: tr. lobotectalis. The tracts are supposed to be projected on a left lateral plane. It will be noted that, in his interpretation, PLATE occasionally uses designations which are not commonly adopted, and in a few instances (e.g. fornix) uses different notations for different Vertebrate forms. His diagrams, however, provide a useful first approximation to the overall arrangement of the communication channels. It is of interest to compare the following Figures: 140A with 133B, 140B with 135C, 140D with 139C, 141G with 132B and 146A.

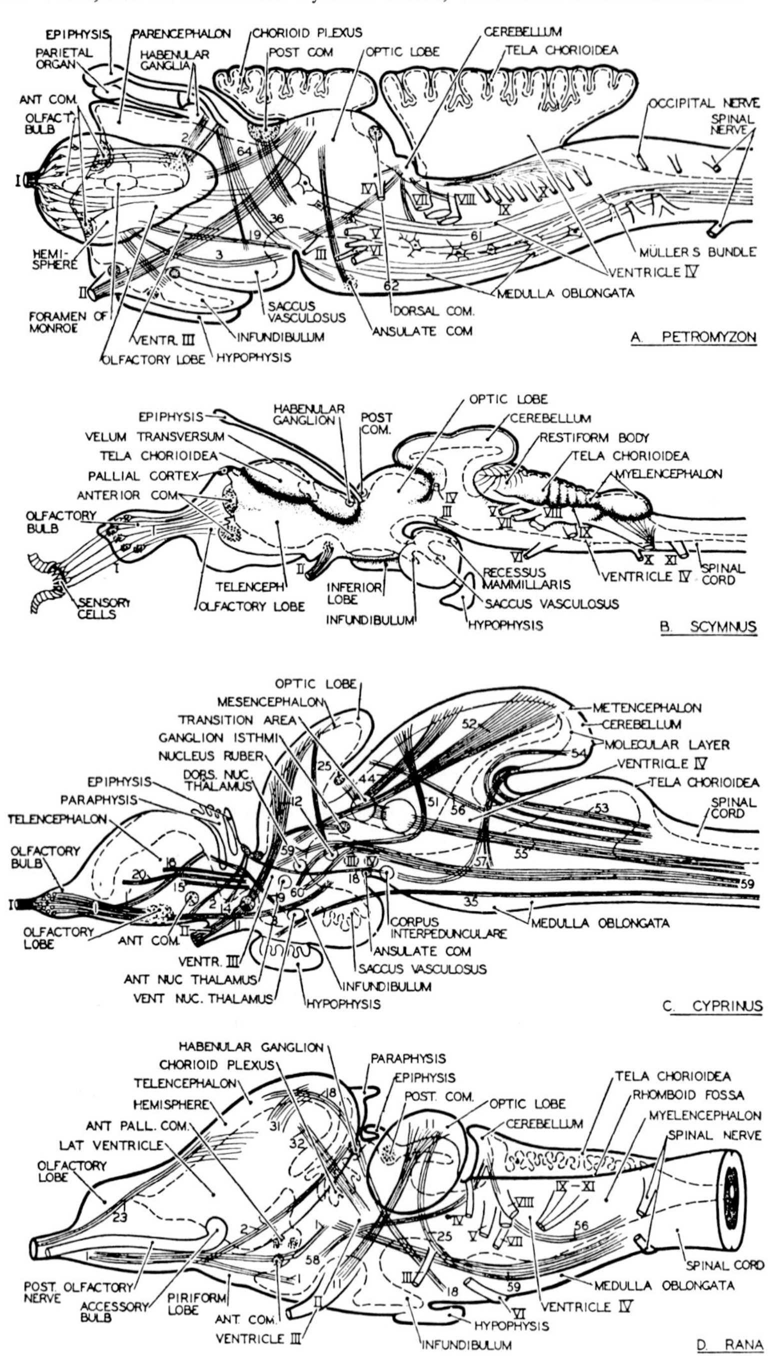

Figure 140. Legend see p. 427.

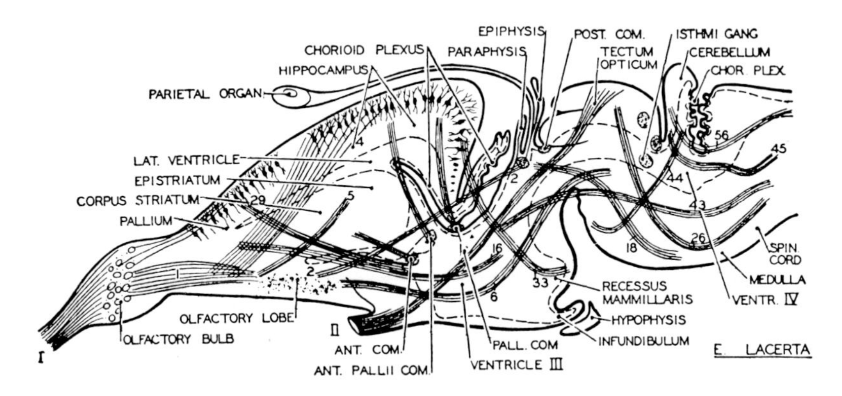

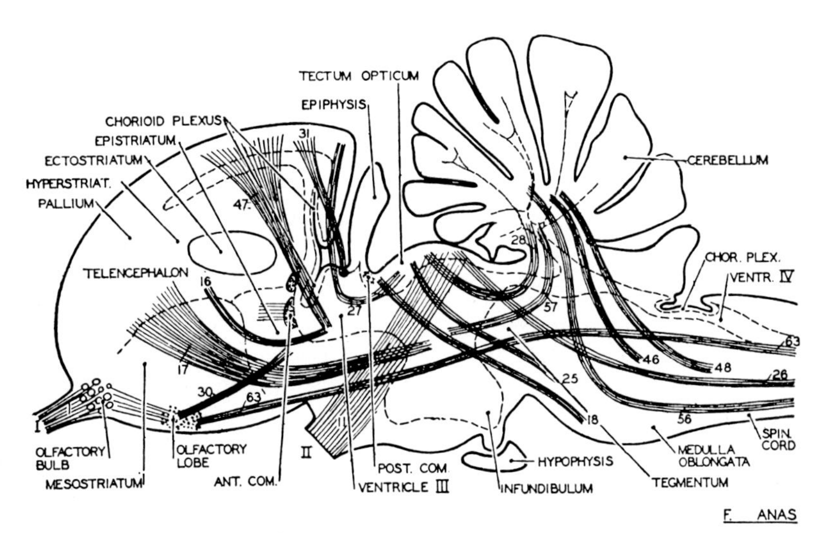

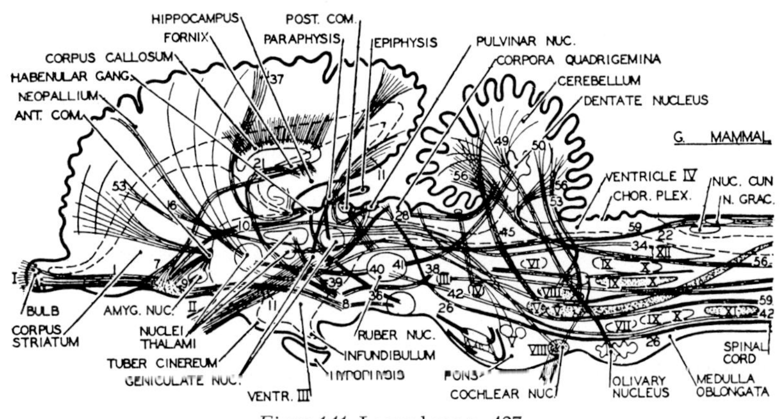

Figure 141. Legend see p. 427.

be subsumed under a sort of superclass Sauropsida. Crocodilia display various features of the neuraxis somewhat resembling those in Birds, while various such features in Lacertilia are, in this respect, more easily comparable with those obtaining in 'lower Mammals'. As regards Chelonia, a few resemblances with Avian as well as with primitive Mammalian neuraxial features become suggested.

The Reptilian telencephalon is, like that of Amphibians, completely inverted and formed by rostral paired evagination as well as by caudal paired evagination, which latter results in a true polus posterior of lobus hemisphaericus. In contradistinction to all Gnathostome Anamnia, however, the early ontogenetic rostral unpaired evagination becomes almost completely reduced, so that a telencephalon impar disappears, except for the median space dorsally to commissura anterior and between the two inverventricular foramina. These features are common to all three classes of Amniota, and are, moreover, and strangely enough, also manifested in Cyclostomes.[25]

The Reptilian olfactory bulb is stalked in numerous forms, but may also be contiguous with lobus hemisphaericus, as e.g. commonly in Chelonia.

The lobus hemisphaericus is relatively larger than in all Anamnia,[26] and more complexly differentiated. A most important morphologic feature, again common to all classes of Amniota, is the '*internation*' or 'introversion' of the D_1 component, which joins the basal B_1 and B_2 components. All Amniota become thus characterized by a secondarily enlarged basis, and a morphologically reduced, 'secondary' pallium. Despite this 'reduction' however, the secondary Amniote pallium greatly expands, culminating, *qua* extension, in Primates, Proboscideans and Cetaceans.

The secondary pallium displays a cerebral cortex which is more differentiated than that obtaining in diverse Anamnia and formed by complete or near complete exhaustion of the periventricular matrix. There is a lateral cortical plate essentially derived from D_2, a dorsal

---

[25] The disappearance of the telencephalon impar is related to the configuration of the lamina terminalis as evolving in the course of ontogeny. Cf. e.g. Figures 62, p.173, 63, p.174, and 65, p.179 of chapter VI, volume 3/II.

[26] An exception, *qua* relative expansion of the telencephalon, but not *qua* degree of differentiation, are here the Amphibian Gymnophiones, whose large, predominantly olfactory telencephalon extends far caudalward and whose well developed olfactory bulbs are each provided with two substantial olfactory nerves.

cortical plate derived from D_2, and a medial ('hippocampal') cortical plate derived from D_3. A 'primordium neocorticis *sive* neopallii' has arisen at the rostral boundary between lateral and dorsal cortical plate.

The zone D_1 provides the epibasal (or hypopallial) complex with diverse 'nuclear' differentations. The B_1 and B_2 zones form a nucleus basilateralis superior and inferior. The lateral basal grisea are much more extensive than the paraterminal medial grisea derived from B_3 and B_4, with grisea basimedialia (inferius and superius). The basilateral and basimedial grisea are covered by a nondescript 'semiparietine' basal cortex ('cortex olfactoria' and cortical amygdaloid nucleus). The caudal components of epibasal and basilateral grisea form an 'amygdaloid complex'.

The diencephalon, discounting the 'progressive' differentation of its grisea, which were described in section 7 of chapter XII of volume 5/I, displays a configuration similar to that obtaining in Amphibians. The thalamus ventralis tends to become relatively reduced.

The mesencephalon is, on the whole, similar to that of Anuran Amphibians, with fairly well developed tectum opticum. In some forms, the torus semicircularis tends to bulge caudalward, suggesting primordia of the Mammalian inferior colliculi. In the mesencephalic tegmentum, a rather well identifiable nucleus ruber is present.

The cerebellum, dealt with in section 7, chapter X of volume 4, is poorly to moderately developed, the Lacertilian one being although relatively slightly larger, rather similar to that of Anuran Amphibians, while the greatest cerebellar expansion is displayed in Crocodilians (cf. Fig. 332, p. 725, vol. 4).

In the oblongata, the lateralis system of Anamnia has become replaced by a rudimentary cochlearis system, which, however, is somewhat better differentiated than that of the Amphibian Anurans and Gymnophiones. The arrangement of the deuterencephalic cranial nerves including nervus accessorius (XI) and hypoglossus (XII) corresponds, more closely than in Amphibians, to that obtaining in Mammals.

Figures 142 and 143 illustrate overall aspects of the Reptilian neuraxis. Comparing Figure 143 with Figure 139B showing the configuration of an Anuran Amphibian neuraxis, it will be noted that, in the Lacertilian oblongata, the ventrally convex rostral rhombencephalic flexure and the dorsally convex cervical flexure are rather prominent. The significance of these flexures, which play a role in stages of on-

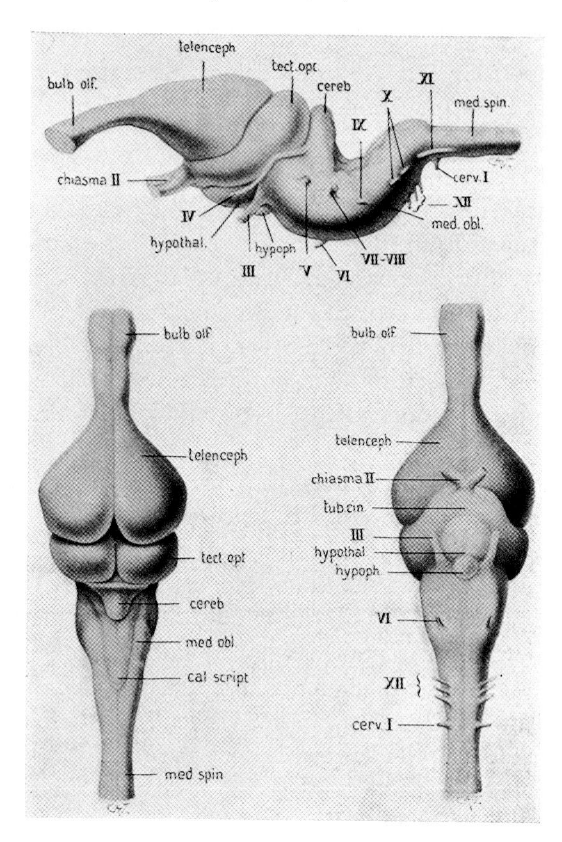

Figure 142. Brain and rostral spinal cord of the Reptilian Lacerta viridis, as seen in lateral, dorsal and ventral view (from FREDERIKSE, 1931).

togeny of the Vertebrate brain, was dealt with and illustrated[27] in section 1B, chapter VI of volume 3/II. These flexures are also conspicuous in adult Amphibian Gymnophiones,[28] and can still be noticed in most Avian brains (cf. Fig. 144A) but tend to disappear in many Mammalian brains, including the Human one (cf. Figs. 145B, 146B, 147A, B, 149A, B).[29] A marked manifestation of said flexures in a Mammalian (Cetacean) brain is illustrated by Figure 148.

---

[27] Cf. e.g. Figures 56D, p.159, 57G, p.164, 66B, p.183, 70, p.194, of volume 3/II.

[28] Cf. Figure 76, p.201 of volume 3/II.

[29] Cf. also Figures 78D and E, p.205 and 206 of volume 3/II.

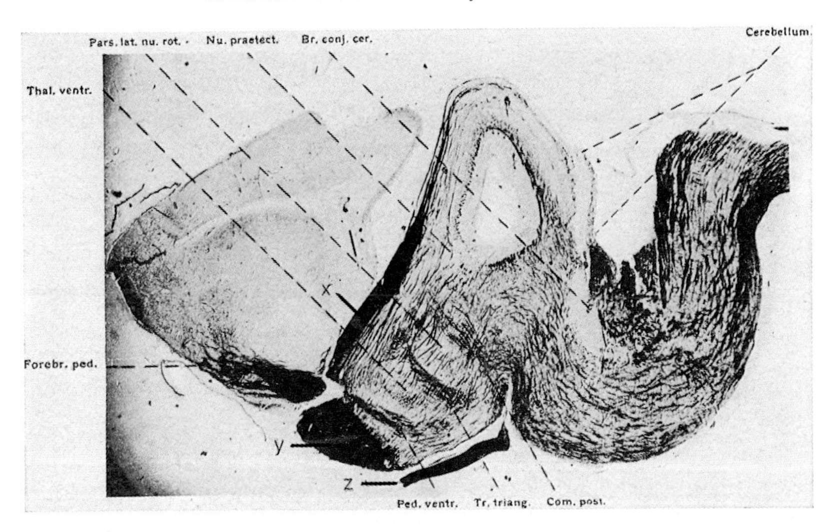

Figure 143. Sagittal section (myelin stain) through the brain of the Reptilian Lacerta viridis (from FREDERIKSE, 1931). Br.conj.cer.: brachium conjunctivum; Com.post.: radiation of posterior commissure; Forebr.ped. and Ped.ventr.: components of basal forebrain bundle; Pars.lat.nu.rot.: lateral portion of nucleus rotundus thalami ventralis; Tr.triang.: tractus triangularis of FREDERIKSE, which corresponds to BECCARI's fasciculus geniculatus descendens, perhaps also including components of that author's fasciculus praetectalis descendens; x: optic tract; y: optic chiasma and supra- respectively postoptic commissures; z: portion of trigeminal nerve (x, y, z have been added).

Figure 141 E depicts, in simplified form, the arrangement of some main communication channels. As regards these latter, both the ascending and descending ones between brain and spinal cord seem better differentiated and more distinct than those in Anamnia. WARNER (1952) has described their progressive myelinization during ontogeny in the Ophidian Natrix sipedon. Besides the descending components of fasciculus longitudinalis medialis, the reticulospinal bundles, and the lateral verstibulospinal tract, tectospinal fibers and a rubrospinal tract seem to be present. There is, however, no definite evidence that descending fibers from telencephalon reach the spinal cord. As regards the descending pathways from the brain stem to the spinal cord in some Reptiles, a recent paper by DONKELAAR (1976) reports some details based on that author's interpretation of data provided by experimental methods. Some taxonomically related differences concerning these pathways in diverse Reptilian forms seem to obtain.

8. Birds

As regards the major subdivisions of the Avian neuraxis, the following main typologic features, many of which are in common with the Sauropsidan Reptilian ones, could be enumerated.

The telencephalon is far more developed and differentiated than that of Reptiles, but shares with this latter and with that of Mammals the essential characteristics of the Amniote telencephalic bauplan which were summarized in section 7.

The small olfactory bulbs are directly contiguous to lobus hemisphaericus, and, because of a pronounced hemispheric bend, in a ventrally displaced position.

In the hemisphere's (secondary) pallium, there is a lateral corticoid layer, mainly originating from lateral components of D_2 with some contributions from D_1. The medially adjacent pallial neighborhood, corresponding to Reptilian primordium neopallii and to neopallium respectively neocortex of Mammals, does not display a cortical structure but is represented by the nuclei diffusi within the so-called *wulst* (or *sagittalwulst*), which is laterally delimited by the groove of the vallecula. The parahippocampal cortex, medially adjacent to nuclei diffusi, corresponds to the dorsal cortical lamina of Reptiles, being derived from the topologic neighborhood D_{2am}. This cortex is highly differentiated and may display a stratification comparable with that of Mammalian neocortex. A medial hippocampal cortex, derived from D_3, is present, and shows some similarities with the precommissural hippocampus of Mammals.

The lateral secondary basis telencephali of Birds, formed by the D_1, B_1 and B_2 complex, is the most developed and extensive Avian telencephalic griseal complex, consisting of diverse epibasal and basal grisea, whose ontogenetic derivation, 'nuclear' components, and nomenclature were discussed in section 6, chapter VI of volume 3/II and in section 9, chapter XIII of volume 5/I. A rostral and a caudal subdivision of these griseal aggregates can be distinguished, the caudal one corresponding to the Mammalian amygdaloid complex. There can be little doubt that the extensive epibasal grisea of the Avian lateral telencephalic basis, although morphologically not homologous to the Mammalian neocortex, are, at least in part, functionally comparable, i.e. in this respect analogous, to the Mammalian neopallium. Concomitantly with the development of these basal ganglia, the Avian lateral forebrain bundle is a massive communication channel, crowded with many

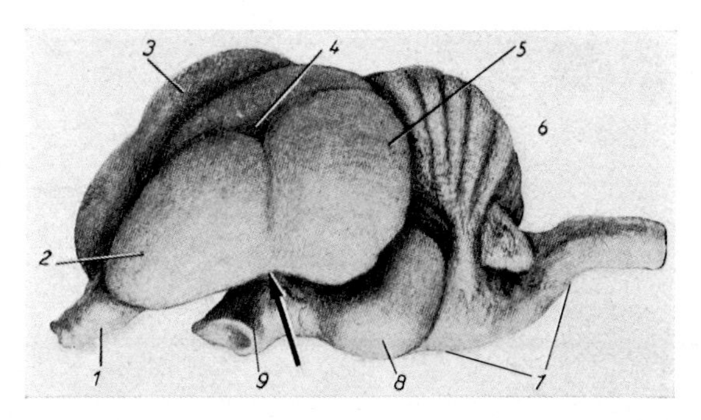

Figure 144 A. Dorsolateral aspect of the brain in a Goose (Anser domesticus), illustrating configuration of an Avian brain (from BECCARI, 1943). 1: olfactory bulb; 2: frontal pole; 3: sagittal wulst; 4: vallecula; 5: occipitotemporal region; 6: cerebellum; 7: oblongata; 8: tectum mesencephali (lobus opticus); 9: optic chiasma. Added arrow indicates depression, corresponding to basal concavity of hemispheric bend, between rostral and 'temporal' regions of lobus hemisphaericus. It is kathomologous to *fossa Sylvii* of Mammals, and somewhat more pronounced than in Reptiles (cf. Fig. 142). In a few Amphibians (Gymnophiones) it is likewise faintly suggested. In the Goose as well as in some other, but not all Birds, a dorsal extension of the basal concavity forms a groove between rostral and caudal portions of lobus hemisphaericus, as clearly recognizable (but not labelled) in the Figure.

components. A peculiar Avian differentiation of one of the channels pertaining to the medial forebrain bundle is the tractus septomesencephalicus. A rather poorly developed, nondescript basal cortical layer is present.

The diencephalon, with a large optic chiasma and optic tract along its basis, is rostrally covered by lobus hemisphaericus, and caudally by tectum opticum. The epithalamus is relatively small, but thalamic and hypothalamic grisea are highly differentiated as described in chapter XII of volume 5/I.

The mesencephalon is characterized by the large optic lobes of the tectum, which are displaced ventrolateralward and remain connected across the midline by a thin roof of the mesencephalic ventricle, representing a stretched median portion of alar plate, in which large cells of nucleus radicis mesencephalicae trigemini are located. As in Reptiles and Mammals, the tegmentum contains a nucleus ruber.

The cerebellum is remarkably well developed and comparable to the Mammalian one, including what could be evaluated as a neocerebellar component (cf. section 8, chapter X, vol. 4). The sulci, lobes, and folia of the Avian cerebellum are morphologically respectively to-

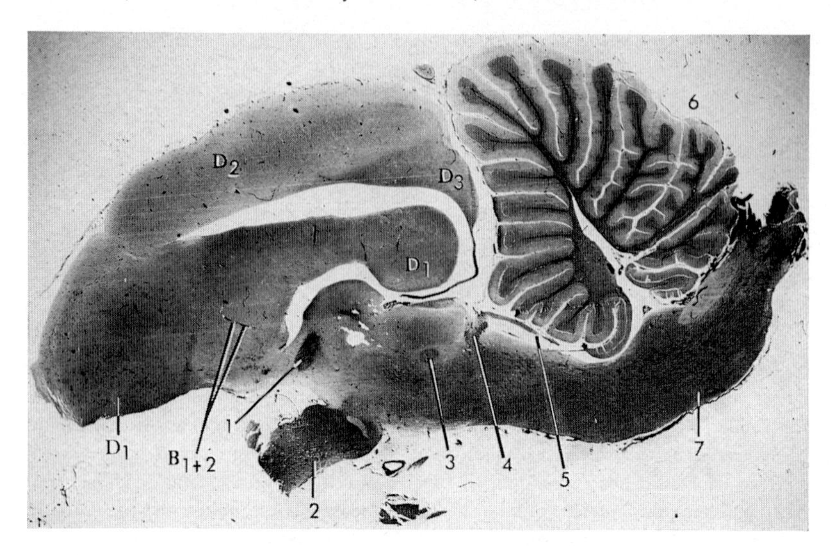

Figure 144B. Sagittal section (myelin stain) through the brain of the Goose. B, D: telencephalic longitudinal griseal zonal system; 1: basal forebrain bundle complex; 2: optic nerve, chiasma, and dorsolaterally adjacent supraoptic commissures; 3: tractus nuclei ovoidalis; 4: fiber system of commissura posterior; 5: thin medial roof of mesencephalon containing nucleus radicis mesencephalicae trigemini; 6: cerebellum; 7: oblongata.

pologically homologous to those in Mammals. This parallel development in two different Vertebrate classes, whose presumptive Reptilian or Reptilian-like ancestors would not have displayed a corresponding differentiation of the cerebellum is an interesting case indicating that detailed morphological homologies cannot always be interpreted to suggest common or close ancestral relationship, but may arise, as it were, by 'parallel' phylogenetic development in 'diverging radiations'.[30]

The oblongata of Birds includes pontine nuclei, which are relatively small, but homologous, although presumably not analogous[31] to those of Mammals. The reticular grisea are highly differentiated, and the cochlear system is more developed than that of Reptiles. The deuterencephalic cranial nerves correspond to those in Reptiles and Mammals, the trigeminal nerve, whose sensory functional significance in Birds

[30] Cf. pp. 146–150, chapter II of volume 1, and pp. 752–755, chapter X of volume 4.

[31] These nuclei presumably receive their telencephalic input from grisea of the basal complex, and not from cerebral cortex, as in Mammals.

appears somewhat diminished, being relatively reduced, but the hypoglossal grisea, some of which innervate the syrinx, are rather complex.

The Avian spinal cord is characterized by the sinus rhomboidalis and a long cervical portion with more segmental nerves than in Mammals. The ascending and descending communication channels are better differentiated than in Reptiles and likewise appear to include a rubrospinal tract. It also seems not unlikely that a direct descending fiber system from telencephalon, analogous[32] to the Mammalian pyramidal tract, may reach spinal levels. Figures 144 A and B illustrate some aspects of the Avian neuraxis, and Figure 141 F depicts the overall arrangement of main communication channels.

9. Mammals

In the class Mammalia, the Vertrebrate telencephalon reaches its highest degree of development as regards structural differentiation, functional capability, and expansion.[33] With respect to its morphologic grundbestandteile, however, the Mammalian telencephalon does not show any deviation from the fundamental Amniote telencephalic bauplan dealt with in section 6, chapter VI of volume 3/II as well as in chapter XIII of volume 5/I, and briefly referred to in section 7 and 8 of the present chapter.

The most conspicuous general typologic feature displayed by the aforementioned progressive development is represented by the expansion of the pallium. Particularly progressive features are manifested by the *neopallium* with its *hexalaminar neocortex*, and by the hippocampal cortex which, except for a rostral region roughly comparable to the Sauropsidan hippocampal cortex, shows a typically Mammalian differentiation into cornu Ammonis and fascia dentata *sive* gyrus dentatus. The parahippocampal cortex likewise displays a considerable progressive development exceeding that in Birds, which latter, nevertheless, especially in some forms, also possess a rather well developed and stratified parahippocampal cortex.

---

[32] Such telencephalic input to spinal cord would presumably originate in grisea of the basal complex, and not in the cerebral cortex (cf. footnote 31).

[33] The degree of expansion *(qua* weight and volume), which culminates in Cetaceans and Proboscideans, is not necessarily correlated with the degree of functional capability respectively of structural differentiation, which culminates in Man.

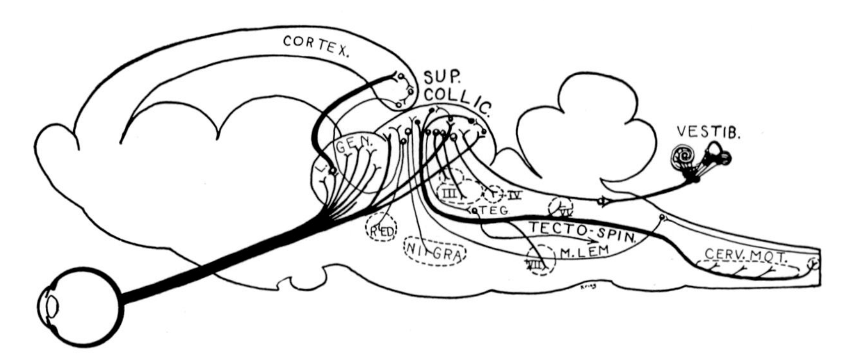

Figure 145 A. Diagram of sagittal section through a 'generalized' Mammalian brain, showing optic and vestibular input channels with some of their connections (from KRIEG, 1966).

The lateral portion of the Mammalian pallium includes the anterior cortex ('prepiriform cortex') of the piriform lobe, which is fairly extensive in macrosmatic forms but relatively reduced in microsmatic ones. This cortex displays a lesser degree of progressive development from its precursor represented by the lateral cortical plate of Reptiles.

As regards the lateral topologic neighborhoods of the basis telencephali, a many-one transformation of the rostral D_1, B_1 and B_2 components has occurred. Instead of the stratified epibasal and basal Reptilian and particularly Avian grisea, a relatively homogeneous *corpus striatum* obtains. A lateral remnant of Reptilian epibasal nucleus is the *claustrum*. Caudally, however, the D_1, B_1 and B_2 components remain differentiated, and form the various cell groups of the *amygdaloid complex*.

A relatively poorly differentiated merogenic (semiparietine) basal cortex, homologous to that of Sauropsidans, Gymnophione Amphibians, Dipnoans, and Selachians is likewise present. It comprises a rostral area which includes the tuberculum olfactorium, and a caudal area representing the cortical amygdaloid 'nucleus'.

The medial neighborhoods of basis telencephali correspond, on the whole, to the paraterminal B_3 and B_4 complex of submammalian forms with inverted telencephalon, and include inferior and superior basimedial grisea pertaining to what is generally designated as the *septal region*.

Concomitantly with the development of neopallium and parahippocampal pallium, an original component of the lateral forebrain bundle has become transformed into the substantial communication channel

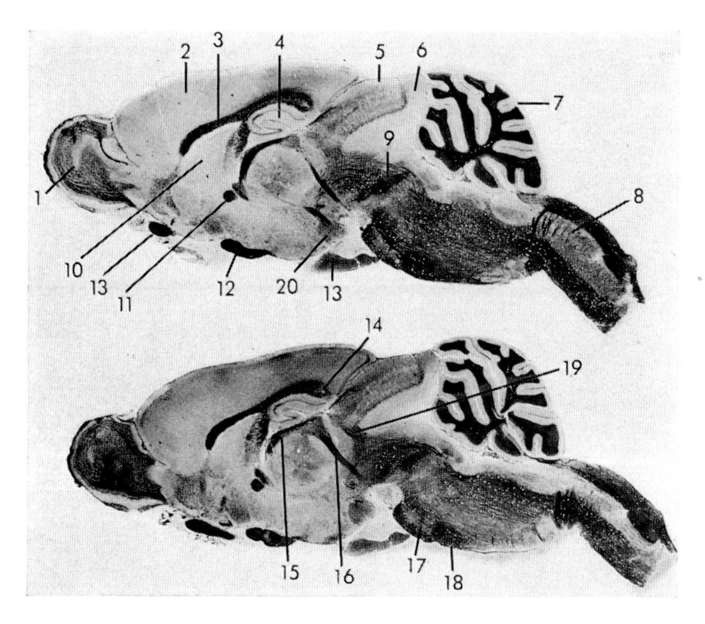

Figure 145 B. Two sagittal sections (myelin stain) through the brain of the Rodent Mouse, showing diverse fiber systems. 1: olfactory bulb; 2: pallium; 3: corpus callosum; 4: hippocampus; 5: superior colliculus; 6: inferior colliculus; 7: cerebellum; 8: oblongata; 9: brachium conjunctivum; 10: paraterminal grisea ('septum'); 11: anterior commissure; 12: optic chiasma, respectively tract, and supraoptic commissures; 13: parts of trigeminal nerve; 14: commissura hippocampi (ventral to splenium of corpus callosum); 15: stria medullaris system; 16: fasciculus retroflexus; 17: pons; 18: trapezoid body; 19: commissura posterior; 20: mammillary body. Above commissura anterior (11) part of fornix can be identified in both sections, also another part of fornix rostroventrally to hippocampus.

provided by the Mammalian *capsula interna* which, in many forms, separates the corpus striatum into nucleus caudatus and putamen. The fornix system has developed as a progressive differentiation of the medial forebrain bundle. The other components of both forebrain bundles become relatively reduced as the Mammalian (secondary) basal forebrain bundle.[34]

Characteristic Mammalian communication channels related to neocortex are *pyramidal* and *corticopontine systems*, moreover the well differentiated afferent sensory pathways[35] to visual, auditory and other sen-

[34] Frequently designated as 'medial forebrain bundle'. In Mammals, however, this contains components pertaining to both lateral and medial forebrain bundles of submammalian forms.

[35] It should, however, be kept in mind that pathways providing all sorts of sensory

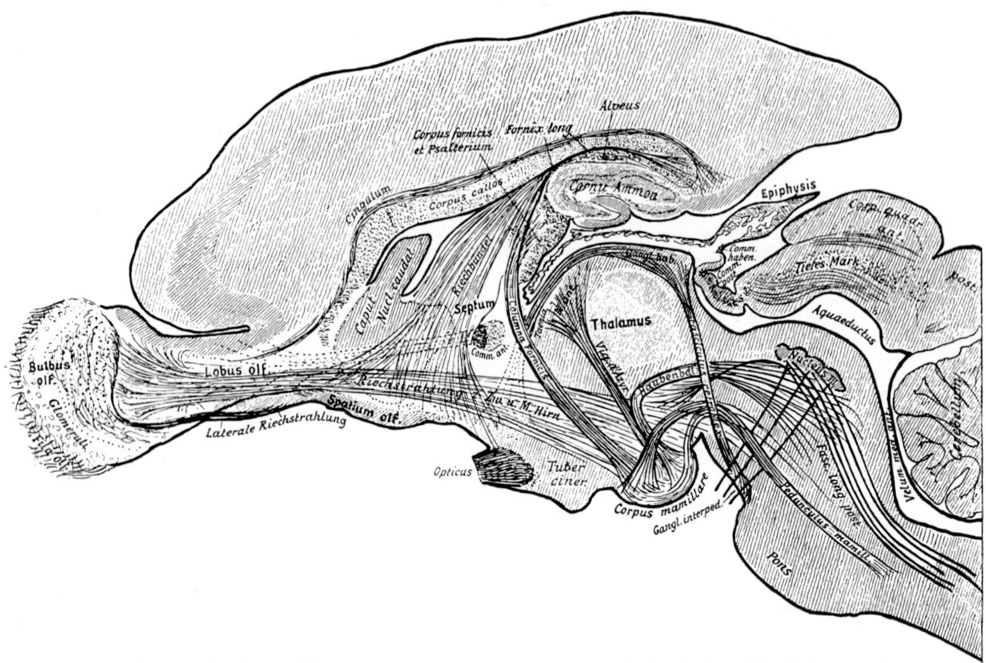

Figure 146 A. Semidiagrammatic sagittal section through the brain of the Lagomorph Rabbit, showing diverse grisea and communication channels (from EDINGER, 1912). Vicq d'Azyr: mammillo-thalamic tract.

sory cortices. These channels run through internal capsule and therewith associated fiber systems (thalamic stalks).

Further details concerning the differentiation of the Prototherian and Metatherian commissura dorsalis *sive* hippocampi into the corpus callosum of Eutheria, the evolution of the lissencephalic into the gyrencephalic fissuration pattern, and the differentiation of the paraterminal grisea can be ommited, in the aspect here under consideration for a concise summary of typologic features. These details were dealt with in the relevant chapters and sections of volume 5/I and volume 3/II.

The configuration of the Mammalian diencephalon becomes characterized by the considerable expansion of the hemispheric stalk, main-

---

input (in addition to the *ab initio* obtaining olfactory one) to telencephalic grisea seem to be present, to a variable degree of development, in all submammalian Vertebrates from Cyclostomes to Sauropsida.

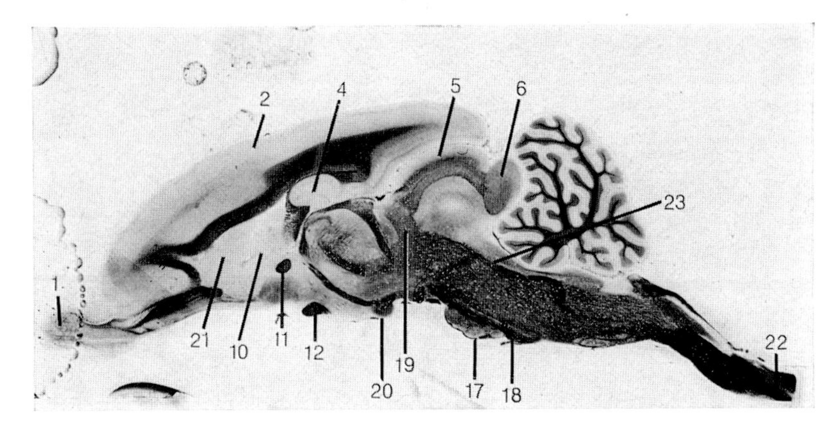

Figure 146 B. Sagittal section (myelin stain) through the brain and rostral spinal cord of the Rabbit. 21: corpus striatum; 22: cervical spinal cord; 23: fasciculus longitudinalis medialis (below end of lead also part of brachium conjunctivum). Other designations as in Figure 145B. Stria medullaris, fasciculus retroflexus, mammillo-thalamic tract, and fornix are easily identified by comparison with Figures 145B and 146A.

ly due to the development of capsula interna. The lateral surface of the diencephalon thereby, as it were, disappears, being reduced to a posterior one, which commonly includes the prominences of lateral and medial geniculate bodies. The thalamus dorsalis, with its relay grisea to neocortex and parahippocampal cortex, and its diverse 'modulating' grisea becomes greatly expanded and differentiated, while the thalamus ventralis becomes relatively much reduced. The hypothalamus displays a complex differentiation particularly also with regard to paraventricular grisea, which also show progressive features in thalamus dorsalis and even thalamus ventralis. The Mammalian *globus pallidus* represents a dorsal rostrolateral hypothalamic griseum which protrudes into the hemispheric stalk, and adjoins the striatal grisea. It is a derivative of the submammalian anterior entopeduncular 'nucleus'.

The mesencephalon is characterized by the paired superior and inferior colliculi, of which the former correspond to the tectum opticum of submammalian Vertebrates, while the latter represent the two caudodorsalward displaced tori semicirculares. This displacement is also, to a lesser degree, manifested in some Squamate Reptilians. The mesencephalic ventricle is relatively reduced, its lumen being the *Sylvian aqueduct* of Human anatomical nomenclature. In addition to a well developed periventricular central gray with fiber tracts, to a substantial

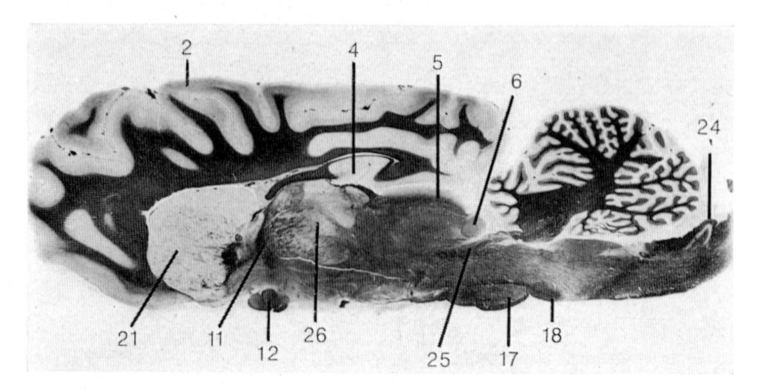

Figure 147A. Sagittal section (myelin stain) through the brain of the Ungulate Sheep. 24: nucleus of descending trigeminal root; 25: brachium conjunctivum; 26: diencephalon. Other designations as in Figures 145B and 146B.

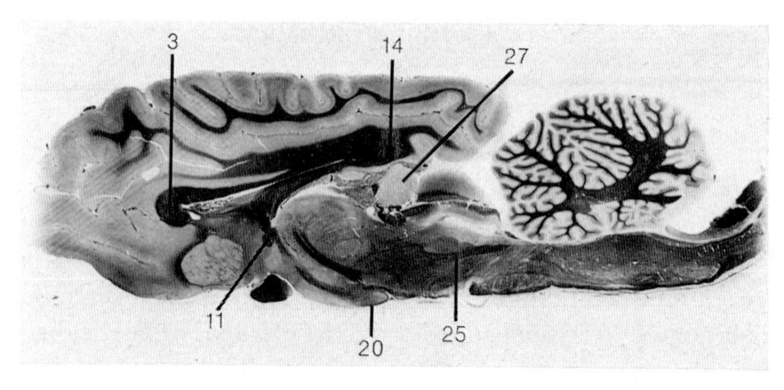

Figure 147B. Sagittal section (myelin stain) through the brain of the Sheep, slightly more medial than Figure 147A. 27: epiphysis. Other designations as in Figures 145B to 147A. The course of fornix and mammillo-thalamic tract, not labelled, can easily be identified.

nucleus ruber and other tegmental cell groups, a conspicuous tegmental derivative, the *substantia nigra*, forms the boundary between tegmentum and basally located communication channels, which comprise the bilateral *pes pedunculi*. These channels are the corticopontine and the pyramidal tracts.

Although some cell groups in the Sauropsidan mesencephalon can be interpreted as a primordium or 'forerunner' of substantia nigra, the distinctive differentiation of this griseum together with the presence of

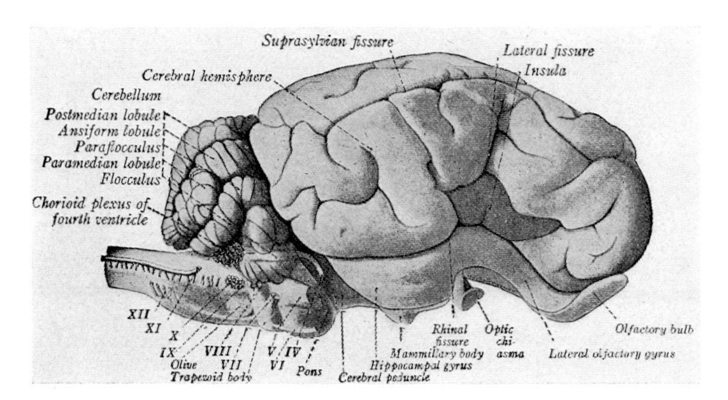

Figure 147C. Lateral view of the Sheep's brain (from RANSON, 1943). The lead 'Cerebral hemisphere' points to the neighborhood of posterior ectosylvian sulcus; the lead 'Lateral fissure' points to boundary of *fossa Sylvii;* 'Lateral olfactory gyrus' is rostral part of anterior piriform lobe; 'Hippocampal gyrus' is posterior portion of piriform lobe; 'Rhinal fissure' is sulcus rhinalis lateralis, the lead points to transition between anterior and posterior portion of the sulcus.

basal pedes pedunculi and median fossa interpeduncularis, are specific configurational features of the Mammalian mesencephalon.

The cerebellum with its three stalks, its hemispheres, its vermis, its paraflocculi and flocculi, and with its cerebellar nuclei, reaches the highest degree of development in Mammals, and was dealt with in chapter X of volume 4. The noteworthy homologies displayed by Mammalian and Avian cerebellum, discussed in the cited chapter, were also briefly mentioned in the preceding section 8.

The rostrobasal portion of the oblongata *sensu latiori* is characterized by the presence of a bulging pons with its brachia pontis forming the middle cerebellar peduncles. In 'lower' Mammals the pons is relatively less voluminous than in most higher ones, and does not cover the entire trapezoid body, whose transverse fibers, pertaining to the cochlearis system, can then be seen on the ventral bulbar surface caudally to pons. The complex nuclear and tract differentiations of the Mammalian oblongata were dealt with in section 10, chapter IX of volume 4. It will here be sufficient to recall *inter alia* the development of superior olivary complex, of the trigeminal system,[36] of the large infer-

[36] The detailed study on the trigeminal system in the Mouse by LORENTE DE NÓ (1922) should here be mentioned. This author also described afferent trigeminal fibers reaching the nucleus loci caerulei.

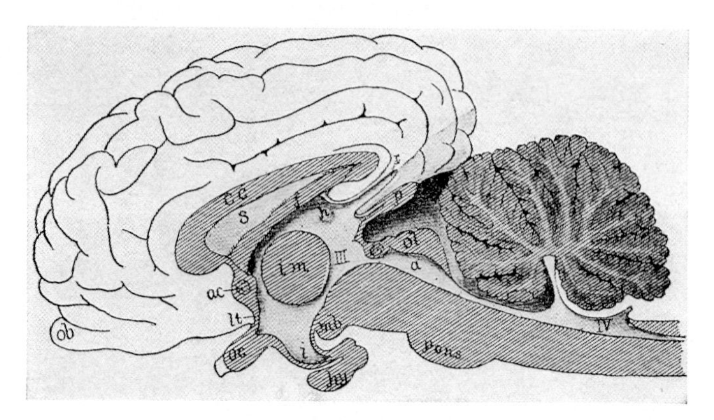

Figure 147 D. Drawing of midsagittal section through the brain of the Ungulate Calf, for comparison with Figures 147A and B of the Ungulate Sheep (after BÜTSCHLI and KINGSLEY, from NEAL and RAND, 1936). a: aquaeduct; ac: anterior commissure; cc: corpus callosum; f: fornix; h: ganglion habenulae; hy: hypophysial complex; i: posterior hypothalamus; im: massa intermedia (commissura mollis) of diencephalon; mb: mammillary body; ob: olfactory bulb; oc: optic chiasma; ol: tectum mesencephali (rostrally to ol: posterior commissure); p: pineal body; r: suprapineal recess; s: septum; III, IV, third and fourth ventricles.

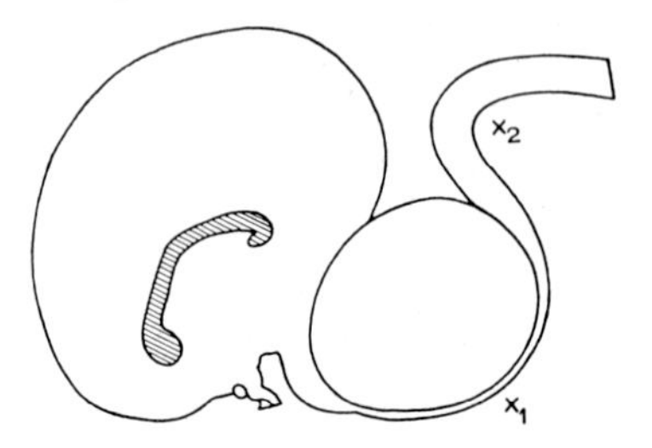

Figure 148. Diagram of midsagittal section depicting pronounced pontine and cervical flexures in the brain of the Cetacean Wale Eubalena australis (from PILLERI, 1964). x_1: pontine flexure basally to cerebellum; x_2: cervical flexure around ventral border of foramen occipitale magnum. The brain is depicted as in situ.

ior olivary complex, and of the peculiar noradrenergic system provided by nucleus loci caerulei. This latter, although apparently recognizable in Birds, but, so far, not identifiable with certainty in Reptiles, seems to represent a griseum of particular functional significance for the Mammalian neuraxis.

As regards the spinal cord, dealt with in chapter VIII of volume 4, a cytoarchitectural dorsoventral stratification or laminar arrangement

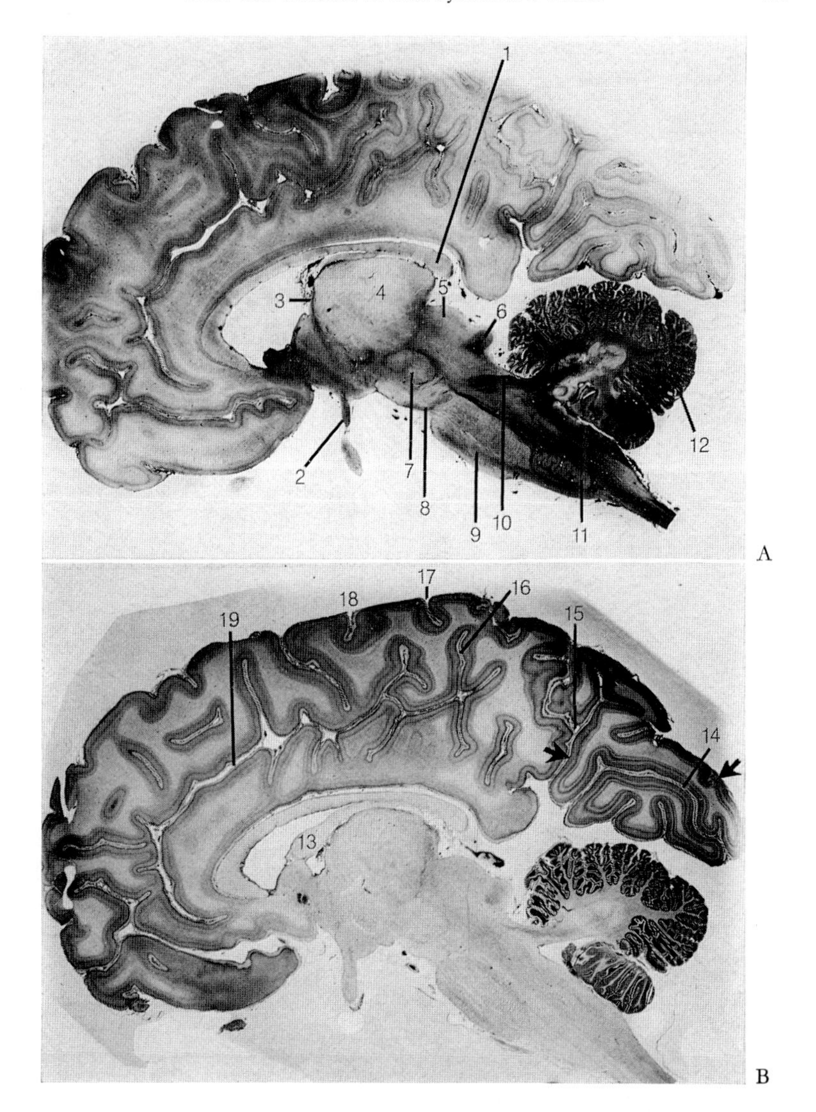

Figure 149 A, B. Sagittal sections, myelin stain (A) and Nissl stain (B), through the brain of a Human newborn. 1: corpus callosum (splenium); 2: optic nerve; 3: stria medullaris thalami; 4: thalamus; 5: superior colliculus; 6: inferior colliculus; 7: nucleus ruber tegmenti; 8: substantia nigra; 9: pons; 10: brachium conjunctivum; 11: inferior olivary nucleus, 12: cerebellum (the nucleus dentatus in its interior is easily identified); 13: fornix; 14: sulcus (fissura) calcarina (the two arrows indicate limits of area striata); 15: sulcus (fissura) parieto-occipitalis; 16: ramus marginalis sulci cinguli; 17: sulcus centralis; 18: sulcus praecentralis; 19: sulcus cinguli. Plane of A is slightly medial to that of B.

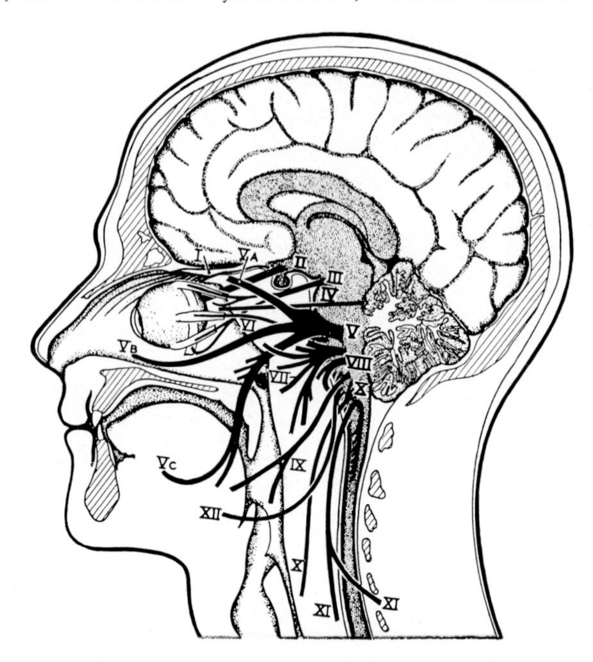

Figure 150 A. Drawing of midsagittal section through Human head, showing brain *in situ*, and indicating overall course of cranial nerves I to XII (from NEAL and RAND, 1936). VA: ophthalmic branch; VB: maxillary branch; VC: mandibular branch of trigeminus.

within the grisea of dorsal and ventral horn is far more conspicuous than its suggestion in submammalian forms. The long ascending and descending tracts seem likewise to be more distinctly differentiated. With respect to the descending ones, this applies particularly to the rubrospinal tract of many Mammalian forms, while the corticospinal tracts represent a specifically Mammalian communication channel. Of interest is here the diversity of said tract's course through dorsal funiculus (e.g. Mouse, Rat), or (e.g. in Man) lateral and ventral funiculus.

Figures 141 G, and 145–149 illustrate diverse typologic features concerning some main communication channels and configurational aspects of the Mammalian neuraxis in a few 'lower', 'intermediate', and 'higher' forms. Figure 150 A, which should be compared with the photographs reproduced as Figures 78 D and E, pp. 205–206 of volume 3/II, depicts the Human brain and cervical spinal cord *in situ*, together with its relationship to the cranial nerves. It is here of interest to note the changes that have resulted with respect to the original

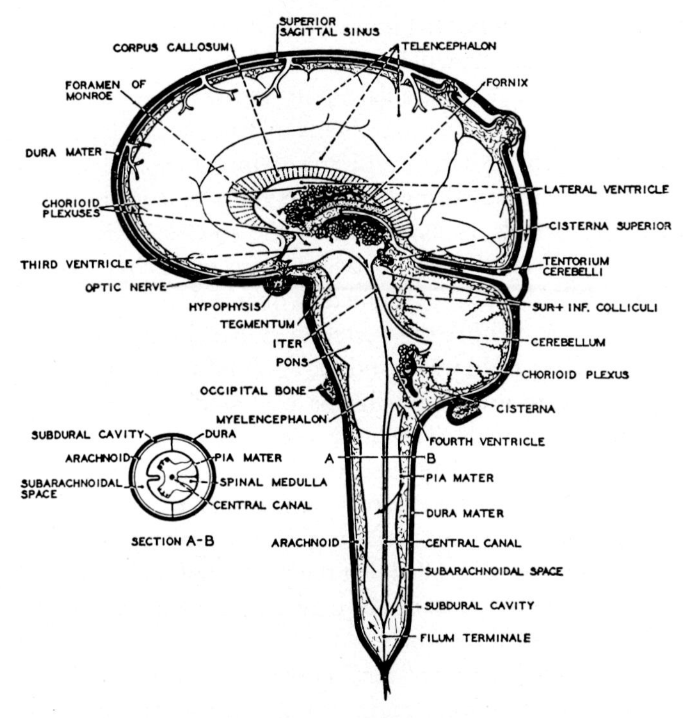

Figure 150 B. Diagram showing relationships of meninges to Human neuraxis (redrawn after RASMUSSEN, 1932, from NEAL and RAND, 1936).

bends (or flexures) displayed by the neuraxis at early ontogenetic stages (cf. Fig. 66B, p. 183, vol. 3/II). These changes are related to the body's upright position. Pontine and cervical flexures have practically disappeared, while the nearly horizontal longitudinal axis of prosencephalon (telencephalon and diencephalon) forms an almost right or even rostrally slightly obtuse angle with the almost vertical longitudinal common axis of mesencephalon, rhombencephalon and spinal cord. This bend roughly corresponds to the rostral limb of original cephalic flexure. An almost straight axis, on the other hand, is displayed in Figure 147D, while an instance of an adult Mammalian neuraxis with pronounced cephalic, pontine and cervical flexures is shown in Figure 148.

The comparative anatomy of the meninges, which still remains controversial as regards various points, and the blood vessels of the Vertebrate neuraxis were dealt with in chapter VI of volume 3/II. In the present context, meant to bring an overall brief summary of salient typologic aspects, reference to Figure 150B, illustrating the Human neuraxis within its meninges, will be sufficient.

It can be seen that the cranial cavity represents a closed box with rigid walls. Since the brain parenchyma may be conceived as essentially incompressible, changes in the volume of intracranial blood would require compensation by displacement or replacement of cerebrospinal fluid. In other words, as expressed by the so-called *Monro-Kellie doctrine*, a reciprocal relationship between brain parenchyma, blood, and cerebrospinal fluid must be assumed to obtain. In contradistinction to the cranial cavity, neither the vertebral subarachnoid, dural, and epidural cavities have rigid walls, and the epidural venous plexuses can be roughly likened to a compressible as well as distensible spongy network of blood channels. The bony vertebral canal, moreover, has a partly membranous wall represented by the elastic ligamenta flava. Thus, ample play for compensatory fluctuations related to intracranial pressure changes is hereby given within the limits of normal conditions. It should also be recalled that the volume of the brain or neuraxial parenchyma is not constant but may e.g. become increased by edema or decrease in dehydration.

The blood requirement of the Mammalian and particularly that of the Human neuraxis is very high. The great susceptibility of the Human brain, and especially of the cerebral cortex, to anoxia is well known. The cardiac output of Man is estimated at about more or less 4.5 liters per minute. Although the brain represents only a rather small fraction of the total body mass, the total blood flow for the entire brain has been estimated at up to nearly one third of the entire cardiac output. This cerebral blood flow, however, may considerably vary in accordance with vasomotor control and numerous additional parameters.

10. Some Phylogenetic Comments

If the typologic configurational features of the neuraxis in the diverse Vertebrate classes are considered from a morphologic, formanalytic, or topologic viewpoint, a remarkable *conservatism* of the overall *bauplan* from Cyclostomes to Mammals, respectively from Petromyzon to Man becomes evident.

In addition, however, the diverse Vertebrate classes, and within these classes, the diverse orders, display, if compared with each other, and arranged in a suitable array, a conspicuous '*progressive*' *differentia-*

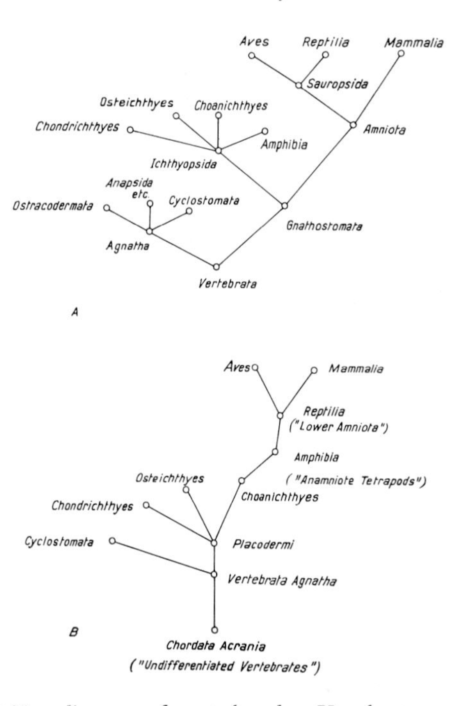

Figure 151 A, B. Two diagrams of posets based on Vertebrate taxonomy. A expresses logical taxonomic relationships (subsumptions). B expresses probable phylogenetic relationships based on morphologic taxonomic features. Despite the differences, some relationships are common to both diagrams.

tion of the neuraxis characterized by one-many and in some instances by many-one transformations of the invariant fundamental formal components.

On the basis of present-day concepts, it can be assumed that the conservative or 'archetypical' aspect of the central nervous system's bauplan represents a constraint, firmly anchored in the macromolecular mechanisms of the genome. The various progressive as well as some obtaining regressive changes displayed in the Vertebrate series are thereby restrained to remain, by homomorphic topologic transformations,[37] within the given spatially configurated pattern, i.e. within the given bauplan.

---

[37] Cf. pp. 62–68 of volume 3/II. Further details on this topic are also elaborated in section 10, chapter II, and in chapter III of volume 1.

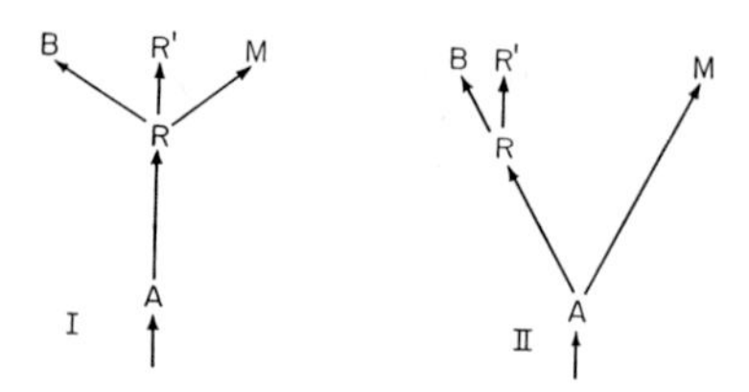

Figure 151 C. Two diagrams indicating possible origins of Mammalian phylogenetic 'radiation' either from Reptilian forms (I) or directly from Amphibian ones (II). A: Amphibians; B: Birds; M: Mammals; R, R′: primitive respectively more advanced Reptiles.

With regard to the features manifested by the differentation of the *grundbestandteile* and *formbestandteile* included in the common overall bauplan of the Vertebrate neuraxis, the aforementioned arrangement in a logically ordered array represents an ascending sequence with a greatest common lower bound, and with diverse branches ending in a least upper bound. These sequences correspond to the partially ordered sets (or *posets*) of mathematics (Fig. 151 A, B). Such arrays, series, or posets strongly suggest transformations along lines of genetically related transspecific evolution, but, unfortunately, this circumstantial evidence remains, despite its very high degree of probability, ambiguous with respect to details, since the observable transforms do not indicate unique operands. As pointed out on p. 144, section 9, chapter II of volume 1, it is thus quite possible, although by no means certain, that the Mammalian phylogenetic 'radiation' originated directly from Amphibian rather than from Reptilian forms (cf. Fig. 151 C).

It will be recalled that the concept of organic evolution, already proposed by the classical Greeks (Anaximander, floruit ca. 500 B.C., Anaxagoras, 500–428 B.C.; Empedocles, 484–424 B.C.), became firmly established somewhat over 100 years ago, predominantly through the pioneering efforts of Charles Darwin (1809–1882) and Ernst Haeckel (1834–1919).

Darwin (1859) elaborated on the significance of *variability* in combination with *natural selection* 'aided in an important manner by the inherited effects of the use and disuse of parts; and in an unimportant manner, that is, in relation to adaptive structures, whether past of present, by the direct action of external conditions, and by variations which seem to us in our ignorance to arise spontaneously'.

In his treatise of 1859, however, Darwin did not elaborate on the origin of life, nor did he propose genealogic trees or sequences

(Stammbäume) in order to explain the taxonomic systems of plants and animals, nor did he apply his conclusions to the origin of Man.

On the other hand, HAECKEL, inspired by DARWIN's classical treatise on the origin of species, provided the morphologic basis of organic evolution in the two volumes of his great work 'Generelle Morphologie der Organismen' (1866), with the subtile '*Allgemeine Grundzüge der organischen Formen-Wissenschaft, mechanisch begründet durch die von Charles Darwin reformirte Deszendenz-Theorie*'. In the first volume, dedicated to CARL GEGENBAUR, he dealt with the '*Allgemeine Anatomie der Organismen*'. The second volume, dedicated to DARWIN, GOETHE, and LAMARCK, was concerned with the '*Allgemeine Entwicklungsgeschichte der Organismen*'.

HAECKEL, who coined many of the now commonly used terms, such as *phylogeny*, *ontogeny*, *ecology*, etc., postulated *abiogenesis*, that is to say the origin of life from non-living, prebiotic physico-chemical reactions, stressing the role played by the carbon atom *(Karbogentheorie*, cf. vol. 3/I, p. 650). He established the significance of the morphologic pattern as providing the relevant clues for evidence of phylogenetic descent, supplemented by evidence from ontogenetic development. In this connection, he formulated the important *gastraea-theory*, based on the significance of the gastrula, and propounded his '*biogenetic law*' by which MECKEL's '*law of reduction*' became re-formulated in terms of phylogeny, such that the observable ontogenetic sequences could be evaluated as evolutionary processes suitable for a cautious comparison with the phylogenetic sequences inferred from the comparison of adult morphologic features. Like DARWIN, and for that matter, like LAMARCK, HAECKEL assumed the heredity of acquired characters.

HAECKEL, moreover, was the first who elaborated genealogic sequences *(Stammbäume)* based on the relevant morphologic features in combination with the then available data of geologic science (Fig. 152A) and including the origin of Man.[38] In his treatise '*Natürliche Schöpfungsgeschichte*' (1868) HAECKEL subsequently reviewed and summarized his conclusions, which DARWIN (1871) corroborated in his second great work 'The descent of Man, and selection in relation

[38] In this respect, a morphologically less detailed theory of organic evolution from abiogenesis to plants and animals including Man had been elaborated about 1809 by LAMARCK in his '*Philosophie zoologique*'. More cautiously, such evolution was also about 1790 suggested by KANT, and generalized evolutionary concepts had been expressed by others (cf. section 4, chapter II, vol. 1).

to sex'.[39] HAECKEL's accomplishment in outlining the first phylogenetic 'trees' or 'radiations' based on substantial morphologic, embryologic, taxonomic and palaeontologic evidence was rejected with contemptuous derision by contemporary leading scientists such as W. HIS (1831–1904), R. VIRCHOW (1821–1902, a teacher of HAECKEL), and E. DUBOIS REYMOND (1818–1896). VIRCHOW maintained that '*es ist ganz gewiss, dass der Mensch nicht vom Affen oder von irgendeinem anderen Thiere abstammt*'. DUBOIS REYMOND stated that the phylogenetic trees '*etwa so viel werth sind, wie in den Augen der historischen Kritik die Stammbäume homerischer Helden*' (cf. HEBERER, 1968).

The recent volume edited by HEBERER (1968) and entitled '*Der gerechtfertigte Haeckel*' brings, in addition to a selection from HAECKEL's important writings, a detailed account of the heated controversies concerning HAECKEL's views in particular, and organic evolution in general. It can easily be seen that the just mentioned noted contemporary scientists were, in this respect, quite amiss and that, in almost all essential questions (abiogenesis, gastraea-theory, phylogenetic sequences or '*Stammbäume*', biogenetic law, and the descent of Man), HAECKEL, whose outstanding accomplishments became largely ignored in the course of the 20th century, must be considered completely vindicated. Various weaknesses in HAECKEL's biologic concepts were mainly related to the less advanced knowledge available at his time, particularly also with regard to the hereditary mechanisms, which seem to preclude an inheritance of acquired characters in the manner assumed by LAMARCK, DARWIN and HAECKEL. The weaknesses inherent in the philosophical views elaborated by HAECKEL resulted from his failure to realize the *epistemologic significance of mind*. These latter weaknesses, however, are no more substantial than those displayed by the writings of numerous famed professional philosophers, and, moreover, are not relevant to the biologic problems here under consideration.

For his vindication and rehabilitation of ERNST HAECKEL, whose outstanding accomplishments were shabbily passed over by the pandits celebrating the 'Centennial Celebration' of DARWIN's 'Origin of Species' ('Evolution after Darwin', 3 vols., SOL TAX, ed., 1960), HEBERER (1968) deserves considerable credit. Summarizing HAECKEL's

---

[39] DARWIN (1871) stated here that, if said work had appeared before the work on his own essay, he should probably never have completed it, and added: 'Almost all the conclusions at which I have arrived I find confirmed by this naturalist, whose knowledge on many points is much fuller than mine.'

contributions, most of which were violently opposed by leading contemporary scientists, the following could be stressed.[39a]

(1) The valid argumentations in favor of abiogenesis, which latter is now widely accepted, and his thereto related carbogen theory, pointing out the significance of the carbon atom. (2) The distinction between plasmodome and plasmophage organisms, of which the former (namely plants synthetizing organic carbon compounds) provide food for the latter. This has some bearing on the possibility that plants, such as algae, producing oxygen by photosynthesis, provided an atmosphere suitable for plasmophage animal life and thus preceded the appearance of this latter. (3) The recognition that non-nucleated (in present-day terminology *Procaryote)* organisms, the '*Moneres*', preceded nucleated *(Eucaryote)* '*Protista*'. (4) The apparently earliest recognition (1866) that the cell nucleus, in contradistinction to the cytoplasm, was concerned with the transmission of hereditary characters.[39b] (5) The distinction and valid formulation of the three different and significant sets of morphologic data supporting the concept of phylogeny, namely, (a) *comparative anatomy of adult recent forms*, (b) *paleontology, and* (c) *ontogeny* which, to a relevant degree, recapitulates phylogeny *(biogenetic law).*

As regards this latter, it seems evident that, by the mechanisms of heredity, offsprings not only resemble their adult progenitors but also ontogenetically evolve in a similar manner. It appears thus fully justified to assume that, despite substantial changes *(cenogenesis)*, which HAECKEL duly emphasized, fundamental ontogenetic mechanisms remain firmly anchored or encoded in the genome and persist throughout phylogenetic transformations, thereby providing evidence for phylogenetic relationships. In his Latin formulation of the biogenetic law, HAECKEL stated: '*Ontogenesis summarium (vel recapitulatio) est phylogeneseos, tanto integrius quanto hereditate palingenesis conservatur, tanto minus integrum quanto adaptatione cenogenesis introducitur.*' (6) *The gastraea theory*, postulating that the gastrula represents a fundamental stage in animal

---

[39a] With regard to the mutual-admiration-coteries participating in said 'Centennial Celebration', one is reminded of the aphorism expressed by LA BRUYÈRE (1645–1696) in his chapter '*Des Jugements*': '*Du même fond dont on néglige un homme de mérite, l'on sait encore admirer un sot*'.

[39b] It should also here be recalled that, under HAECKEL's sponsorship, his disciple OSCAR HERTWIG was, in 1875, the first to follow the (Echinoderm, Echinoid) sperm nucleus to its union with the egg nucleus, thus finally solving the riddle of fertilization (cf. vol. 3/I, chapter V, section 2, p. 103, footnote 32).

phylogeny. The cenogenetic 'discogastrulae' or blastoderms of animals with polylecithal ova are easily understood as secondarily 'adapted' to the large amount of 'inert' yolk. (7) The fully justified first attempts at establishing the then much ridiculed phylogenetic *genealogic trees (Stammbäume)* which HAECKEL stressed as being tentative and subject to revisions depending on new data. With relatively minor modifications, HAECKEL's phylogenetic trees, based on comparative anatomy, paleontology and ontogeny, are now generally accepted (cf. further below, Figs. 152A and B). (8) The valid argumentations concerning the origin of Man from preceding Simian forms. HAECKEL postulated an intermediate form *('missing link')* for which he coined the term *Pithecanthropus* (1866) before DUBOIS discovered, in 1894, remains of *Pithecanthropus erectus* in Java. (9) The concepts of *epacme*, *acme*, *paracme (anaplasia, metaplasia, cataplasia)*, i.e. development, modification, degeneration *(Aufbildung, Umbildung Rückbildung)* as general trends or 'natural laws' displayed by organic life and obtaining in both ontogenetic and phylogenetic evolution.

As an aside, it is perhaps of interest to recall SCHOPENHAUER's views on evolution, which were expressed in the first half of the 19th century, preceding the writings of DARWIN and HAECKEL. Discounting SCHOPENHAUER's reification or hypostatization of '*will*' for all physical actions respectively interactions, many of his opinions are in full accordance with subsequent biologic theories. He assumed life to have originated by abiogenesis *('generatio aequivoca')*, and considered it as 'certain', that '*Thiere früher als Menschen, Fische früher als Landthiere, Pflanzen auch früher als diese, das Unorganische vor allem Organischen dagewesen ist; dass folglich die ursprüngliche Masse eine lange Reihe von Veränderungen durchzugehen gehabt, bevor das erste Auge sich öffnen konnte*' (*Welt als Wille und Vorstellung*, I, p. 66).

SCHOPENHAUER was familiar with *Meckel's* '*law of reduction*' and, although rather vaguely, anticipated its subsequent interpretation as biogenetic law by HAECKEL. This is evidenced by SCHOPENHAUER's following statement which also implies a concept somewhat similar to mutation: '*Die Batrachier führen vor unseren Augen ein Fischleben, ehe sie ihre eigene volkommene Gestalt annehmen; und nach einer jetzt ziemlich allgemein anerkannten Bemerkung, durchgeht eben so jeder Fötus successive die Formen der unter seiner Species stehenden Klassen, bis er zur eigenen gelangt. Warum sollte nun nicht jede neue und höhere Art dadurch entstanden seyn, dass diese Steigerung der Fötusform ein Mal noch über die Form der ihn tragenden Mutter um eine Stufe hinausgegangen ist? – Es ist die einzige rationelle, d. h.*

vernünftigerweise denkbare Entwicklungsart der Species, die sich ersinnen lässt.'

'*Wir haben aber diese Steigerung uns zu denken nicht als in einer einzigen Linie, sondern in mehreren nebeneinander aufsteigenden*' (V, p. 168).

'*Wir wollen es uns nicht verhehlen, dass wir danach die ersten Menschen uns zu denken hätten als in Asien vom Pongo (dessen Junges Oran-Utan heisst), und in Afrika vom Schimpanse geboren, wiewohl nicht als Affen, sondern sogleich als Menschen*' (V, p. 169). SCHOPENHAUER thus believed in sudden jumps, as it were in 'macromutations' and in Orangoid and Schimpansoid Human origin, somewhat in a manner as still recently claimed by KURZ (cf. chapter XIV, p. 108). Yet, despite various weaknesses of this sort, these early speculations by a philosopher are rather remarkable.

Concerning the morphologic *bauplan*, designated by SCHOPENHAUER as '*das anatomische Element*', this author remarked: '*Wir müssen daher annehmen, dass dies anatomische Element theils auf der Einheit und Identität des Willens zum Leben überhaupt beruhe, theils darauf, dass die Urformen der Thiere eine aus der anderen hervorgegangen sind und daher der Grundtypus des ganzen Stammes beibehalten wurde*' (Parerga, Bd. 2, § 91). The second statement is in full accordance with the phylogenetic significance of the morphologic pattern stressed by HAECKEL, GEGENBAUR, T.H. HUXLEY and others. The first statement could be reworded to mean that the significance of morphologic pattern is a manifestation of the overall fundamental orderliness correlating, despite the obtaining substantial factor of chance or randomness (disorder, entropy), all aspects of physical interactions.

As regards present-day theories concerning the mechanisms of evolution, the following causal factors are generally assumed to obtain: gene mutation, changes in chromosome structure and number, genetic recombination, changes involving cytoplasmic rather than nuclear structures, natural selection, reproductive isolation, migration of individuals from one population to another, and hybridization (cf. e.g. RENSCH, 1956, 1968; STEBBINS, 1966). Particular emphasis is thereby laid on '*chance*' manifested by random[40] gene mutations, whose results

[40] 'Random' and 'chance' are rather slippery terms. The former subsumes lack of definable orderliness or rule, lack of specifiable direction, and lack of 'purpose'. Chance subsumes a purposeless 'happening of events' without ascertainable cause, or as the result of unknown interactions, or without, in any given respectively assumed causal aspect, intelligible correlation with another event or set of events. It is not logically justifiable dogmatically to equate chance, unpredictability, or uncertainty with indeterminism.

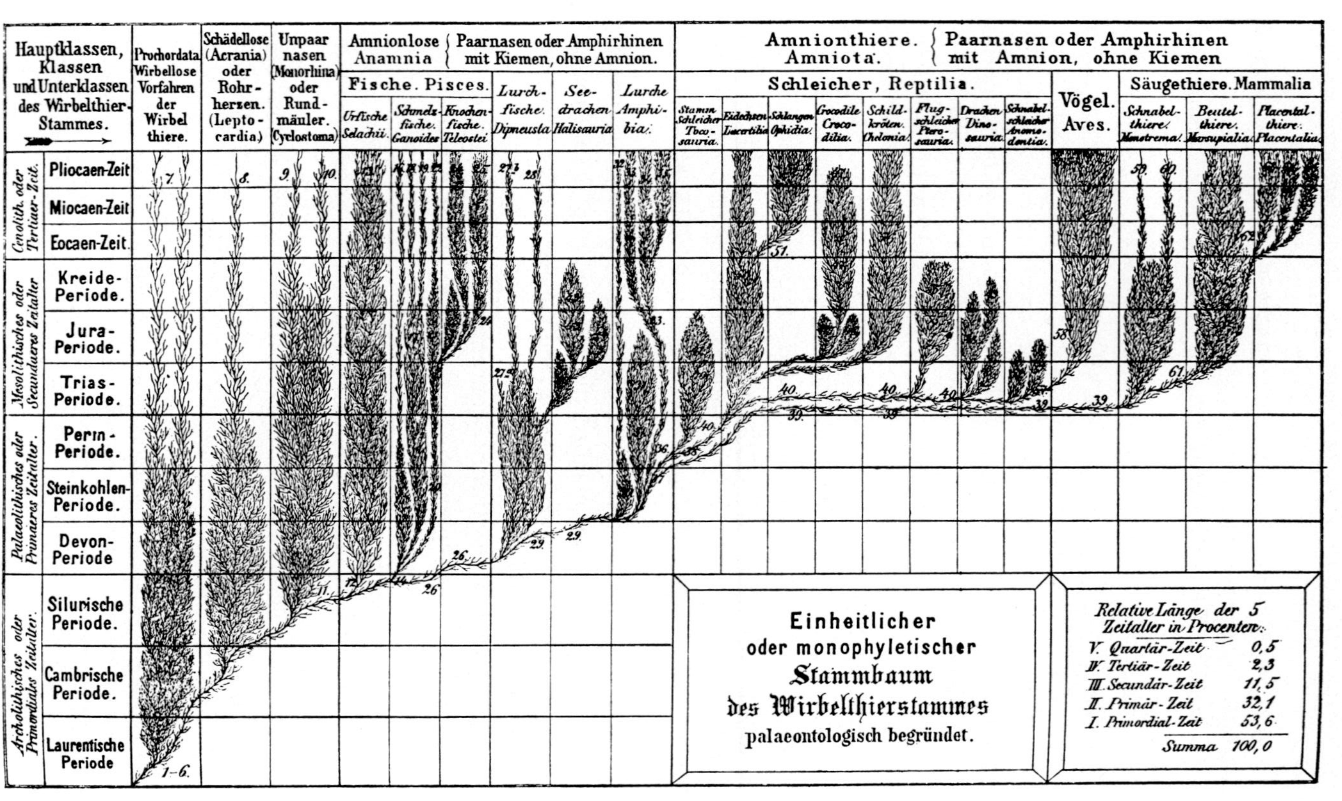

Figure 152 A. Genealogic tree of the Vertebrate phylum as proposed by HAECKEL in 1874 (from HEBERER, 1968).

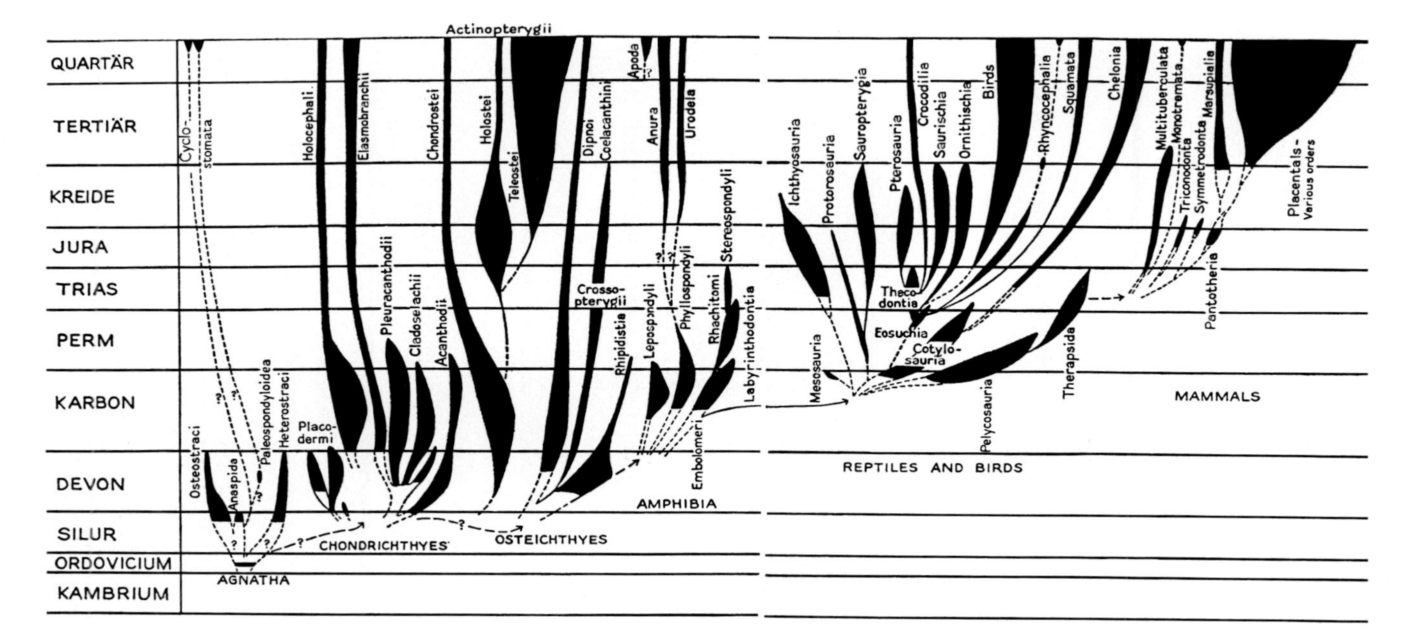

Figure 152 B. Genealogic tree of the Vertebrate phylum as elaborated by ROMER in 1937 (from HEBERER, 1968).

become constrained by *natural selection* respectively favored by '*adaptation*'.

At present, macromolecular biologists generally assume that the genome provides, by strictly one-way transmission of information, a 'blueprint', that is to say, function rules, for the development of the phenotype. These function rules, although essentially fixed, appear to be flexible in their detailed application to particular circumstances. A feedback from phenotype to genome or genotype is denied, thereby excluding the inheritance of acquired characters. At most, 'adaptation' to environment, and related changes in the behavioral pattern are presumed to provide new '*selective pressures*'. Adherents of this mutation-selection theory believe that, by further elucidation of still insufficiently known details, postulated by said theory, a complete causal explanation of phylogenetic evolution can be given. Be that as it may, and regardless whether non-inheritance or a restricted and qualified, not yet ascertained inheritance of some sorts of acquired characters obtain, phylogenetic evolution, as first systematically outlined by HAECKEL on the basis of *morphologic* respectively *taxonomic* changes, is no longer seriously in doubt. It is here of interest to compare Figure 152A, depicting HAECKEL's early genealogic tree of Vertebrates (1874), with Figure 152B, illustrating a contemporary concept of this 'family tree' as elaborated by ROMER (1937). The similarity of both concepts is striking, and the differences, related to an increased amount of available paleontologic data, can be assessed as irrelevant in the aspect here under consideration. It is, moreover, evident that the relevant clues for phylogenetic interpretations are based on *morphologic features* displayed by the fossil and by the recent forms.

In this respect, it is also important to realize that the presently much emphasized results of macromolecular biology, developmental mechanics, and genetics, have, so far, not given any intelligible causal explanation whatsoever for the development, formation, maintenance, and evolution of organic configurational patterns.[41] The nature, distribution, and action of the pertinent forces (physico-chemical interactions) remain, except for vague and essentially meaningless conjectures,[42] entirely unknown. It seems, nevertheless, quite reasonable to

---

[41] Cf. also the comments in chapter II of volume 1, and in section 1A, chapter VI of volume 3/II, particularly p.30 and 73.

[42] Cf. e.g. the recent *Ciba Symposium* (1975) on 'Cell patterning', and the publications by MOROWITZ (1968), v. BERTALANFFY (1969), and OXNARD (1973).

assume that these events are strictly causal, respectively determined in accordance with physicochemical orderliness, and do not require any postulate of 'vitalistic', 'teleologic', or 'teleonomic' principles.

With regard to the *configurational aspects*, a cautious interpretation of HAECKEL's biogenetic law provides considerable evidence permitting phylogenetic interpolations and extrapolations where the morphology of adult recent and fossil organisms does not supply sufficient clues. Authors attempting to deny the significance of HAECKEL's 'law' (or better: 'rule') stress that phylogenetic and ontogenetic evolution are separate biological processes with entirely different time scales, and, moreover, displaying an extremely complex relationship to one another, which precludes a simple understanding of evolutionary mechanisms and sequences through study of ontogenetic mechanisms. Evolution does not follow a set of ordered laws, and chance-based mechanisms are very important (cf. e.g. BOCK, 1969).

To these objections it can be replied that as mentioned above, the relevant form-giving or configurational 'mechanisms' of both phylogeny[43] and ontogeny, despite the available data of genetics and macromolecular biology, still remain entirely unknown, while the formal sequences of ontogeny are well known, and those of phylogeny have been inferred with a reasonable degree of probability concerning many main lines of evolution. Although the molecular components of biological systems display chance-based mechanisms, that is to say tendencies toward 'randomness' or 'disorder', said systems, nevertheless, because of numerous poorly understood constraints or parameters, tend toward order, as e.g. expressed in configurational or topologic orderliness, displaying invariants which characterize both ontogeny and phylogeny. By physicochemical interactions ('forces'), 'living masses', namely, arrays of multiplying cells, are set in motion, becoming displaced and molded into orderly organic patterns.

Without going into further details concerning the ambiguous meanings of 'chance' and 'randomness' it could here be said that the obtaining mechanisms (including 'mutations') operate *partly* on what

---

[43] REMANE (1956) justly stated: '*Das Ergebnis der bisherigen Versuche, die Triebkräfte der Phylogenie zu enträtseln, ist also recht gering. Kombinations- und Mutationsphänomene sind zweifellos an der Umbildung der Organismen beteiligt, wir kennen recht genau die Mechanismen, die zur Artspaltung führen, wir kennen die Grundlagen der Rassenbildungen, von Rückbildungserscheinungen, Symmetrieänderungen usw., an die organisatorische Umbildung der Lebewesen führen sie aber noch nicht heran.*' More recent comments by REMANE on '*offene Probleme der Evolution*' can be found in his contribution to the volume '*Evolution*' edited by SCHARF (1975).

may be called the 'principle of trial and error', which, however, is kept within certain limits by poorly understood and poorly definable intrinsic 'properties'[43a] and 'constraints'. 'Trial and error' would here be e.g. undirected, i.e. by 'random' mutations, whose effectiveness and successions, nevertheless, would become 'guided' or 'directed' by said intrinsic constraints as well as by extrinsic, environmental ones. The evolutionary process might thus be compared to a *game of chance with loaded dice*. Although unpredictable it can, and in my opinion should, be regarded as strictly determinate, causal, [43b] and 'mechanistic' in the wider sense.

The inadequacies of the contemporary '*Neo-Darwinian*' theory of evolution based on present-day concepts of the transmission of genetic information and the occurrence of 'random' mutations were pointed out on pp. 128–138 in section 7, chapter II of volume 1. In this respect, it is of interest that competent mathematicians have expressed their doubts in a symposium edited by MOORHEAD and KAPLAN (1967), which subsequently came to my attention.

Thus, EDEN (1967), with respect to the element of 'randomness' which is claimed to provide the mutational variation upon which evolution is claimed to depend, stated that 'no currently existing formal language can tolerate random changes in the symbol sequences which express its sequences. Meaning is almost invariably destroyed. Any changes must be syntactically lawful ones'. EDEN presumes that 'what one might call "genetic grammaticality" has a deterministic explanation and does not owe its stability to selection pressure acting on random variations'.

SCHÜTZENBERGER (1967) emphasizes that the contemporary 'dogma' of genetic information implies an extremely special net of derivability relations representing a 'syntactic topology', requiring algorithms in which the very concept of syntactic correctness has been incorporated. Organisms, however, 'are related by another topology which simply results from their being physical objects in space-time. Although this second topology is far harder to formalize, it is the basis of systematics, and it is objectively studied when observing the developmental effects of variations in the milieu.' This, the cited author calls 'phenotypic topology'.

[43a] Cf. 'Mind and Matter' (K., 1961), § 74, pp. 425–426, concerning unpredictable 'emerging' properties.

[43b] It can be maintained that the outcome of a game of chance by throwing dice, although unpredictable, is strictly causal and determinate.

SCHÜTZENBERGER stresses, with respect to *Neo-Darwinian theory*, the present lack of a conceivable mechanism which would insure within a relevant range 'the faintest amount of matching between the two above mentioned topologies'. He therefore believes 'that an entirely new set of rules is needed to obtain the sort of correspondence which is assumed to hold (one way – DARWIN, or the other – LAMARCK) between neighboring phenotypes and which is needed in similar types of evolutions. If these new principles, or deductions from old ones, were to be postulated, it would seem then a subsidiary point to discuss how much of random mutations and selections are at work in conjunction with them.'

The attempts of the *Neo-Darwinian* biologists participating in the symposium (e.g. EISELEY, MAYR, MEDAWAR, WADDINGTON, WALD) at answers to the challenge by the mathematicians remained vague, weak, and rather unconvincing. Yet, WADDINGTON conceded that realistic models would need to take account of what he called main elements of a theory of phenotypes involving canalized processes of development, namely the heritability of developmental responses to environmental stimuli, and a principle of 'archetypes', that is to say inbuilt characteristics of an evolving group which determine the directions in which evolutionary change is especially easy. The interested reader is referred to the various papers included in the cited symposium and is left to draw his own conclusions.

In accordance with my own views it could be said that the configurational changes representing the formal course of evolution with its phylogenetic or genealogic trees, as elaborated in the pioneering contributions of HAECKEL, GEGENBAUR, T.H. HUXLEY, and others on the basis of strictly *morphologic* data provided by comparative anatomy, paleontology, and ontogeny, are far more satisfactorily established and understood than the 'causal' aspects of that evolution. There is little doubt that, with respect to this 'causal' aspect, DARWINIAN variability (e.g. mutations) and natural selection (*Malthusian parameters* and environment) represent very substantial factors. Yet, all hitherto propounded 'causal' theories of evolution remain incomplete and unsatisfactory. This, however, should not justify the introduction of teleologic, vitalistic, theologic or similar supernatural factors, and a strictly mechanistic (in the wider sense) interpretation or 'explanation' of organic evolution may still be postulated. I would essentially agree with EDEN's (1967) comment 'that the principal task of the evolutionist is to discover and examine mechanisms which constrain the variations of phenotypes to a very small class and to relegate the notion of randomness to a minor and non-crucial role'. I would here add that the mechanisms of evolution, despite considerable randomness or 'chance', nevertheless constrain the variations of phenotypes to proceed along particular (strictly determined) directions.

Discounting the completely unknown details of these formative mechanisms, the ontogenetic configurational sequences can be ascertained, and the phylogenetic ones, closely related to a 'natural' taxonomic order, can be inferred. Thus, as regards the evolution of the longitudinal zonal systems in deuterencephalon and spinal cord, in diencephalon and telencephalon, and the evolution of the cerebral cortex, the validity of HAECKEL's biogenetic rule was, as I believe, corroborated by my own investigations.

With regard to other Vertebrate organic configurations, the branchial and aortic arches, the development of the heart, of the urogenital system with its unfolding of pronephros, mesonephros, metanephros and associated ducts provide other examples. From a morphological viewpoint, and despite the numerous complicating factors, it can be maintained that a formal recapitulation of presumably ancestral stages becomes displayed. This opinion is shared by competent contemporary biologists, including AUTRUM, MÄGDEFRAU, MÖHRES, REMANE, RENSCH, and ZIMMERMANN, as cited by HEBERER (1968).[44]

With respect to phylogenetic lines of descent suggested by paleontology, relatively complete sequences have been found in some more or less coherent layers of deposits. Such series were recorded, *inter alia*, for Brachiopods, Snails, Ammonoids, Trilobites, and certain Mammals. It is true that the obtained results are only partly satisfactory, because little is known concerning important environmental factors (cf. e.g. RENSCH, 1968).

It will here be sufficient to illustrate two sequences, of which one,

---

[44] Thus, REMANE is quoted to state: '*Eine Parallelität zwischen Ontogenese und Phylogenese existiert. Der Kern des Biogenetischen Grundgesetzes besteht nach wie vor.*' This author estimates that said rule is valid in 70 to 90 per cent of its application. RENSCH is quoted to state: '*Fast jedes Stadium der Embryogenese und der Jugendentwicklung eines Lebewesens hat eine heuristische Bedeutung für die Erschliessung der Stammesgeschichte, denn die frühen Stadien sind zumeist konservativer als die späteren und sind dadurch Hinweise auf den stammesgeschichtlichen Wandel.*' (Quoted after HEBERER, 1968). MÄGDEFRAU and ZIMMERMANN, moreover, are *botanists* stressing the validity of HAECKEL's rule for plant phylogeny. The reader may be referred to my own comments in section 6, chapter III, pp. 224–254 of volume 1.

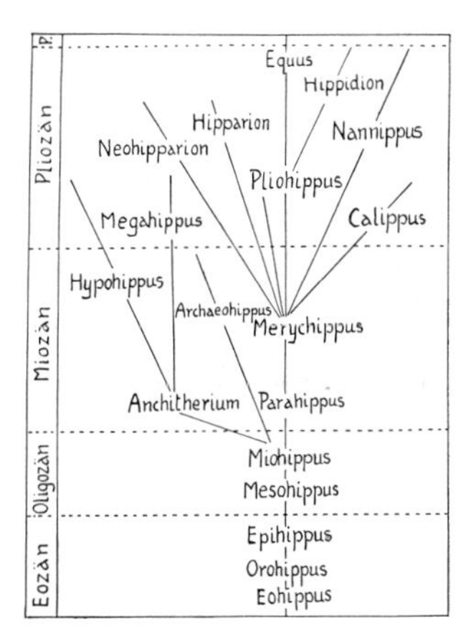

Figure 153 A. Assumed phylogenetic tree of Equidae, based on paleontologic evidence and involving a sequence of perhaps 60 million years from Eohippus to recent Equus (from RENSCH, 1956).

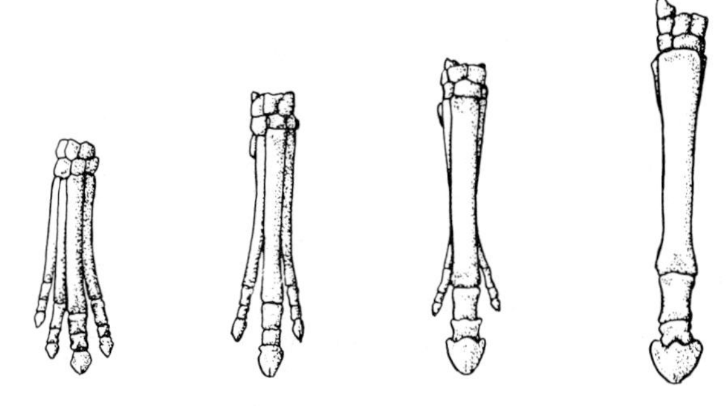

Figure 153 B. Evolution of forefoot in the phylogenetic series of Equidae (after various authors, from RENSCH, 1956). In left-right sequence: Hyracotherium *sive* Eohippus (Eocene), Mesohippus (Oligocene), Merychippus (Miocene), Equus (Pleistocene).

concerning Equine evolution, is particularly well documented, and the other one, concerning Hominid evolution, is of special import.

Figures 153 A–D illustrate the presumptive phylogenetic radiation, beginning in the Eocene epoch, perhaps 45 to 50 million years ago, with the small Mammalian Hyracotherium (Eohippus), and culminating in the recent Horse. The changes, as e.g. displayed by bones of the anterior extremity (Fig. 153 B), by increase of size, including that of brain (Figs. 153 C, D) appear to indicate a direction, with side lines or 'branches' of a 'progressing' and essentially 'irreversible' evolutionary

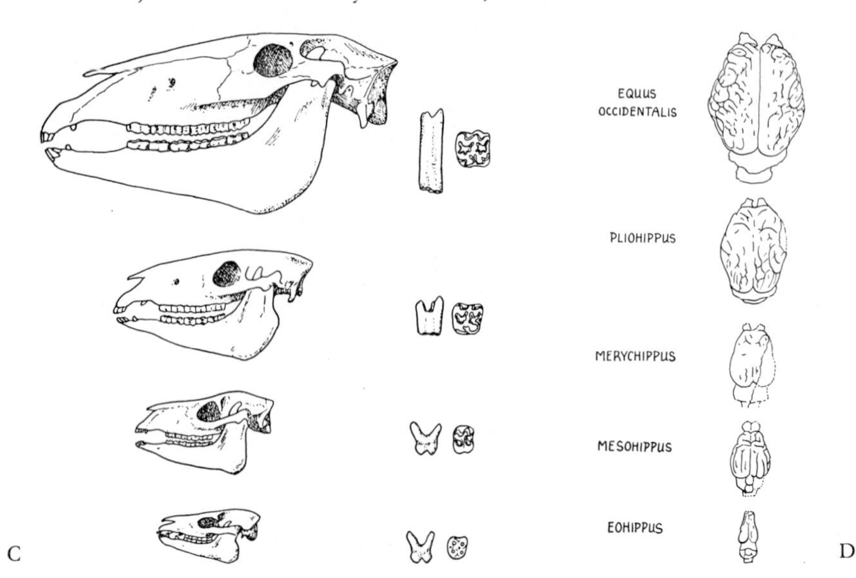

Figure 153 C. Evolution of skull and molars in the phylogenetic series of Equidae (from Rensch, 1956). In ascending sequence: Eohippus, Mesohippus, Merychippus, Equus.

Figure 153 D. Progressive evolution of brain in the phylogenetic series of Equidae (after T. Edinger, 1948, from Rensch, 1956). Pliohippus pertains to Pliocene Epoch (about 12 to 1 million years ago, and preceding the Pleistocene).

process. Such seemingly directed evolution has been designated as '*orthogenesis*'.[45] The 'directing' parameters, however, which constrain fluctuations or randomness of the subsequent changes, and may finally result in some sort of 'stabilization' or 'fixation', can be evaluated as devoid of teleologic implications. Although largely unknown, some such parameters may be intrinsic to the mechanisms of ontogenetic developmental processes, and, as it were, determined from within, whereby the course of evolution, representing a reaction between 'living substance' and environment, is not in its entirety directly and passively shaped by the environmental mold (Herrick, 1920). Other parameters are believed to be 'stable selective factors'. Yet, the highly unsatisfactory and vague state of all causal explanations based on the presently available knowledge should be admitted. The poorly under-

[45] Cf. e.g. p. 150, section 10, chapter II of volume 1. Orthogenesis is also dealt with in a paper by Herrick (1920).

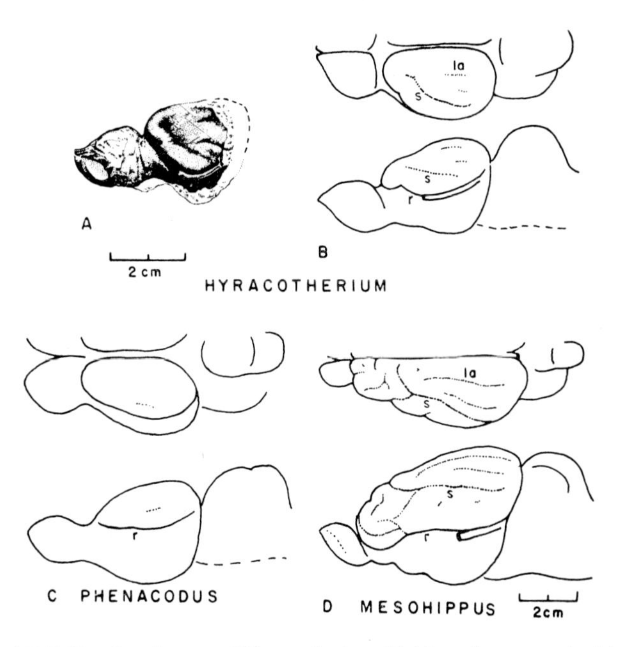

Figure 153 E. Fossil endocasts of Hyracotherium (Eohippus) compared with that of its contemporary Condylarth Phenacodus and with that of Mesohippus according to RADINSKY (from RADINSKY, 1976). la: lateral sulcus; r: sulcus rhinalis lateralis; s: suprasylvian sulcus.

stood problems of 'complex adaptations' in evolving populations were recently reviewed by FRAZZETTA (1975).

In the particular case of Equine phylogeny, the diagram of Figure 153 A can be reasonably interpreted as probably indicating a coherent, continuous line of evolution. As regards the transformations of the Horse brain, the cranial endocasts shown in Figure 153 D display not only the increase in size, but an early transition from lissencephalic to gyrencephalic surface pattern of the hemispheres. This Figure also illustrates the significance of paleoneurology, of which TILLY EDINGER (1929, 1948) is one of the main founders, for phylogenetic interpretations (cf. also Fig. 397, p. 734, vol. 3/II). It should, however, be added that the endocast of Eohippus (Hyracotherium) described by T. EDINGER (1948) is, according to RADINSKY, actually an endocast of the likewise Eocene Condylarth Phenacodus. Newly prepared endocasts of Eohippus, the 'oldest Horse' and one of the earliest Perissodactyls, are said to reveal a relatively larger brain, with a more expand-

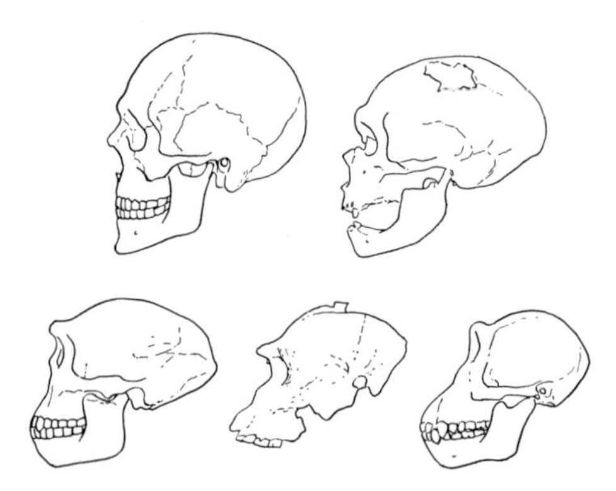

Figure 154 A. Array of primate skulls displaying transitions from Ape to Man, and, although not representing a direct genealogic lineage, presumably roughly corresponding to the Primate skull's actual phylogenetic evolution (from RENSCH, 1956). From lower right to upper left: Chimpanzee, South African Paranthropus (Australopithecine), Java Pithecanthropus, Neanderthal Man (La Chapelle aux Saints), recent European.

ed neocortex, than existed in the Condylarth ancestors of Perissodactyls. According to this interpretation, the brain of Hyracotherium, about 50 million years ago, had suprasylvian, ectolateral, and lateral sulci, but its frontal lobe was still poorly developed (cf. Fig. 153E). Be that as it may, the discrepancy between TILLY EDINGER's and RADINSKY's interpretations of fossil remnants and endocasts, as well as the diversified opinions, concerning details of Hominid evolution and based on palaeontologic evidence, clearly demonstrate the uncertainties inherent in attempts at specific phylogenetic interpretations.

Concerning the skeletal system, Figure 153B illustrates, with regard to phylogenetic changes, the sort of many-one transformation representing the exclusive type, characterized by extrinsic diminution or elimination of closed neighborhoods, as contrasted with the inclusive type, whereby open neighborhoods, without being eliminated as such, become fused (cf. above section 1, p. 389).

Paleontologic Mammalian series comparable to those suggesting the evolution of the Horse, although perhaps somewhat less complete, have been recorded as regards Elephants, Carnivores, especially Ursids, and the extinct Ungulate Titanotheres.

Figure 154A, which illustrates stages of *Hominid evolution*, depicts a series somewhat comparable to that of Figure 153C. In contradistinction to this latter, however, the given array of Primate skulls cannot be interpreted to indicate a direct line of descent, as e.g. tentatively traced,

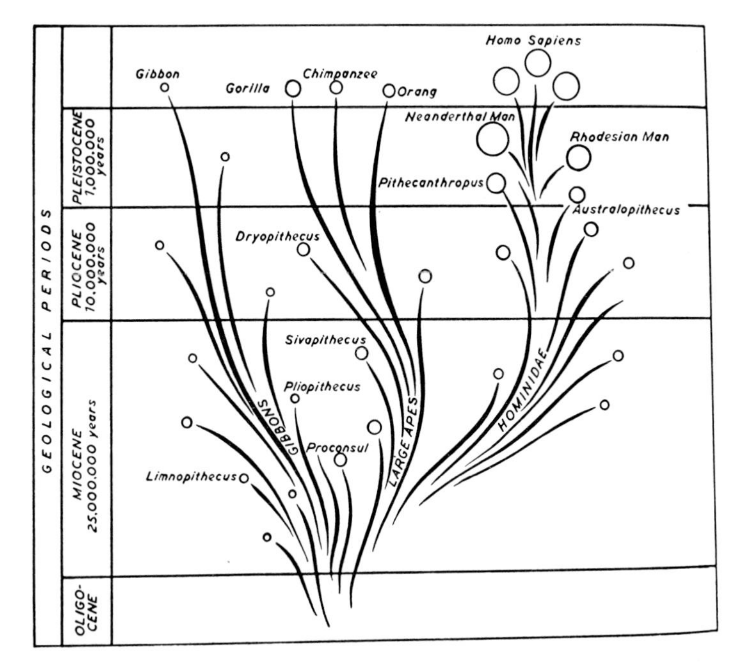

Figure 154B. Diagram indicating hypothetic phylogenetic relationship of Hominidae and Anthropoid Apes as interpreted by LE GROS CLARK (from LE GROS CLARK, 1960). The circles are intended to represent approximate differences in the relative size of the brain. It is here assumed that the brain did not undergo the expansion characteristic for recent Man until the early or middle Pleistocene, perhaps 800 000 to 600 000 years ago (roughly 20 000 Human generations). The author's size indication for the Neanderthal Man's brain is perhaps somewhat exaggerated.

according to LE GROS CLARK (1960), in the diagram of Figure 154B. It should here be added that the results of this[46] and other attempts still remain rather uncertain, although most likely conforming, regardless of unsettled details, to the overall course of Hominid evolution. The now firmly established term '*Pithecanthropus*', initially much ridiculed, was already coined, about 100 years ago, by ERNST HAECKEL, as stated above.

In this connection, BOLK's theory concerning the principle of 'fetalization' or 'neoteny', briefly discussed on p. 108, section 5 of chapter XIV, should be recalled. There has been considerable controversy

[46] In contradistinction to the diagram of LE GROS CLARK I would presume that the Chimpanzee rather than the Orang stands closer to Man in the genealogic tree.

about the exact definition and application of the relevant terms fetalization, neoteny, paedogenesis, paedomorphosis, proterogenesis, and retardation. Thus, it has been claimed the *fetalization* (retention of some intermediate ontogenetic feature) and *neoteny (sensu stricto:* attainment of sexual maturity in Urodele Amphibia previously to, or without, metamorphosis) are entirely different processes which are not comparable (cf. e.g. STARCK, 1962). *Sensu latiori*, however, the retention of juvenile characteristics, as justly pointed out by BOLK, can be conceived to display features obtaining, as abstract 'invariants' namely as retention combined with progression in the various processes subsumed under the above-mentioned disputed terms. Despite the numerous uncertainties inherent in all phylogenetic speculations, BOLK's theory, regardless of terminologic controversies, may nevertheless be evaluated as indicating a significant factor in Hominid evolution. Taken *sensu latiori*, the terms 'fetalization' or 'neoteny' can be upheld as unobjectionable.

With respect to the phylogenetic evolution of Mammals and presumably also of numerous submammalian Vertebrate and other animal and vegetal organisms, the faunal and floral isolations related to the now generally recognized *continental drift* appear to have played an important role.

Although a few previous authors, impressed by the jigsaw-like fitting outlines of African and South-American coasts, had suggested a splitting of these continents, ALFRED WEGENER (1880–1930), since about 1912, was the first to propound a systematic elaboration of the *theory of continental drift* (1915). This concept was not accepted by most geologists and became practically ignored until revived by evidence obtained in the period after midcentury.

Said evidence was provided by explorations of oceanic floors, by observations of paleomagnetism, and by remnants of the Permian Reptile Lystrosaurus, found in Antarctica. This Reptile, known to have lived in Africa, India and China, could not possibly have crossed oceanic barriers.

It is now assumed, in overall conformity with WEGENER's theory concerning '*Verschiebung der Kontinente*', that at the transition from Palaeozoic to Mesozoic eras[47] (Permian-Triassic periods), about 200 million years ago, our planet's land areas formed a single continent, *Pangaea*, surrounded by a single ocean, *Panthalassa*.

---

[47] A short review of the relevant geologic and paleontologic aspects, omitting the theory of continental drift, was given in section 6, pp. 104–115, chapter II of volume 1.

There is, moreover, some evidence that this continent was initially entirely submerged, and that its emergence from a primordial ocean, covering the entire globe, occurred in the Precambrian, about 3–4 billion years ago. This emergence of a continent from beneath the primordial sea may have been a slow process, occupying most of the Precambrian time (cf. e.g. HARGRAVES, 1976).

During the Mesozoic era, at the Jurassic-Cretaceous periods, about 135 million years ago, a northern landmass, Laurasia, seems to have split from a southern one, Gondwana, from which, in turn, Antarctica and India split off, with India heading toward Laurasia. Perhaps in the lower Cretaceous, Laurasia split into North-America and Eurasia, while South-America split off from Gondwana, whose remainder became Africa. During the Cretaceous, Madagascar began to separate from Africa, and Australia-New Guinea may have started to drift away from Antarctica.

In the Palaeocene epoch of the Cenozoic era, perhaps 65 million years ago, the North-Atlantic and Indian Oceans took shape, the South-Atlantic widened, and India was still floating toward Eurasia, which it joined, possibly in the Miocene epoch, some 20 million years ago, thrusting up the Himalaya range. North- and South-America likewise became joined in the Tertiary period. During this period, concomitantly with collisions of the African and Eurasian continents, the Mediterranean sea repeatedly dried up and became again flooded. One such major flooding within that sequence seems to have occurred in the middle Pliocene, about 5 million years ago.

Present-day elaborations of WEGENER's old concept of '*Kontinentalschollen*' are known as the *plate tectonic theory* and postulate that the still shifting crust of the earth represents a mosaic of perhaps about 20 'plates', about 30 to 100 miles thick. These plates,[48] slowly sliding over a hot, semiplastic layer *('asthenosphere')* carry the continents and ocean basins with them. Earthquakes and volcanic eruptions occur along some boundaries, 'faults' or 'cracks' between plates, while other edges slide beneath an opposing plate. Polar wandering is supposed to have taken place in connection with these events, and the present Sahara may have been at the South Pole about 450 million years ago, in the Palaeozoic era, at the Cambrian period.

---

[48] E.g. Eurasian plate, African plate, American plate, Indo-Australian plate, Antarctic plate, Pacific plate, etc.

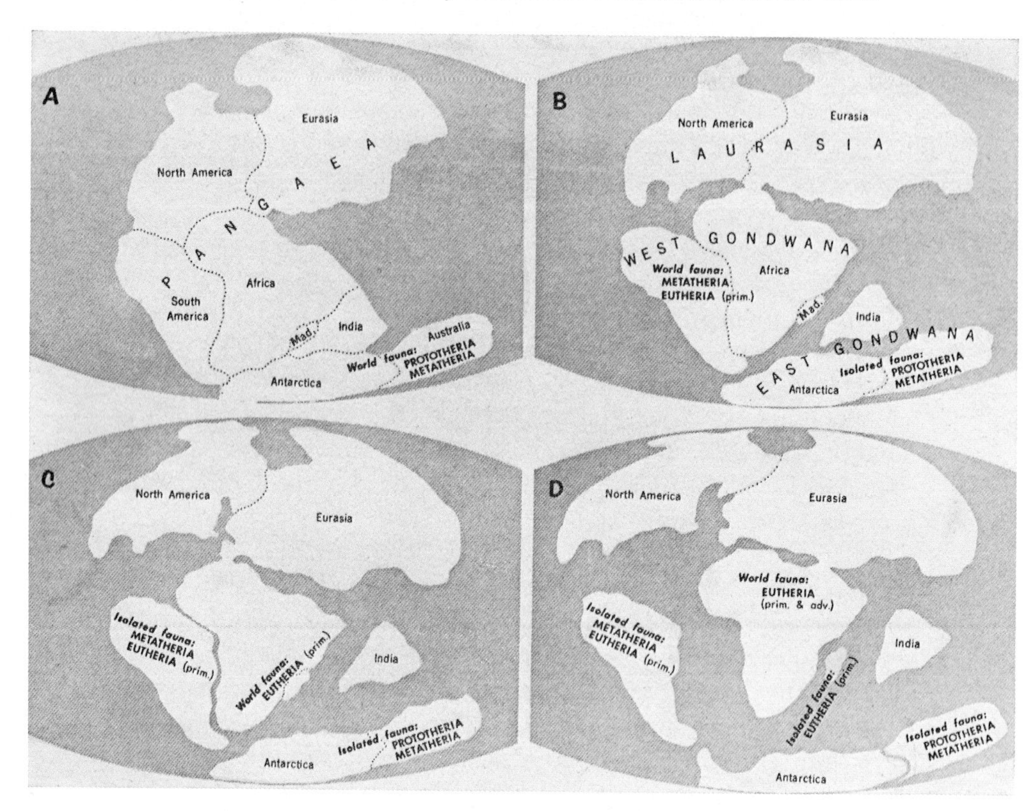

Figure 155. Hypothetic maps illustrating the breakup of Pangaea (simplified after DIETZ and HOLDEN, from FOODEN, 1972). A: at end of Palaeozzoic (ca. 200 million years ago); B: near end of Triassic (perhaps 165 million years ago); C: at upper Jurassic or lower Cretaceous (perhaps roughly 135 million years ago); D: during Cretaceous Period (135 to 75 million years ago).

Figures 155 A–D illustrate current concepts of the breakup of Pangaea,[49] and indicate some of the implications for Mammalian phylogeny. The aboriginal land Mammal faunas in Australia and New Guinea (Prototherians and Metatherians), in South-America (Metatherians and Eutherians), and Madagascar (Eutherians only) can be considered to represent successively detached samples of the evolving world Mammal fauna as it existed when each of these land masses became

[49] Although the maps drawn up by WEGENER and by the subsequent diverse authors show various differences in detail, overall agreement as to the main outlines and sequences can be said to obtain.

faunally isolated from the rest of the world as a result of the progressive fragmentation of Pangaea (FOODEN, 1972). Isolation of aboriginal Prototherians and Metatherians in Australia and New Guinea may date, according to the cited author, from the Upper Jurassic-Lower Cretaceous; isolation of aboriginal Metatherians and Eutherians in South America may date from the Middle Cretaceous – Upper Cretaceous; isolation of aboriginal Eutherians in Madagascar could date from the Paleocene-Eocene.

In concluding the present chapter's comments on phylogeny, it could be said that, seen from the viewpoint of the fictional but indispensable materialistic semantic model, the origin of 'life' represents an inevitable 'disease' or 'putrefaction' of matter, intrinsically related to properties of the carbon atom, as well as to properties of cyanamide compounds and of the nucleic acid macromolecules.[50]

The evolution of prebiotic and biotic forms, again, led, in Metazoan organisms, to the evolution of nervous systems and to the origin of transitory consciousness phenomena. In these latter, and as seen from the idealistic epistemologic viewpoint, the intrinsically incomprehensible aspects of the orderliness x, devoid of space and time characteristics, became manifested (or 'symbolized') as a multiplicity of spatiotemporal phenomena occurring in a large diversity of discrete ('private') consciousness sets or manifolds.

Reverting to the fully justified but fictional materialistic viewpoint, evolution can be conceived as a purposeless, i.e. meaningless but strictly causal and thereby rigidly determined *cosmic game of chance*, in which, as it were, the dice are loaded in favor of the highly 'improbable' origin of organic life, such that the whole subsequent phylogenetic development displays a predestined spatial and temporal pattern manifesting itself along the time coordinate.[51]

The rigidly predetermined course of the future is, causally as well as logically, correlated with the uniqueness of the present and the irrevocability of the past:

---

[50] Cf. pp. 650f., section 8, chapter V of volume 3/I.

[51] These conclusions from the rules of evolution, quite in contradiction to views expressed by DOBZHANSKY (1953), were pointed out in the monograph 'Mind and Matter' (K., 1961, p. 273). The zoologist RENSCH (1968, 1971) has likewise, in somewhat similar manner, attempted to formulate relevant rules of evolution. I disagree, however, with his panpsychistic concept of consciousness, based on a 'cryptomaterialistic' interpretation of the 'extramental world' (that is to say of what I prefer to designate as the unknown and unknowable 'orderliness x').

'The Moving Finger writes; and having writ,
Moves on: nor all your Piety nor Wit
Shall lure it back to cancel half a Line,
Nor all your Tears wash out a Word of it.'
(FITZGERALD's *Omar* LXXI, 1879).

'Fate, show thy force. Ourselves we do not owe:
What is decreed must be; and be this so!'
(SHAKESPEARE, *Twelfth Night*, I, 5).

11. References to Chapter XVI

ALBERS, R.W.; SIEGEL, G.J., and KATZMANN, R. (eds.): Basic neurochemistry (Little, Brown, Boston 1972).

ASHBY, W. ROSS: An introduction to cybernetics (Wiley, New York 1956, 1957).

ASHBY, W. ROSS: Design for a brain; 1st ed., 2nd ed. (Wiley, New York 1952, 1960).

BAER, C.E. v.: Über Entwicklungsgeschichte der Thiere (Beobachtung und Reflexion). 2 vols. (Bornträger, Königsberg 1828, 1837).

BARKER, L.F.: The nervous system and its constituent neurones (Stechert, New York 1901).

BARR, M.L.: The human nervous system; 2nd ed. (Harper & Row, Hagerstown 1974).

BARRON, D.H. and MATTHEWS, B.H.C.: Intermittent conduction in the spinal cord. J. Physiol., Lond. *85:* 73–103 (1935).

BARRON, D.H. and MATTHEWS, B.H.C.: The interpretation of potential changes in the spinal cord. J. Physiol., Lond. *92:* 276–321 (1938).

BECCARI, N.: Neurologia comparata (Sansoni, Firenze 1943).

BERTALANFFY, L. v.: General system theory. Foundations, development, applications (Braziller, New York 1969).

BOCK, W.J.: Evolution by orderly law (book review). Science *164:* 684–685 (1969).

BOYDEN, A.: Homology and analogy. Science *164:* 455–456 (1969).

BRADLEY, P.B. (ed.): Methods in brain research (Wiley, New York 1975).

BUDDENBROCK, W. VON: The senses (University of Michigan Press, Ann Arbor 1958).

BURR, H.S.: The central nervous system of Orthagoriscus mola. J. comp. Neurol. *45:* 33–128 (1928).

CAMPBELL, C.B.G and HODOS, W.: The concept of homology and the evolution of the nervous system. Brain Behav. Evol. *3:* 353–367 (1970). Ciba Foundation Symposium 29: Cell patterning (Elsevier, New York 1974).

CLARA, M.: Das Nervensystem des Menschen; 3. Aufl. (Barth, Leipzig 1959).

CLARK, W.E. LE GROS: The antecedents of man. An introduction to the evolution of primates (Quadrangle Books, Chicago 1960).

CONRAD, L.C.A. and PFAFF, D.W.: Axonal projections of medial preoptic and anterior hypothalamic nuclei. Science *190:* 1112–1114 (1975).

COWAN, M. and CUÉNOD, M. (eds.): The use of axonal transport for studies of neuronal connectivity (Elsevier, New York 1975).

Cowan, W. M.; Gottlieb, D. I.; Hendrickson, A. E.; Price, J. L., and Woolsey, T. A.: The autoradiographic demonstration of axonal connections in the central nervous system. Brain Res. *37:* 21–51 (1972).

Craigie, E. H.: An introduction to the finer anatomy of the central nervous system based upon that of the albino Rat (Blakiston, Philadelphia 1925).

Creps, E. S.: Time of neuron origin in preoptic and septal areas of the mouse. An autoradiographic study. J. comp. Neurol. *157:* 161–243 (1974).

Crosby, E. C.; Humphrey, T., and Lauer, E. W.: Correlative anatomy of the nervous system (MacMillan, New York 1962).

Darwin, C. R.: The origin of species (Murray, London 1859; also Modern Library, New York n.d.).

Darwin, C. R.: The descent of man, and selection in relation to sex (Murray, London 1871; also Modern Library, New York n.d.).

Davidson, N.: Neurotransmitter amino acids (Academic Press, New York 1976).

Dejerine, J.: Anatomie des centres nerveux. 2 vols. (Rueff, Paris 1895, 1901).

Dexler, H.: Zur Anatomie des Zentralnervensystems von Elephas indicus. Arb. neurol. Inst. Univ. Wien *15:* 137–181 (1907).

Dobzhansky, T.: The genetic basis of evolution. Scientific American Reader, pp. 293–308 (Simon & Schuster, New York 1953).

Donkelaar, H. J. Ten: Descending pathways from the brain stem to the spinal cord in some reptiles. I. Origin. II. Course and site of termination. J. comp. Neurol. *167:* 421–463 (1976).

Eden, M.: Inadequacies of Neo-Darwinian evolution as a scientific theory; in Moorhead and Kaplan Mathematical challenges to the Neo-Darwinian interpretation of evolution, pp. 5–19, 109–111 (Wistar Institute Press, Philadelphia 1967).

Edinger, L.: Über das Gehirn von Myxine glutinosa. Phys. Abh. Kgl. preuss. Akad. Wiss. (1906).

Edinger, L.: Einführung in die Lehre vom Bau und den Verrichtungen des Nervensystems (Vogel, Leipzig 1912).

Edinger, T.: Die fossilen Gehirne (Springer, Berlin 1929).

Edinger, T.: Evolution of the horse brain. Mem. geol. Soc. America, No. 25 (1948).

Elliott, H. C.: Textbook of neuroanatomy (Lippincott, Philadelphia 1963).

Elliott, H. C.: The shape of intelligence. The evolution of the human brain (Scribner's, New York 1969).

Flechsig, P.: Anatomie des menschlichen Gehirns und Rückenmarks auf myelogenetischer Grundlage (Thieme, Leipzig 1920).

Fooden, J.: Breakup of Pangaea and isolation of relict mammals in Australia, South America, and Madagascar. Science *175:* 894–898 (1972).

Frazzetta, T. H.: Complex adaptations in evolving populations (Sinauer, Sunderland, Mass. 1975).

Frederikse, A.: The lizard's brain. An investigation of the histologic structure of the brain of Lacerta vivipara (Callenbach, Nijkerk 1931).

Gaupp, E.: Eckers und Wiedersheims Anatomie des Frosches. 2 Abt. (Vieweg, Braunschweig 1899).

Gerlach, J.: Über das Gehirn von Protopterus annectens. Ein Beitrag zur Morphologie des Dipnoerhirnes. Anat. Anz. *75:* 311–406 (1933).

Gerlach, J.: Beiträge zur vergleichenden Morphologie des Selachierhirnes. Anat. Anz. *96:* 79–165 (1947).

GILBERT, C.D. and KELLY, J.P.: The projection of cells in different layers of the cat's visual cortex. J. comp. Neurol. *163:* 81–106 (1975).

GÜNERAL, I.: Atlas cerebri humani (Uycan, Istanbul 1972).

HAECKEL, E.: Generelle Morphologie der Organismen. 2 vols. (Reimer, Berlin 1866).

HAECKEL, E.: Natürliche Schöpfungsgeschichte; 1., 4., 9. Aufl. (Reimer, Berlin 1868, 1874, 1898).

HARGRAVES, R.B.: Precambrian geologic history. Continents grew and emerged from beneath the primordial sea. Science *193:* 363–371 (1976).

HEBERER, G. (ed.): Der gerechtfertigte Haeckel (Fischer, Stuttgart 1968).

HEIER, P.: Fundamental principles in the structure of the brain. A study of the brain of Petromyzon fluviatilis (Hakan, Lund 1948; also Acta anat., suppl. VI).

HERRICK, C.J.: Irreversible differentiation and orthogenesis. Science *51:* 621–625 (1920).

HERRICK, C.J.: Morphogenesis of the brain. J. morph. *54:* 233–258 (1933).

HERRICK, C.J.: The brain of the tiger salamander Ambystoma tigrinum (University of Chicago Press, Chicago 1948).

HERTWIG, R.: Lehrbuch der Zoologie; 10. Aufl. (Fischer, Jena 1912).

HODOS, W.: The comparative study of brain-behavior relationship; in GOODMAN and SCHEIN Birds: brain and behavior, pp. 15–25 (Academic Press, New York 1974).

HOLMGREN, N. and HORST, C.J. VAN DER: Contributions to the morphology of the brain of Ceratodus. Acta zool. *6:* 59–165 (1925).

HOOGENBOOM, K.J.H.: Das Gehirn von Polyodon folium Lacep. Z. mikr.-anat. Forsch. *18:* 311–392 (1929).

HULTKRANTZ, J.W.: Gehirnpräparation mittels Zerfaserung. Anleitung zum makroskopischen Studium des Gehirns (Springer, Berlin 1929).

JACOBOWITZ, D.M. and PALKOVITS, M.: Topographic atlas of catecholamine and acetylcholinesterase-containing neurons in the rat brain. I. Forebrain (telencephalon, diencephalon). II. Hindbrain (mesencephalon, rhombencephalon). J. comp. Neurol. *157:* 13–42 (1974).

JACOBSHAGEN, E.: Allgemeine vergleichende Formenlehre der Tiere (Klinkhardt, Leipzig 1925).

JACOBSHAGEN, E.: Zur Reform der allgemeinen vergleichenden Formenlehre der Tiere (Fischer, Jena 1927).

JACOBSHAGEN, E.: Die dynamische Erklärung der tierischen Konstruktionen und die vergleichende Morphologie. Anat. Anz. *65:* 314–319 (1928).

JANSEN, J.: The brain of Myxine glutinosa. J. comp. Neurol. *49:* 359–507 (1930).

JANSEN, J. and JANSEN, J.K.S.: The nervous system of Cetacea; in ANDERSON The biology of marine mammals, chapter 7, pp. 175–252 (Academic Press, New York 1968).

JELGERSMA, G.: Atlas anatomicum cerebri humani (Scheltema & Holkema, Amsterdam n.d.).

JOHNSTON, J.B.: The brain of Acipenser. Zool. Jb. *15:* 59–260 (1901).

JOHNSTON, J.B.: The brain of Petromyzon. J. comp. Neurol. *12:* 87–106 (1902).

KAHLE W.: Studien über die Matrixphasen und die örtlichen Reifungsunterschiede im embryonalen menschlichen Hirn. Dtsch. Z. Nervenheilk. *166:* 273–302 (1951).

KAPPERS, C.U.A.: The structure of the Teleostean and Selachian brain. J. comp. Neurol. *16:* 1–109 (1906).

KAPPERS, C.U.A.: Untersuchungen über das Gehirn von Amia calva und Lepidosteus osseus. Abh. Senkenberg, naturforsch. Ges. *30:* 449–500 (1907).

KAPPERS, C.U.A.: Die vergleichende Anatomie des Nervensystems der Wirbeltiere und des Menschen. 2 vols. (Bohn, Haarlem 1920, 1921).

KAPPERS, C.U.A.: The evolution of the nervous system in invertebrates, vertebrates and man (Bohn, Haarlem 1929).

KAPPERS, C.U.A.: Anatomie comparée du système nerveux, particulièrement de celui des mammifères et de l'homme. Avec la collaboration de E. STRASBURGER (Masson, Paris 1947).

KAPPERS, C.U.A. und CARPENTER, F.W.: Das Gehirn von Chimaera monstrosa. Folia neurobiol. *5:* 127–160 (1911).

KAPPERS, C.U.A. und HAMMER, E.: Das Zentralnervensystem des Ochsenfrosches (Rana catesbyana). Psych. neurol. Bladen (Feestbundel *Winkler*)*:* 368–415 (1918).

KAPPERS, C.U.A.; HUBER, G.C., and CROSBY, E.C.: The comparative anatomy of the nervous system of vertebrates including man. 2 vols. (Macmillan, New York 1936).

KARTEN, H.J. and HODOS, W.: A stereotaxic atlas of the brain of the pigeon (Columba livia) (The Johns Hopkins Press, Baltimore 1967).

KELLER, H.: The story of my life (Doubleday, New York 1954).

KÖNIG, J.F.R. and KLIPPEL, R.A.: The rat brain: a stereotaxic atlas of the forebrain and lower parts of the brain stem (Williams & Wilkins, Baltimore 1963).

KRIEG, W.J.S.: Functional neuroanatomy; 3rd ed. (Brain Books, Evanston 1966).

KRIEG, W.S.: Interpretative atlas of the monkey brain (Brain Books, Evanston 1975).

KUHLENBECK, H.: Zur Morphologie des Gymnophionengehirns. Jena. Z. Naturw. *58:* 453–484 (1922).

KUHLENBECK, H.: Vorlesungen über das Zentralnervensystem der Wirbeltiere (Fischer, Jena 1927).

KUHLENBECK, H.: Die Grundbestandteile des Endhirns im Lichte der Bauplanlehre. Anat. Anz. *67:* 1–50 (1929a).

KUHLENBECK, H.: Über die Grundbestandteile des Zwischenhirnbauplans der Anamnier. Morph. Jb. *63:* 50–95 (1929b).

KUHLENBECK, H.: Über die Grundbestandteile des Zwischenhirnbauplans bei Reptilien. Morph. Jb. *66:* 244–317 (1931).

KUHLENBECK, H.: Über die Grundbestandteile des Zwischenhirnbauplans der Vögel. Morph. Jb. *77:* 61–109 (1936).

KUHLENBECK, H.: The human diencephalon. A summary of development, structure, function, and pathology (Karger, Basel 1954).

KUHLENBECK, H.: Brain and consciousness. Some prolegomena to an approach of the problem (Karger, Basel 1957).

KUHLENBECK, H.: Mind and matter. An appraisal of their significance for neurologic theory (Karger, Basel 1961).

KUHLENBECK, H.: Gehirn und Bewusstsein (Translated by Prof. J. GERLACH and Dr. U. PROTZER). Erfahrung und Denken. Schriften zur Förderung der Beziehungen zwischen Philosophie und Einzelwissenschaften, vol. 39 (Duncker & Humblot, Berlin 1973).

KUHLENBECK, H. and NIIMI, K.: Observations on the morphology of the brain in the Holocephalian Elasmobranchs Chimaera and Callorhynchus. J. Hirnforsch. *11:* 265–314 (1969).

LIM, R.K.S.; LIU, C.N., and MOFFITT, R.L.: A stereotaxic atlas of the dog's brain (Thomas, Springfield 1960).

LORENTE DE NÓ, R.: Contribución al conocimiento del nervio trigemino. Libro en honor de *D.S.R.y Cajal*, vol. II, pp. 13–30 (1922).

LUCCHI, M.R. DE; DENNIS, B.J., and ADEY, W.R.: A stereotaxic atlas of the chimpanzee brain (Cambridge University Press, London 1965).

LUDWIG, E. and KLINGLER, J.: Atlas cerebri humani (Karger, Basel 1956).

MARBURG, O.: Mikroskopisch-topographischer Atlas des menschlichen Zentralnervensystems; 3. Aufl. (Deuticke, Leipzig 1927).

MARGOLIASH, E.: Homology: a definition. Science *163:* 127 (1969).

MASTERTON, R.B.; CAMPBELL, C.B.G.; BITTERMAN, M.E. and HOTTON, N., eds.: Evolution of the brain and behavior in vertebrates. Papers from a conference, Tallahassee, Fla., Feb. 1973 (Halsted/Wiley, New York 1976). Quoted from Science 196, No. 4288, 1977.

MASTERTON, R.B.; HODOS, W., and JERISON, H., eds.: Evolution, brain and behavior. Persistent problems (Halsted/Wiley, New York 1976). Quoted from Science 196, No. 4288, 1977.

MATTHEWS, B.H.C.: Impulses leaving the spinal cord by dorsal roots. J. Physiol, Lond. *81:* 29–31 (1934).

MILLOT, J. et ANTHONY, J.: Anatomie de Latimeria chalumnae, vol. II. Système nerveux et organes des sens (Editions du Centre national de la Recherche scientifique, Paris 1965).

MIURA, R.: Über die Differenzierung der Grundbestandteile im Zwischenhirn des Kaninchens. Anat. Anz. *77:* 310–406 (1933).

MOORHEAD, P.S. and KAPLAN, M.M. (eds.): Mathematical challenges to the Neo-Darwinian interpretation of evolution (Wistar Institute Press, Philadelphia 1967).

MOROWITZ, H.J.: Energy flow in biology. Biological organization as a problem in thermal physics (Academic Press, New York 1968).

MYERS, R.D.: Handbook of drug and chemical stimulation of the brain. Behavioral, pharmacological and physiological aspects (Van Nostrand, New York 1974).

NATHAN, P.W.: The gate-control theory of pain – a critical review. Brain *99:* 123–158 (1976).

NAUTA, H.J.W.; PRITZ, M.B., and LASEK, R.J.: Afferents in the rat caudoputamen studied with horseradish peroxidase. An evaluation of a retrograde neuro-anatomical research method. Brain Res. *67:* 219–239 (1974).

NEAL, H.V. and RAND, H.W.: Comparative anatomy (Blakiston, Philadelphia 1936).

NICHOLSON, H.: On the presence of ganglion cells in the third and sixth nerves of man. J. comp. Neurol. *37:* 31–36 (1924).

NIIMI, K.: Shinkei kaibôgaku (Neuroanatomy, Japanese) (Asakura Shoten, Tôkyô 1976).

OBERSTEINER, H.: Anleitung beim Studium der nervösen Zentralorgane im gesunden und kranken Zustande (Deuticke, Leipzig 1912).

OXNARD, C.: Form and pattern in human evolution. Some mathematical, physical, and engineering approaches (University of Chicago Press, Chicago 1973).

PAPEZ, J.W.: Comparative neurology (Crowell, New York 1929).

PATTEN, B.M.: The embryology of the pig; 2nd ed. (Blakiston, Philadelphia 1931).

PEARSON, R.: The Avian brain (Academic Press, London 1972).

PEARSON, R., and PEARSON, L.: The vertebrate brain (Academic Press, New York 1976). Quoted from Science 196. No. 4292, 1977.

PEELE, T.L.: The neuroanatomic basis for clinical neurology; 2nd ed. (McGraw Hill, New York 1961).

PILLERI, G.: Morphologie des Gehirns des 'Southern-Right Wale' Eubalaena australis (Cetacea, Balaenidae, Mysticeti). Acta zool. *45:* 245–272 (1964).

RADINSKY, L.: Oldest horse brains: more advanced than previously realized. Science *194:* 626–627 (1976).

RANSON, S.W.: The anatomy of the nervous system from the standpoint of development and function; 7th ed., 10th ed. revised by S.L. CLARK (Saunders, Philadelphia 1943, 1961).

RASMUSSEN, A.T.: The principal nervous pathways; 1st and 2nd ed. (Macmillan, New York 1932, 1941).

REMANE, A.: Die Grundlagen des natürlichen Systems, der vergleichenden Anatomie und der Phylogenetik; 2. Aufl. (Akad. Verlagsges., Leipzig 1956).

RENSCH, B.: Tatsachen und Probleme der Evolution; in RENSCH Vom Unbelebten zum Lebendigen, pp. 198–221 (Enke, Stuttgart 1956).

RENSCH, B.: Biophilosophie auf erkenntnistheoretischer Grundlage. Panpsychistischer Identismus (Fischer, Stuttgart 1968).

RENSCH, B.: Biophilosophy. Translated by C.A.M. SYM (Columbia University Press, New York 1971).

ROMER, A.S.: Man and the vertebrates (University of Chicago Press, Chicago 1937).

ROSENBLUETH, A.: The transmission of nerve impulses at neuroeffector junctions and peripheral synapses (Wiley, New York 1950).

RUSTIONI, A.: Spinal neurons project to the dorsal column nuclei of Rhesus monkeys. Science 196: 656–658 (1977).

SAITO, T.: Über das Gehirn des japanischen Flussneunauges (Entosphenus japonicus Martens). Folia anat. japon. *8:* 189–263 (1930).

SCHARF, J.H. (ed.): Evolution. Nova Acta Leopoldina vol. 42 (1975).

SCHMATOLLA, E.: Retino-tectal course of optic nerves in cyclopic and synophthalmic Zebrafish embryos. Anat. Rec. *180:* 377–383 (1974).

SCHOPENHAUER, A.: Sämmtliche Werke (1818–1851), ed. *Grisebach* (Reclam, Leipzig n.d.); also ed. *Hübscher*, 7 vols. (Brockhaus, Leipzig 1946–1950).

SCHÜTZENBERGER, M.P.: Algorithms and the Neo-Darwinian theory of evolution; in MOORHEAD and KAPLAN Mathematical challenges to the Neo-Darwinian interpretation of evolution, pp. 73–80, 121 (Wistar Institute Press, Philadelphia 1967).

SIMPSON, G.G.: The meaning of evolution (Yale University Press, New Haven 1967).

SINGER, M. and YAKOVLEV, P.I.: The human brain in sagittal sections (Thomas, Springfield 1954).

SNIDER, R.S. and LEE, J.C.: Stereotaxic atlas of the monkey brain Macaca mulatta (University of Chicago Press, Chicago 1962).

SNIDER, R.S. and NIEMER, T.: A stereotaxic atlas of the cat brain (University or Chicago Press, Chicago 1962).

SNOW, P.J.; ROSE, P.K., and BROWN, A.G.: Tracing axons and axon collaterals of spinal neurons using intracellular injection of horseradish peroxidase. Science *191:* 312–313 (1976).

SOURY, J.: Le système nerveux central (Naud, Paris 1899).

SPIEGEL, E.A. and WYCIS, H.T.: Stereoencephalotomy (thalamotomy and related procedures). I. Methods and stereotaxic atlas of the human brain (Grune & Stratton, New York 1952).

STARCK, D.: Wandlungen des Homologiebegriffes. Zool. Anz. *145:* suppl., pp. 957–969 (1950).

STARCK, D.: Der heutige Stand des Fetalisationproblems (Parey, Hamburg 1962).

STEBBINS, G.L.: Processes of organic evolution (Prentice-Hall, Englewood 1966).

STEPHAN, H.: Allocortex; in v. MÖLLENDORFF and BARGMANN Handb. d. mikr. Anat. d. Menschen, vol. IV/9 (Springer, Berlin 1975).

STERZI, C.: Il sistema nervoso centrale dei Vertebrati, vol. 2, Selaci (Draghi, Padova 1909).

TALAIRACH, J.; DAVID, M.; TOURNOUX, P.; CORREDOR, H. et KVASINA, T.: Atlas d'anatomie stéréotaxique (Masson, Paris 1957).

TAX, S. (ed.): Evolution after Darwin. 3 vols. (University of Chicago Press, Chicago 1960).

TIENHOVEN, A. VAN and JUHASZ, L.P.: The chicken telencephalon, diencephalon and mesencephalon in stereotaxic coordinates. J. comp. Neurol. *118:* 185–197 (1962).

TOWER, D.B.: Neurochemistry – one hundred years, 1875–1975. Collective review. Ann. Neurol. *1:* 2–36 (1977).

VILLIGER, E.: Gehirn und Rückenmark. Leitfaden für das Studium der Morphologie und des Faserverlaufs; 7. Aufl. (Engelmann, Leipzig 1920).

WARNER, F.J.: The myelinization of the central nervous system of the American water snake (Natrix sepedon). Trans. zool. Soc. Lond. *27:* 307–348 (1952).

WEGENER, A.: Die Entstehung der Kontinente und Ozeane (Vieweg, Braunschweig 1915).

WINKLER, C. and POTTER, A.: An anatomical guide to experimental researches on the rabbit's brain (Versluys, Amsterdam 1911).

WINKLER, C. and POTTER, A.: An anatomical guide to experimental researches on the cat's brain (Versluys, Amsterdam 1914).

WINTER, W.P.; WALSH, K.A., and NEURATH, H.: Homology as applied to proteins. Science *162:* 1433 (1968).

ZEMAN, W. and INNES, J.R.M.: *Craigie's* neuroantomy of the rat, revised and expanded (Academic Press, New York 1963).

Addenda to Volume 5/II

A. To Chapter XIV

The brain of HELMHOLTZ, referred to on p. 102, 105, and 122, was investigated by HANSEMANN in a paper which accidentally remained unquoted in the text and omitted from the List of References. It is herewith added to the bibliography: HANSEMANN, D.: Über das Gehirn von Hermann von Helmholtz (mit 2 Tafeln). Z. Psychol. Physiol. Sinnesorg. *20:* 1–12 (1899).

B. To Chapter XV

With respect to concepts of cortical localization in 'lower Mammals', reviewed in section 10, chapter XIII of vol. 5/I, and to the cortical map of the Rabbit, dealt with on p. 263, section 4 of chapter XV, a recent paper by TOWNS *et al.* should be mentioned, which discusses, on the basis of corticocortical fiber connections, investigated with degeneration techniques, the assumed presence, in the occipital (visual) cortex, of a dorsal visual area 1 and of a ventral visual area 2, the adjacent boundary zones of both of which are presumed to be binocular. The fiber connections elucidated by the cited authors appear reasonably well substantiated, but the added interpretation of cytoarchitectural findings in general agreement with the mapping by M. ROSE (1931), namely as displaying three rostrocaudal strips of visual cortex, designated, in dorsobasal sequence, as peristriate, striate, and occipital cortex cannot be considered supported by valid evidence. The authors' cytoarchitectural illustrations appear, in this regard, unconvincing. I would maintain that the Rabbit's visual, i.e. 'occipital cortex' in my terminology, is represented by a cytoarchitecturally essentially homogeneous field which, regardless of probable differences in details of input and output fiber distribution or of synaptology, cannot be parcellated into sufficiently distinctive architectural subregions. This visual cortex, in turn, is continuous, through blurred nondescript gradients, with parahippocampal cortex dorsomediocaudally, parietal cortex rostrally and temporal cortex basally. The cited paper, containing numerous references to apposite recent studies, whose results are difficult to reconcile with the well-documented experimental findings of LASHLEY in the Rat, is nevertheless of interest (TOWNS, L.C.; GIOLLI, R.A., and HASTE, D.A.: Corticocortical fiber connections of the rabbit visual cortex. A fiber degeneration study. J. comp. Neurol. *173:* 537–559, 1977).

Corrigenda to Volume 5/I

P. 443, line 18 from top, read: HARRIS, G.W., and DONOVAN, B.T., instead of: DONAVAN.
P. 641, footnotes 136, line 3–4 from top, read: more likely, instead of: most likely.

Directions for Use of the Subject and Authors Indexes

1. General

The author and subject indexes are listed separately. The numbers which follow the individual names and terms denote: (1) the number of the volume, recognizable by *italic* type and a subsequent colon, and (2) the pages of the respective volumes on which the names and terms are found. The numbers of footnotes and legends (and figures) are not mentioned in the index, but only the numbers of the pages on which they are found. Footnotes, legends and references are denoted by the capital letters F, L and R, respectively.

2. Index of Subjects

Considering the broad and comprehensive background of the entire work as well as the many basic questions that are dealt with, it becomes obvious that the index of subjects does not only contain terms from comparative morphology and biology of the nervous system but also from numerous other sciences, primarily natural sciences as well as mathematics and philosophy. In order to prevent ambiguity and confusion in the text the author has adopted not only the common terminologies as they have been suggested by the various professional committees but has also used special terms of his own which he selected by considering problems of logic and semantics. Therefore, for the index of subjects a definite nomenclature system cannot be applied. Synonyms are differentiated by attributes added in parentheses, e.g. nucleus (cell) and nucleus (griseum). When the terms consist of several single words that can be subordinated to one another, the most important word is listed in the index and the subordinates follow it in alphabetic order, the key word being replaced by dashes. Consequently when looking for terms of this kind the main word has to be found before the compound word, e.g. Sulcus, – tubero-infundibularis. In some cases the index is in tabular form, e.g. Diencephalon, comparative anatomy. The index of authors can also be used to find subject information if the required terms are sufficiently defined by names, e.g. Lorenzini's ampullae, vena Galeni. Emphasis of a page number by *italic* print indicates essential or comprehensive information, extensive treatment of facts or terms or historical or semantic explanations of words or terms.

For further subject information the numerous references of the author to preceding and following pages and volumes of the whole work may be used.

3. Index of Authors

The last names of the authors are followed by the initial of their first name(s). This is to differentiate between authors with identical last names and does not appear when the authors are not mentioned in the chapter references or with some historical personalities, e.g. Galenus, Rolando, Sylvius.

Subject Index to Volumes 1–5

Table I Diencephalon, comparative anatomy

	Cyclo-stomes	Sela-chians	Ganoids	Tele-osts	Lati-meria	Dipno-ans	Amphi-bians	Rep-tiles	Birds	Mammals
	138	160	175	182	202	206	217	246	270	292
Sulci	142	167	180		L204	210	218	248	*270*	*3/II:* 445
Habenular ganglion	143 155	161	175	183		207	219 231	250	276	298 305
Habenular commissure	143 160	161 174	182	201	L204	213	620	647	681	747
Thalamus dorsalis	143 155	164	175	183	L204	209 213	220 232	253	*276*	*299*
Thalamus ventralis	144 156	164	175	186	L204	211 216	220 233	258	281	301 361
Genic. lat	148	164		186			220 221	256	*278* 283	353 364
Hypothalamus	148 158	165	177	*187*		210 216	221 *235* *245*	*261*	*284*	*301* *366*
Nucleus praeopticus magnocellularis	158	167	177	191			245	262	286	303
Fiber connections	149	170	180	197	206	211	224	263	*287*	335
Fasciculus retroflexus	151	174	180	200	206	217	229 245	267	290	335 343

The page numbers are related to volume 5/I unless otherwise stated. Genic. lat. = Dorsal respectively ventral lateral geniculate nucleus. Other designations self-explanatory

Table II Nuclei, cranial nerves

Nerve	Cyclostomes	Selachians	Ganoids	Teleosts	Latimeria	Dipnoans	Amphibians	Reptiles	Birds	Mammals
III		834	858	858	866	870	883	911	930	959 *961*
IV		834	858	858	866	870	883	911	931	951 *962*
Vs	358 375	393	409	419	434	437	454	475	493	528 587
Vm	361 380	394	409	428	434	440	459	478	495	537 587
Vd	358	393	409	419	434	437	454	475	494	531
Vme	358	829	409	843	865	869	455 878	900	920	949
VI	362 F301	394	412	429	436	440	459 F301	480	500	541
VIIs	358	393	409	421	434	437	457	475	494	534
VIIm	361	394	409	429	434	440	459	478	495	538
VIII	377 351 (lat)	390 F390 (lat)	406 (lat)	415 (lat)	434 (lat)	437 (lat)	454		485	
VIIIc							453	467	485	512

VIIIv	352 355	391	407	415		L710	449	471	F486 487	520
Deiters	356	393 402		418			454	472	490	521
IXs	358	394	409	421	434	437	457	475	494	534
IXm	361 380	395	409	429	434	440	459	478	497	539
Xs	358 377	394	409	421	434	437	457	475	495	534
Xm	361	395	412 409	429	434	440	459	478	497	539 540
XI		395 F395					459	479	498	539
spo	350 361 381	397	412	429	436	440				
XII							459	480 474	501	541

The page numbers are related to volume 4. The Roman numerals denote the cranial nerves. m = Motor, me = mesencephalic root, d = radix descendens, c = cochlear nucleus, v = vestibular nucleus, Deiters = Deiters' nucleus, s = sensory, spo = spinooccipital nuclei.

Table III Telencephalon, comparative anatomy

	Cyclo- stomes	Sela- chians	Ganoids	Tele- osts	Lati- meria	Di- pnoans	Amphi- bians	Rep- tiles	Birds	Mammals
Bulbus olfactorius	543	554	572	572	590	592 599	502	624	650	500
Tractus olfactorius	544 559		582		590		616	642	674	518
Lobus hemisphaericus	543 549	555	573	570 574	590	593 601	603	625	*650*	*686*
Pallium	543 549	555	573	*576*	590		*603*	*625*	655	691
Hippoc. (prim.)	550	558					603	626	*656*	*693*
Basis	543 549	556	574	576	590	594	605	636	*666* 672	*698*
Communication channels	544 551	562	582	596	590		620	641	674	*727*
Forebrain bundle lateral		562		582			618	644	675	731
medial				582			618	646	680	733
basal	544 552	564	566	586						738 (secondary)
Stria medullaris	544		582	583	591	597	619	646	681	739
Stria terminalis								646	680	790
Commissura anterior	545	565	584	584	590	598	619	647	681	745
Commissura pallii (hippocampi)	545 552	565		584			619	647	581	730 742
Lesions				587			620	649	583	*747*

The page numbers are related to volume 5/I. Hippoc. (prim.) = Hippocampus respectively primordium hippocampi, Lesions = experimental lesions, (secondary) = incomplete homology.

Authors Index to Volumes 1–5